Advances in Mechanics and Mathematics

Volume 29

For further volumes:
http://www.springer.com/series/5613

Jan Awrejcewicz

Classical Mechanics

Dynamics

Springer

Jan Awrejcewicz
Department of Automation & Biomechanics
Łódź University of Technology
Łódź, Poland

ISSN 1571-8689 ISSN 1876-9896 (electronic)
ISBN 978-1-4614-3739-0 ISBN 978-1-4614-3740-6 (eBook)
DOI 10.1007/978-1-4614-3740-6
Springer New York Heidelberg Dordrecht London

Library of Congress Control Number: 2012938706

Printed on acid-free paper

Springer is part of Springer Science+Business Media (www.springer.com)

Preface

This is the second of three books devoted to classical mechanics. The first book, entitled *Classical Mechanics: Statics and Kinematics*; the third, being coauthored by Z. Koruba, is entitled *Classical Mechanics: Applied Mechanics and Mechatronics*. All three books reference each other, and hence they are highly recommended to the reader. In this book dynamical and advanced mechanics problems are stated, illustrated, and discussed, and a few novel concepts, in comparison to standard textbooks and monographs. Apart from being addressed to a wide spectrum of graduate and postgraduate students, researchers, and instructors from the fields of mechanical and civil engineering, this volume is also intended to be used as a self-contained body of material for applied mathematicians and physical scientists and researchers.

In Chap. 1 the dynamics of a particle and system of particles, as well as rigid-body motion about a point, are studied. Section 1.1 is focused on particle dynamics. First, Newton's second law of motion is revisited. Classification of forces is carried out, and Newton's second law is formulated in cylindrical, spherical, and polar coordinates. Forward and inverse dynamics problems are defined and analyzed. The dynamics of a particle subjected to the action of a particular excitation from the previously classified forces is studied. Governing second-order ordinary differential equations (ODEs) are derived and then solved. An illustrative example of particle motion along an ellipse is provided. The laws of conservation of momentum, angular momentum, kinetic energy, and total (mechanical) energy are introduced, illustrated, and examined. In Sect. 1.2, the fundamental laws of the collection of particles (discrete or continuous) are introduced and studied. In the case of momentum conservation of a continuous mechanical system, two important theorems are formulated. Then the conservation of the center of gravity of either a discrete or continuous mechanical system is described. Essential corollaries and principles are formulated, and an illustrative example is provided. Next, the conservation of the angular (kinetic) momentum of a discrete mechanical system is considered. Five important definitions are introduced including Köning's systems, and Köning's theorem is formulated. A kinetic-energy formula is derived. Next, the conservation of angular momentum of a discrete (lumped) mechanical system is studied.

A theorem regarding the necessary and sufficient condition for the existence of the first integral of angular momentum is formulated and proved. In addition, two examples are provided. In what follows the formulation of the law of conservation of kinetic energy is given. Body motion about a point is analyzed in Sect. 1.3.

Mathematical and physical pendulums are studied in Chap. 2. In Sect. 2.1 second-order ODE governing the dynamics of a mathematical pendulum is derived and then explicitly solved for two different sets of initial conditions. In addition, a mathematical pendulum oscillating in a plane rotating with constant angular velocity is analyzed. A physical pendulum is studied in Sect. 2.2. Again a governing dynamics equation is formulated. In the case of a conservative system mechanical energy of the physical pendulum is also derived. Section 2.3 concerns the plane dynamics of a triple physical pendulum. Initially, three second-order ODEs governing the dynamics of a triple pendulum are derived, and then they are presented in matrix notation. Since the obtained differential equations are strongly non-linear, they are then solved numerically. In particular, periodic, quasiperiodic, and chaotic motions are illustrated and discussed. Furthermore, the dynamic reactions in pendulum joints are determined and monitored.

In Chap. 3 static and dynamic problems of discrete mechanical systems are discussed. In Sect. 3.1 the constraints and generalized coordinates are defined. That is, geometric, kinematic (differential), and rheonomic (time-dependent), as well as holonomic and non-holonomic, constraints are illustrated and analyzed through several examples. Furthermore, unilateral and bilateral constraints are introduced and explained using two illustrative examples. Possible and ideal virtual displacements are also introduced and further examples are provided. Variational principles of Jourdain and Gauss are introduced in Sect. 3.2, and their direct application to static problems is illustrated through two examples. The general equations of statics, as well as the stability of equilibrium configurations of mechanical systems embedded in a potential force field, are considered in Sect. 3.3. Important theorems as well as four principles are formulated, and two examples illustrating theoretical considerations are provided. In Sect. 3.4 the Lagrange equations of the second and first kind are rigorously derived. Both discrete and continuous mechanical systems are considered, and some particular cases of the introduced various constraints are analyzed separately. Five illustrative examples are given. In Sect. 3.5 properties of Lagrange's equation, i.e., covariance, calibration invariance, kinetic-energy form, non-singularity, and the least action principle, are briefly described. The first integrals of the Lagrange systems are derived and discussed in Sect. 3.6. Cyclic coordinates are introduced, and two theorems are formulated. Routhian mechanics is briefly introduced in Sect. 3.7. Using the Legendre transformation, we derive Routh's equations. Next, in Sect. 3.8 the cyclic coordinates are discussed, and their validity is exhibited through examples. A three-degree-of-freedom manipulator serving as an example of rigid-body kinetics is studied in Sect. 3.9. First, physical and mathematical models are introduced, then the Denavit–Hartenberg notation is applied, and the obtained differential equations are solved numerically. Furthermore, the results of numerical simulations are discussed on the basis of an analysis of three different cases, and some conclusions are formulated.

Chapter 4 is devoted to classical equations of mechanics. Section 4.1 concerns Hamiltonian mechanics, Sect. 4.2 describes methods of solution of the Euler–Lagrange equations, Sect. 4.3 deals with Whittaker equations, Sect. 4.4 concerns Voronets and Chaplygin equations, and, finally, Sect. 4.5 includes both a derivation and discussion of Appell's equations. In Sect. 4.1.1 the canonically conjugate variables are introduced, and Hamilton's canonical equations are derived. An example is added for clarification. In Sect. 4.1.2 the Jacobi–Poisson theorem is formulated and proved, followed by an introduction to the Poisson bracket. Canonical transformations are discussed in Sect. 4.1.3, where six theorems are also given. Non-singular canonical transformations and guiding functions are introduced in Sect. 4.1.4, whereas the Jacobi method and the Hamilton–Jacobi equations are presented in Sect. 4.1.5. Then two particular cases of Hamilton–Jacobi equations are considered in Sect. 4.1.6, i.e., the Hamilton–Jacobi equations for cyclic coordinates and conservative systems. In Sects. 4.2.1–4.2.4 solutions of the Euler–Lagrange equations are presented. Section 4.2.2 includes definitions of weak and strong minima and Euler's theorem. The Bogomol'nyi equation and decomposition are briefly stated in Sect. 4.2.3, whereas the Bäcklund transformation is described in Sect. 4.2.4 supplemented with two examples. Whittaker's equations are derived in Sect. 4.3, whereas the Voronets and Chaplygin equations are formulated in Sect. 4.4. Applications of Chaplygin's equations are presented as an example of a homogeneous disk rolling on a horizontal plane. The Appell equations, followed by an example, are derived in Sect. 4.5.

Classical impact theory is introduced and illustrated in Chap. 5. Basic concepts of phenomena associated with impact are presented in Sect. 5.1. The fundamental laws of an impact theory such as conservation of momentum and angular momentum are given in Sect. 5.2, where two theorems are also formulated. A particle's impact against an obstacle is studied in Sect. 5.3, and the physical interpretation of impact is given in Sect. 5.4. Next (Sect. 5.5) the collision of two balls in translational motion is analyzed. In particular, kinetic-energy loss during collision is estimated through the introduced restitution coefficient. In Sect. 5.6 a collision of two rigid bodies moving freely is studied, and in Sect. 5.7 a center of percussion is defined using as an example the impact of bullet against a compound pendulum.

Chapter 6 deals with vibrations of mechanical systems. After a short introduction multi-degree-of-freedom mechanical systems are studied. In Sect. 6.2 linear and non-linear sets of second-order ODEs are derived from Lagrange's equations. Classification and properties of mechanical forces are presented for linear systems in Sect. 6.3. Subsequently, the dissipative, gyroscopic, conservative, and circulatory forces are illustrated and discussed in general and using an example. In Sect. 6.4 small vibrations of linear one-degree-of-freedom mechanical systems are presented. Both autonomous and non-autonomous cases are considered, and (contrary to standard approaches) solutions are determined for homogenous/non-homogenous equations using the notion of complex variables. Amplitude and phase responses are plotted in two graphs, allowing for direct observation of the influence of damping magnitude on the amplitude-frequency and phase-frequency plots. In particular, resonance and non-resonance cases are discussed. In addition, transverse vibrations of a disk mounted on a flexible steel shaft modeled by two second-order linear ODEs are analyzed, and the critical speeds of the shaft are defined. Both cases, i.e., with and without damping, are studied. The phenomenon of shaft self-centering is illustrated. Finally, a one-degree-of-freedom system driven by an arbitrary time-dependent force is studied using Laplace transformations. An illustrative example is given. A one-degree-of-freedom non-linear autonomous and conservative system is studied in Sect. 6.5. It is shown step by step how to obtain its period of vibration. In addition, the so-called non-dimensional Duffing equation is derived. One-degree-of-freedom systems excited in a piecewise linear or impulsive fashion are studied in Sect. 6.6. It is shown how to find the corresponding solutions.

The dynamics of planets is briefly studied in Chap. 7. In the introduction, Galileo's principle of relativity is discussed. Second-order vector ODEs are presented and the homogeneity of space and time are defined. Potential force fields are introduced in Sect. 7.2, whereas Sect. 7.3 is devoted to the analysis of two-particle dynamics. Total system energy, momentum, and angular momentum are derived. In addition, a surface integral and the Laplace vector are defined explicitly, and their geometrical interpretations are given. First and second cosmic velocities are defined, among others. Kepler's three laws are revisited.

The dynamics of variable mass systems is studied in Chap. 8. After a short introduction, the change in the quantity of motion and angular momentum is described in Sect. 8.2. Then an equation of motion of a particle of variable mass (the Meshcherskiy equation) is derived. Two Tsiolkovsky problems are studied in Sect. 8.4. Finally, in Sect. 8.4.1 an equation of motion of a body with variable mass is derived and studied. Two illustrative examples are given.

Body and multibody dynamics are studied in Chap. 9. First, in Sect. 9.1, the rotational motion of a rigid body about a fixed axis is introduced. In Sect. 9.2 Euler's dynamic equations are derived, and the so-called Euler case is analyzed. Poinsot's geometric interpretation of rigid-body motion with one fixed point is illustrated. The roles of a polhode and a herpolhode are discussed. In Sect. 9.3 the dynamics of a rigid body about a fixed point in the gravitational field is studied. The Euler, Lagrange, and Kovalevskaya cases, where first integrals have been found, are also briefly described. The general free motion of a rigid body is analyzed in Sect. 9.4. In Sect. 9.5, the motion of a homogenous ball on a horizontal plane in the gravitational field with Coulomb friction is modeled and analyzed. Equations of motion are derived and then solved. The roles of angular velocities of spinning and rolling and the associated roles of the rolling and spinning torques are illustrated and discussed. Section 9.6 deals with the motion of a rigid body of convex surface on a horizontal plane. Equations of motion are supplemented by the Poinsot equation, and the dynamic reaction is derived. Dynamics of a multibody system coupled by universal joints is studied in Sect. 9.7. Equations of motion are derived using Euler's angles and Lagrange equations of the second kind. Conservative vibrations of a rigid body supported elastically in the gravitational field are analyzed in Sect. 9.8, and one illustrative example is provided. Wobblestone dynamics is studied in Sect. 9.9. The Coulomb–Contensou friction model is first revisited, and the importance of the problem is exhibited emphasizing a lack of a correct and complete solution of the stated task in Sect. 9.9.1. Three vectorial equations of motion are derived, followed by a tenth scalar equation governing the perpendicularity condition. Several numerical simulation results are presented. Next, a hyperbolic tangent approximation of friction spatial models are introduced and discussed in Sect. 9.9.2. The advantages and disadvantages of the introduced approximation versus the Padé approximations are outlined. A few numerical simulation results are given.

Stationary motions of a rigid body and their stability is studied in Chap. 10. It includes problems related to stationary conservative dynamics (Sect. 10.1) and invariant sets of conservative systems (Sect. 10.2).

A geometrical approach to dynamical problems is the theme of Chap. 11. In Sect. 11.1, the correspondence between dynamics and a purely geometrical approach through the Riemannian space concept is derived. It is shown how dynamical problems, supported by a configuration space and the Jacobi metric, are reduced to an equation of geodesic deviation known also as the Jacobi–Levi–Civita (JLC) equation (Sect. 11.3). Next, in Sects. 11.4 and 11.5, the Jacobi metric on a configuration space is rigorously defined, the JLC equation is derived, and then it is rewritten in geodesic coordinates. Finally, a two-degree-of-freedom mechanical system is used to illustrate the theoretical background introduced earlier in Sect. 11.6.

Finally, it is rather impossible nowadays to write a comprehensive book on classical mechanics and include an exhaustive bibliography related to classical mechanics. Therefore, this volume and the two related books mentioned earlier, are rather located in standard classical mechanics putting emphasis on some important topics being rarely mentioned in published literature on mechanics.

Furthermore, in the particular case of dynamics/dynamical systems, there is a vast number of books that are either devoted directly to classical dynamics or that include novel branches of dynamics like stability problems, bifurcational behavior, or deterministic chaos. Although the latter material is beyond the book contents, the reader may be acquainted with my authored co-authored books/monographs devoted to the mentioned subjects, i.e., *Bifurcation and Chaos in Simple Dynamical Systems*, J. Awrejcewicz (World Scientific, Singapore, 1989); *Bifurcation and Chaos in Coupled Oscillators*, J. Awrejcewicz (World Scientific, Singapore, 1991); *Bifurcation and Chaos: Theory and Application*, J. Awrejcewicz (Ed.) (Springer, New York, 1995); *Nonlinear Dynamics: New Theoretical and Applied Results*, J. Awrejcewicz (Ed.) (Akademie Verlag, Berlin, 1995); *Asymptotic Approach in Nonlinear Dynamics: New Trends and Applications*, J. Awrejcewicz, I.V. Andrianov, and L.I. Manevitch (Springer, Berlin, 1998); *Bifurcation and Chaos in Nonsmooth Mechanical Systems*, J. Awrejcewicz and C.-H. Lamarque (World Scientific, Singapore, 2003); *Nonlinear Dynamics of a Wheeled Vehicle*, J. Awrejcewicz and R. Andrzejewski (Springer, Berlin, 2005); *Smooth and Nonsmooth High Dimensional Chaos and the Melnikov-Type Methods*, J. Awrejcewicz and M.M. Holicke (World Scientific, Singapore, 2007); *Modeling, Simulation and Control of Nonlinear Engineering Dynamical Systems: State of the Art, Perspectives and Applications*, J. Awrejcewicz (Ed.) (Springer, Berlin, 2009).

The dynamics of continuous mechanical systems governed by PDEs and completely omitted in this book is widely described in the following books/monographs authored/co-authored by the present author: *Nonclassical Thermoelastic Problems in Nonlinear Dynamics of Shells*, J. Awrejcewicz and V.A. Krysko (Springer, Berlin, 2003); *Asymptotical Mechanics of Thin Walled Structures: A Handbook*, J. Awrejcewicz, I.V. Andrianov, and L.I. Manevitch (Springer, Berlin 2004); *Nonlinear Dynamics of Continuous Elastic Systems*, J. Awrejcewicz, V.A. Krysko and A.F. Vakakis (Springer, Berlin, 2004); *Thermodynamics of Plates and Shells*, J. Awrejcewicz, V.A. Krysko, and A.V. Krysko (Springer, Berlin, 2007); *Chaos in Structural Mechanics*, J. Awrejcewicz and V.A. Krysko (Springer, Berlin, 2008); *Nonsmooth Dynamics of Contacting Thermoelastic Bodies*, J. Awrejcewicz and Yu. Pyryev (Springer, New York, 2009).

The author wishes to express his thanks to P. Dąbek and M. Kamierczak for their help in the book preparation.

Many helpful comments and suggestions given by the author's coworkers from the Łódź University of Technology, Poland, and in particular the contributions of K. Januszkiewicz, J. Mrozowski, and G. Kudra are gratefully acknowledged.

It should also be emphasized that a part of Chaps. 3 and 9 was prepared together with G. Kudra, and Chap. 11 was written in cooperation with D. Sendkowski.

Finally, I acknowledge the Humboldt Foundation Award and the hospitality of Professor P. Hagedorn during my stay at Fraunhofer Institute in Darmstadt, Germany, as well as the support of the Master Award of the Polish Foundation of Science (FNP) during the final stage of the book's preparation.

Łódź and Darmstadt Jan Awrejcewicz

Contents

Chapter 1
Particle Dynamics, Material System Dynamics and Rigid-Body Motion About a Point

1.1 Dynamics of a Particle

1.1.1 Newton's Second Law of Motion

Let us return to Chap. 1 of [1], where Newton's laws were introduced. The second law formulates the relation between the acceleration of motion of a particle and the force acting on it. However, it turns out that only in certain cases is the force **F** acting on this particle independent of the kinematic parameters of motion of the particle. Let us emphasize that in this chapter all masses are considered constant.

In the general case Newton's second law takes the form

$$m\frac{\mathrm{d}^2\mathbf{r}}{\mathrm{d}t^2} = \mathbf{F}(\mathbf{r}, \mathbf{v}, t), \tag{1.1}$$

which emphasizes that the force depends on the position of a particle defined by a radius vector as well as on the velocity of motion of this particle (and sometimes even on acceleration [2], which, however, will not be considered in this book) and on time. Equation (1.1) is a second-order non-linear differential equation.

Figure 1.1a presents the free motion of a particle and the vectors of force and acceleration. In turn, in Fig. 1.1b are shown the distribution of acceleration and forces acting on the particle in natural coordinates.

Introducing the Cartesian coordinates and projecting the vectors occurring in (1.1) onto the axes of the system we obtain

$$\begin{aligned} m\ddot{x}_1 &= F_1\left(x_1, x_2, x_3, \dot{x}_1, \dot{x}_2, \dot{x}_3, t\right),\\ m\ddot{x}_2 &= F_2\left(x_1, x_2, x_3, \dot{x}_1, \dot{x}_2, \dot{x}_3, t\right),\\ m\ddot{x}_3 &= F_3\left(x_1, x_2, x_3, \dot{x}_1, \dot{x}_2, \dot{x}_3, t\right). \end{aligned} \tag{1.2}$$

J. Awrejcewicz, *Classical Mechanics: Dynamics*, Advances in Mechanics and Mathematics 29, DOI 10.1007/978-1-4614-3740-6_1,

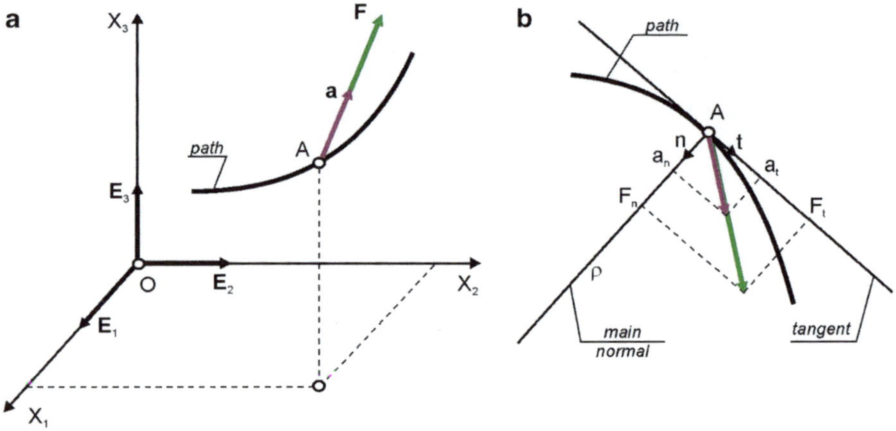

Fig. 1.1 Motion of particle in Cartesian coordinates (**a**) and in natural coordinates (**b**)

According to previous calculations and Fig. 1.1b, in the *natural coordinates* (see Chap. 4.4, Sect. 4, of [1] for more details) we can resolve the acceleration and force into *normal* and *tangential components*, that is,

$$\mathbf{F} = F_t\mathbf{t} + F_n\mathbf{n},$$
$$\mathbf{a} = \mathrm{a}_t\mathbf{t} + \mathrm{a}_n\mathbf{n}, \tag{1.3}$$

and according to Newton's second law we have

$$m\frac{\mathrm{d}v}{\mathrm{d}t} = F_t,$$
$$m\frac{v^2}{\rho} = F_n. \tag{1.4}$$

Let us consider the dynamics of a particle moving on a prescribed fixed plane curve (Fig. 1.2).

In this case, the motion of the particle is not free motion, which can be illustrated by the motion (without any resistance to motion) of a small ball inside a bent tube. Let the curve along which the particle moves be described by the equation $f(x_1, x_2) = 0$ in the coordinate system OX_1X_2. The equation of motion in the vector form reads

$$m\mathbf{a} = \mathbf{F} + \mathbf{N}. \tag{1.5}$$

Projecting vectors onto the coordinate axes we obtain

$$m\ddot{x}_1 = F_{x_1} + N\cos(\mathbf{N}, \mathbf{E}_1),$$
$$m\ddot{x}_2 = F_{x_2} + N\cos(\mathbf{N}, \mathbf{E}_2). \tag{1.6}$$

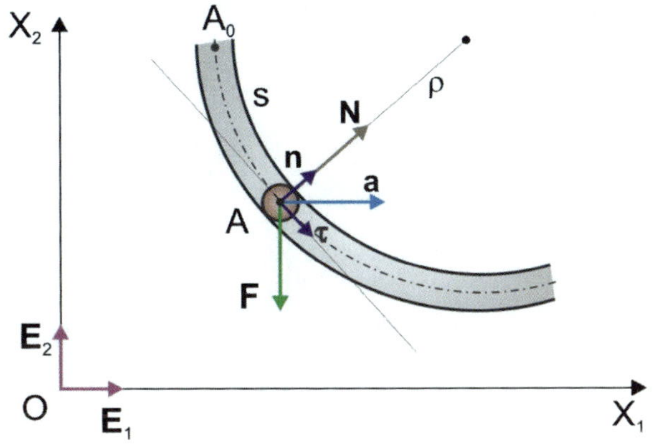

Fig. 1.2 Motion of a particle on a plane OX_1X_2 along prescribed plane curve

The force **N** has the direction of a normal to the prescribed curve, and the direction cosines are described by formulas known from differential geometry:

$$\cos(\mathbf{N}, \mathbf{F}_i) = \frac{1}{f^*}\frac{\partial f}{\partial x_i}, \quad f^* = \sqrt{\left(\frac{\partial f}{\partial x_1}\right)^2 + \left(\frac{\partial f}{\partial x_2}\right)^2}. \tag{1.7}$$

Equations (1.6), after taking into account (1.7), take the form

$$m\ddot{x}_i = F_{x_i} + \lambda\frac{\partial f}{\partial x_1}, \quad i = 1, 2, \tag{1.8}$$

where $\lambda = N/f^*$. In those equations the unknowns are λ, $\ddot{x}_1$, $\ddot{x}_2$ [the third equation is the equation of constraints $f(x_1, x_2) = 0$]. Solving this system of differential-algebraic equations we can determine the three desired quantities and, subsequently, the normal force $N = \lambda f^*$.

The vectors in (1.5) can also be projected onto the axes of the natural coordinate system (τ, n), where the sense of $\boldsymbol{\tau}$ is in accordance with the sense of a ball's motion relative to its initial position A_0. We obtain

$$\begin{aligned} ma\cos(\mathbf{a}, \boldsymbol{\tau}) &= F_\mathrm{t}, \\ ma\cos(\mathbf{a}, \mathbf{n}) &= F_\mathrm{n} + N, \end{aligned} \tag{1.9}$$

where F_t and F_n are projections of the force **F** onto tangent and normal directions.

Because

$$\begin{aligned} a\cos(\mathbf{a}, \boldsymbol{\tau}) &= \frac{\mathrm{d}^2 s}{\mathrm{d}t^2}, \\ a\cos(\mathbf{a}, \mathbf{n}) &= \frac{v^2}{\rho}, \end{aligned} \tag{1.10}$$

after taking into account (1.10) in (1.9) we get

$$m\frac{\mathrm{d}^2 s}{\mathrm{d}t^2} = F_\mathrm{t},$$

$$m\frac{v^2}{\rho} = F_\mathrm{n} + N. \tag{1.11}$$

The preceding equations are called *Euler's equations* of the constrained motion of a particle.

Integrating the first of (1.11) and assuming that $F_\mathrm{t} = \mathrm{const}$ we have

$$m\frac{\mathrm{d}s}{\mathrm{d}t} = F_\mathrm{t} t + C_2,$$

$$ms(t) = \frac{F_\mathrm{t} t^2}{2} + C_2 t + C_1. \tag{1.12}$$

Let $v(0) = v_0$, $s(0) = 0$; then we have $C_1 = 0$, $C_2 = mv_0$, and in view of that, $v(t) = \frac{F_\mathrm{t}}{m}t + v_0$. From the second of (1.11) we determine $N = N(t)$. From the example of the application of natural coordinates it is seen that by the appropriate choice of coordinates we can significantly simplify (or complicate) a solution with regard to the mathematical model describing the problem.

One may introduce the following classification of forces:

(1) $\mathbf{F} = \mathrm{const}$. The examples of a force like this can be the gravity force or the friction force. Let us note that both these forces are only approximately constant. In the former case the motion should take place in the proximity of Earth, whereas in the latter case the friction can depend on many parameters, and in many cases it cannot be treated as constant [1,3].

(2) $\mathbf{F} = \mathbf{F}(t)$. The force depends on time, and the dependency has a variety of forms. In the case of the oscillations of discrete (lumped) systems, such a dependency is most often a harmonic, periodic, or quasiperiodic function [3]. In the case of the vibrations of continuous systems such as beams, plates, or shells, the dependency of force on time is often assumed in the form of step, rectangular, triangular, or impulse excitations [4]. Another example can be the force of attraction in a magnetic field because it is dependent on the field strength, which may vary in time.

(3) $\mathbf{F} = \mathbf{F}(\mathbf{r})$. The dependency of a force on the position in the case of the gravitational attraction of particles of masses m_1 and m_2 can be expressed as the relation $F = G(m_1 m_2)/r^2$, where r is the distance between the particles and G is a constant. We deal with a similar case if to the particle we connect a spring (of negligible mass) whose stiffness, that is, the dependency of the spring load on its deflection, is a non-linear function.

(4) $\mathbf{F} = \mathbf{F}(\mathbf{v})$. We deal with this case if a particle is moving in a liquid or gaseous medium. This dependency most often describes damping and is either linear

(viscous damping) or non-linear and proportional to v^2, but it may also assume a linear negative value, which leads to self-excited motion of a particle [3].

Note that in real mechanical systems combinations of these forces is possible.

Apart from rectangular coordinates one may use curvilinear coordinates for description of particle dynamics. According to earlier considerations (Chap. 4 of [1]) equations of motion take the form

(1) In cylindrical coordinates

$$\begin{aligned} m\left(\ddot{r} - r\dot{\Theta}^2\right) &= F_r, \\ m\left(r\ddot{\Theta} + 2\dot{r}\dot{\Theta}\right) &= F_\Theta, \\ m\ddot{z} &= F_z, \end{aligned} \tag{1.13}$$

where $\mathbf{F} = F_r\mathbf{e}_r + F_\Theta\mathbf{e}_\Theta + F_Z\mathbf{E}_3$;

(2) In spherical coordinates

$$\begin{aligned} m\left(\ddot{R} - R\dot{\phi}^2 - R\dot{\Theta}^2\sin^2\phi\right) &= F_R, \\ m\left(R\ddot{\phi} + 2\dot{R}\dot{\phi} - R\dot{\Theta}^2\sin\phi\cos\phi\right) &= F_\phi, \\ m\left(R\ddot{\Theta}\sin\phi + 2\dot{R}\dot{\Theta}\sin\phi + 2R\dot{\Theta}\dot{\phi}\cos\phi\right) &= F_\Theta, \end{aligned} \tag{1.14}$$

where $\mathbf{F} = F_R\mathbf{e}_R + F_\phi\mathbf{e}_\phi + F_\Theta\mathbf{e}_\Theta$;

(3) In polar coordinates on a plane

$$\begin{aligned} m\left(\ddot{r} - r\dot{\Theta}^2\right) &= F_r, \\ m\left(r\ddot{\Theta} + 2\dot{r}\dot{\Theta}\right) &= F_\Theta. \end{aligned} \tag{1.15}$$

This equation was obtained from (1.13) after setting $z = \text{const}$.

1.1.2 Classifying Dynamics Problems

The dynamics of a particle deals with two classes of problems, *forward problems* and *inversed problems*. In the first case the motion of a particle is known, and the force that causes the motion is desired. In the second case, the force is known, and the motion of a particle is to be determined.

Forward dynamics problems, according to its name and the definition introduced earlier, do require knowledge of the vector $\mathbf{r} = \mathbf{r}(t)$, and the same $\dot{\mathbf{r}} = \dot{\mathbf{r}}(t)$ and $\ddot{\mathbf{r}} = \ddot{\mathbf{r}}(t)$.

According to (1.1) we have

$$m\left(\ddot{x}_1(t)\mathbf{E}_1 + \ddot{x}_2(t)\mathbf{E}_2 + \ddot{x}_3(t)\mathbf{E}_3\right) = F_1\mathbf{E}_1 + F_2\mathbf{E}_2 + F_3\mathbf{E}_3, \tag{1.16}$$

which allows for the determination of projections of vector **F** from the equations

$$m\ddot{x}_i = F_i, \quad i = 1, 2, 3. \tag{1.17}$$

Vector **F** is defined by the magnitude

$$F = \sqrt{F_1^2 + F_2^2 + F_3^2}, \tag{1.18}$$

and the direction cosines of angles α_i formed by the vector with the axes X_i are as follows:

$$\cos\alpha_i = \frac{F_i}{F}, \quad i = 1, 2, 3. \tag{1.19}$$

The inverse dynamics problem is more complicated. In order to determine the motion of a particle knowing the right-hand side of (1.1), one should integrate the differential equation (1.2) twice. Then we obtain a general solution of the system of equations (1.2). In order to uniquely determine the motion of the system, one should choose from the obtained whole family of solutions only those that are consistent with the initial conditions adopted earlier, that is, one should solve the so-called Cauchy problem.

In order to determine uniquely the motion of a particle, one should know its displacement and velocity at the initial time instant t_0, that is,

$$\begin{aligned}\mathbf{r}(t_0) &= x_{10}\mathbf{E}_1 + x_{20}\mathbf{E}_2 + x_{30}\mathbf{E}_3,\\ \dot{\mathbf{r}}(t_0) &= \dot{x}_{10}\mathbf{E}_1 + \dot{x}_{20}\mathbf{E}_2 + \dot{x}_{30}\mathbf{E}_3,\end{aligned} \tag{1.20}$$

where $x_i(t_0) = x_{i0}$, $\dot{x}_i(t_0) = \dot{x}_{i0}$, $i = 1, 2, 3$.

Because each of equations (1.2) is a second-order differential equation, in order to integrate it twice, the introduction of two constants of integration is required.

Since there are three equations, we have to determine six constants C_i. Then, the solutions take the form

$$\begin{aligned}x_i &= x_i\,(t, C_1, C_2, C_3, C_4, C_5, C_6),\\ \dot{x}_i &= \dot{x}_i\,(t, C_1, C_2, C_3, C_4, C_5, C_6).\end{aligned} \tag{1.21}$$

Next, according to (1.20) and (1.21) for $t = t_0$ we have

$$\begin{aligned}x_{i0} &= x_i\,(t_0, C_1, C_2, C_3, C_4, C_5, C_6),\\ \dot{x}_{i0} &= \dot{x}_i\,(t_0, C_1, C_2, C_3, C_4, C_5, C_6), \quad i = 1, 2, 3.\end{aligned} \tag{1.22}$$

The preceding system of algebraic equations consisting of six equations allows us to determine six unknown constants $C_i = C_i(x_{10}, x_{20}, x_{30}, \dot{x}_{10}, \dot{x}_{20}, \dot{x}_{30})$.

Substituting the constants C_i determined in that way into (1.21), we obtain the particular solutions of the form

$$x_i = x_i\left(t, x_{10}, x_{20}, x_{30}, \dot{x}_{10}, \dot{x}_{20}, \dot{x}_{30}\right). \tag{1.23}$$

Only in a few cases does the inverse dynamics problem allow one to determine solutions in analytical form. Systems of ordinary differential equation (1.2) are often strongly non-linear and their solutions are not known.

There exists, however, a whole range of approximate methods allowing for the determination of the desired solutions. These methods include numerical methods, analytical asymptotic methods, and mixed numerically analytical methods. If the ordinary differential equations are linear with constant coefficients, their formal solution does not present difficulties.

This problem is a broad one, and the reader can find more information, for instance, in the monographs of the author and his coworkers [5–10].

1.1.3 Particle Motion Under the Action of Simple Forces

Let us now consider special cases of exciting forces, which were classified at the beginning of Sect. 1.1.1. If the force acting on a particle is constant $\mathbf{F} = \text{const}$, the from (1.2) we obtain

$$\ddot{x}_i = \frac{1}{m} F_i. \tag{1.24}$$

Following the first integration of the three preceding equations, we have

$$\dot{x}_i = \frac{1}{m} F_i t + C_i. \tag{1.25}$$

Integrating (1.25) we obtain

$$\begin{aligned} x_1 &= \frac{F_1}{2m} t^2 + C_1 t + C_4, \\ x_2 &= \frac{F_2}{2m} t^2 + C_2 t + C_5, \\ x_3 &= \frac{F_3}{2m} t^2 + C_3 t + C_6. \end{aligned} \tag{1.26}$$

In order to uniquely determine the motion, the initial conditions should be taken into account:

$$x_i(0) = x_{i0}, \quad \dot{x}_i(0) = v_{i0}, \tag{1.27}$$

where $t_0 = 0$ was assumed. Substituting $t = 0$ into (1.26) and taking into account (1.27) we obtain the following values of the constants:

$$C_i = v_{i0}, \quad C_{i+3} = x_{i0}, \quad i = 1, 2, 3. \tag{1.28}$$

Taking into account (1.28) in (1.26), the equations of motion of the particle take the form

$$x_i(t) = \frac{F_i}{2m}t^2 + v_{0i}t + x_{i0}. \tag{1.29}$$

Introducing new variables $\xi_i(t) = x_i(t) - x_{i0}$, we eventually obtain

$$\begin{aligned} \xi_1(t) &= \frac{F_1}{2m}t^2 + v_{01}t, \\ \xi_2(t) &= \frac{F_2}{2m}t^2 + v_{02}t, \\ \xi_3(t) &= \frac{F_3}{2m}t^2 + v_{03}t. \end{aligned} \tag{1.30}$$

Setting $F_i = 0$ in (1.30) we obtain equations of uniform rectilinear motion of a particle, which is in agreement with Newton's fist law.

If the force acting on a particle has the form $\mathbf{F} = \mathbf{F}(t)$, then from (1.2) we obtain

$$\ddot{x}_i = \frac{F_i(t)}{m}, \quad i = 1, 2, 3. \tag{1.31}$$

Let us note that each of the three equations in (1.31) is independent of the others. The solutions are sought in an analogous way to that presented previously. As a result of integration we obtain

$$x_i = \frac{1}{m}\int_0^t \left[\int_0^\tau F_i(\xi)\mathrm{d}\xi\right]\mathrm{d}\tau + v_{0i}t + x_{i0}, \tag{1.32}$$

where the initial conditions (1.27) were assumed.

Let us note that, if we assume F_i to be constant in (1.32), we obtain (1.29). In the general case the solutions (1.32) depend on the form of the function $F_i(t)$. If these relationships are given in the form of elementary functions and the integration can be successfully conducted twice, we obtain the explicit analytical solution of the form (1.32), that is, an exact solution.

If the force depends only on position, that is,

$$\mathbf{F} = \mathbf{F}(\mathbf{r}) = F_1(x_1, x_2, x_3)\mathbf{E}_1 + F_2(x_1, x_2, x_3)\mathbf{E}_2 + F_3(x_1, x_2, x_3)\mathbf{E}_3,$$

then the problem is not an easy one. According to (1.2), equations of motion take the form

$$\begin{aligned}\ddot{x}_1 &= \frac{1}{m} F_1(x_1, x_2, x_3),\\ \ddot{x}_2 &= \frac{1}{m} F_2(x_1, x_2, x_3),\\ \ddot{x}_3 &= \frac{1}{m} F_3(x_1, x_2, x_3).\end{aligned} \tag{1.33}$$

However, now each of the equations depends on the solutions of two other equations. In the general case the solution of the system of equations (1.33) is not known. However, in two special cases the solutions can be determined.

The first case occurs when the relationships $F_i(x_1, x_2, x_3)$ are linear, that is, we have

$$F_i(x_1, x_2, x_3) = k_{i1}x_1 + k_{i2}x_2 + k_{i3}x_3, \quad i = 1, 2, 3. \tag{1.34}$$

Equations (1.33) take the form

$$\ddot{\mathbf{x}}(t) = \mathbf{A}\mathbf{x}(t), \tag{1.35}$$

where

$$\mathbf{A} = \frac{1}{m}\begin{bmatrix} k_{11} & k_{12} & k_{13} \\ k_{21} & k_{22} & k_{23} \\ k_{31} & k_{32} & k_{33} \end{bmatrix}, \tag{1.36}$$

$$\mathbf{x} = \begin{bmatrix} x_1(t) \\ x_2(t) \\ x_3(t) \end{bmatrix}. \tag{1.37}$$

The system of second-order differential equations of the form (1.35) is linear, and its solution can be determined analytically [3]. We deal with the second case where an exact solution can be obtained when a particle moves in rectilinear motion. Then, for a description of the motion a single axis is enough instead of a system of axes.

Such a case was discussed, for instance, in [3]. Let the equation of motion have the form

$$m\ddot{y} = F(y) \tag{1.38}$$

for the initial conditions

$$y(0) = y_0, \quad \dot{y}(0) = v_0. \tag{1.39}$$

Let us note that

$$\ddot{y} = \frac{\mathrm{d}\dot{y}}{\mathrm{d}y}\frac{\mathrm{d}y}{\mathrm{d}t} = v\frac{\mathrm{d}v}{\mathrm{d}y}, \tag{1.40}$$

where $v = \dot{y}$.

Substituting expression (1.40) into (1.38) we have

$$mvdv - F(y)\mathrm{d}y = 0, \tag{1.41}$$

and after integration

$$m\int_{v_0}^{\dot{y}} vdv = \int_{y_0}^{y} F(\eta)\mathrm{d}\eta \tag{1.42}$$

we obtain

$$m\frac{\dot{y}^2 - v_0^2}{2} = \int_{y_0}^{y} F(\eta)\mathrm{d}\eta. \tag{1.43}$$

From the preceding equation we obtain

$$\dot{y} \equiv \frac{\mathrm{d}y}{\mathrm{d}t} = \pm\sqrt{v_0^2 + \frac{2}{m}\int_{y_0}^{y} F(\eta)\mathrm{d}\eta}, \tag{1.44}$$

and separating the variables we have

$$t = \int_{y_0}^{y} \frac{\mathrm{d}\xi}{\pm\sqrt{v_0^2 + \frac{2}{m}\int_{y_0}^{\xi} F(\eta)\mathrm{d}\eta}} + t_0. \tag{1.45}$$

From (1.45) it follows that unexpectedly we obtained the solution in the form of the inverse function of $y(t)$, that is, $t = t(y)$. If we assume that the relationship $F(y)$ has the form enabling the integrations described in (1.45), then the relationship $t(y)$ can usually be represented through elementary functions. If the relationship $F(y) = ky$ is linear, then additionally it is possible to find the relationship $y = y(t)$.

The last of the considered cases is associated with the relationship $\mathbf{F} = \mathbf{F}(\mathbf{v})$. Similar to the case of $\mathbf{F} = \mathbf{F}(\mathbf{r})$, the problem cannot be generally solved in exact form. However, if the relationship is linear, the solution can be obtained in analytical form.

Let us consider one special case of the rectilinear motion of a particle, that is,

$$m\ddot{y} = F(\dot{y}) \tag{1.46}$$

or

$$m\frac{\mathrm{d}v}{\mathrm{d}t} = F(v). \tag{1.47}$$

Following separation of the variables and integration we obtain

$$t = \int \frac{m}{F(\xi)} \mathrm{d}\xi + C_1. \tag{1.48}$$

On the assumption that as a result of integration we will succeed in determining the function $t = t(v)$, and also in determining $v = v(t)$, it is easy to carry out the integration for the second time obtaining

$$y(t) = \int v(t, v_0) \mathrm{d}t + C_2. \tag{1.49}$$

If this method fails, then we can proceed in a different way [2]. One should multiply (1.47) by dy obtaining

$$m \frac{\mathrm{d}y}{\mathrm{d}t} \mathrm{d}v = F(v) \mathrm{d}y, \tag{1.50}$$

and separating the variables

$$\mathrm{d}y = \frac{mv}{F(v)} \mathrm{d}v. \tag{1.51}$$

Integrating the preceding equation we obtain

$$y(v) = m \int \frac{\xi}{F(\xi)} \mathrm{d}\xi + C_2. \tag{1.52}$$

At the end of the calculations in this section it should be emphasized that the subject of our analysis was the simple dependencies of the force on particular parameters, such as time, position, and velocity, that occurred separately. The problem becomes more complicated if they occur together.

As was already mentioned, the *forward dynamics problems* for a particle consist in the determination of the force producing the motion, on the assumption that the mass of the particle is known and the equation of the trajectory of the particle motion is known as well.

While solving problems of this class, one often applies the so-called *method of kinetostatics*. One makes use of d'Alembert's principle through introduction of the inertia forces $\mathbf{F}_B = -m\mathbf{a}$, where $\mathbf{a}$ is the acceleration (kinetics), and then one takes into consideration the equilibrium of forces acting on a particle (statics).

Example 1.1. A particle A of mass m moves along a path that is an ellipse of the form $x_1^2 a^{-2} + x_2^2 b^{-2} = 1$ (Fig. 1.3). The acceleration of the moving particle is parallel to the axis OX_2. Determine the relationship $\mathbf{F}(t)$, where $\mathbf{F}$ is the force acting on the particle. One should assume the following initial conditions: $\mathbf{r}(0) = b\mathbf{E}_2$, $\dot{\mathbf{r}}(0) = v_0\mathbf{E}_1$.

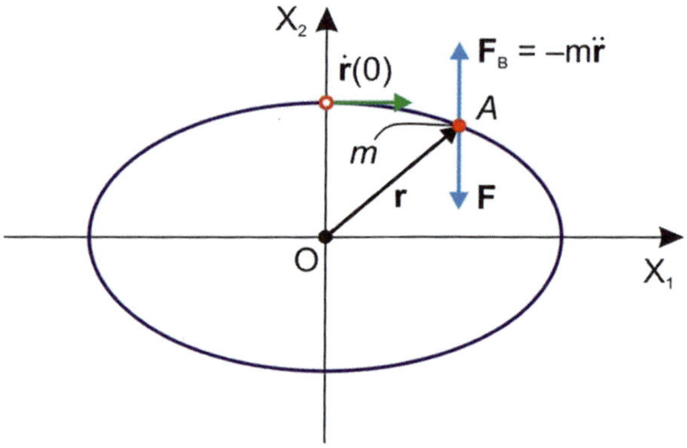

Fig. 1.3 Motion of a particle along an ellipse

According to the conditions of the problem we have

$$m\ddot{\mathbf{r}} = \mathbf{F},$$

that is,

$$m\left(\ddot{x}_1\mathbf{E}_1 + \ddot{x}_2\mathbf{E}_2\right) = F_1\mathbf{E}_1 + F_2\mathbf{E}_2,$$

and hence we obtain the following scalar equations:

$$m\ddot{x}_1 = \mathbf{F}_1, \quad m\ddot{x}_2 = \mathbf{F}_2.$$

Because the acceleration $\ddot{\mathbf{r}} \parallel \mathbf{E}_2$, we have $F_1 = 0$. Thus we have

$$\dot{x}_1 = \text{const};$$

following integration, and taking into account the initial condition $x_1 = v_0 t$, and from the equation of an ellipse we find

$$x_2 = \pm b\sqrt{1 - \frac{v_0^2 t^2}{a^2}},$$

where the plus sign means that the particle is located on the upper half of the ellipse.

Because $F_1 = 0$, the determination of $\mathbf{F}$ boils down to the determination of F_2, that is, to the determination of the acceleration $\ddot{x}_2$. From the last relationship we have

$$\dot{x}_2 = \frac{b\left(-\frac{2v_0^2 t}{a^2}\right)}{2\sqrt{1 - \frac{v_0^2 t^2}{a^2}}} = -\frac{bv_0^2 t}{a^2}\left(1 - \frac{v_0^2 t^2}{a^2}\right)^{-\frac{1}{2}},$$

$$\ddot{x}_2 = -\frac{bv_0^2}{a^2}\frac{\left[\sqrt{1-\frac{v_0^2t^2}{a^2}}-\left(\frac{-\frac{v_0^2t^2}{a^2}}{\sqrt{1-\frac{v_0^2t^2}{a^2}}}\right)\right]}{1-\frac{v_0^2t^2}{a^2}}$$

$$= -\frac{bv_0^2}{a^2}\frac{\left[\left(1-\frac{v_0^2t^2}{a^2}\right)+\frac{v_0^2t^2}{a^2}\right]}{\left(\sqrt{1-\frac{v_0^2t^2}{a^2}}\right)^3} = -\frac{b^4v_0^2}{a^2x_2^3},$$

and eventually we obtain

$$\mathbf{F} = F_2\mathbf{E}mb = m\ddot{x}_2\mathbf{E}mb_2 = -\frac{mb^4v_0^2}{a^2x_2^3}\mathbf{E}mb_2.$$

In the end, let us observe that zero horizontal acceleration of the particle (constant horizontal velocity v_0) and "start" from point (O, b) mean that the particle can move only on the ellipse quadrant, and its vertical velocity at point (a, O) will be infinitely large. □

1.1.4 Law of Conservation of Momentum

The law of conservation of momentum of a particle is also known as Newton's second law or Euler's first law. According to the previous considerations and taking into account the introductory definitions given in [1] we will analyze the motion of a particle in the adopted Cartesian coordinate system of basis $\{\mathbf{E}_1, \mathbf{E}_2, \mathbf{E}_3\}$. We will describe the position of particle A by the radius vector $\mathbf{r} = \overrightarrow{OA}$, where O is the origin of the coordinate system. The integral form of the law of conservation of momentum is expressed by the formula

$$\mathbf{P}(t) - \mathbf{P}(t_0) = \int_{t_0}^{t} \mathbf{F}(\tau)\mathrm{d}\tau, \tag{1.53}$$

where $\mathbf{F}$ denotes the vector of force acting on the particle.

If we assume that vector $\mathbf{P}(t)$ is differentiable with respect to time, then differentiating both sides of (1.53) we have

$$\mathbf{F} = \dot{\mathbf{P}} = \frac{\mathrm{d}\mathbf{P}}{\mathrm{d}t}. \tag{1.54}$$

The preceding equation can be written in the form

$$\mathrm{d}\mathbf{P} = \mathbf{F}\mathrm{d}t, \tag{1.55}$$

where $\mathbf{F}\mathrm{d}t$ is the *elementary impulse of a force*.

Let us note [2] that the elementary impulse of a force depends on the force vector and on time, that is, it contains more information about the external action on the particle than the notion of force used so far.

On the other hand, $\mathbf{P}$ denotes the momentum of a particle, that is,

$$\mathbf{P} = m\mathbf{v} = m\dot{\mathbf{r}}. \tag{1.56}$$

Comparing (1.55) and (1.56) we obtain

$$m\mathrm{d}\mathbf{v} = \mathbf{F}\mathrm{d}t. \tag{1.57}$$

The preceding formula has the following interpretation.

An infinitesimal increment of momentum of a particle is equal to the elementary impulse of forces acting on the particle.

Let us note that from (1.57) we obtain Newton's second law

$$m\frac{\mathrm{d}\mathbf{v}}{\mathrm{d}t} = \mathbf{F}, \tag{1.58}$$

which justifies the equivalence of the two names of the discussed law.

Because the elementary impulse of a force and elementary momentum are both vectors, one may add them by means of sums or integrals (see next section). Integrating both sides of (1.57) we obtain

$$\int\limits_{v_1}^{v_2} \mathrm{d}(m\mathbf{v}) = \int\limits_{t_1}^{t_2} \mathbf{F}\mathrm{d}t, \tag{1.59}$$

where $m = \text{const}$ was assumed. From the preceding equation we get

$$\Delta\mathbf{P} = \mathbf{J}_{12}, \tag{1.60}$$

where

$$\Delta\mathbf{P} = \mathbf{P}_2 - \mathbf{P}_1, \quad \mathbf{P}_2 = m\mathbf{v}_2, \quad \mathbf{P}_1 = m\mathbf{v}_1,$$

$$\mathbf{J}_{12} = \int\limits_{t_1}^{t_2} \mathbf{F}\mathrm{d}t. \tag{1.61}$$

The formula (1.60) expresses the law of conservation of momentum in time interval $(t_2 - t_1)$. From the formula it also follows that $\mathbf{P}_2 = \mathbf{P}_1 + \mathbf{J}_{12}$. This means that if we know momentum vector $\mathbf{P}_1$ and the impulse [i.e., we know the relation $\mathbf{F} = \mathbf{F}(t)$, and we are able to calculate integral (1.61)], then after geometric addition of these two vectors we obtain momentum vector $\mathbf{P}_2$.

If $\mathbf{F} = \text{const}$, then from (1.61) we get

$$\mathbf{J}_{12} = \mathbf{F}(t_2 - t_1). \tag{1.62}$$

Let us also note that if there is no force acting on a particle ($\mathbf{F} = \mathbf{0}$), then according to (1.56) and (1.57) we have

$$\mathbf{P} \equiv m\mathbf{v} = \text{const}, \tag{1.63}$$

which means that the momentum of a particle is constant.

1.1.5 Laws of Conservation of the Kinematic Quantities of a Particle

We say that an arbitrary kinematic quantity is conserved during the motion of a particle if it is constant during this motion.

Such quantities are very important in mechanics and are called *integrals of motion*. It turns out that in the case of particle motion we are dealing with the laws of conservation of three kinetic quantities. We will successively present these laws below.

1. *The law of conservation of momentum*

According to (1.53) and (1.56), the momentum $\mathbf{P}(t)$ is conserved (constant) during the time of motion $t - t_0$ if $\int_{t_0}^{t} \mathbf{F}(\tau)\mathrm{d}\tau = 0$. This takes place when $\mathbf{F}(\tau) = \mathbf{0}$.

There exists also the second variant of the law of conservation of momentum of a particle. That is, we can consider the case of the conservation of the projection of momentum vector $\mathbf{P}$ onto an arbitrary direction represented by vector $\mathbf{b}(t)$, that is, we demand that

$$\frac{\mathrm{d}}{\mathrm{d}t}(\mathbf{P} \circ \mathbf{b}) = 0. \tag{1.64}$$

This means that certain components of vector $\mathbf{P}$ in the adopted basis of a vector space are conserved. From the preceding equation we obtain

$$\dot{\mathbf{P}} \circ \mathbf{b} + \mathbf{P} \circ \dot{\mathbf{b}} = \mathbf{F} \circ \mathbf{b} + \mathbf{P} \circ \dot{\mathbf{b}} = 0. \tag{1.65}$$

From formula (1.65) it follows that if $\mathbf{F} \circ \mathbf{b} + \mathbf{P} \circ \dot{\mathbf{b}} = 0$, then the quantity $\mathbf{P} \circ \mathbf{b}$ is conserved.

If a particle moves in a gravitational field, then the components of the vector $\mathbf{P} = m\mathbf{v}$ in the directions $\mathbf{E}_1$ and $\mathbf{E}_2$ are conserved because $\mathbf{F} = -mg\mathbf{E}_3$. Another example can be that of a particle hitting a smooth plane. This time the components of vector $\mathbf{P}$ parallel to the aforementioned plane are conserved.

2. *The law of conservation of angular momentum*
Taking an arbitrary point O as fixed we will observe the motion of a particle of mass m. At time instant t the particle has a velocity $\mathbf{v}$ and momentum $\mathbf{P} = m\mathbf{v}$ and is acted upon by a force $\mathbf{F}(t)$.

Definition 1.1. The moment of the vector of a particle's momentum with respect to point O is called the angular momentum of the particle with respect to point O ($\mathbf{K}_O$).

According to the preceding definition we have

$$\mathbf{K}_O = \mathbf{r} \times \mathbf{P} = \mathbf{r} \times m\mathbf{v}, \tag{1.66}$$

where $\mathbf{r} = \mathbf{r}(t)$ is a radius vector of the previously analyzed particle of the tail at point O. Differentiating (1.66) with respect to time we have

$$\dot{\mathbf{K}}_O = \dot{\mathbf{r}} \times m\mathbf{v} + \mathbf{r} \times m\dot{\mathbf{v}} = \mathbf{v} \times m\mathbf{v} + \mathbf{r} \times \mathbf{F} = \mathbf{r} \times \mathbf{F}. \tag{1.67}$$

Equation (1.67) takes the form

$$\frac{\mathrm{d}\mathbf{K}_O}{\mathrm{d}t} \equiv \dot{\mathbf{K}}_O = \mathbf{M}_O, \tag{1.68}$$

where

$$\mathbf{M}_O = \mathbf{r} \times \mathbf{F} \tag{1.69}$$

and $\mathbf{M}_O$ is a moment of force $\mathbf{F}$ about pole O.

The equation obtained enables us to formulate the following conclusion.

The rate of change of angular momentum of a particle with respect to a given fixed pole (point O) is equal to the moment of forces acting on the particle with respect to that pole.

Multiplying both sides of (1.68) by $\mathrm{d}t$ we obtain

$$\mathrm{d}\mathbf{K}_O = \mathbf{M}_O \mathrm{d}t, \tag{1.70}$$

which enables us to formulate the following conclusion.

An elementary increment of the angular momentum vector is equal to an elementary impulse of a moment of force.

By conducting calculations analogous to those regarding the conservation of momentum presented previously, one may integrate both sides of (1.70), obtaining

$$\Delta\mathbf{K}_O = \mathbf{K}_{O2} - \mathbf{K}_{O1} = \int_{t_1}^{t_2} \mathbf{M}_O(t)\mathrm{d}t, \tag{1.71}$$

where $\mathbf{K}_{Oi} = \mathbf{K}_O(t_i)$, $i = 1, 2$. Equation (1.71) enables us to formulate the following conclusion.

The increment of the angular momentum of a particle in a certain time interval is equal to the impulse of a moment of force acting on the particle during that time interval.

If $\mathbf{M}_O = \text{const}$, then from (1.71) we obtain

$$\mathbf{K}_{O2} - \mathbf{K}_{O1} = \mathbf{M}_O(t_2 - t_1). \tag{1.72}$$

If $\mathbf{M}_O = \mathbf{0}$, then according to (1.70), $\mathbf{K}_O = \text{const} = \mathbf{C}$. That enables us to formulate the law of conservation of angular momentum.

The angular momentum of a particle with respect to an arbitrary fixed pole is constant if the moment of forces acting on the particle with respect to that pole equals zero.

According to (1.66) we have

$$\begin{aligned}
\dot{\mathbf{K}}_O &= \frac{\mathrm{d}}{\mathrm{d}t}(\mathbf{r} \times m\mathbf{v}) = m\frac{\mathrm{d}}{\mathrm{d}t}(\mathbf{r} \times \mathbf{v}) \\
&= m\frac{\mathrm{d}}{\mathrm{d}t}\begin{vmatrix} \mathbf{E}_1 & \mathbf{E}_2 & \mathbf{E}_3 \\ x_1 & x_2 & x_3 \\ \dot{x}_1 & \dot{x}_2 & \dot{x}_3 \end{vmatrix} = m\frac{\mathrm{d}}{\mathrm{d}t}(x_2\dot{x}_3 - x_3\dot{x}_2)\,\mathbf{E}_1 \\
&\quad + m\frac{\mathrm{d}}{\mathrm{d}t}(\dot{x}_1 x_3 - x_1\dot{x}_3)\,\mathbf{E}_2 + m\frac{\mathrm{d}}{\mathrm{d}t}(x_1\dot{x}_2 - x_2\dot{x}_1)\,\mathbf{E}_3,
\end{aligned} \tag{1.73}$$

and according to (1.69) we obtain

$$\begin{aligned}
\mathbf{M}_O &= \begin{vmatrix} \mathbf{E}_1 & \mathbf{E}_2 & \mathbf{E}_3 \\ x_1 & x_2 & x_3 \\ F_1 & F_2 & F_3 \end{vmatrix} = (x_2F_3 - x_3F_2)\,\mathbf{E}_1 \\
&\quad + (F_1x_3 - x_1F_3)\,\mathbf{E}_2 + (x_1F_2 - x_2F_1)\,\mathbf{E}_3.
\end{aligned} \tag{1.74}$$

Finally, from (1.68) [taking into account (1.73) and (1.74)] we obtain

$$\begin{aligned}
m\frac{\mathrm{d}}{\mathrm{d}t}(x_2\dot{x}_3 - x_3\dot{x}_2) &= x_2F_3 - x_3F_2, \\
m\frac{\mathrm{d}}{\mathrm{d}t}(x_3\dot{x}_1 - x_1\dot{x}_3) &= x_3F_1 - x_1F_3, \\
m\frac{\mathrm{d}}{\mathrm{d}t}(x_1\dot{x}_2 - x_2\dot{x}_1) &= x_1F_2 - x_2F_1.
\end{aligned} \tag{1.75}$$

For the case $\mathbf{M}_O = \mathbf{0}$, $\mathbf{K}_O = C_1\mathbf{E}_1 + C_2\mathbf{E}_2 + C_3\mathbf{E}_3$ and from (1.75) we obtain

$$K_{O1} \equiv m(x_2\dot{x}_3 - x_3\dot{x}_2) = C_1,$$
$$K_{O2} \equiv m(x_3\dot{x}_1 - x_1\dot{x}_3) = C_2,$$
$$K_{O3} \equiv m(x_1\dot{x}_2 - x_2\dot{x}_1) = C_3, \tag{1.76}$$

where C_i, $i = 1, 2, 3$ are constants.

We obtained, then, the first integrals of the equations of motion, which are called the *integrals of angular momentum.*

We encounter the conservation of angular momentum in two cases, namely, either when the angular momentum vector is completely conserved or when one of the angular momentum vector components is conserved. The former case takes place if $\mathbf{F}\|\mathbf{r}$. Additionally, the angular momentum is conserved when $\mathbf{F} = 0$ (the trivial case). Then from (1.67) it follows that $\mathbf{K}_O = \text{const} = \mathbf{C}$. Such force $\mathbf{F}$ we call a *central force* and its example can be a gravitational force of attraction. Then, according to (1.66), we have $\mathbf{r}(t) \times m\mathbf{v}(t) = \mathbf{C}$. Vector $\mathbf{C}$ is constant over time, that is, vectors $\mathbf{r}$ and $\mathbf{v}$ form one plane, and thus the particle motion takes place on a plane. If we treat an arbitrary vector $\mathbf{b}(t)$ as a component of vector $\mathbf{K}_O$, then the second case of the angular momentum conservation takes place when $\mathbf{K}_O \circ \mathbf{b} = \text{const}$. Then we have

$$\dot{\mathbf{K}}_O \circ \mathbf{b} + \mathbf{K}_O \circ \dot{\mathbf{b}} = \mathbf{r} \times \mathbf{F} \circ \mathbf{b} + \mathbf{K}_O \circ \dot{\mathbf{b}} = 0, \tag{1.77}$$

where formula (1.67) was used.

3. *The law of conservation of kinetic energy*

Before we formulate the law of conservation of the kinetic energy of a particle, we will introduce certain basic notions associated with work, power, and efficiency.

Figure 1.4 presents the elementary displacement $\mathrm{d}\mathbf{r}$ of particle A caused by the action of force $\mathbf{F}$.

The elementary work $\mathrm{d}W$ of force $\mathbf{F}$ during the elementary displacement $\mathrm{d}\mathbf{r}$ we define as a scalar product of the form

$$\mathrm{d}W = \mathbf{F} \circ \mathrm{d}\mathbf{r} = \mathbf{F} \circ \frac{\mathrm{d}\mathbf{r}}{\mathrm{d}t}\mathrm{d}t = \mathbf{F} \circ \mathbf{v}\mathrm{d}t. \tag{1.78}$$

Although $\mathrm{d}W$ denotes the total differential, one should bear in mind that generally the elementary work is not a total differential of any function [2]. From formula (1.78) it is seen that the work is a scalar and can be expressed both through the position vector and through the vector of velocity of a particle. A unit of work is the joule ($1J = N \cdot m$).

Because a particle moves along a path, we have $|\mathrm{d}\mathbf{r}| = \mathrm{d}s$, and from (1.78) we obtain

$$\mathrm{d}W = F\mathrm{d}s\cos\alpha. \tag{1.79}$$

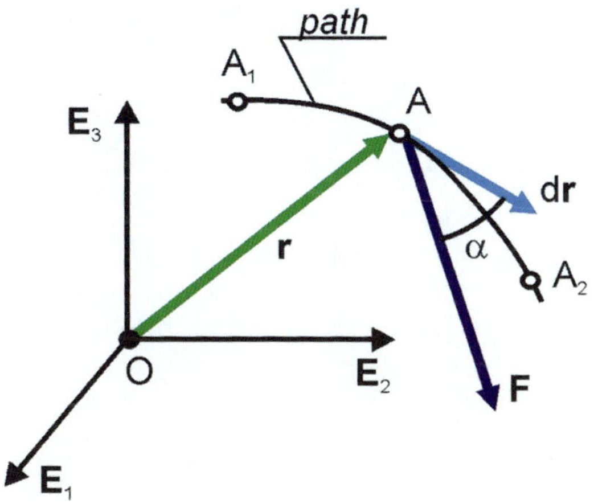

Fig. 1.4 Elementary displacement of particle A

The measure of mechanical motion is kinetic energy $\frac{mv^2}{2}$ and momentum $m\mathbf{v}$, and the measure of force action is the impulse $\mathbf{J}$ and work of force W.

Theorem 1.1. *The work of a resultant force on a certain displacement is equal to the algebraic sum of works of each of the component forces on this displacement.*

Proof. By assumption we have

$$\mathbf{F}_r = \mathbf{F}_1 + \mathbf{F}_2 + \cdots + \mathbf{F}_N,$$

and the work on displacement $\mathbf{dr}$ is equal to

$$\delta W = \mathbf{F}_r \circ \mathbf{dr} = \mathbf{F}_1 \circ \mathbf{dr} + \mathbf{F}_2 \circ \mathbf{dr} + \cdots + \mathbf{F}_N \circ \mathbf{dr},$$

and after integration we obtain

$$W = \int_{A_1}^{A_2} \mathbf{F}_r \circ \mathbf{dr} = \int_{A_1}^{A_2} \mathbf{F}_1 \circ \mathbf{dr} + \int_{A_1}^{A_2} \mathbf{F}_2 \circ \mathbf{dr} + \cdots + \int_{A_1}^{A_2} \mathbf{F}_N \circ \mathbf{dr},$$

that is,

$$W = W_1 + W_2 + \cdots + W_N,$$

which completes the proof. □

Theorem 1.2. *The work of a force constant with regard to the magnitude and direction on a resultant displacement is equal to the sum of works of this force on each of the components of the displacement.*

Proof. Let point A under the action of force $\mathbf{F}$ undergo consecutive displacements $\mathbf{r}_n$ such that

$$\mathbf{r} = \mathbf{r}_1 + \mathbf{r}_2 + \cdots + \mathbf{r}_N.$$

The work done by force $\mathbf{F}$ is equal to

$$W = \mathbf{F} \circ \mathrm{d}\mathbf{r} = \mathbf{F} \circ \mathrm{d}\mathbf{r}_1 + \mathbf{F} \circ \mathrm{d}\mathbf{r}_2 + \cdots + \mathbf{F} \circ \mathrm{d}\mathbf{r}_N = W_1 + W_2 + \cdots + W_N,$$

which completes the proof. □

Until now, we have used the notion of elementary work. If under the action of the force $\mathbf{F} = \mathbf{F}(t)$ the particle changed its position from point A_1 to point A_2 in the time interval $t_2 - t_1$, then the work done by this force would be equal to the sum of the elementary works, which can be written as

$$W_{\widehat{A_1 A_2}} = \int\limits_{\widehat{A_1 A_2}} \mathbf{F} \circ \mathrm{d}\mathbf{r} = \int\limits_{\widehat{A_1 A_2}} F \cos\alpha \mathrm{d}s, \tag{1.80}$$

where $\widehat{A_1 A_2}$ denotes the arc connecting points A_1 and A_2 along the curve of the path of the particle.

If the force $\mathbf{F} = \text{const}$, then one of the axes of the Cartesian coordinate system can be taken as parallel to $\mathbf{F}$. Then we obtain

$$\begin{aligned} W_{\widehat{A_1 A_2}} &= \int\limits_{\widehat{A_1 A_2}} (F_1\mathbf{E}_1 + F_2\mathbf{E}_2 + F_3\mathbf{E}_3) \circ (\mathrm{d}x_1\mathbf{E}_1 + \mathrm{d}x_2\mathbf{E}_2 + \mathrm{d}x_3\mathbf{E}_3) \\ &= \int\limits_{X_{1A_1}}^{X_{1A_2}} F_1 \mathrm{d}x_1 = F_1\left(x_{1A_2} - x_{1A_1}\right), \end{aligned} \tag{1.81}$$

where axis $\mathbf{E}_1 \parallel \mathbf{F}$, that is, $F_2 = F_3 = 0$.

If we consider the work of the force $m\mathbf{g}$ in the gravitational field, then, on the assumption that $\mathbf{E}_3 \parallel \mathbf{g}$, and $\mathbf{E}_3$ and $\mathbf{g}$ have opposite senses (i.e., $\mathbf{E}_3 \circ \mathbf{g} = -g$), we have

$$W_{\widehat{A_1 A_2}} = -mg\left(x_{3A_2} - x_{3A_1}\right) = mg\left(x_{3A_1} - x_{3A_2}\right). \tag{1.82}$$

If one takes the force

$$\mathbf{F} = k\mathbf{r}, \tag{1.83}$$

where k is a stiffness coefficient of a massless elastic element, then

$$\begin{aligned} W_{\widehat{A_1 A_2}} &= -k \int\limits_{\widehat{A_1 A_2}} \mathbf{r} \circ \mathrm{d}\mathbf{r} \\ &= -k \int\limits_{r_{A_1}}^{r_{A_2}} (x_1\mathrm{d}x_1 + x_2\mathrm{d}x_2 + x_3\mathrm{d}x_3) = \frac{k}{2}\left(\mathbf{r}_{A_1}^2 - \mathbf{r}_{A_2}^2\right). \end{aligned} \tag{1.84}$$

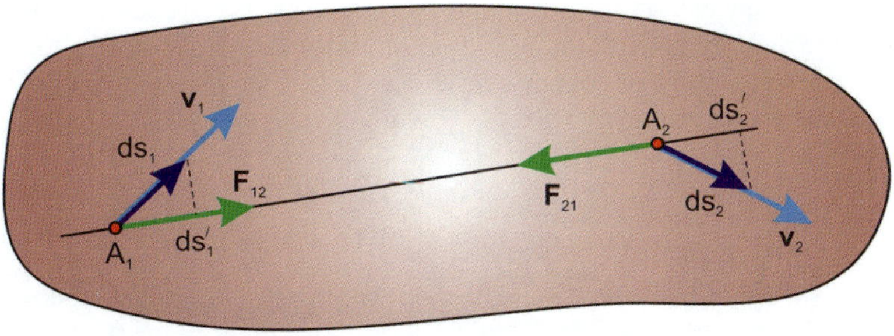

Fig. 1.5 Work of internal forces on example of two points of a rigid body

Let us note that during calculation of the integrals used for estimation of the work, the time $t_2 - t_1$ over which the work was done was not exploited. It follows that the work does not depend on time. However, from our everyday experience we know that the time in which the work is done is of great importance. Therefore, we introduce the notion of power.

Work of internal forces in a rigid body
Let us consider two arbitrary points of a rigid body A_1 and A_2 (Fig. 1.5) subjected to the action of internal forces F_{12} and F_{21}, where according to Newton's third law $\mathbf{F}_{12} = -\mathbf{F}_{21}$.

Let the vectors of velocities respectively at points A_1 and A_2 be equal to $\mathbf{v}_1$ and $\mathbf{v}_2$. Displacements of those points at the time $\mathrm{d}t$ are equal to $\mathrm{d}s_1 = v_1\mathrm{d}t$ and $\mathrm{d}s_2 = v_2\mathrm{d}t$. Projections of velocities on the line A_1A_2 are equal, so the projections $\mathrm{d}s_1^{'}$ and $\mathrm{d}s_2^{'}$ are also equal. The work done by forces $\mathbf{F}_{12}$ and $\mathbf{F}_{21}$ is equal to

$$F_{12}\mathrm{d}s_1\cos(\mathbf{F}_{12}, \mathbf{v}_1) + F_{21}\mathrm{d}s_2\cos(\mathbf{F}_{21}, \mathbf{v}_2) = F_{12}\mathrm{d}s_1' - F_{12}\mathrm{d}s_2' = 0.$$

Considering all points we have

$$\delta W = \sum_{n=1}^{r} \delta W_n = 0,$$

that is, the sum of works of internal forces on an arbitrary displacement of a rigid body is equal to zero.

The following ratio of elementary work done by a force to the time of its duration we call the power

$$N = \frac{\mathrm{d}W}{\mathrm{d}t} \equiv \dot{W}. \tag{1.85}$$

According to (1.78) we obtain

$$N = \mathbf{F} \circ \mathbf{v}. \tag{1.86}$$

It is easy to notice that when a machine is working, part of the power is consumed to overcome the resistance to motion, friction, self-heating, wear, etc. As a result as the machine's output we obtain power that is smaller than the power initially supplied. The ratio of these two powers we call efficiency:

$$\eta = \frac{W_e}{W_{in}} = \frac{N_e}{N_{in}}, \tag{1.87}$$

where the subscript e denotes the effective work or power and the subscript in denotes the input work or power. The efficiency varies in the range $0 \leq \eta \leq 1$.

The law of conservation of kinetic energy (work) is described by a simple formula:

$$\dot{T} = N, \tag{1.88}$$

where $T = \frac{1}{2}m\mathbf{v} \circ \mathbf{v}$ and the power of the force is described by formula (1.86). Let us note that force $\mathbf{F}$ plays here an important role, although it does not occur explicitly in (1.88).

Equation (1.88) is validated, since we have

$$\dot{T} = \frac{1}{2}m\frac{\mathrm{d}}{\mathrm{d}t}(\mathbf{v} \circ \mathbf{v}) = m\dot{\mathbf{v}} \circ \mathbf{v} = \mathbf{F} \circ \mathbf{v}. \tag{1.89}$$

We will also demonstrate the validity of the following statement.

The differential of the kinetic energy of a particle is equal to the elementary work done by the force acting on the particle.

From Newton's second law we have

$$m\frac{\mathrm{d}\mathbf{v}}{\mathrm{d}t} \circ \mathrm{d}\mathbf{r} = \mathbf{F} \circ \mathrm{d}\mathbf{r}. \tag{1.90}$$

The right-hand side of (1.90) is $\mathbf{F} \circ \mathrm{d}\mathbf{r} = \mathrm{d}W$, and its left-hand side is

$$m\mathrm{d}\mathbf{v} \circ \frac{\mathrm{d}\mathbf{r}}{\mathrm{d}t} = m\mathbf{v} \circ \mathrm{d}\mathbf{v} = \mathrm{d}\left(\frac{m\mathbf{v}^2}{2}\right) = \mathrm{d}T, \tag{1.91}$$

and in view of that,

$$\mathrm{d}T = \mathrm{d}W. \tag{1.92}$$

If a particle moving along a certain path was located at point A_1 at time instant t_1, and at point A_2 at instant t_2, then, integrating (1.92), we obtain

$$\int_{v_1}^{v_2} \mathrm{d}T = \int_{\widehat{A_1 A_2}} \mathrm{d}W \tag{1.93}$$

or

$$T_2 - T_1 = W_{\widehat{A_1 A_2}}, \tag{1.94}$$

where

$$T_i = \frac{m}{2} v_i^2, \quad i = 1, 2. \tag{1.95}$$

Equation (1.94) leads to the formulation of the following conclusion.

The increment of the kinetic energy of a particle is equal to the work done by the forces acting on the path traveled by the particle.

4. *The law of conservation of total energy (mechanical energy)*
Before this law is formulated, we will introduce certain basic notions regarding a potential force field [2]. If for a certain force field (i.e., the region of a space within which forces act on a particle) described by the scalar function

$$V = V(\mathbf{r}, t), \tag{1.96}$$

which depends on the position of the particle and on time, the force acting on the particle described by the radius vector is equal to

$$\mathbf{F} = -\mathrm{grad} V(\mathbf{r}, t), \tag{1.97}$$

then such a field we call a *non-stationary potential field* and function $V(\mathbf{r}, t)$ we call the *potential*. If $V = V(\mathbf{r})$, then the field is called a *stationary field*.

According to Definition (1.97) we have

$$\mathbf{F} = -\frac{\partial V}{\partial \mathbf{r}} = -\frac{\partial V}{\partial x_i} \mathbf{E}_i. \tag{1.98}$$

Introducing the definition of the total differential of the potential as

$$\mathrm{d}V = \frac{\partial V}{\partial \mathbf{r}} \circ \mathrm{d}\mathbf{r} = \sum_{i=1}^{3} \frac{\partial V}{\partial x_i} \mathrm{d}x_i, \tag{1.99}$$

and taking into account (1.98), we obtain

$$\mathrm{d}V = -\mathbf{F} \circ \mathrm{d}\mathbf{r} = -\sum_{i=1}^{3} F_i \mathrm{d}x_i. \tag{1.100}$$

The elementary work of a potential force is a total differential of the potential of the field, but with a minus sign. This means that the positive work done in the potential field is accompanied by a decrease in potential value.

If the force field is a conservative field, the potential

$$V = V(\mathbf{r}). \tag{1.101}$$

In turn, the work of the potential force during displacement of a particle under the action of this force from a given point of the field $V(x_1, x_2, x_3)$ to the point of reference $V_0(x_{10}, x_{20}, x_{30})$, that is, $E_p = V - V_0$, is called the potential energy of a potential force. If $V_0 = 0$, then $E_p = V$.

The field is conservative if the following conditions are satisfied:

(1) The work of a force on a closed path S in a conservative field is equal to

$$\oint_S \mathbf{F} \circ \mathrm{d}\mathbf{r} = \oint_S F_i \mathrm{d}x_i = \oint_S \frac{\partial V}{\partial x_i} \mathrm{d}x_i = 0. \tag{1.102}$$

(2) The curl at every point of the conservative field

$$\operatorname{rot} \mathbf{F}(\mathbf{r}) = \mathbf{0}, \tag{1.103}$$

that is,

$$\begin{aligned}\operatorname{rot} \mathbf{F}(\mathbf{r}) = \nabla \times \mathbf{F} &= \begin{vmatrix} \mathbf{E}_1 & \mathbf{E}_2 & \mathbf{E}_3 \\ \frac{\partial}{\partial x_1} & \frac{\partial}{\partial x_2} & \frac{\partial}{\partial x_3} \\ F_1 & F_2 & F_3 \end{vmatrix} \\ &= \left(\frac{\partial F_3}{\partial x_2} - \frac{\partial F_2}{\partial x_3}\right)\mathbf{E}_1 + \left(\frac{\partial F_1}{\partial x_3} - \frac{\partial F_3}{\partial x_1}\right)\mathbf{E}_2 + \left(\frac{\partial F_2}{\partial x_1} - \frac{\partial F_1}{\partial x_2}\right)\mathbf{E}_3 \\ &= \mathbf{0}, \end{aligned} \tag{1.104}$$

which means that the conservative field is irrotational.

Three scalar equations following from (1.104) are called *Schwartz conditions*, and if they are satisfied, then the examined field is called the *potential field*.

Figure 1.6 illustrates the notions introduced previously connected with the potential force field.

Particle A in the figure lies on the surface over which the potential has a constant value $V(x_1, x_2, x_3) = C$. Such a surface is called an *equipotential surface*. The marked gradient grad $V(x_{1A}, x_{2A}, x_{3A})$ is a vector normal to the equipotential surface at the point occupied by particle A. Its sense is taken toward increasing values of the potential.

The lines (paths) of forces are perpendicular to the equipotential surfaces (since force F has the opposite sense to the vector grad $V \perp V(x_1, x_2, x_3) = C$).

Below we present some examples of potential fields. One of them is the gravitational field of Earth.

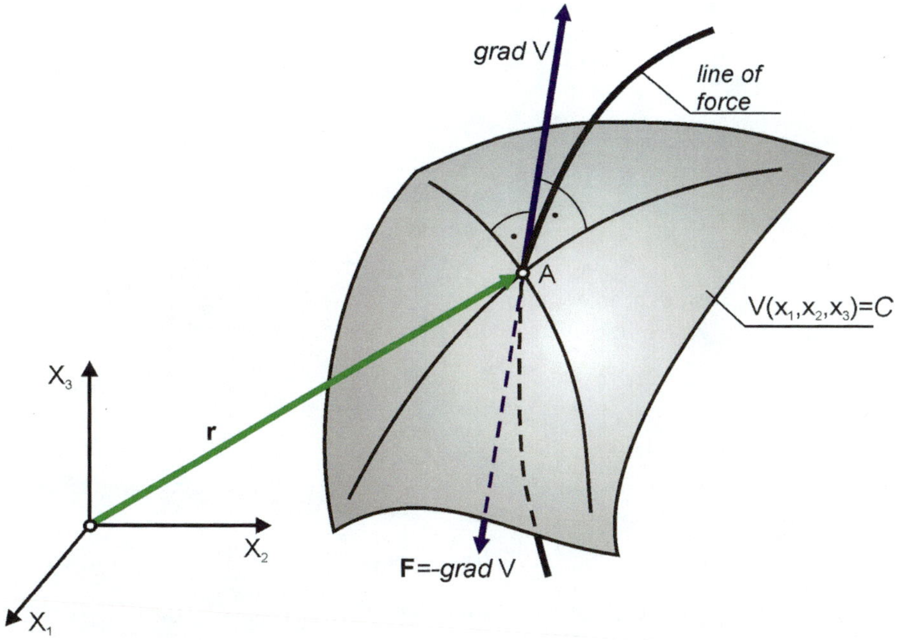

Fig. 1.6 Force acting on particle $A(\mathbf{r})$ situated in a potential force field

The force acting on a particle of mass m is equal to

$$\mathbf{F} = 0 \cdot \mathbf{E}_1 + 0 \cdot \mathbf{E}_2 + mg \cdot \mathbf{E}_3, \tag{1.105}$$

where the system of coordinates was introduced in such a way that $\mathbf{g} \parallel \mathbf{E}_3$ and $\mathbf{g} \circ \mathbf{E}_3 = -g$. From (1.105) it follows that $F_3 = -mg$.

We determine the total differential based on formula (1.100). It is equal to

$$\mathrm{d}V = -F_3 \mathrm{d}x_3 = d(mgx_3), \tag{1.106}$$

and integrating both sides of this equation we obtain

$$V = mgx_3 + C_3, \tag{1.107}$$

where C_3 is a constant of integration.

Assuming $V(0) = 0$ we determine the constant $C_3 = 0$. The equation $V = mgx_3$ determines the family of equipotential planes parallel to the plane OX_1X_2. The lines of forces are perpendicular to these planes, and the potential V increases with increasing x_3. We call such a conservative field a *uniform field*.

Now we will show the difference between the *uniform field* and the *non-uniform field* on an example of the gravitational field of Earth.

In the case of the *uniform field* we have

$$F_1 = F_2 = 0, \qquad F_3 = -mg, \tag{1.108}$$

where m is the mass of a particle.

According to (1.106) we have

$$V_1(x_3) = -\int_0^{x_3} (-mg)\mathrm{d}x_3 = mgx_3 + C_1, \tag{1.109}$$

where C_1 is a constant of integration.

In turn, in the case of the *non-uniform field* the force of gravity is equal to

$$F_3(x_3) = -G\frac{mM}{(R + x_3)^2}, \tag{1.110}$$

where R is the radius of Earth and the origin of the coordinate system lies on Earth's surface, that is,

$$F_3(0) = -mg. \tag{1.111}$$

From (1.110) and taking into account (1.111) we obtain

$$G = \frac{gR^2}{M}, \tag{1.112}$$

that is,

$$F_3(x_3) = -\frac{mgR^2}{(R + x_3)^2}. \tag{1.113}$$

The potential of the non-uniform field is equal to

$$\begin{aligned} V_2(x_3) &= -\int F_3(x_3)\mathrm{d}x_3 + C_2 = mg\int \frac{\mathrm{d}x_3}{(R + x_3)^2} + C_2 \\ &= -mgR^2\frac{1}{R + x_3} + C_2. \end{aligned} \tag{1.114}$$

Let

$$V_1(0) = V_2(0) = 0, \tag{1.115}$$

then from (1.109) we have $C_1 = 0$, and from (1.114) we obtain $C_2 = mgR$.

Eventually we have

$$\begin{aligned} V_1 &= mgx_3, \\ V_2 &= mgR\left(1 - \frac{R}{R + x_3}\right). \end{aligned} \tag{1.116}$$

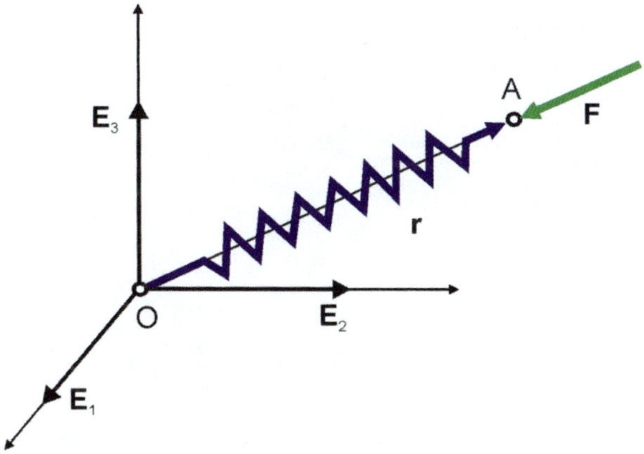

Fig. 1.7 Example of a central force

Let us consider now the case of the so-called *central field*. The direction of a central force at the point occupied by A is the same as the direction of the position vector of particle A, and the sense of the force is opposite to the sense of the position vector. An example of the central force can be the force acting in a spring (Fig. 1.7).

According to Fig. 1.7, we have

$$\mathbf{F} = -K(r)\hat{\mathbf{r}} = -K(r)\frac{\mathbf{r}}{r}, \tag{1.117}$$

where $\hat{\mathbf{r}}$ is the unit vector of the axis OA.

We obtain the total differential from formulas (1.100) and (1.117):

$$\mathrm{d}V = \frac{K(r)}{r}\mathbf{r} \circ \mathrm{d}\mathbf{r} = \frac{K(r)}{r} r\hat{\mathbf{r}} \circ \mathrm{d}\mathbf{r} = K(r)\mathrm{d}r. \tag{1.118}$$

We obtain the potential of the central force after integrating (1.118)

$$V = \int K(r)\mathrm{d}r + C. \tag{1.119}$$

It is not always possible to obtain the analytical form of a solution because it depends on the form of the non-linear function $K(r)$. If we are dealing with the linear case, that is, $K(r) = kr$, then from formula (1.119) we obtain

$$V = \frac{1}{2}kr^2 + C. \tag{1.120}$$

Eventually, assuming $V(0) = 0$, we have

$$V = \frac{1}{2}kr^2 = \frac{1}{2}k\left(x_1^2 + x_2^2 + x_3^2\right). \tag{1.121}$$

The equipotential surface is, then, a sphere of radius $\sqrt{\frac{2V}{k}}$, where $V = \text{const}$. Lines of action of forces are determined by the radii of this sphere.

An example of a central field is the universal gravitational field. If we denote masses of planets by m_1 and m_2, then for this field we have

$$F(r) = G\frac{m_1 m_2}{r^2} \tag{1.122}$$

and the potential

$$V = Gm_1 m_2 \int \frac{\mathrm{d}r}{r^2} = -\frac{Gm_1 m_2}{r} + C. \tag{1.123}$$

Assuming a value of $C = 0$ for a singular point $r = 0$, we have $V(0) = -\infty$. This means that the gravitational force of attraction of two planets (treated as point masses) tends to infinity for $r \to 0$. Now we will demonstrate that any central field is a potential field.

Multiplying (1.117) by sides by $\mathbf{E}_i$ we have

$$F_i = -\frac{F(r)x_i}{r}, \qquad i = 1, 2, 3. \tag{1.124}$$

The first Schwartz condition described by (1.104) has the form

$$\frac{\partial F_3}{\partial x_2} - \frac{\partial F_2}{\partial x_3} = 0. \tag{1.125}$$

We successively calculate

$$\begin{aligned}
\frac{\partial F_3}{\partial x_2} &= -\left(\frac{\partial F}{\partial r}\frac{\partial r}{\partial x_2}\frac{x_3}{r} + F(r)x_3\frac{\partial}{\partial x_2}\left(\frac{1}{r}\right)\right) \\
&= -\left(\frac{\partial F}{\partial r}\frac{x_2}{r}\frac{x_3}{r} - F(r)x_3\frac{x_2}{r}\frac{1}{r^2}\right) = -\left(\frac{\partial F}{\partial r}\frac{x_2 x_3}{r^2} - F(r)\frac{x_2 x_3}{r^3}\right), \qquad (1.126) \\
\frac{\partial F_2}{\partial x_3} &= -\left(\frac{\partial F}{\partial r}\frac{\partial r}{\partial x_3}\frac{x_2}{r} + F(r)x_2\frac{\partial}{\partial x_3}\left(\frac{1}{r}\right)\right) \\
&= -\left(\frac{\partial F}{\partial r}\frac{x_3}{r}\frac{x_2}{r} - F(r)x_2\frac{x_3}{r}\frac{1}{r^2}\right) = -\left(\frac{\partial F}{\partial r}\frac{x_2 x_3}{r^2} - F(r)\frac{x_2 x_3}{r^3}\right). \qquad (1.127)
\end{aligned}$$

Comparing (1.126) and (1.127) with one another, we can see that (1.125) is satisfied. In a similar way it is possible to prove the two remaining Schwartz conditions.

Now we will take up the analysis of the law of conservation of total energy of a particle

$$E = T + V, \tag{1.128}$$

where T is the kinetic energy and V a potential energy of the particle. We will divide the forces $\mathbf{F}$ acting on the particle into conservative forces $\mathbf{F}^c$ described by formula (1.98) and non-conservative forces $\mathbf{F}^n$.

Exploiting the calculations conducted earlier, according to (1.89), we have

$$\dot{T} = \mathbf{F} \circ \mathbf{v} = (\mathbf{F}^c + \mathbf{F}^n) \circ \mathbf{v} = -\frac{\partial V}{\partial \mathbf{r}} \circ \mathbf{v} + \mathbf{F}^n \circ \mathbf{v} = -\dot{V} + \mathbf{F}^n \circ \mathbf{v}. \tag{1.129}$$

From (1.128) and (1.129) we obtain

$$\dot{E} = \mathbf{F}^n \circ \mathbf{v}. \tag{1.130}$$

The obtained equation enables us to formulate the following conclusion.

The total energy of a particle is conserved, provided that the non-conservative forces perform no work, that is, $\mathbf{F}^n \circ \mathbf{v} = 0$.

Forces in the potential field are conservative. In view of that, $E = \text{const}$, and the law of conservation of the total energy of a particle can be formulated in the following form.

The total energy of a particle moving in a potential conservative field has a constant value.

Example 1.2. A particle of mass m moves on the inside surface of a cylinder of radius ρ whose generatrix is parallel to the vector of acceleration of gravity $\mathbf{g}$. On the assumption that the inside surface is perfectly smooth, determine the equation of motion of the particle and the reaction force of the cylinder.

Introducing the Cartesian coordinate system (Fig. 1.8), the initial conditions of motion are as follows: $x_1(0) = \rho$, $x_2(0) = 0$, $x_3(0) = 0$, $\dot{x}_1(0) = 0$, $\dot{x}_2(0) = v_0 \cos\alpha$, $\dot{x}_3(0) = v_0 \sin\alpha$, where α means the angle between the tangent to the trajectory and the plane OX_1X_2.

From Newton's second law for the particle it follows that

$$m\ddot{x}_1 = R_{x_1}, \quad m\ddot{x}_2 = R_{x_2}, \quad m\ddot{x}_3 = mg.$$

Let us integrate the third equation of the obtained system of equations. As a result we obtain

$$\dot{x}_3 = gt + C_1, \quad x_3 = \frac{gt^2}{2} + C_1 t + C_2,$$

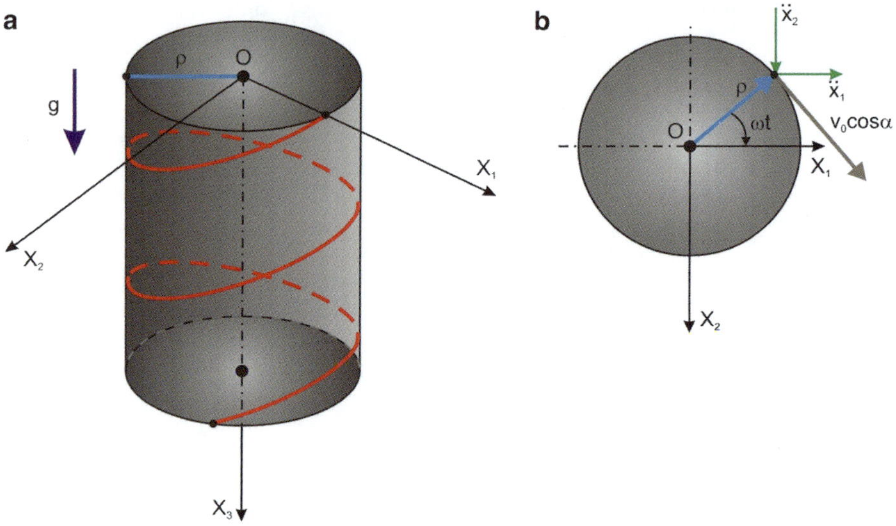

Fig. 1.8 Motion of a particle on the smooth inside surface of a cylinder (**a**) and motion of particle projection onto the plane OX_1X_2 (**b**)

and taking into account the initial conditions we have

$$v_0 \sin\alpha = C_1, \quad C_2 = 0.$$

Eventually,

$$x_3(t) = \frac{gt^2}{2} + v_0 t \sin\alpha.$$

Now, let us make use of the law of conservation of total energy of a particle moving in the potential field.

From (1.128) it follows that

$$\frac{mv_0^2}{2} = \frac{m}{2}(\dot{x}_1^2 + \dot{x}_2^2 + \dot{x}_3^3) - mgx_3.$$

In the preceding equation, $x_3(t)$ is known, and in view of that we obtain

$$\begin{aligned}\dot{x}_1^2 + \dot{x}_2^2 &= v_0^2 - \dot{x}_3^2 + 2gx_3 \\ &= v_0^2 - (gt + v_0 \sin\alpha)^2 + 2g\left(\frac{gt^2}{2} + v_0 t \sin\alpha\right) \\ &= v_0^2 - v_0^2 \sin^2\alpha = v_0^2(1 - \sin^2\alpha) = v_0^2 \cos^2\alpha.\end{aligned}$$

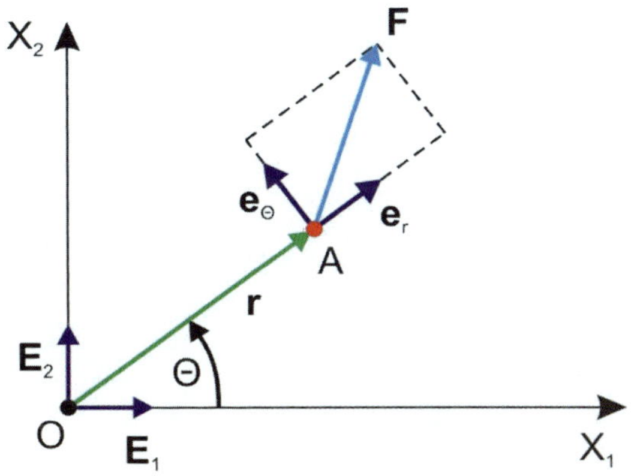

Fig. 1.9 Motion of particle A in plane OX_1X_2

Let us note that the equation just obtained has the following physical interpretation. It describes the motion of the projection of a particle onto the plane OX_1X_2. That is, the projection of a particle moving on the inside surface of the cylinder moves with a constant velocity $v_0 \cos\alpha$ along a circle of radius ρ (Fig. 1.8b). Because of this we can determine the remaining components of motion $x_1(t)$ and $x_2(t)$ from the following equations:

$$x_1 = \rho \cos\omega t, \quad x_2 = -\rho \sin\omega t,$$

where

$$\omega = \rho^{-1} v_0 \cos\alpha.$$

From the preceding equations we obtain the components of acceleration

$$\ddot{x}_1 = -\omega^2 x_1, \qquad \ddot{x}_2 = -\omega^2 x_2.$$

The reaction of the surface of cylinder R is caused by the acceleration of motion of the particle and is equal to

$$R = m\sqrt{\ddot{x}_1^2 + \ddot{x}_2^2} = m\omega^2\sqrt{x_1^2 + x_2^2} = m\omega^2\rho = m\frac{v_0^2\cos^2\alpha}{\rho}. \qquad \square$$

1.1.6 Particle Motion in the Central Field

In the case of planar motion of a particle it is convenient to make use of the polar coordinates in the calculations (Fig. 1.9).

Using the derived relationships [see (4.195) and (4.196), Chap. 4 of [1]] for the polar system we obtain

$$\begin{aligned}\dot{\mathbf{r}} &= \dot{r}\mathbf{e}_r + r\dot{\Theta}\mathbf{e}_\Theta,\\ \ddot{\mathbf{r}} &= \left(\ddot{r} - r\dot{\Theta}^2\right)\mathbf{e}_r + \left(r\ddot{\Theta} + 2\dot{r}\dot{\Theta}\right)\mathbf{e}_\Theta.\end{aligned} \tag{1.131}$$

From Newton's second law we obtain

$$m\ddot{\mathbf{r}} = \mathbf{F}, \tag{1.132}$$

and taking into account (1.131) in formula (1.132) we have

$$\begin{aligned}m\left(\ddot{r} - r\dot{\Theta}^2\right) &= F_r,\\ m\left(r\ddot{\Theta} + 2\dot{r}\dot{\Theta}\right) &= F_\Theta.\end{aligned} \tag{1.133}$$

The definition of the central field implies that $F_\Theta = 0$, that is, from the second equation of (1.133) we obtain

$$\frac{1}{r}\frac{\mathrm{d}}{\mathrm{d}t}\left(r^e\dot{\Theta}\right) \equiv \frac{1}{r}\left(2r\dot{r}\dot{\Theta} + r^2\ddot{\Theta}\right) = 0, \tag{1.134}$$

which leads to the following relationship:

$$r^2\dot{\Theta} = \mathrm{const} \equiv C. \tag{1.135}$$

We obtain the velocity of the particle from the first equation of (1.131). It is equal to

$$v^2 = \dot{r}^2 + r^2\dot{\Theta}^2 = \left(r^2\dot{\Theta}\right)^2\left(\left(\frac{\dot{r}}{r^2\dot{\Theta}}\right)^2 + \frac{1}{r^2}\right).$$

Because

$$\left(\frac{\dot{r}}{r^2\dot{\Theta}}\right)^2 = \left(\frac{\left(\frac{\mathrm{d}r}{\mathrm{d}t}\right)}{r^2\frac{\mathrm{d}\Theta}{\mathrm{d}t}}\right)^2 = \left(\frac{\mathrm{d}\left(\frac{1}{r}\right)}{\mathrm{d}\Theta}\right)^2,$$

we have

$$v^2 = C^2\left[\frac{1}{r^2} + \left(\frac{\mathrm{d}\left(\frac{1}{r}\right)}{\mathrm{d}\Theta}\right)^2\right]. \tag{1.136}$$

From the first equation of (1.133) we find

$$\begin{aligned} F_r &= m\left(\ddot{r} - r\dot{\Theta}^2\right) = -m\left(r\dot{\Theta}^2 - \ddot{r}\right) \\ &= -m\left(r^2\dot{\Theta}\right)^2\left[\frac{r\dot{\Theta}^2}{\left(r^2\dot{\Theta}\right)^2} - \frac{\frac{\mathrm{d}}{\mathrm{d}t}\left(\frac{\mathrm{d}r}{\mathrm{d}t}\right)}{\left(r^2\frac{\mathrm{d}\Theta}{\mathrm{d}t}\right)^2}\right] \\ &= -\frac{mC^2}{r^2}\left(\frac{1}{r} - \frac{\mathrm{d}(\mathrm{d}r)}{r^2\mathrm{d}\Theta^2}\right), \end{aligned}$$

where the following relationships were used:

$$\frac{\mathrm{d}\Theta}{\mathrm{d}t}\frac{\mathrm{d}\Theta}{\mathrm{d}t} = \left(\frac{\mathrm{d}\Theta}{\mathrm{d}t}\right)^2,$$

$$\frac{\mathrm{d}}{\mathrm{d}t}\left(\frac{\mathrm{d}\left(\frac{1}{r}\right)}{\mathrm{d}t}\right) = \frac{\mathrm{d}}{\mathrm{d}t}\left(-\frac{1}{r^2}\frac{\mathrm{d}r}{\mathrm{d}t}\right) = -\frac{1}{r^2}\frac{\mathrm{d}}{\mathrm{d}t}\left(\frac{\mathrm{d}r}{\mathrm{d}t}\right).$$

Eventually

$$F_r = -\frac{mC^2}{r^2}\left[\frac{1}{r} + \frac{\mathrm{d}^2\left(\frac{1}{r}\right)}{\mathrm{d}\Theta^2}\right]. \tag{1.137}$$

The formulas obtained serve as means of the direct determination of the velocity (1.136) and the force (1.137) of a particle in the central field and are called *Binet's*[1] *formulas*.

Now we will show that the angular momentum of a particle in the central field is conserved. Since we have

$$\mathbf{M} = \mathbf{r} \times \mathbf{F}_r = \mathbf{r} \times \mathbf{e}_r F_r = \mathbf{0}, \tag{1.138}$$

and because

$$\frac{\mathrm{d}\mathbf{K}}{\mathrm{d}t} = \mathbf{r} \times m\dot{\mathbf{r}} = \mathbf{M}, \tag{1.139}$$

the angular momentum of the particle is conserved

$$\mathbf{K} = \mathbf{r} \times m\dot{\mathbf{r}} = \mathrm{const.} \tag{1.140}$$

From (1.140) it follows that **K** changes neither the magnitude nor the direction. This means that the motion always takes place in one plane perpendicular to the angular momentum vector **K** and determined by vectors **r** and $\dot{\mathbf{r}}$.

---

[1]Jacques Binet (1786–1856), French mathematician.

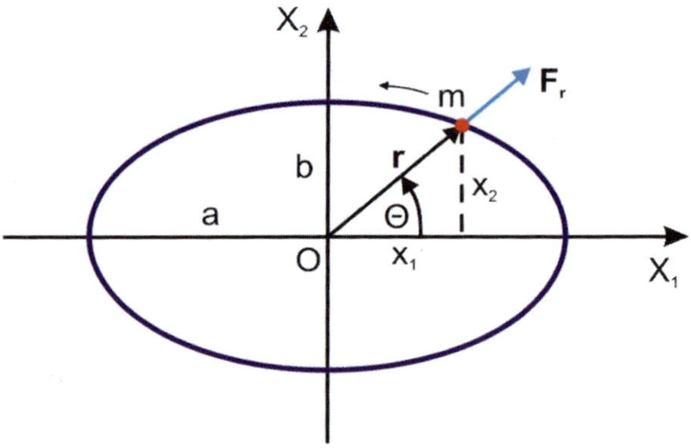

Fig. 1.10 Motion of particle along ellipse of semiaxes a and b

Example 1.3. Determine the magnitude of the central force $\mathbf{F}_r$, provided that the particle acted upon by this force moves along an ellipse of semiaxes a and b (Fig. 1.10).

Substituting $x_1 = r\cos\Theta$, $x_2 = r\sin\Theta$ into the equation of the ellipse

$$\frac{x_1^2}{a^2} + \frac{x_2^2}{b^2} = 1,$$

we obtain

$$\frac{\cos^2\Theta}{a^2} + \frac{\sin^2\Theta}{b^2} = \frac{1}{r^2},$$

and after transformation

$$\begin{aligned}\frac{1}{r^2} &= \frac{b^2\cos^2\Theta + a^2\left(1-\cos^2\Theta\right)}{a^2b^2} = \frac{-\left(a^2-b^2\right)\cos^2\Theta + a^2}{a^2b^2}\\ &= \frac{1}{b^2}\left[1 - \frac{(a^2-b^2)}{a^2}\cos^2\Theta\right].\end{aligned}$$

Eventually

$$\frac{1}{r} = \frac{1}{b}\sqrt{1-e^2\cos^2\Theta}\,, \quad e = \frac{1}{a}\sqrt{a^2-b^2}.$$

In order to make use of Binet's formulas one needs the following derivatives:

$$\frac{\mathrm{d}\left(\frac{1}{r}\right)}{\mathrm{d}\Theta} = \frac{1}{b}\frac{\mathrm{d}}{\mathrm{d}\Theta}\left(\sqrt{1-e^2\cos^2\Theta}\right) = \frac{1}{b}\frac{e^2\sin 2\Theta}{2\sqrt{1-e^2\cos^2\Theta}},$$

$$\frac{\mathrm{d}^2\left(\frac{1}{r}\right)}{\mathrm{d}\Theta^2} = \frac{e^2}{2b}\frac{\mathrm{d}}{\mathrm{d}\Theta}\left(\frac{\sin 2\Theta}{\sqrt{1-e^2\cos^2\Theta}}\right)$$

$$= \frac{e^2}{2b}\left[\frac{2\cos 2\Theta\sqrt{1-e^2\cos^2\Theta} - \frac{e^2\sin^2 2\Theta}{2\sqrt{1-e^2\cos^2\Theta}}}{1-e^2\cos^2\Theta}\right]$$

$$= \frac{e^2}{2b}\left[\frac{2\cos 2\Theta\left(1-e^2\cos^2\Theta\right) - \frac{1}{2}e^2\sin^2 2\Theta}{\sqrt{1-e^2\cos^2\Theta}\,(1-e^2\cos^2\Theta)}\right].$$

According to formula (1.137) we have

$$F_r = -\frac{mC^2}{r^2}\left[\frac{\sqrt{1-e^2\cos^2\Theta}}{b} + \frac{2e^2\cos 2\Theta\left(1-e^2\cos^2\Theta\right) - \frac{1}{2}e^4\sin^2 2\Theta}{2b\sqrt{1-e^2\cos^2\Theta}\,(1-e^2\cos^2\Theta)}\right]$$

$$= -\frac{mC^2}{r^2}\left[\frac{2\left(1-e^2\cos^2\Theta\right)^2 + 2e^2\cos 2\Theta\left(1-e^2\cos^2\Theta\right) - \frac{1}{2}e^4\sin^2 2\Theta}{2b\sqrt{1-e^2\cos^2\Theta}\,(1-e^2\cos^2\Theta)}\right]$$

$$= -\frac{mC^2}{r^2}\frac{L}{2b\sqrt{1-e^2\cos^2\Theta}\,(1-e^2\cos^2\Theta)},$$

where

$$L = 2\left(1 - 2e^2\cos^2\Theta + e^4\cos^4\Theta + e^2\cos^2\Theta - e^4\cos^4\Theta\right.$$
$$\left. - e^2\sin^2\Theta + e^4\sin^2\Theta\cos^2\Theta - e^4\sin^2\Theta\cos^2\Theta\right)$$
$$= 2\left(1-e^2\right).$$

Finally,

$$F_r = -\frac{mC^2}{r^2}\frac{\left(1-e^2\right)r^3}{b^4} = -\frac{mC^2\left(1-e^2\right)}{b^4}r,$$

because

$$\sqrt{1-e^2\cos^2\Theta}\left(1-e^2\cos^2\Theta\right) = \frac{b}{r}\frac{b^2}{r^2}.$$

Owing to the obtained sign of the force, it is evident that the moving particle is attracted to the center of the ellipse. □

Example 1.4. A particle of mass m moves along a circle of radius R under the action of force $\mathbf{F_r}$ directed toward point O' located on the circle (Fig. 1.11). Determine the velocity of the particle and the magnitude of the force as a function of radius vector $\mathbf{r}$, provided that for $r = 2R$ the particle A has velocity v_0.

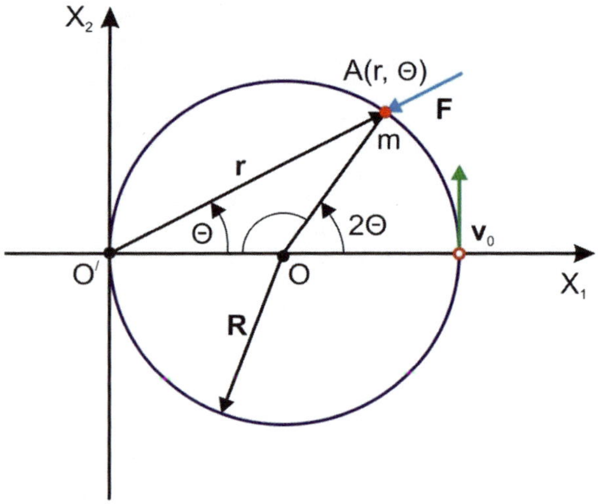

Fig. 1.11 Motion of a particle along a circle

The radius vector of particle A is described by the equation

$$\mathbf{r} = (R + R\cos 2\Theta)\,\mathbf{E}_1 + R\sin 2\Theta\mathbf{E}_2,$$

hence we get

$$\begin{aligned} r^2 &= 2R^2\cos 2\Theta + R^2 + R^2\cos^2 2\Theta + R^2\sin^2 2\Theta = 2R^2\,(1+\cos 2\Theta)\\ &= 2R^2\left(1+\cos^2\Theta - \sin^2\Theta\right) = 4R^2\cos^2\Theta, \end{aligned}$$

that is, the equation of a circle in the polar coordinate system takes the form

$$r = 2R\cos\Theta.$$

In order to make use of Binet's formulas we calculate

$$\begin{aligned} \frac{\mathrm{d}\left(\frac{1}{r}\right)}{\mathrm{d}\Theta} &= \frac{1}{2R}\frac{\mathrm{d}\left(\frac{1}{\cos\Theta}\right)}{\mathrm{d}\Theta} = \frac{1}{2R}\frac{\sin\Theta}{\cos^2\Theta},\\ \frac{\mathrm{d}^2\left(\frac{1}{r}\right)}{\mathrm{d}\Theta^2} &= \frac{1}{2R}\left[\frac{\cos^3\Theta + 2\sin^2\Theta\cos\Theta}{\cos^4\Theta}\right]\\ &= \frac{1}{2R}\left[\frac{\cos^2\Theta + 2\sin^2\Theta}{\cos^3\Theta}\right]. \end{aligned}$$

Fig. 1.12 Motion of particle *A* under action of force $\mathbf{F}_r = \mathbf{F}(\mathbf{r})$

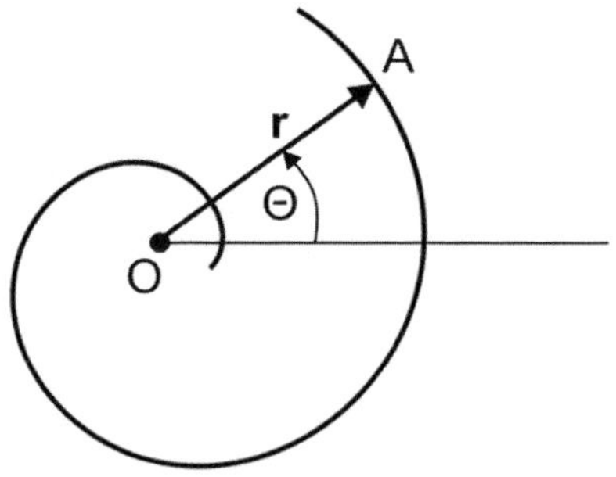

From (1.136) we have

$$v^2 = C^2 \left[\frac{1}{4R^2 \cos^2 \Theta} + \frac{\sin^2 \Theta}{4R^2 \cos^4 \Theta} \right]$$

$$= \frac{C^2}{4R^2} \left[\frac{\cos^2 \Theta + \sin^2 \Theta}{\cos^4 \Theta} \right] = \frac{4C^2 R^2}{r^4}.$$

According to the conditions of the problem

$$C = r^2 \dot{\Theta} = (2R)^2 \frac{v_0}{2R} = 2R v_0,$$

and hence the velocity of particle *A* reads

$$v = \sqrt{\frac{16R^4 v_0^2}{r^4}} = \frac{4R^2}{r^2} v_0.$$

In turn, from (1.137) we obtain the magnitude of the central force

$$F_r = -\frac{mC^2}{2Rr^2} \left[\frac{\cos^2 \Theta}{\cos^3 \Theta} + \frac{\cos^2 \Theta + 2\sin^2 \Theta}{\cos^3 \Theta} \right]$$

$$= -\frac{mC^2}{Rr^2} \frac{1}{\cos^3 \Theta} = -\frac{8mC^2 R^2}{r^5} = -\frac{32mR^4 v_0^2}{r^5}. \qquad \square$$

Example 1.5. Particle *A* of mass m moves around a fixed center O under the action of the force $\mathbf{F}_r = \mathbf{F}(\mathbf{r})$ directed along the radius vector of the particle (Fig. 1.12). Determine the force $\mathbf{F}_r$ and the path of the particle on condition that the velocity of the particle is equal to $v = C_1/r$, where C_1 is a constant.

According to (1.131) the velocity of the particle is equal to

$$v = \sqrt{\dot{r}^2 + r^2\dot{\Theta}^2} = \frac{C_1}{r}.$$

Hence, taking into account relation (1.135) we obtain

$$\dot{r}^2 + r^2\frac{C^2}{r^4} = \frac{C_1^2}{r^2}.$$

We transform the preceding equation into the form

$$\frac{\mathrm{d}r}{\mathrm{d}t} = \frac{1}{r}\sqrt{C_1^2 - C^2}.$$

Let us note that

$$\frac{\mathrm{d}r}{\mathrm{d}t} = \frac{\mathrm{d}r}{\mathrm{d}\Theta}\frac{\mathrm{d}\Theta}{\mathrm{d}t} = \frac{\mathrm{d}r}{\mathrm{d}\Theta}\frac{C}{r^2},$$

and hence we have

$$\frac{\mathrm{d}r}{\mathrm{d}\Theta} = r\frac{\sqrt{C_1^2 - C^2}}{C},$$

and we obtain

$$\ln\frac{r}{C_2} = \Theta\sqrt{\left(\frac{C_1}{C}\right)^2 - 1},$$

where C_2 is a constant. Eventually, the equation of the path along which particle A travels has the form

$$r = C_2 \mathrm{e}^{\left(\Theta\sqrt{\left(\frac{C_1}{C}\right)^2 - 1}\right)},$$

and it is the logarithmic spiral equation.

In order to make use of Binet's formulas (1.137) we calculate the derivatives

$$\frac{\mathrm{d}\left(\frac{1}{r}\right)}{\mathrm{d}\Theta} = \left[-\frac{1}{C_2}\sqrt{\left(\frac{C_1}{C}\right)^2 - 1}\right]\mathrm{e}^{-\Theta\sqrt{\left(\frac{C_1}{C}\right)^2 - 1}},$$

$$\frac{\mathrm{d}^2\left(\frac{1}{r}\right)}{\mathrm{d}\Theta^2} = \frac{1}{C_2}\left[\left(\frac{C_1}{C}\right)^2 - 1\right]\mathrm{e}^{-\Theta\sqrt{\left(\frac{C_1}{C}\right)^2 - 1}} = \frac{1}{r}\left[\left(\frac{C_1}{C}\right)^2 - 1\right],$$

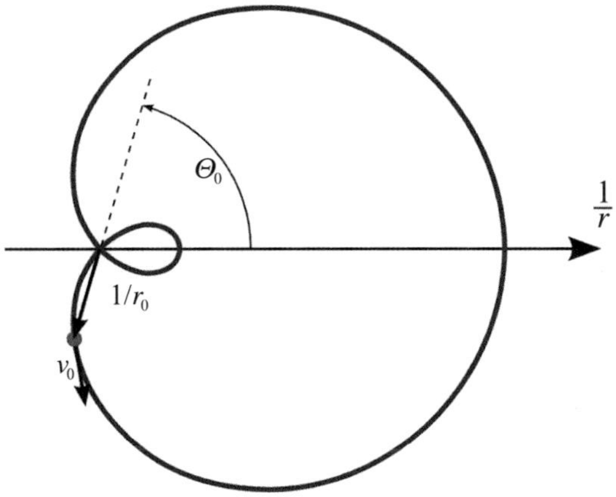

Fig. 1.13 Path of particle motion determined using numerical methods

and according to formula (1.137) we have

$$F_r = -\frac{mC^2}{r^2}\left[\frac{1}{r} + \frac{\mathrm{d}^2\left(\frac{1}{r}\right)}{\mathrm{d}\Theta^2}\right] = -\frac{mC^2}{r^3}\frac{C_1^2}{C^2} = -\frac{mC_1^2}{r^3}.$$

□

Example 1.6. Determine the path along which moves a particle under the action of the force $F_r = mv_0^2R^2/r^2$ for the following initial conditions:

$$r(0) = R, \;\; v(0) = v_0, \;\; \sphericalangle(\mathbf{r}, \mathbf{v}_0) = \Theta(t = 0) = \arctan a^{-1}.$$

From formula (1.135) we obtain

$$C = rr\dot{\Theta} = r(0)v_0 = Rv_0,$$

and from formula (1.137) we have

$$\frac{1}{r} + \frac{\mathrm{d}^2\left(\frac{1}{r}\right)}{\mathrm{d}\Theta^2} = -1.$$

The preceding non-homogeneous differential equation has the following solution:

$$\frac{1}{r} = A\cos\Theta + B\sin\Theta - 1.$$

The numerical solution of the differential equation for the initial conditions $a = 5$, $R = 2$ is presented in Fig. 1.13. □

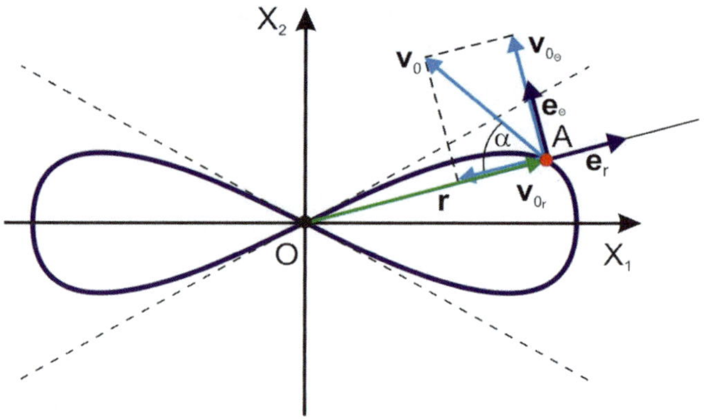

Fig. 1.14 Motion of particle along lemniscate of Bernoulli

Example 1.7. A particle of mass m in the central field moves along the lemniscate of Bernoulli[2] of the equation $r^2 = a\cos 2\Theta$, where r is the position vector of the particle attached at the center of a central force field. For the initial time instant of motion it is assumed that $r(0) = r_0$, $v(0) = v_0$ and $\sphericalangle(\mathbf{r}, \mathbf{v}_0) = \alpha$. Determine the central force acting on the particle (Fig. 1.14).

According to formula (1.131) the components of the velocity of particle A are equal to $v_r = \dot{r}$, $v_\Theta = r\dot{\Theta}$, and taking into account the initial time instant we have $v_\Theta = r\dot{\Theta}(0) = v_0 \sin\alpha$.

According to (1.134) and (1.135) we obtain

$$r^2(0)\dot{\Theta}(0) = r_0 v_0 \sin\alpha = C,$$

and the we represent lemniscate of Bernoulli in the following form:

$$\frac{1}{r} = \frac{1}{\sqrt{a}} (\cos 2\Theta)^{-\frac{1}{2}}.$$

We calculate successively

$$\frac{\mathrm{d}\left(\frac{1}{r}\right)}{\mathrm{d}\Theta} = \frac{1}{\sqrt{a}} \frac{\sin 2\Theta}{(\cos 2\Theta)^{\frac{3}{2}}},$$

[2]Jacob Bernoulli (1654–1705), distinguished mathematician in Bernoulli family (he described the lemniscate properties in 1694).

$$\frac{\mathrm{d}^2\left(\frac{1}{r}\right)}{\mathrm{d}\Theta^2} = \frac{1}{\sqrt{a}}\left[\frac{2\cos 2\Theta\,(\cos 2\Theta)^{\frac{3}{2}} + 3\,(\cos 2\Theta)^{\frac{1}{2}}\sin^2 2\Theta}{(\cos 2\Theta)^3}\right]$$
$$= \frac{2}{\sqrt{a}}(\cos 2\Theta)^{-\frac{1}{2}} + \frac{3}{\sqrt{a}}\tan^2 2\Theta\,(\cos 2\Theta)^{-\frac{1}{2}}$$
$$= \frac{1}{\sqrt{a}}(\cos 2\Theta)^{-\frac{1}{2}}\left(2 + 3\tan^2 2\Theta\right) = \frac{1}{r}\left(2 + 3\tan^2 2\Theta\right).$$

From the second of Binet's formulas (1.137) it follows that

$$F_r = -\frac{mC^2}{r^2}\left(\frac{1}{r} + \frac{\mathrm{d}^2\left(\frac{1}{r}\right)}{\mathrm{d}\Theta^2}\right) = -\frac{mr_0^2v_0^2\sin^2\alpha}{r^2}\frac{1}{r}\left[1 + 2 + 3\tan^2 2\Theta\right]$$
$$= -\frac{3mr_0^2v_0^2\sin^2\alpha}{r^3}\left[1 + \tan^2 2\Theta\right] = -\frac{3mr_0^2v_0^2\sin^2\alpha}{r^3}\left[\frac{\sin^2 2\Theta + \cos^2 2\Theta}{\cos^2 2\Theta}\right]$$
$$= -\frac{3mr_0^2v_0^2\sin^2\alpha}{r^3}\frac{1}{\cos^2 2\Theta} = -\frac{3mr_0^2v_0^2a^2\sin^2\alpha}{r^7},$$

and the minus sign denotes that the particle is attracted to the field center, that is, to point O. □

1.2 Fundamental Laws of Dynamics of a Mechanical System

1.2.1 Introduction

A (discrete or continuous) group of particles, isolated from the environment, is called a material system (see also Chap. 1 of [1]). Such a system can consist of a finite or infinite number of particles.

A group of particles in which the position or motion of every particle depends on the positions and motions of the othe particles is called a system of particles (SOP).

A SOP whose motion is not limited (is limited) by constraints we call a free (constrained) SOP. An example of a free SOP is the Solar System, where planets are treated as particles and move along orbits only under the action of forces acting on them. An example of a constrained SOP is any mechanism or machine.

The main vector of all internal forces $\mathbf{F}^i = \mathbf{0}$, that is, $\sum_{n=1}^{N} F_{x_{ni}} = 0, i = 1, 2, 3$. The main moments of all internal forces with respect to an arbitrary point $\mathbf{M}_0^i$, that is, $\sum_{n=1}^{N} M_{0ni} = 0, i = 1, 2, 3$.

In this book more attention will be devoted to discrete (lumped) systems. The reader will find more information regarding the dynamics of continuous systems in, for example, monographs of the author and his coworkers [3–10].

Material systems can be divided into *free systems* and *constrained systems.*

Free systems have no limitations imposed on the motion of any particle of the system. The particles of such systems are characterized by the possibility of displacement in an arbitrary direction and by velocities that are determined only by the initial conditions and forces acting on the particles. However, the motions of these particles are mutually dependent on each other since the force acting on one particle may depend on the velocities and displacements of other particles. In contrast, constrained systems are characterized by the fact that on all or some part of the particles limitations of motion are imposed. Those limitations are called *constraints.*

Let us now consider a discrete material system (DMS) composed of a finite number N of particles n and let every particle have mass m_n.

The dynamics of such a system can be analyzed based on Newton's laws presented in Sect. 1.1. It can be demonstrated on the basis of Newton's third law that all *internal forces*, that is, the forces coming only from the other particles of the system, cancel out each other (they form a closed system of vectors). The forces acting on the material system that come from the particles not belonging to the investigated system are called *external forces.*

Similar considerations also apply to the continuous material system (CMS), where we assume that a mass element dm is acted upon by the force $\mathbf{F}$dm (here $\mathbf{F}$ denotes a mass density of force) [11, 12].

In the case of DMS, according to Newton's second law, we have

$$m_n \mathbf{a}_n = \mathbf{F}_n^e + \mathbf{F}_n^i, \tag{1.141}$$

where $\mathbf{a}_n$ denotes the vector of acceleration of particle n of mass m_n whose position is described by vector $\mathbf{r}_n$. The force $\mathbf{F}_n^e$ ($\mathbf{F}_n^i$) denotes a resultant force of all external forces (internal forces) applied to particle n.

Our aim is the determination of the motion of the DMS, that is, the motion of all its particles $n = 1, 2, \ldots, N$. This boils down to the integration of the system of second-order differential equations (1.141), since $\mathbf{a}_n = \ddot{\mathbf{r}}_n$. For large N this problem is very complicated and often its solution is not feasible. However, it turns out that there exist certain characteristics (functions) of the considered mechanical system that are dependent on the velocity and position of its particles (and often also on time), which remain constant despite changes in $\mathbf{r}_n(t)$ and $\dot{\mathbf{r}}_n(t)$. If such a function exists and is constant, we call it the *first integral of the system of differential equations* (1.141). The integration of all equations of system (1.141) can be replaced with the analysis of certain quantities characteristic of the system, such as *quantity of motion*, *angular momentum (moment of momentum)*, and *kinetic energy of a system.*

The material appearing in this section (and also in this chapter) might be extended by the reader by [13–21].

1.2.2 *Law of Conservation of Momentum*

According to the previous considerations, let an arbitrary element dm (whose position is described by position vector $\mathbf{r}$ in a certain inertial Cartesian coordinate system) of CMS be acted upon by the resultant vectors of internal $\mathbf{F}^i$ and external $\mathbf{F}^e$ forces. Then, by Newton's second law, we have

$$\mathrm{d}m\frac{\mathrm{d}^2\mathbf{r}}{\mathrm{d}t^2} = \mathbf{F}^i + \mathbf{F}^e. \quad (1.142)$$

Integrating the preceding equation over the whole mass m of the CMS we obtain

$$\int\limits_m \frac{\mathrm{d}^2\mathbf{r}}{\mathrm{d}t^2}\mathrm{d}m = \int\limits_m \left(\mathbf{F}^e + \mathbf{F}^i\right)\mathrm{d}m. \quad (1.143)$$

According to Newton's third law the internal forces cancel out each other. Therefore we obtain

$$\int\limits_m \frac{\mathrm{d}^2\mathbf{r}}{\mathrm{d}t^2}\mathrm{d}m = \int\limits_m \mathbf{F}^e\mathrm{d}m. \quad (1.144)$$

Because we are considering a system whose mass is constant over time, (1.144), we can transform it into the form

$$\frac{\mathrm{d}}{\mathrm{d}t}\int\limits_m \mathbf{v}\mathrm{d}m = \frac{\mathrm{d}\mathbf{P}}{\mathrm{d}t} = \int\limits_m \mathbf{F}^e\mathrm{d}m, \quad (1.145)$$

where

$$\mathbf{v} = \frac{\mathrm{d}\mathbf{r}}{\mathrm{d}t}, \quad \mathbf{P} = \int\limits_m \mathbf{v}\mathrm{d}m, \quad (1.146)$$

and $\mathbf{P}$ is called the vector of momentum of a CMS.

One obtains similar relationships for a DMS consisting of N isolated particles, but the integral symbol in (1.145) should be replaced with the sum symbol. In this case, for DMS, (1.145) assumes the form [it is the sum of (1.141)]

$$\frac{\mathrm{d}}{\mathrm{d}t}\left(\sum_{n=1}^{N} m_n\mathbf{v}_n\right) = \frac{\mathrm{d}\mathbf{P}}{\mathrm{d}t} = \sum_{n=1}^{N}\mathbf{F}_n^e. \quad (1.147)$$

Because also in this case the internal forces cancel out each other ($\sum_{n=1}^{N}\mathbf{F}_n^i = 0$), we have

$$\mathbf{v}_n = \frac{\mathrm{d}\mathbf{r}_n}{\mathrm{d}t}, \quad \mathbf{P} = \sum_{n=1}^{N} m_n\mathbf{v}_n, \quad (1.148)$$

where now $\mathbf{P}$ is the vector of momentum of the DMS.

If we have a constrained DMS, then, after its release from constraints, (1.147) takes the form

$$\frac{\mathrm{d}\mathbf{P}}{\mathrm{d}t} = \sum_{n=1}^{N} \mathbf{F}_n^e + \sum_{n=1}^{N} \mathbf{F}_n^R, \tag{1.149}$$

where it was assumed that reaction forces $\mathbf{F}_n^R$ act on every particle n. Considerations regarding the constrained CMS will be omitted here.

Equations (1.147) and (1.149) enable us to state the following theorems.

Theorem 1.3. *The rate of change of the vector of momentum (quantity of motion) of a free mechanical system is equal to the geometric sum of external forces acting on the system.*

Theorem 1.4. *The rate of change of the vector of momentum (quantity of motion) of a CMS is equal to the geometric sum of external forces and reaction forces acting on the system.*

Equation (1.149) can be written in the equivalent form

$$\mathrm{d}\mathbf{P} = \sum_{n=1}^{N} \mathbf{F}_n^e \mathrm{d}t + \sum_{n=1}^{N} \mathbf{F}_n^R \mathrm{d}t, \tag{1.150}$$

which allows for formulation of the following conclusion.

An elementary increment of momentum of a mechanical system is equal to the sum of elementary impulses of external forces and reaction forces acting on the system.

The following integral is called the impulse of a force acting over time interval $t - t_0$:

$$\mathbf{J} = \int_{t_0}^{t} \mathbf{F}(\tau)\mathrm{d}\tau, \tag{1.151}$$

where in the case of a DMS the force $\mathbf{F}$ is discrete, and in the case of a CMS the vector of force $\mathbf{F}$ denotes the mass density of force. In the latter case vector $\mathbf{J}$ is called the mass density of impulse of a force.

Integrating (1.151) we obtain

$$\mathbf{P}(t) - \mathbf{P}(t_0) = \Delta\mathbf{P} = \int_{t_0}^{t} \left(\int_{m} \mathbf{F}^e \mathrm{d}m \right) \mathrm{d}\tau = \mathbf{J}. \tag{1.152}$$

The relationship just obtained can be formulated in the following way.

The change of momentum (quantity of motion) of a CMS during time tt_0 is equal to the action of the impulse of a force during that time interval.

In the case of a DMS, relationship (1.152) takes the form

$$\Delta\mathbf{P} = \int_{t_0}^{t} \left(\mathbf{F}^e + \mathbf{F}^R\right) \mathrm{d}\tau, \tag{1.153}$$

where

$$\mathbf{F}^e = \sum_{n=1}^{N} \mathbf{F}_n^e, \quad \mathbf{F}^R = \sum_{n=1}^{N} \mathbf{F}_n^R,$$

and $\mathbf{F}^e$ ($\mathbf{F}^R$) is the main vector of external forces (reaction forces).

Equation (1.153) enables us to formulate the following conclusion.

An elementary increment of momentum of a DMS during time $t - t_0$ is equal to the sum of impulses of the main vector of external forces and of the main vector of reaction forces acting during that time.

1.2.3 Law of Motion of Center of Mass

Let us consider further a DMS consisting of N particles n of masses $m_n, n \in [1, N]$, whose positions are described by position vectors $\mathbf{r}_n$. Point C of the system is called the center of mass of the DMS (its position is described by the radius vector $\mathbf{r}_C$ and the total mass of the DMS is concentrated there, i.e., $m = \sum_{n=1}^{N} m_n$), provided that the following equation is satisfied:

$$m\mathbf{r}_C = \sum_{n=1}^{N} m_n \mathbf{r}_n. \tag{1.154}$$

Although the notion of the center of mass was introduced on the example of the DMS, it can also be applied to liquid, gaseous, rigid and flexible material systems, or those subjected to the action of various fields of forces (Chap. 1 of [1]).

In the case of the CMS, at first, one should divide the considered system into mass elements Δm_n and for each of them determine $\mathbf{r}_{C_n}$. Next, one should increase the number of mass elements up to infinity by the transition $\Delta m_n \to \mathrm{d}m_n$ and in the limit determine the position of the mass center, which is presented in the following equation:

$$\mathbf{r}_C = \frac{\lim\limits_{\Delta m_n \to 0} \sum\limits_{n=1}^{N} \Delta m_n \mathbf{r}_n}{m} = \frac{\int_m \mathbf{r} \mathrm{d}m}{m}. \tag{1.155}$$

Assuming that the considered material system is located in the uniform gravitational field it is possible to define the notion of gravity center of both a DMS and a CMS.

Multiplying both sides of (1.154) as well as the numerator and denominator of (1.155) by the acceleration of gravity $\mathbf{g}$, we obtain

$$\mathbf{Q} \circ \mathbf{r}_C = \sum_{n=1}^{N} \mathbf{Q}_n \circ \mathbf{r}_n, \tag{1.156}$$

$$\mathbf{Q} \circ \mathbf{r}_C = \int_m \mathbf{r} \circ \mathrm{d}\mathbf{Q}, \tag{1.157}$$

where $\mathbf{Q} = m\mathbf{g}$, $\mathbf{Q}_n = m_n\mathbf{g}$.

Formulas (1.156) and (1.157) were obtained on the assumption that the mass center and the gravity center are coincident.

The notion of the mass center has a more general meaning since it does not depend on the force field in which the system is located (for example, the notion of the center of gravity loses its meaning when vector $\mathbf{g}$ does not exist).

Let us assign to every particle n of a DMS the velocity $\mathbf{v}_n$. The momentum of the DMS is equal to

$$\mathbf{P} = \sum_{n=1}^{N} m_n \mathbf{v}_n = \sum_{n=1}^{N} m_n \frac{\mathrm{d}\mathbf{r}_n}{\mathrm{d}t} = \frac{\mathrm{d}}{\mathrm{d}t} \sum_{n=1}^{N} (m_n \mathbf{r}_n) = m \frac{\mathrm{d}\mathbf{r}_C}{\mathrm{d}t} = m\mathbf{v}_C. \tag{1.158}$$

In the preceding transformations, $\dot{m} = 0$ was assumed, and (1.154) was used. Similar considerations can be carried out regarding the CMS. Equation (1.158) allows for the formulation of the following conclusion.

The vector of momentum of a material system is equal to the product of the vector of velocity of system mass center and its total mass.

Differentiating (1.158) with respect to time we obtain

$$\frac{\mathrm{d}\mathbf{P}}{\mathrm{d}t} = m \frac{\mathrm{d}^2\mathbf{r}_C}{\mathrm{d}t^2} \equiv m\mathbf{a}_C, \tag{1.159}$$

and taking into account (1.148) we have

$$m\mathbf{a}_C = \sum_{n=1}^{N} \mathbf{F}_n^e + \sum_{n=1}^{N} \mathbf{F}_n^R. \tag{1.160}$$

Equation (1.160) is the expression for Newton's second law for particle C (*mass center of a system*). It enables us to formulate the following conclusion.

The mass center is a special point at which the total mass of a material system is concentrated, on which act all vectors of external forces and reactions, and which moves according to Newton's second law.

Instead of observing all N particles of a material system, it suffices to observe the motion of the mass center of the system, which is very advantageous. However, this knowledge is not sufficient to determine the motion of other particles of the system.

The principle of motion of the mass center is obtained from (1.159) and (1.160), setting $\mathbf{a}_C = \mathbf{0}$, which means that $\mathbf{v}_C = \text{const}$, and this is stated below.

If the sum of external forces and reaction forces acting on a material system is equal to zero, then the mass center either remains at rest or is in uniform rectilinear motion.

A material system on which no external forces and no reaction forces act is called an *isolated system (free system).*

It follows that for an isolated system the velocity of its mass center is constant.

From (1.159) we obtain

$$\mathbf{P} = m\mathbf{v}_C = \text{const} \tag{1.161}$$

or

$$C_1\mathbf{E}_1 + C_2\mathbf{E}_2 + C_3\mathbf{E}_3 = m\left(c_1\mathbf{E}_1 + c_2\mathbf{E}_2 + c_3\mathbf{E}_3\right),$$

where $C_i = P_{x_i}$ and $c_i = v_{cx_i}$, and C_i and c_i $(i = 1, 2, 3)$ are arbitrary constants.

In other words, by projecting the quantity of motion vector onto the axes of the system $OX_1X_2X_3$ we obtain three first integrals of motion. Therefore, the motion of the mass center C is uniquely defined.

Example 1.8. Two boys wearing ice skates are standing on an ice rink (a perfectly smooth horizontal surface) at a distance x from each other. One of them throws a ball of mass m, which is caught by the other after time t. What is the velocity with which the boy starts to move after throwing the ball if he has the mass M?

Because we neglect friction, the horizontal component of the reaction force produced by the weight of the boy is equal to zero. According to (1.160) there are no external forces and no reactions along the horizontal direction, that is, (1.161) is valid. If at the initial time instant of motion the boy throwing the ball was at rest, then $\mathbf{P} = \mathbf{v}_C = \mathbf{0}$. Differentiating (1.154) we have

$$(m + M)\,\dot{x}_C = Mv - m\frac{x}{t} = 0,$$

because the horizontal component of the velocity of mass center of the ball is equal to x/t, and the velocities of the boy v and the ball have opposite senses.

From the preceding equation we obtain

$$v = \frac{m}{M}\frac{x}{t}.$$

Finally, let us note that along the vertical direction the momentum of the boy–ball system is not conserved. □

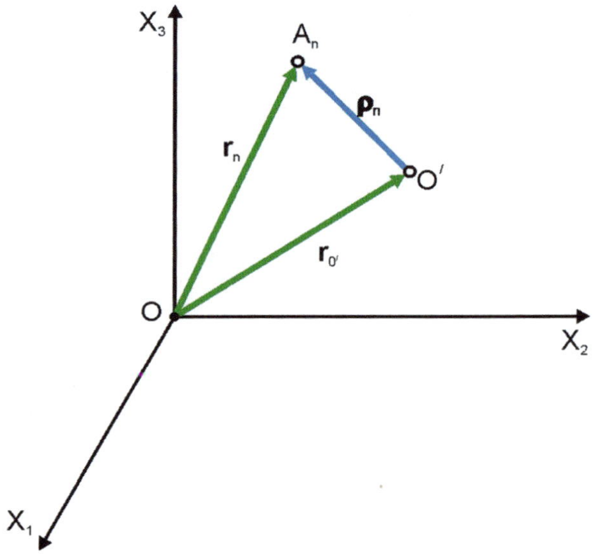

Fig. 1.15 Schematic leading to the definition of the moment of quantity of motion of material point A_n and of a DMS

1.2.4 Moment of Quantity of Motion (Angular Momentum)

Figure 1.15 shows the material point A_n of a DMS, point O' (called the center), and the adopted Cartesian coordinate system $OX_1X_2X_3$ [12].

Vectors $\mathbf{r}_{O'}$ and $\mathbf{r}_n$ respectively describe the positions of points O' and A_n in the system $OX_1X_2X_3$, and vector $\boldsymbol{\rho}_n$ is the radius vector of point A_n with respect to the adopted center O'.

Definition 1.2. We call the following vector the moment of quantity of motion (angular momentum) of a material point A_n with respect to a point O' (center):

$$\mathbf{K}_{nO'} = \boldsymbol{\rho}_n \times m_n \mathbf{v}_n. \tag{1.162}$$

Definition 1.3. A projection of the moment of quantity of motion of a material point A_n on an axis, where the moment of quantity of motion of the point is determined with respect to an arbitrarily chosen point (center) on the aforementioned axis (because the projection is independent of the choice of the center), is called the moment of quantity of motion (angular momentum) of material point A_n with respect to that axis.

Definition 1.4. The following vector is called the main moment of quantity of motion of a DMS (angular momentum of a DMS) with respect to a point O' (center):

$$\mathbf{K}_{O'} = \sum_{n=1}^{N} \boldsymbol{\rho}_n \times m_n \mathbf{v}. \tag{1.163}$$

Definition 1.5. A projection of the main moment of quantity of motion on an axis with respect to an arbitrarily chosen point (center) lying on this axis is called the main moment of quantity of motion of a DMS (angular momentum of a DMS) with respect to that axis.

According to the introduced definitions it can be easily seen that the angular momentum changes with the change of the center.

Let us choose in Fig. 1.15 another point (the center) O'' (not drawn) and let the position of point A_n be described by the radius vectors $\overrightarrow{O'A_n}$ and $\overrightarrow{O''A_n}$. The moment of quantity of motion of a DMS with respect to O'' is equal to

$$\begin{aligned}\mathbf{K}_{O''} &= \sum_{n=1}^{N}\left(\overrightarrow{O''A_n} \times m_n\mathbf{v}_n\right) = \sum_{n=1}^{N}\left(\overrightarrow{O''O'} + \overrightarrow{O'A_n}\right) \times m_n\mathbf{v}_n \\ &= \sum_{n=1}^{N}\left(\overrightarrow{O'A_n} \times m_n\mathbf{v}_n\right) + \sum_{n=1}^{N}\left(\overrightarrow{O''O'} \times m_n\mathbf{v}_n\right) \\ &= \mathbf{K}_{O'} + \overrightarrow{O''O'} \times \mathbf{P},\end{aligned} \tag{1.164}$$

where $\mathbf{P}$ is the momentum of the DMS [see (1.158)].

The preceding equation describes the relation between the vectors of the quantity of motion of a DMS calculated with respect to distinct centers O' and O''.

Definition 1.6. The motion of material points A_n of a DMS with respect to the coordinate system of origin at the DMS mass center C and moving in translational motion (the so-called König[3] system) is called the motion of a DMS with respect to its mass center C.

Let $\mathbf{v}_C$ denote the absolute velocity of the mass center C (that is, in the system $OX_1X_2X_3$), $\mathbf{v}_n$ the absolute velocity of point A_n, and $\mathbf{v}_n^r$ the relative velocity of point A_n with respect to point C.

Point A_n moves in composite motion, and its velocity is equal to

$$\mathbf{v}_n = \mathbf{v}_C + \mathbf{v}_n^r. \tag{1.165}$$

The relative angular momentum of a DMS with respect to its mass center (the point C) is equal to

$$\mathbf{K}_C^r = \sum_{n=1}^{N} \boldsymbol{\rho}_n^r \times m_n\mathbf{v}_n^r, \tag{1.166}$$

where now $\boldsymbol{\rho}_n^r(\mathbf{v}_n^r)$ is the radius vector (velocity) of point A_n in the König system with respect to point C.

[3]Johann Samuel König (1712–1757), German mathematician.

The absolute angular momentum of a DMS with respect to point C, taking into account (1.165) and (1.166), is equal to

$$\begin{aligned}\mathbf{K}_C &= \sum_{n=1}^{N} \boldsymbol{\rho}_n^r \times m_n \mathbf{v}_n = \sum_{n=1}^{N} \boldsymbol{\rho}_n^r \times m_n \left(\mathbf{v}_C + \mathbf{v}_n^r\right) \\ &= \sum_{n=1}^{N} \left(m_n \boldsymbol{\rho}_n^r\right) \times \mathbf{v}_C + \sum_{n=1}^{N} \left(\boldsymbol{\rho}_n^r \times m_N \mathbf{v}_n^r\right) \\ &= m\boldsymbol{\rho}_C^r \times \mathbf{v}_C + \sum_{n=1}^{N} \left(\boldsymbol{\rho}_n^r \times m_n \mathbf{v}_n^r\right) = \sum_{n=1}^{N} \left(\boldsymbol{\rho}_n^r \times m_n \mathbf{v}_n^r\right) = \mathbf{K}_C^r,\end{aligned} \tag{1.167}$$

because $\boldsymbol{\rho}_C^r = \mathbf{0}$ (point C lies at the origin of the König system).

One may conclude from (1.167) that the absolute angular momentum of a DMS with respect to the mass center of system C is equal to the relative angular momentum of the DMS with respect to point C, that is, $\mathbf{K}_C = \mathbf{K}_C^r$.

From (1.164), and setting $O' = C$, it follows that

$$\begin{aligned}\mathbf{K}_{O''} &= \mathbf{K}_C + \overrightarrow{O''C} \times \mathbf{P} = \mathbf{K}_C + \overrightarrow{O''C} \times \sum_{n=1}^{N} m_n \mathbf{v}_n^r \\ &= \mathbf{K}_C + \overrightarrow{O''C} \times m\mathbf{v}_C^r = \mathbf{K}_C,\end{aligned} \tag{1.168}$$

because $\mathbf{v}_C^r = \mathbf{0}$.

If the center O'' was chosen arbitrarily, then the angular momentum of the system with respect to the mass center of the system (point C) would be identical for each of the points of the DMS and would equal $\mathbf{K}_C$.

1.2.5 Kinetic Energy of a DMS and a CMS

The kinetic energy of a DMS is equal to

$$T = \frac{1}{2} \sum_{n=1}^{N} m_n \mathbf{v}_n^2. \tag{1.169}$$

Taking into account (1.165) in relation (1.169) we obtain

$$\begin{aligned}T &= \frac{1}{2}\sum_{n=1}^{N} m_n\left(\mathbf{v}_C + \mathbf{v}_n^r\right)^2 \\ &= \frac{1}{2}\left(\sum_{n=1}^{N} m_n\right)\mathbf{v}_C^2 + \sum_{n=1}^{N}\left(m_n\mathbf{v}_n^r\right)\cdot\mathbf{v}_C + \frac{1}{2}\sum_{n=1}^{N}\left(m_n(\mathbf{v}_n^r)^2\right) \\ &= \frac{1}{2}m\mathbf{v}_C^2 + m\mathbf{v}_C^r\cdot\mathbf{v}_C + \frac{1}{2}\sum_{n=1}^{N} m_n(\mathbf{v}_n^r)^2 \\ &= \frac{1}{2}m\mathbf{v}_C^2 + \frac{1}{2}\sum_{n=1}^{N} m_n(\mathbf{v}_n^r)^2, \end{aligned} \tag{1.170}$$

because $\mathbf{v}_C^r = \mathbf{0}$.

The obtained result can be presented in the form of the following theorem.

Theorem 1.5 (König theorem). *The kinetic energy of a DMS is equal to the sum of the kinetic energy of a material point located at the mass center of the DMS and of mass equal to the sum of masses of all points of the DMS and the kinetic energy associated with the motion of the DMS with respect to its mass center.*

Let us now consider the motion of a CMS (a rigid body) with respect to the fixed point O. Let $O'X_1'X_2'X_3'$ be the body system such that $O' = O$, and let the instantaneous angular velocity of the rigid body $\boldsymbol{\omega}$ have the direction of the l axis determined by direction cosines.

In Chap. 3 of [1] it was shown that

$$T = \frac{1}{2}I_l\omega^2, \tag{1.171}$$

where I_l denotes the mass moment of inertia of a body with respect to an axis l, and that

$$\begin{aligned}T &= \frac{1}{2}\left(I_{X_1'}\omega_1'^2 + I_{X_2'}\omega_2'^2 + I_{X_3'}\omega_3'^2\right) \\ &\quad - I_{X_1'X_2'}\omega_1'\omega_2' - I_{X_1'X_3'}\omega_1'\omega_3' - I_{X_2'X_3'}\omega_2'\omega_3', \end{aligned} \tag{1.172}$$

where $I_{X_i'}$ and $I_{X_{ij}'}$ denote the *mass moments of inertia* and the *products of inertia* of the body with respect to the system $O'X_1'X_2'X_3'$, and in turn ω_i' are projections of vector $\boldsymbol{\omega}$ onto axes of this coordinate system, that is, $\omega_i' = \boldsymbol{\omega}\circ\mathbf{E}_i'$.

If $O'X_1'$, $O'X_2'$, and $O'X_3'$ are the principal axes of inertia of a body for point O', then (1.172) takes the form

$$T = \frac{1}{2}\left(I_{X_1'}\omega_1'^2 + I_{X_2'}\omega_2'^2 + I_{X_3'}\omega_3'^2\right). \tag{1.173}$$

If a rigid body rotates with a constant velocity $\boldsymbol{\omega}$ about one of the principal axes, for example, the X_1' axis, then we have $\boldsymbol{\omega} = \mathbf{E}_1'\omega_1'$ and from (1.172) we obtain

$$T = \frac{1}{2} I_{X_1'} \omega_1'^2. \tag{1.174}$$

Let us now describe the moment of quantity of motion of a rigid body with one point fixed O. Let us introduce the body system $O'X_1'X_2'X_3'$, where $O' = O$. The position vector of material point A_n in the body system is $\mathbf{r}_n'$, and we assume that point A_n has the following coordinates in the body system: $\mathbf{r}_n' \circ \mathbf{E}_1' = x_{1n}'$, $\mathbf{r}_n' \circ \mathbf{E}_2' = x_{2n}'$, $\mathbf{r}_n' \circ \mathbf{E}_3' = x_{3n}'$. The vector of instantaneous angular velocity of the body is equal to

$$\boldsymbol{\omega} = \omega_1'\mathbf{E}_1' + \omega_2'\mathbf{E}_2' + \omega_3'\mathbf{E}_3'. \tag{1.175}$$

The angular momentum of the body with respect to point O' is equal to

$$\begin{aligned}\mathbf{K}_{O'} &= \sum_{n=1}^{N} \mathbf{r}_n' \times m_n \mathbf{v}_n = \sum_{n=1}^{N} \mathbf{r}_n' \times m_n \left(\boldsymbol{\omega} \times \mathbf{r}_n'\right) \\ &= \sum_{n=1}^{N} m_n \mathbf{r}_n' \times \left(\boldsymbol{\omega} \times \mathbf{r}_n'\right).\end{aligned} \tag{1.176}$$

Replacing the vector triple product with scalar products we obtain

$$\begin{aligned}\mathbf{K}_{O'} &= \sum_{n=1}^{N} m_n \mathbf{r}_n'^2 \boldsymbol{\omega} - \sum_{n=1}^{N} m_n \left(\boldsymbol{\omega} \circ \mathbf{r}_n'\right) \mathbf{r}_n' \\ &= \sum_{n=1}^{N} m_n \left(x_{1n}'^2 + x_{2n}'^2 + x_{3n}'^2\right) \boldsymbol{\omega} \\ &\quad - \sum_{n=1}^{N} m_n \left(\omega_1' x_{1n}' + \omega_2' x_{2n}' + \omega_3' x_{3n}'\right) \mathbf{r}_n'.\end{aligned} \tag{1.177}$$

Projections of the vector of angular momentum of the body onto axes $O'X_i'$ are equal to

$$K_{O'X_i} = \mathbf{K}_{O'} \circ \mathbf{E}_i'. \tag{1.178}$$

In the case of the $O'X_1'$ axis from (1.177) we obtain

$$\begin{aligned}K_{O'X_1} &= \mathbf{K}_{O'} \circ \mathbf{E}_1' = \sum_{n=1}^{N} m_n \left(x_{1n}'^2 + x_{2n}'^2 + x_{3n}'^2\right) \omega_1' \\ &\quad - \sum_{n=1}^{N} m_n \left(\omega_1' x_{1n}' + \omega_2' x_{2n}' + \omega_3' x_{3n}'\right) x_{1n}'\end{aligned}$$

$$= \sum_{n=1}^{N} m_n \left(x'^2_{2n} + x'^2_{3n}\right) \omega'_1 - \sum_{n=1}^{N} m_n \left(x'_{1n} x'_{2n}\right) \omega'_2$$

$$- \sum_{n=1}^{N} m_n (x'_{1n} x'_{3n}) \omega'_3 = I_{X'_1} \omega'_1 - I_{X'_1 X'_2} \omega'_2 - I_{X'_1 X'_3} \omega'_3, \quad (1.179)$$

where the definitions of principal moments of inertia and products of inertia introduced in Chap. 3 of [1] were used.

Eventually, writing the projections on two remaining axes we obtain an equation in matrix form

$$\mathbf{K}_{O'} = \mathbf{I}\boldsymbol{\omega}, \quad (1.180)$$

where

$$\mathbf{I} = \begin{bmatrix} I_{X'_1} & -I_{X'_1 X'_2} & -I_{X'_1 X'_3} \\ -I_{X'_1 X'_2} & I_{X'_2} & -I_{X'_2 X'_3} \\ -I_{X'_1 X'_3} & -I_{X'_2 X'_3} & I_{X'_3} \end{bmatrix}. \quad (1.181)$$

If the axes of the system $O'X'_1X'_2X'_3$ coincide with the principal axes of inertia of the body, then matrix (1.181) is the diagonal matrix of the form

$$\mathbf{I} = \begin{bmatrix} I_{X'_1} & 0 & 0 \\ 0 & I_{X'_2} & 0 \\ 0 & 0 & I_{X'_3} \end{bmatrix}. \quad (1.182)$$

If vector ω lies on the $O'X'_3$ axis, then $\omega'_1 = \omega'_2 = 0$, and from (1.180) we obtain

$$K_{O'X'_1} = -I_{X'_1 X'_3} \omega'_3, \quad K_{O'X'_2} = -I_{X'_2 X'_3} \omega'_3, \quad K_{O'X'_3} = -I_{X'_3} \omega'_3. \quad (1.183)$$

From (1.180) and (1.183) it follows that vectors $\mathbf{K}_{O'}$ and $\boldsymbol{\omega}$ do not lie on the same axis. From (1.183) it follows that it can take place if the axis of rotation (i.e., direction of vector ω) is the principal axis of inertia of the body (then we have $\mathbf{K}_{O'X'_3} = I_{X'_3} \boldsymbol{\omega}'_3$).

Taking into account (1.172), (1.180) and (1.181), the following relationship holds true:

$$T = \frac{1}{2} (\mathbf{K}_{O'} \circ \boldsymbol{\omega}). \quad (1.184)$$

Let us note that because the kinetic energy $T > 0$, the angle between vectors $\mathbf{K}_{O'}$ and $\boldsymbol{\omega}$ is an acute angle.

1.2.6 Law of Conservation of Angular Momentum

According to Fig. 1.15 the angular momentum of a DMS with respect to the center O' is described by (1.163), where the coordinate system introduced in that figure is the inertial one.

Differentiating (1.163) with respect to time we obtain

$$\begin{aligned}\frac{\mathrm{d}\mathbf{K}_{O'}}{\mathrm{d}t} &= \sum_{n=1}^{N} \dot{\boldsymbol{\rho}}_n \times m_n \mathbf{v}_n + \sum_{n=1}^{N} \boldsymbol{\rho}_n \times m_n \mathbf{a}_n \\ &= \sum_{n=1}^{N} \dot{\boldsymbol{\rho}}_n \times m_n \mathbf{v}_n + \sum_{n=1}^{N} \boldsymbol{\rho}_n \times \left(\mathbf{F}_n^e + \mathbf{F}_n^R\right) \\ &= \sum_{n=1}^{N} \dot{\boldsymbol{\rho}}_n \times m_n \mathbf{v}_n + \mathbf{M}_{O'}^e + \mathbf{M}_{O'}^R. \end{aligned} \tag{1.185}$$

According to Fig. 1.15 we have

$$\mathbf{r}_n = \mathbf{r}_{O'} + \boldsymbol{\rho}_n, \tag{1.186}$$

and differentiating (1.186) with respect to time we obtain

$$\dot{\boldsymbol{\rho}}_n = \dot{\mathbf{r}}_n - \dot{\mathbf{r}}_{O'} = \mathbf{v}_n - \mathbf{v}_{O'}. \tag{1.187}$$

From (1.185) we obtain

$$\begin{aligned}\dot{\mathbf{K}}_{O'} &= \sum_{n=1}^{N} (\mathbf{v}_n - \mathbf{v}_{O'}) \times m_n \mathbf{v}_n + \mathbf{M}_{O'}^e + \mathbf{M}_{O'}^R \\ &= \left(\sum_{n=1}^{N} m_n \mathbf{v}_n\right) \times \mathbf{v}_{O'} + \mathbf{M}_{O'}^e + \mathbf{M}_{O'}^R \\ &= m\mathbf{v}_C \times \mathbf{v}_{O'} + \mathbf{M}_{O'}^e + \mathbf{M}_{O'}^R, \end{aligned} \tag{1.188}$$

where $\mathbf{v}_C$ is the velocity of the mass center of the DMS, $\mathbf{M}_{O'}^e$ is the main moment of external forces with respect to point O', and $\mathbf{M}_{O'}^R$ is the main moment of reactions with respect to point O' and during transformations was also used differentiation of (1.154) with respect to time, and hence $\mathbf{v}_{O'} \neq \mathbf{0}$. However, if we take point O' as stationary ($O' = O$), then from (1.188) we obtain

$$\dot{\mathbf{K}}_O = \mathbf{M}_O^e + \mathbf{M}_O^R. \tag{1.189}$$

Let us observe that origin O of the coordinate system was chosen arbitrarily and only in a special case will it coincide with the mass center of the considered system.

Equation (1.189) allows for the statement of the law of conservation of angular momentum of an SOP.

The time derivative of angular momentum of a material system with respect to an arbitrary fixed pole O is equal to the sum of main moments of external forces and reaction forces with respect to that point.

Let us note that if in formula (1.188) an arbitrary non-stationary point O' is chosen at point C, then we obtain $m\mathbf{v}_C \times \mathbf{v}_C = \mathbf{0}$, and then formula (1.188) takes the form

$$\dot{\mathbf{K}}_C = \mathbf{M}_C^e + \mathbf{M}_C^R. \qquad (1.190)$$

A comparison of the obtained relationship with (1.189) leads to the conclusion that the change in the system's angular momentum about the fixed point O and about mass center C is the same.

The preceding calculations can be also represented in integral form.

According to (1.189) we have

$$\Delta\mathbf{K}_O = \mathbf{K}_O(t) - \mathbf{K}_O(t_0) = \int_{t_0}^{t} \left(\mathbf{M}_O^e + \mathbf{M}_O^R\right) \mathrm{d}t. \qquad (1.191)$$

The integral occurring on the right-hand side of (1.191) is called an *impulse (action) of external forces and reactions* in time interval $t - t_0$.

One may conclude from (1.191) that the increment of the vector of the system's angular momentum with respect to a stationary pole in a finite time interval is equal to the impulse of moments of external forces and reaction forces about that point in the same time interval.

If we are dealing with an isolated system ($\mathbf{M}_O^e = \mathbf{M}_O^R = \mathbf{0}$), then we obtain the following law of conservation of angular momentum.

The angular momentum of an isolated material system with respect to an arbitrary stationary pole is constant and equal to

$$\mathbf{K}_O(t) = \text{const} \qquad (1.192)$$

or

$$\mathbf{K}_O = C_1\mathbf{E}_1 + C_2\mathbf{E}_2 + C_3\mathbf{E}_3,$$

where C_i, $i = 1, 2, 3$ *are arbitrary constants.*

In a system where, during motion, $\mathbf{M}_O^e = \mathbf{M}_O^R = \mathbf{0}$, (1.192) remains valid during the motion of point O' with respect to the system mass center C, on the condition that the relation $\mathbf{r}_{O'} = C_1\mathbf{r}_C(t) + C_2$ holds true, where C_1 and C_2 are constants. Following differentiation we get $\mathbf{v}_{O'} = C_1\mathbf{v}_C$, and substitution into (1.188) proves the validity of the preceding observation.

Now let us take an arbitrary stationary axis l passing through the system mass center.

According to relation (1.188) we have

$$\dot{\mathbf{K}}_l = \mathbf{M}_l^e + \mathbf{M}_l^R, \tag{1.193}$$

and on the assumption that during motion of the DMS we have $\mathbf{M}_l^e = \mathbf{M}_l^R = \mathbf{0}$, we obtain the following first integral:

$$\mathbf{K}_l = \text{const.} \tag{1.194}$$

Theorem 1.6. *The necessary and sufficient condition for the existence of the first integral* (1.194) *is that projections of the velocity of mass center of the system and of the velocity of an arbitrary point O' lying on the l axis onto a plane perpendicular to this axis are parallel to each other during the motion of the system.*

Proof. Let us take a unit vector of the l axis as $\boldsymbol{l} = \mathbf{E}_3$ and multiply (scalar multiplication) (1.188) by sides by $\mathbf{E}_3 = \text{const}$ obtaining

$$\frac{\mathrm{d}\left(\mathbf{K}_{O'} \circ \mathbf{E}_3\right)}{\mathrm{d}t} = m\left(\mathbf{v}_C \times \mathbf{v}_{O'}\right) \circ \mathbf{E} + \left(\mathbf{M}_{O'}^e + \mathbf{M}_{O'}^R\right) \circ \mathbf{E}_3.$$

□

Because $\mathbf{K}_{O'} \circ \mathbf{E}_3 = K_l$ and $(\mathbf{M}_{O'}^e + \mathbf{M}_{O'}^R) \circ \mathbf{E}_3 = M_l^e + M_l^R = 0$, we have $\mathbf{K}_l = \text{const}$ if and only if $(\mathbf{v}_C \times \mathbf{v}_{O'}) \circ \mathbf{E}_3 \equiv 0$.

Since

$$\mathbf{v}_C \times \mathbf{v}_{O'} = \begin{vmatrix} \mathbf{E}_1 & \mathbf{E}_2 & \mathbf{E}_3 \\ v_{1C} & v_{2C} & v_{3C} \\ v_{1O'} & v_{2O'} & v_{3O'} \end{vmatrix}$$

$$= \mathbf{E}_1\left(v_{2C}v_{3O'} - v_{3C}v_{2O'}\right) - \mathbf{E}_2\left(v_{1C}v_{3O'} - v_{3C}v_{1O'}\right) + \mathbf{E}_3\left(v_{1C}v_{2O'} - v_{2C}v_{1O'}\right),$$

we have

$$\left(\mathbf{v}_C \times \mathbf{v}_{O'}\right) \circ \mathbf{E}_3 = v_{1C}v_{2O'} - v_{2C}v_{1O'} = 0.$$

Following transformation we have

$$\frac{v_{1C}}{v_{2C}} = \frac{v_{1O'}}{v_{2O'}}.$$

The obtained relationship proves that the projections of vectors $\mathbf{v}_C$ and $\mathbf{v}_{O'}$ onto a plane perpendicular to the l axis are parallel.

In the end, let us sum up basic knowledge regarding the conservation of momentum and angular momentum. In the case of choice of an arbitrary pole, the state of a material system is determined by the main vector of momentum $\mathbf{P}$ and main moment of angular momentum $\mathbf{K}_O$.

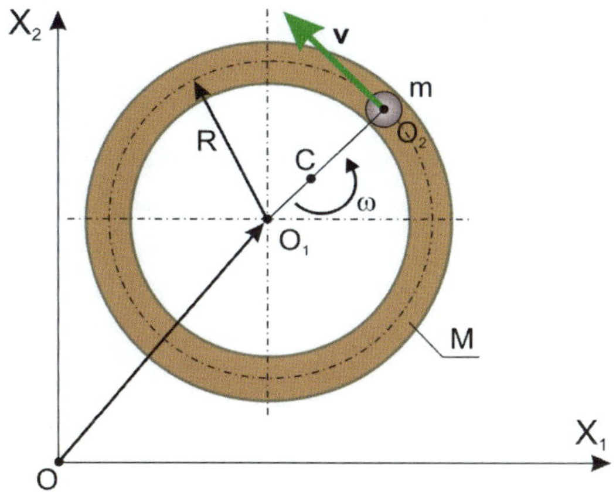

Fig. 1.16 Motion of a mechanical system composed of a small ball (particle) of mass m and a metal rim of mass M

External forces and reaction forces acting on the material system are determined by the main vectors of these forces $\mathbf{F}^e$ and $\mathbf{F}^R$, and main moments $\mathbf{M}_O^e$ and $\mathbf{M}_O^R$ with respect to the chosen point O.

The laws of conservation of momentum and angular momentum are expressed by the following simple equations:

$$\dot{\mathbf{P}} = \mathbf{F}^e + \mathbf{F}^R, \tag{1.195}$$

$$\dot{\mathbf{K}}_O = \mathbf{M}_O^e + \mathbf{M}_O^R. \tag{1.196}$$

Example 1.9. On a horizontal smooth surface lies a metal rim of mass M and radius R with a groove along which a particle of mass m moves with velocity constant as to the magnitude $\mathbf{v} = const$ (Fig. 1.16). Determine the motion of the bodies (the rim and the particle).

The mass center C of the system lies on the segment connecting points O_1 and O_2.

According to Fig. 1.16 we have

$$(m+M)\mathbf{r}_C = M \cdot \overrightarrow{OO_1} + m \cdot \overrightarrow{OO_2}, \qquad \left|\overrightarrow{O_1C}\right| + \left|\overrightarrow{CO_2}\right| = R,$$

$$\mathbf{r}_C = \overrightarrow{OO_1} + \overrightarrow{O_1C} = \overrightarrow{OO_2} + \overrightarrow{O_2C}.$$

Following transformation we obtain

$$(m+M)\cdot\left(\overrightarrow{OO_1} + \overrightarrow{O_1C}\right) = M \cdot \overrightarrow{OO_1} + m \cdot \overrightarrow{OO_2},$$

that is,

$$(m+M)\cdot\overrightarrow{O_1C} = m \cdot \left(\overrightarrow{OO_2} - \overrightarrow{OO_1}\right).$$

In a similar way we calculate

$$(m+M)\cdot\left(\overrightarrow{OO_2}+\overrightarrow{O_2C}\right)=M\cdot\overrightarrow{OO_1}+m\cdot\overrightarrow{OO_2},$$

that is,

$$(m+M)\cdot\overrightarrow{CO_2}=M\cdot\left(\overrightarrow{OO_2}-\overrightarrow{OO_1}\right).$$

Finally, we have

$$O_1C=\frac{m}{m+M}R,$$
$$CO_2=\frac{M}{m+M}R.$$

The calculations leading to the determination of the position of the mass center simplify significantly following the introduction of the local coordinate system of one axis directed along segment O_1O_2 and the origin at point O_1.

The angular momentum of the system is equal to zero about the axis parallel to the OX_3 axis and passing through point C. Then we have

$$I_C\omega+mCO_2\left(v+\omega CO_2\right)=0.$$

The quantity I_C is a moment of inertia of the rim with respect to the CX_3 axis, which, according to the parallel axis theorem, is equal to

$$I_C=MR^2+M\left(O_1C\right)^2,$$

and substituting into the previous equation we find

$$\omega=-\frac{v}{R}\frac{m(m+M)}{M^2+3mM+2m^2}.$$

□

Example 1.10. Figure 1.17 shows a disk of mass m and radius R rolling on a horizontal plane along a straight line (without slip, air resistance, or rolling resistance). The mass center C of the disk is located at distance r from its geometric center O. Derive the equation of motion of the disk on the assumption that the mass moment of inertia of the disk with respect to the axis perpendicular to it and passing through point C is equal to I_C.

Let us introduce an angle $\psi=\psi(t)$ formed between a line perpendicular to the horizontal surface and passing through the geometric center of the disk and the segment OC. Figure 1.17 shows two vectors of the forces causing the motion of the disk, that is, the vector of the gravity force and the vector of the reaction force at point B of two components (vertical and horizontal).

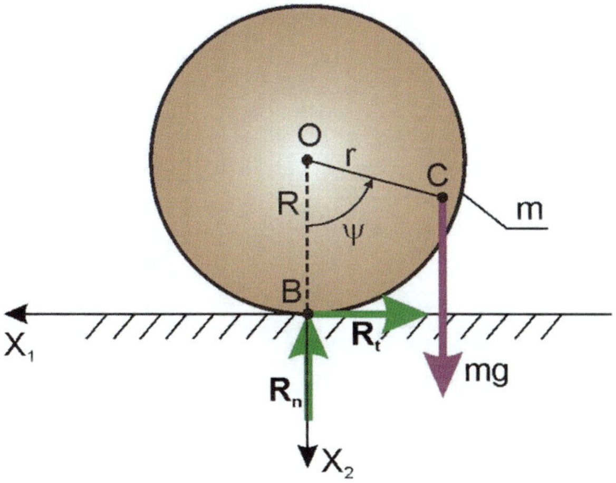

Fig. 1.17 A disk rolling without slip on a horizontal surface

The vector

$$\mathbf{v}_B = \left(\dot{\boldsymbol{\Psi}} \times \overrightarrow{OB}\right) = \begin{vmatrix} \mathbf{E}_1 & \mathbf{E}_2 & \mathbf{E}_3 \\ 0 & 0 & \dot{\Psi} \\ 0 & -R & 0 \end{vmatrix} = \mathbf{E}_1 R\dot{\Psi},$$

where $\mathbf{v}_B$ denotes the velocity of the point moving on the straight line determined by the "track" left by the disk on the horizontal surface.

In the introduced absolute right-handed coordinate system $BX_1X_2X_3$ vector $\overrightarrow{BC}$ is given by

$$\overrightarrow{BC} = -r \sin \Psi \mathbf{E}_1 - (R - r \cos \Psi)\mathbf{E}_2,$$

hence

$$\begin{aligned} \left|\overrightarrow{BC}\right|^2 &= r^2 \sin^2 \Psi + (R - r \cos \Psi)^2 \\ &= r^2 \sin^2 \Psi + r^2 \cos^2 \Psi + R^2 - 2rR \cos \Psi = r^2 + R^2 - 2rR \cos \Psi. \end{aligned}$$

The angular velocity of the disk has the form

$$\boldsymbol{\omega} = \dot{\Psi}\mathbf{E}_3,$$

whereas the velocity of the mass center with respect to point B is equal to

$$\begin{aligned} \mathbf{v}_C = \boldsymbol{\omega} \times \overrightarrow{BC} &= \begin{vmatrix} \mathbf{E}_1 & \mathbf{E}_2 & \mathbf{E}_3 \\ 0 & 0 & \dot{\Psi} \\ -r \sin \Psi & -R + r \cos \Psi & 0 \end{vmatrix} \\ &= \mathbf{E}_1 \left(\dot{\Psi}(R - r \cos \Psi)\right) - \mathbf{E}_2 \dot{\Psi} r \sin \Psi. \end{aligned}$$

From (1.188) we obtain

$$\dot{\mathbf{K}}_B = \mathbf{M}_B^e + \mathbf{M}_B^R + m\mathbf{v}_C \times \mathbf{v}_B.$$

Let us note that

$$m\mathbf{v}_C \times \mathbf{v}_B = m \begin{vmatrix} \mathbf{E}_1 & \mathbf{E}_2 & \mathbf{E}_3 \\ \dot{\Psi}(R - r\cos\Psi) & -\dot{\Psi}r\sin\Psi & 0 \\ R\dot{\Psi} & 0 & 0 \end{vmatrix}$$
$$= \mathbf{E}_1 \cdot 0 + \mathbf{E}_2 \cdot 0 + Rr\dot{\Psi}^2 \sin\Psi \mathbf{E}_3.$$

The angular momentum of the disk with respect to point B is equal to

$$\mathbf{K}_B = \left[I_C + m(BC)^2\right]\dot{\Psi}\mathbf{E}_3,$$

hence

$$\frac{\mathrm{d}\mathbf{K}_B}{\mathrm{d}t} = \frac{\mathrm{d}}{\mathrm{d}t}\left[I_C + m\left(r^2 + R^2 - 2rR\cos\Psi\right)\right]\dot{\Psi}\mathbf{E}_3$$
$$= \left\{\left[I_C + m\left(r^2 + R^2 - 2rR\cos\Psi\right)\right]\ddot{\Psi} + 2mrR\dot{\Psi}^2\sin\Psi\right\}\mathbf{E}_3.$$

In turn, we have $\mathbf{M}_B^R = \mathbf{0}$ and

$$\mathbf{M}_B^e = \overrightarrow{BC} \times (\mathbf{E}_2 mg) = \begin{vmatrix} \mathbf{E}_1 & \mathbf{E}_2 & \mathbf{E}_3 \\ -r\sin\Psi & -R + r\cos\Psi & 0 \\ 0 & mg & 0 \end{vmatrix} = -mgr\sin\Psi\mathbf{E}_3.$$

Eventually, we obtain the following second-order non-linear differential equation:

$$\left[I_C + m\left(r^2 + R^2 - 2rR\cos\Psi\right)\right]\ddot{\Psi} + mrR\dot{\Psi}^2\sin\Psi + mgr\sin\Psi = 0. \quad \square$$

1.2.7 Law of Conservation of Kinetic Energy

From (1.90) and (1.91) we obtain

$$\mathrm{d}\left(\frac{m_n v_n^2}{2}\right) = \mathbf{F}_n^{e^0}\mathrm{d}\mathbf{r}_n + \mathbf{F}_n^{R^0}\mathrm{d}\mathbf{r}_n + \mathbf{F}_n^{i^0}\mathrm{d}\mathbf{r}_n. \tag{1.197}$$

The preceding calculations are valid for every $n \in [1, N]$, therefore, adding N (1.197), we obtain

$$dT = dW^e + dW^i + dW^r, \tag{1.198}$$

$$T = \sum_{n=1}^{N} \frac{m_n v_n^2}{2}, \quad dW^e = \sum_{n=1}^{N} \mathbf{F}_n^e \circ d\mathbf{r}_n,$$

$$dW^i = \sum_{n=1}^{N} \mathbf{F}_n^i \circ d\mathbf{r}_n, \quad dW^R = \sum_{n=1}^{N} \mathbf{F}_n^R \circ d\mathbf{r}_n, \tag{1.199}$$

where the symbol $d(\cdot)$ denotes the elementary work, which in general is not a differential of any function [2].

Equation (1.198) allows for the formulation of the law of conservation of kinetic energy.

An elementary increment of the kinetic energy is equal to the sum of elementary works of external forces, internal forces, and reaction forces that occur in a given system.

Let us note that, despite the the fact that the system of internal forces is equivalent to zero, the work done by those forces does not have to be equal to zero.

Since, if on two arbitrary points that interact with one another, the same vectors of forces but of opposite senses act, then we can imagine a material medium of properties dependent on the direction of displacement of these points, and thus the displacements can be different, and consequently the works done by the forces are not equal. In the case of a rigid body, the works are equal to zero because there are no displacements between the points of a rigid body.

If all forces acting on a material system result from the accumulated potential, then

$$\mathbf{F}_n^e = -\mathrm{grad}_n V, \tag{1.200}$$

and the work increment

$$dW = \sum_{n=1}^{N} \mathbf{F}_n^e \circ d\mathbf{r}_n = -dV. \tag{1.201}$$

Neglecting the work of the internal forces and reaction forces (a rigid system, where the distances between particles are constant and where we are dealing with ideal constraints) from (1.198) and (1.201) we obtain

$$dT = -dV, \tag{1.202}$$

and following integration we obtain the so-called *integral of energy* of the form

$$T_2 - T_1 = V_1 - V_2. \tag{1.203}$$

From the obtained equation it can be seen that in the potential field the total energy E is conserved because

$$E \equiv T_1 + V_1 = T_2 + V_2. \tag{1.204}$$

Such a material system is called a *conservative system*.

1.3 Motion About a Point

1.3.1 Kinetic Energy, Ellipsoid of Inertia, and Angular Momentum

Recall that the motion about a point of a rigid body is the motion of that body with one point fixed.

In Chaps. 1 and 3 of [1] we introduced the notions of kinetic energy and momentum of a material point. For the purpose of analysis of motion about a point of a rigid body, we need to understand the associated notions of kinetic energy, matrix of inertia, and angular momentum, which implies taking into account the angular velocity of the body $\boldsymbol{\omega}$.

Figure 1.18 presents a rigid body, its instantaneous axis of rotation l, the vector of angular momentum of the body $\mathbf{K}_O$ measured with respect to the center O of motion about a point, and the momentum of a body element of mass $\mathrm{d}m$, that is, $v\mathrm{d}m$.

Let the body rotate about the fixed axis l with angular velocity $\boldsymbol{\omega}$, and let the distance of the element of mass $\mathrm{d}m$ from the rotation axis be denoted by ρ.

The *kinetic energy* of the rotating body is equal to

$$T = \frac{1}{2}\int_m \mathbf{v} \circ \mathbf{v}\mathrm{d}m = \frac{1}{2}\int_m v^2\mathrm{d}m = \frac{\omega^2}{2}\int_m \rho^2\mathrm{d}m = \frac{\omega^2}{2}I_l \tag{1.205}$$

because $v = \rho\omega$ and the unit vector of the l axis is denoted by $\boldsymbol{l}$ (Fig. 1.18). From (1.205) it follows that $I_l = \int_m \rho^2\mathrm{d}m$ is the mass moment of inertia of the body rotating about the l axis.

However, if the axis of rotation is not fixed, that is, it changes position with respect to the body under consideration, then the moment of inertia measured with respect to this instantaneous axis of rotation also undergoes a continuous change.

In the non-stationary coordinate system $OX_1'X_2'X_3'$ we have

$$\mathbf{v} = \boldsymbol{\omega} \times \mathbf{r} = \mathbf{E}_1'\left(\omega_2'r_3' - \omega_3'r_2'\right) + \mathbf{E}_2'\left(\omega_3'r_1' - \omega_1'r_3'\right) + \mathbf{E}_3'\left(\omega_1'r_2' - r_1'\omega_2'\right). \tag{1.206}$$

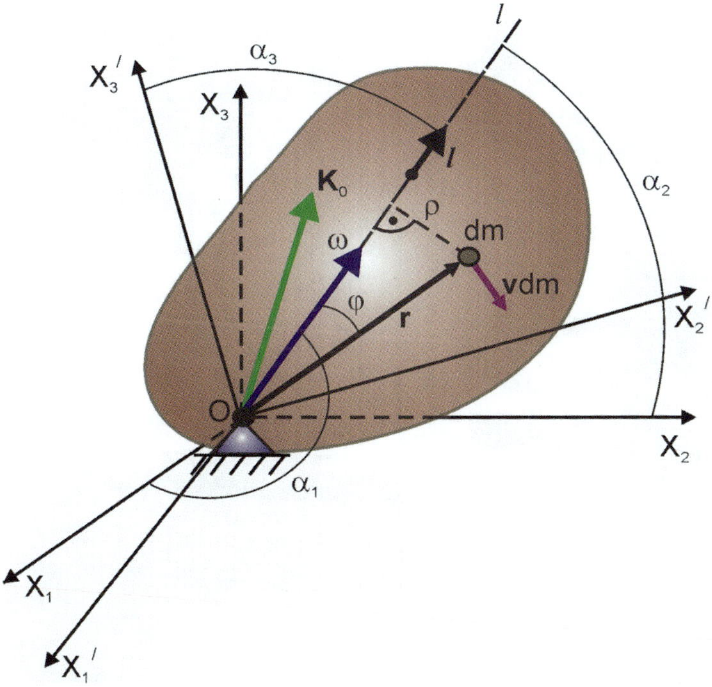

Fig. 1.18 Motion about a point of a rigid body and the adopted stationary coordinate system $OX_1X_2X_3$

Let us note that the preceding result can be obtained through the following linear transformation:

$$\mathbf{v} = \left([r'][\omega']\right) = \begin{bmatrix} 0 & r_3' & -r_2' \\ -r_3' & 0 & r_1' \\ r_2' & -r_1' & 0 \end{bmatrix} \begin{bmatrix} \omega_1' \\ \omega_2' \\ \omega_3' \end{bmatrix}, \tag{1.207}$$

and in turn

$$\begin{aligned} \mathbf{v} \circ \mathbf{v} &= \left(\begin{bmatrix} 0 & r_3' & -r_2' \\ -r_3' & 0 & r_1' \\ r_2' & -r_1' & 0 \end{bmatrix} \begin{bmatrix} \omega_1' \\ \omega_2' \\ \omega_3' \end{bmatrix} \right) \cdot \left(\begin{bmatrix} 0 & r_3' & -r_2' \\ -r_3' & 0 & r_1' \\ r_2' & -r_1' & 0 \end{bmatrix} \begin{bmatrix} \omega_1' \\ \omega_2' \\ \omega_3' \end{bmatrix} \right) \\ &= \begin{bmatrix} \omega_1' \\ \omega_2' \\ \omega_3' \end{bmatrix}^T \begin{bmatrix} 0 & r_3' & -r_2' \\ -r_3' & 0 & r_1' \\ r_2' & -r_1' & 0 \end{bmatrix}^T \begin{bmatrix} 0 & r_3' & -r_2' \\ -r_3' & 0 & r_1' \\ r_2' & -r_1' & 0 \end{bmatrix} \begin{bmatrix} \omega_1' \\ \omega_2' \\ \omega_3' \end{bmatrix}. \end{aligned} \tag{1.208}$$

Let us note that

$$
\begin{aligned}
[\boldsymbol{\omega}']^T[\mathbf{r}']^T &= \left[\omega_1' \; \omega_2' \; \omega_3'\right]\begin{bmatrix} 0 & -r_3' & r_2' \\ r_3' & 0 & -r_1' \\ -r_2' & r_1' & 0 \end{bmatrix} \\
&= \begin{bmatrix} \omega_2' r_3' - r_2'\omega_3' \\ -\omega_1' r_3' + r_1'\omega_3' \\ \omega_1' r_2' - r_1'\omega_2' \end{bmatrix}^T \equiv \left[v_1 \; v_2 \; v_3\right]
\end{aligned}
\tag{1.209}
$$

and

$$
\left[r'\right]\left[\omega'\right] = \begin{bmatrix} \omega_2' r_3' - r_2'\omega_3' \\ -\omega_1' r_3' + r_1'\omega_3' \\ \omega_1' r_2' - r_1'\omega_2' \end{bmatrix} \equiv \begin{bmatrix} v_1 \\ v_2 \\ v_3 \end{bmatrix}. \tag{1.210}
$$

According to (1.208) we have

$$
\begin{aligned}
I^V &= \begin{bmatrix} 0 & -r_3' & r_2' \\ r_3' & 0 & -r_1' \\ -r_2' & r_1' & 0 \end{bmatrix}\begin{bmatrix} 0 & r_3' & -r_2' \\ -r_3' & 0 & r_1' \\ r_2' & -r_1' & 0 \end{bmatrix} \\
&= \begin{bmatrix} I^V_{X_1'} & -I^V_{X_1'X_2'} & -I^V_{X_1'X_3'} \\ -I^V_{X_2'X_1'} & I^V_{X_2'} & -I^V_{X_2'X_3'} \\ -I^V_{X_3'X_1'} & -I^V_{X_3'X_2'} & I^V_{X_3'} \end{bmatrix},
\end{aligned}
\tag{1.211}
$$

where

$$
\begin{aligned}
&I^V_{X_1'} = \left(r_2'\right)^2 + \left(r_3'\right)^2, \quad I^V_{X_1'X_2'} = r_1'r_2', \quad I^V_{X_1'X_3'} = r_1'r_3', \\
&I^V_{X_2'X_3'} = r_2'r_3', \quad I^V_{X_2'} = \left(r_1'\right)^2 + \left(r_3'\right)^2, \quad I^V_{X_3'} = \left(r_1'\right)^2 + \left(r_2'\right)^2.
\end{aligned}
\tag{1.212}
$$

Taking into account (1.208) and (1.211) in (1.205) we obtain

$$
T = \frac{1}{2}\boldsymbol{\omega}'^T\left(\int_m I^V \mathrm{d}m\right)\boldsymbol{\omega}' = \frac{1}{2}\boldsymbol{\omega}'^T\mathbf{I}'\boldsymbol{\omega}', \tag{1.213}
$$

where $\mathbf{I}'$ is the matrix (tensor) of mass moments of inertia of the body in the body system.

Let the position of the l axis in an arbitrary Cartesian coordinate system $OX_1X_2X_3$ be described by angles α_1, α_2, and α_3 (Fig. 1.18).

The desired *moment of inertia* is equal to

$$
I_l = \int_m \rho^2 \mathrm{d}m. \tag{1.214}
$$

Because

$$r^2 = r_1^2 + r_2^2 + r_3^2, \tag{1.215}$$

$$\cos^2\alpha_1 + \cos^2\alpha_2 + \cos^2\alpha_3 = 1, \tag{1.216}$$

$$\begin{aligned}
\rho^2 &= r^2\sin^2\varphi = r^2 - (\mathbf{r}\circ\mathbf{e})^2 \\
&= r^2 - (r_1\cos\alpha_1 + r_2\cos\alpha_2 + r_3\cos\alpha_3)^2 \\
&= r_1^2\left(1-\cos^2\alpha_1\right) + r_2^2\left(1-\cos^2\alpha_2\right) + r_3^2\left(1-\cos^2\alpha_3\right) - 2r_1r_2\cos\alpha_1\cos\alpha_2 \\
&\quad - 2r_2r_3\cos\alpha_2\cos\alpha_3 - 2r_1r_3\cos\alpha_1\cos\alpha_3 \\
&= r_1^2\left(\cos^2\alpha_2 + \cos^2\alpha_3\right) + r_2^2\left(\cos^2\alpha_1 + \cos^2\alpha_3\right) \\
&\quad + r_3^2\left(\cos^2\alpha_1 + \cos^2\alpha_2\right) - 2r_1r_2\cos\alpha_1\cos\alpha_2 \\
&\quad - 2r_2r_3\cos\alpha_2\cos\alpha_3 - 2r_1r_3\cos\alpha_1\cos\alpha_3 \\
&= \left(r_2^2 + r_3^2\right)\cos^2\alpha_1 + \left(r_1^2 + r_3^2\right)\cos^2\alpha_2 + \left(r_1^2 + r_2^2\right)\cos^2\alpha_3 \\
&\quad - 2r_1r_2\cos\alpha_1\cos\alpha_2 - 2r_2r_3\cos\alpha_2\cos\alpha_3 - 2r_1r_3\cos\alpha_1\cos\alpha_3,
\end{aligned} \tag{1.217}$$

substituting (1.217) into (1.214) we obtain

$$\begin{aligned}
I_l &= I_{X_1}\cos^2\alpha_1 + I_{X_2}\cos^2\alpha_2 + I_{X_3}\cos^2\alpha_3 - 2I_{X_1X_2}\cos\alpha_1\cos\alpha_2 \\
&\quad -2I_{X_2X_3}\cos\alpha_2\cos\alpha_3 - 2I_{X_1X_3}\cos\alpha_1\cos\alpha_3.
\end{aligned} \tag{1.218}$$

From the obtained formula it follows that if we know the position of the l axis described by the angles α_1, α_2, α_3, and if we know the moments of inertia and the products of inertia with respect to the adopted coordinate system, we can determine the moment of inertia of the body with respect to the l axis.

The kinetic energy described by formula (1.213) can be also represented in the form

$$T = \frac{1}{2}\boldsymbol{\omega}\circ\mathbf{I}\boldsymbol{\omega}, \tag{1.219}$$

where $\mathbf{I}$ denotes the *inertia tensor*, which is symmetrical [see (1.211)]. On the diagonal of the inertia tensor successively lie the moments of inertia with respect to the axes OX_1, OX_2, and OX_3, and outside the diagonal lie the products of inertia.

We can represent the kinetic energy of the body given by (1.205) as

$$\begin{aligned}
T &= \frac{1}{2}\int\limits_m \mathbf{v}\circ(\boldsymbol{\omega}\times\mathbf{r})\mathrm{d}m = \frac{1}{2}\int\limits_m (\mathbf{r}\times\mathbf{v})\circ\boldsymbol{\omega}\mathrm{d}m \\
&= \frac{1}{2}\boldsymbol{\omega}\circ\int\limits_m(\mathbf{r}\times\mathbf{v})\mathrm{d}m = \frac{1}{2}\boldsymbol{\omega}\circ\mathbf{K},
\end{aligned} \tag{1.220}$$

where $\mathbf{K}$ is the angular momentum of the body with respect to point O.

During transformations of (1.220) the following commutativity property of the scalar product was used:

$$(\boldsymbol{a} \times \mathbf{b}) \circ \mathbf{c} = \boldsymbol{a} \circ (\mathbf{b} \times \mathbf{c}) = \mathbf{c} \circ (\boldsymbol{a} \times \mathbf{b}).$$

From a comparison of (1.219) with (1.220) it follows that the angular momentum is given by the formula

$$\mathbf{K} = \mathbf{I}\boldsymbol{\omega}. \tag{1.221}$$

Because vector $\boldsymbol{\omega}$ is directed along the instantaneous axis of rotation, knowing the inertia tensor along the l axis and the magnitude of the angular velocity $\boldsymbol{\omega}$ it is possible to calculate the angular momentum of the body with respect to point O. If we associate unit vector $\boldsymbol{l}$ with the l axis, then the angular velocity vector is given by $\boldsymbol{\omega} = \omega\, \boldsymbol{l}$. According to (1.219) we have

$$T = \frac{1}{2}\omega \circ \mathbf{I}\omega = \frac{\omega^2}{2}\boldsymbol{l} \circ \mathbf{I}\boldsymbol{l}. \tag{1.222}$$

In turn the definition of kinetic energy measured with respect to the l axis is given by (1.205).

Comparing (1.222) with (1.205) we obtain

$$I_l = \boldsymbol{l} \circ \mathbf{I}\boldsymbol{l}. \tag{1.223}$$

The obtained equation means that after arbitrarily choosing the unit vector $\boldsymbol{l}$ while we know the tensor of inertia $\mathbf{I}\boldsymbol{l}$, we can calculate the mass moment of inertia with respect to the axis defined by unit vector $\boldsymbol{l}$.

Let $\mathbf{S}$ be the rotation matrix such that the vectors of angular momentum $\mathbf{K}$ and of angular velocity $\boldsymbol{\omega}$, following rotation of the axes of the coordinate system $OX_1X_2X_3$ to the axes $OX_1'X_2'X_3'$, in the new coordinates take the form

$$\boldsymbol{\omega}' = \mathbf{S}\boldsymbol{\omega},$$
$$\mathbf{K}' = \mathbf{S}\mathbf{K}, \tag{1.224}$$

and substituting (1.224) into (1.221) we obtain

$$\mathbf{S}^T\mathbf{K}' = \mathbf{I}\mathbf{S}^T\boldsymbol{\omega}'. \tag{1.225}$$

Next, premultiplying by $\mathbf{S}$, we eventually obtain

$$\mathbf{K}' = \mathbf{I}'\boldsymbol{\omega}', \tag{1.226}$$

where the inertia tensor associated with the axes $OX_1'X_2'X_3'$ reads

$$\mathbf{I}' = \mathbf{S}\mathbf{I}\mathbf{S}^T. \tag{1.227}$$

Equation (1.227) describes the relationship between the inertia tensors expressed in the coordinates $OX_1X_2X_3$ and $OX_1'X_2'X_3'$. The property $\mathbf{S}^T\mathbf{S} = \mathbf{E}$, where $\mathbf{E}$ is the identity matrix, was already used earlier during transformations. Equation (1.227) describes matrices $\mathbf{I}'$ and $\mathbf{I}$ as *orthogonally similar* to one another.

Because according to the previous calculations, matrix $\mathbf{I}$ is symmetric and positive definite, so by introducing the rotation matrix $\mathbf{S}$ such that the axes of the coordinate system $OX_1'X_2'X_3'$ become coincident with the principal axes of inertia of the body, we obtain as a result a diagonal matrix of the form

$$\mathbf{I}' = \begin{bmatrix} \lambda_1 & 0 & 0 \\ 0 & \lambda_2 & 0 \\ 0 & 0 & \lambda_3 \end{bmatrix}. \tag{1.228}$$

The kinetic energy of the system expressed through these principal axes of inertia is equal to

$$T = \frac{1}{2}\left(\lambda_1\omega_{X_1'}^2 + \lambda_2\omega_{X_2'}^2 + \lambda_3\omega_{X_3'}^2\right). \tag{1.229}$$

Let us recall that vector $\boldsymbol{a}$, which satisfies the condition below, is called the *eigenvector of matrix I*:

$$\mathbf{I}\boldsymbol{a} = \lambda\boldsymbol{a}. \tag{1.230}$$

The preceding system is a system of linear algebraic equations of the form

$$(\mathbf{I} - \lambda\mathbf{E})\,\boldsymbol{a} = 0. \tag{1.231}$$

It has a non-zero solution if its characteristic determinant is equal to zero, that is,

$$\det(\mathbf{I} - \lambda\mathbf{E}) = 0. \tag{1.232}$$

The preceding equation leads to the determination of the eigenvalues λ_1, λ_2, and λ_3 – real numbers – which can be proved.

Moreover, let us note that

$$\begin{aligned} \det(\mathbf{I} - \lambda\mathbf{E}) &= \det\left(\mathbf{S}^T\mathbf{I}'\mathbf{S} - \lambda\mathbf{S}^T\mathbf{E}\mathbf{S}\right) \\ &= \left(\det\mathbf{S}^T\right)\det\left(\mathbf{I}' - \lambda\mathbf{E}\right)\det\mathbf{S} = \det\left(\mathbf{I}' - \lambda\mathbf{E}\right), \end{aligned} \tag{1.233}$$

which means that matrices of inertia expressed in different coordinate systems have the same eigenvalues.

References

1. J. Awrejcewicz, *Classical Mechanics: Statics and Kinematics.* (Springer, New York, 2012)
2. Z. Osiński, *General Mechanics.* (PWN, Warsaw, 1994)
3. J. Awrejcewicz, *Oscillations of Lumped Deterministic Systems.* (WNT, Warsaw, 1997)

4. J. Awrejcewicz, V.A. Krysko, *Oscillations of Continuous Systems.* (WNT, Warsaw, 2000)
5. J. Awrejcewicz, I.V. Andrianov, L.I. Manevitch, *Asymptotic Approaches in Nonlinear Dynamics: New Trends and Applications* (Springer, Berlin, 1998)
6. J. Awrejcewicz, V.A. Krysko, *Nonclassical Thermoelastic Problems in Nonlinear Dynamics of Shells* (Springer, Berlin, 2003)
7. J. Awrejcewicz, I. Andrianov, L.I. Manevitch, *Asymptotical Mechanics of Thin-Walled Structures* (Springer, Berlin, 2004)
8. J. Awrejcewicz, V.A. Krysko, A.V. Krysko, *Thermo-Dynamics of Plates and Shells* (Springer, Berlin, 2007)
9. J. Awrejcewicz, V.A. Krysko, *Introduction to Asymptotic Methods* (Chapman & Hall/CRC Press, New York, 2006)
10. J. Awrejcewicz, V.A. Krysko, A.V. Krysko, *Chaos in Structural Mechanics* (Springer, Berlin, 2008)
11. V.E. Zhuravlev, *Fundamentals of Theoretical Mechanics* (Fizmatlit, Moscow, 2001)
12. A.D. Markeev, *Theoretical Mechanics* (Nauka, Moscow, 1990)
13. G.R. Fowles , *Analytical Mechanics*, 3rd edn. (Holt, Rinehart and Winston, New York, 1977)
14. N.V. Butenin, N.A. Fufayev, *Introduction to Analytical Mechanics* (Nauka, Moscow, 1991)
15. S.T. Thornton, J.B. Marion, *Classical Dynamics of Particles and Systems* (Saunders College Publishers, New York, 1995)
16. A.P. Arya, *Introduction to Classical Mechanics* (Addison-Wesley, Reading, MA, 1998)
17. L.N. Hand, J.D. Finch, *Analytical Mechanics* (Cambridge University Press, Cambridge, 1998)
18. T.W. Kibble, F.H. Berkshire, *Classical Mechanics*, 5th edn. (Imperial College Press, Danvers, 2004)
19. J.R. Taylor, *Classical Mechanics* (University Science Books, Mill Valley, CA, 2005)
20. P.W. Johnson, *Classical Mechanics with Applications* (World Scientific, Singapore, 2010)
21. R.D. Gregory, *Classical Mechanics* (Cambridge University Press, Cambridge, 2011)

Chapter 2
Mathematical and Physical Pendulum

2.1 The Mathematical Pendulum

A particle of mass m connected by a rigid, weightless rod (or a thread) of length l to a base by means of a pin joint that can oscillate and rotate in a plane we call a *mathematical pendulum* (Fig. 2.1).

Let us resolve the gravity force into the component along the axis the rod and the component perpendicular to this axis, where both components pass through the particle of mass m. The normal component does not produce the particle motion. The component tangent to the path of the particle, being the arc of a circle of radius l, is responsible for the motion. Writing the equation of moments about the pendulum's pivot point we obtain

$$ml^2\ddot{\varphi} + mgl\sin\varphi = 0, \tag{2.1}$$

where ml^2 is a mass moment of inertia with respect to the pivot point.

From (2.1) we obtain

$$\ddot{\varphi} + \alpha^2 \sin\varphi = 0, \tag{2.2}$$

where

$$\alpha = \sqrt{\frac{g}{l}}\left[\frac{\text{rad}}{\text{s}}\right].$$

If it is assumed that we are dealing only with small oscillations of the pendulum, the relationship $\sin\alpha \approx \alpha$ holds true and (2.2) takes the form

$$\ddot{\varphi} + \alpha^2\varphi = 0. \tag{2.3}$$

It is the second-order linear differential equation describing the circular motion of a particle (Sect. 4.2 of [1]). Let us recall that its general solution has the form

$$\varphi = \phi\sin(\alpha t + \Theta_0), \tag{2.4}$$

J. Awrejcewicz, *Classical Mechanics: Dynamics*, Advances in Mechanics and Mathematics 29, DOI 10.1007/978-1-4614-3740-6_2,

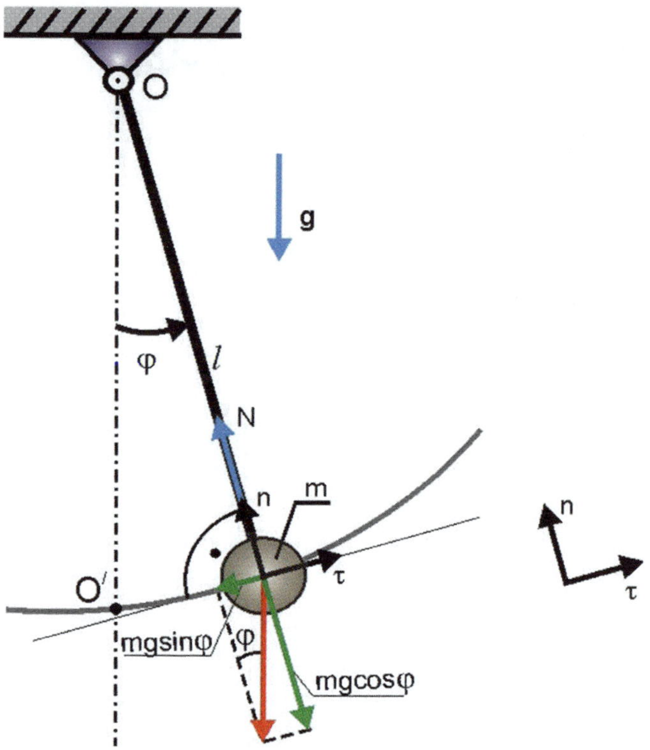

Fig. 2.1 Mathematical pendulum

which means that we are dealing with small harmonic oscillations of the period $T = \frac{2\pi}{\alpha} = 2\pi\sqrt{\frac{l}{g}}$. Let us note that in the case of small oscillations of a pendulum, their period does not depend on the initial angle of deflection of the pendulum but exclusively on the length of pendulum l. We say that such a motion is isochronous. That observation, however, is not valid for the big initial angle of deflection. Such a conclusion can be drawn on the basis of the following calculations (see also [2, 3]).

Let us transform (2.2) into the form

$$\begin{aligned} \dot{\varphi} &= \gamma, \\ \dot{\gamma} &= -\alpha^2 \sin\varphi. \end{aligned} \tag{2.5}$$

Let us note that

$$\dot{\gamma} = \frac{\mathrm{d}\gamma}{\mathrm{d}t} = \frac{\mathrm{d}\gamma}{\mathrm{d}\varphi}\frac{\mathrm{d}\varphi}{\mathrm{d}t} = \frac{\mathrm{d}\gamma}{\mathrm{d}\varphi}\gamma = \frac{1}{2}\frac{\mathrm{d}}{\mathrm{d}\varphi}\left(\gamma^2\right), \tag{2.6}$$

and substituting (2.6) into the second equation of system (2.5) we obtain

$$\frac{\mathrm{d}}{\mathrm{d}\varphi}\left(\gamma^2\right) = -2\alpha^2 \sin\varphi. \tag{2.7}$$

Separating the variables and integrating we obtain

$$\dot{\varphi}^2 = 2\alpha^2 \cos\varphi + 2C, \tag{2.8}$$

where C is the integration constant.

Let us emphasize that we were able to conduct the integration thanks to the fact that the investigated system is conservative (it was assumed that the medium in which the vibrations take place introduced no damping). Equation (2.8) is the *first integral* of the non-linear differential equation (2.2) since it relates the functions $\varphi(t)$ and $\dot{\varphi}(t)$. In other words, it is the non-linear equation of reduced order with respect to the original equation (2.2).

Let us introduce the following initial conditions: $\varphi(0) = \varphi_0$, $\dot{\varphi}(0) = \dot{\varphi}_0$, and following their substitution into (2.8), in order to determine the integration constant, we obtain

$$2C = \dot{\varphi}_0^2 - 2\alpha^2 \cos\varphi_0. \tag{2.9}$$

From (2.8) we obtain

$$\dot{\varphi} = \pm\sqrt{2\left(C + \alpha^2 \cos\varphi\right)}, \tag{2.10}$$

where C is given by (2.9).

The initial condition $\dot{\varphi}_0$ determines the selection of the sign in formula (2.10). If $\dot{\varphi}_0 > 0$, then we select a plus sign, and if $\dot{\varphi}_0 < 0$, then we select a minus sign. If $\dot{\varphi}_0 = 0$, then the choice of the sign in front of the square root should agree with the sign of acceleration $\ddot{\varphi}(0)$.

Following separation of the variables in (2.10) and integration we have

$$t = \pm\int_{\varphi_0}^{\varphi} \frac{\mathrm{d}\varphi}{\sqrt{2\left(C + \alpha^2 \cos\varphi\right)}}. \tag{2.11}$$

Unfortunately, it is not possible to perform the integration of the preceding equation using elementary functions. The preceding integral is called the *elliptic integral*. Equation (2.11) describes the time plot $\varphi(t)$, and the form of the function (the solution) depends on initial conditions as shown subsequently.

Below we will consider two cases of selection of the initial conditions [4].

Case 1. Let us first consider a particular form of the initial condition, namely, let $0 < \varphi_0 < \pi$ and $\dot{\varphi}_0 = 0$. For such an initial condition from (2.9) we obtain

$$C = -\alpha^2 \cos\varphi_0. \tag{2.12}$$

Note that (2.7) indicates that $|C| \le \frac{g}{l}$.

Substituting the constant thus obtained into (2.11) we obtain

$$t = \pm\frac{1}{\alpha}\int_{\varphi_0}^{\varphi}\frac{\mathrm{d}\varphi}{\sqrt{2\left(\cos\varphi - \cos\varphi_0\right)}}. \tag{2.13}$$

The outcome of the actual process determines the sign of the preceding expression, i.e., we have $t \geq 0$. From observations it follows that following introduction of the aforementioned initial condition (or similarly for $-\pi < \varphi_0 < 0$) the angle $\varphi(t)$ decreases, that is, $\cos\varphi_0 > \cos\varphi$; therefore, one should select a minus sign in (2.13). The angle $\varphi(t)$ will be a decreasing function until the second extreme position $\varphi = -\varphi_0$ is attained. From that instant we will be dealing with similar calculations since the initial conditions are determined by the initial angle $-\varphi_0$ and the speed $\dot{\varphi}\left(\frac{T}{2}\right) = 0$, where T is a period of oscillations.

Starting from the aforementioned instant, the angle $\varphi(t)$ will increase from the value $-\varphi_0$ to the value $+\varphi_0$; therefore, in that time interval one should select a plus sign in (2.13).

Note that

$$\begin{aligned}\cos\varphi - \cos\varphi_0 &= 1 - 2\sin^2\frac{\varphi}{2} - \left(1 - 2\sin^2\frac{\varphi_0}{2}\right)\\ &= 2\left(\sin^2\frac{\varphi_0}{2} - \sin^2\frac{\varphi}{2}\right),\end{aligned} \tag{2.14}$$

and hence from (2.13) (minus sign) we obtain

$$t = -\frac{1}{2\alpha}\int_{\varphi_0}^{\varphi}\frac{\mathrm{d}\varphi}{\sqrt{\sin^2\frac{\varphi_0}{2} - \sin^2\frac{\varphi}{2}}} = -\frac{1}{2\alpha}\int_{\varphi_0}^{\varphi}\frac{\mathrm{d}\varphi}{\sin\frac{\varphi_0}{2}\sqrt{1 - \frac{\sin^2\frac{\varphi}{2}}{\sin^2\frac{\varphi_0}{2}}}}. \tag{2.15}$$

For the purpose of further transformations let us introduce a new variable ξ of the form

$$\sin\xi = \frac{\sin\frac{\varphi}{2}}{\sin\frac{\varphi_0}{2}}. \tag{2.16}$$

Differentiating both sides of the preceding equation we obtain

$$\cos\xi\mathrm{d}\xi = \frac{\cos\frac{\varphi}{2}}{2\sin\frac{\varphi_0}{2}}\mathrm{d}\varphi. \tag{2.17}$$

Taking into account that the introduction of the new variable ξ leads also to a change in the limits of integration, i.e., $\xi_0 = \frac{\pi}{2}$ corresponds to φ_0 [see (2.16)], and taking into account relationships (2.16) and (2.17) in (2.15) we obtain

$$t = \frac{1}{2\alpha}\int_{\xi}^{\frac{\pi}{2}}\frac{2\sin\frac{\varphi_0}{2}\cos\xi\mathrm{d}\xi}{\cos\frac{\varphi}{2}\sin\frac{\varphi_0}{2}\sqrt{1 - \sin^2\xi}}$$

$$= \frac{1}{\alpha}\int_{\xi}^{\frac{\pi}{2}} \frac{d\xi}{\sqrt{\frac{(1-\sin^2\xi)\cos^2\frac{\varphi}{2}}{\cos^2\xi}}}$$

$$= \frac{1}{\alpha}\int_{\xi}^{\frac{\pi}{2}} \frac{d\xi}{\sqrt{\frac{\cos^2\frac{\varphi}{2}-(1-\cos^2\xi)\cos^2\frac{\varphi}{2}}{\cos^2\xi}}} = \frac{1}{\alpha}\int_{\xi}^{\frac{\pi}{2}} \frac{d\xi}{\sqrt{\cos^2\frac{\varphi}{2}}}$$

$$= \frac{1}{\alpha}\int_{\xi}^{\frac{\pi}{2}} \frac{d\xi}{\sqrt{1-\sin^2\frac{\varphi_0}{2}\sin^2\xi}}, \tag{2.18}$$

because according to (2.16) we have

$$\sin^2\frac{\varphi}{2} = \sin^2\xi\sin^2\frac{\varphi_0}{2}. \tag{2.19}$$

The change in the angle of oscillations from φ_0 to zero corresponds to the time interval $T/4$, which after using (2.18) leads to the determination of the period of oscillations

$$T = \frac{4}{\alpha}\int_0^{\frac{\pi}{2}} \frac{d\xi}{\sqrt{1-\sin^2\frac{\varphi_0}{2}\sin^2\xi}}. \tag{2.20}$$

Let $x = \sin^2\frac{\varphi_0}{2}\sin^2\xi$; then $|x| < 1$, and the integrand can be expanded in a Maclaurin series about $x = 0$ in the following form:

$$f(x) = \frac{1}{\sqrt{1-x}} = 1 - \frac{1}{2}x + \cdots = 1 - \frac{1}{2}\sin^2\frac{\varphi_0}{2}\sin^2\xi + \cdots. \tag{2.21}$$

Taking into account (2.20) and (2.21) we obtain

$$T = 4\sqrt{\frac{l}{g}}\left(\int_0^{\frac{\pi}{2}} d\xi - \frac{1}{2}\sin^2\frac{\varphi_0}{2}\int_0^{\frac{\pi}{2}}\sin^2\xi d\xi\right)$$

$$= 4\sqrt{\frac{l}{g}}\left(\frac{\pi}{2} - \frac{1}{4}\cdot\frac{\pi}{2}\sin^2\frac{\varphi_0}{2}\right)$$

$$= 2\pi\sqrt{\frac{l}{g}}\left(1-\frac{1}{4}\sin^2\frac{\varphi_0}{2}\right) \approx 2\pi\sqrt{\frac{l}{g}}\left(1-\frac{1}{16}\varphi_0^2\right), \tag{2.22}$$

where in the last transformation the relationship $\sin\frac{\varphi_0}{2} \approx \frac{\varphi_0}{2}$ was used.

Case 2. Now let us consider the case where apart from the initial amplitude φ_0 the particle (bob of pendulum) was given the speed $\dot{\varphi}_0$ big enough that, according to (2.9), the following inequality is satisfied:

$$C = \frac{1}{2}\dot{\varphi}_0^2 - \frac{g}{l}\cos\varphi_0 > \frac{g}{l}. \tag{2.23}$$

Then the radicand in (2.10) is always positive. This means that the function $\varphi(t)$ is always increasing (plus sign) or decreasing (minus sign). The physical interpretation is such that the pendulum rotates clockwise (plus sign) or counterclockwise (minus sign).

It turns out that for certain special values of initial conditions, namely,

$$C = \frac{1}{2}\dot{\varphi}_0^2 - \frac{g}{l}\cos\varphi_0 = \frac{g}{l}, \tag{2.24}$$

we can perform the integration given by (2.11).

From that equation, and taking into account (2.24) and assuming $\dot{\varphi}_0 > 0$, we obtain

$$\begin{aligned} t &= \sqrt{\frac{l}{g}}\int_{\varphi_0}^{\varphi}\frac{\mathrm{d}\varphi}{\sqrt{2(1+\cos\varphi)}} = \sqrt{\frac{l}{g}}\int_{\varphi_0}^{\varphi}\frac{\mathrm{d}\varphi}{\sqrt{2\cos\frac{\varphi}{2}}} = \\ &= \sqrt{\frac{l}{g}}\ln\left[\frac{\tan\left(\frac{\pi}{4}-\frac{\varphi_0}{4}\right)}{\tan\left(\frac{\pi}{4}-\frac{\varphi}{4}\right)}\right], \end{aligned} \tag{2.25}$$

because

$$\sqrt{2(1+\cos\varphi)} = \sqrt{2\left(1+\cos^2\frac{\varphi}{2}-\sin^2\frac{\varphi}{2}\right)} = 2\cos\frac{\varphi}{2}.$$

In (2.25) we have a singularity since when $\varphi \to \pi$, the time $t \to +\infty$. This means that for the initial condition (2.24) the pendulum attains the vertical position for $t \to +\infty$. The foregoing analysis leads to the diagram presented in Fig. 2.2.

As was already mentioned, (2.1) cannot be solved in an analytical way using elementary functions. However, it will be shown that using the natural coordinates $\boldsymbol{\tau}, \mathbf{n}$ one may determine an exact value of reaction $\mathbf{N}$ (Fig. 2.1).

The Euler equations of motion are as follows:

$$\begin{aligned} m\frac{\mathrm{d}^2 s}{\mathrm{d}t^2} &= F_\tau = -mg\sin\varphi, \\ \frac{mv^2}{\rho} &= F_n + N = -mg\cos\varphi + N. \end{aligned} \tag{2.26}$$

Taking into account $s = l\varphi$, the first equation of (2.26) takes the form of (2.2). The second equation of (2.26) yields

$$N = \frac{mv^2}{l} + mg\cos\varphi. \tag{2.27}$$

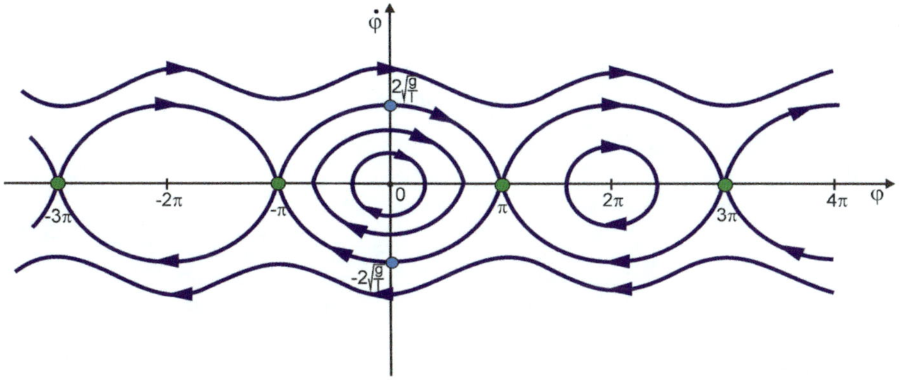

Fig. 2.2 Small oscillations about an equilibrium position, a critical case described by the initial condition (2.24), and pendulum rotations

Observe that

$$\frac{\mathrm{d}v}{\mathrm{d}t} = \frac{\mathrm{d}v}{\mathrm{d}\varphi} \cdot \frac{\mathrm{d}\varphi}{\mathrm{d}t} = \frac{v}{l}\frac{\mathrm{d}v}{\mathrm{d}\varphi}, \tag{2.28}$$

and from (2.2) one obtains

$$\frac{\mathrm{d}\varphi}{\mathrm{d}t} + \alpha^2 \sin\varphi = \frac{\mathrm{d}v}{l\,\mathrm{d}t} + \alpha^2 \sin\varphi = \frac{v}{l^2}\frac{\mathrm{d}v}{\mathrm{d}\varphi} + \alpha^2 \sin\varphi = 0 \tag{2.29}$$

or, equivalently,

$$v\frac{\mathrm{d}v}{\mathrm{d}\varphi} = -gl\sin\varphi. \tag{2.30}$$

Separation of variables and integration give

$$\frac{(v^2 - v_0^2)}{2} = gl(\cos\varphi - \cos\varphi_0) \tag{2.31}$$

or, equivalently,

$$\frac{v^2}{l} = \frac{v_0^2}{l} + 2g(\cos\varphi - \cos\varphi_0). \tag{2.32}$$

Substituting (2.32) into (2.27) yields

$$N = \frac{mv_0^2}{l} + 2mg(\cos\varphi - \cos\varphi_0).$$

A minimum force value can be determined from the equation

$$\frac{\mathrm{d}N}{\mathrm{d}\varphi} = -2mg\sin\varphi = 0, \tag{2.33}$$

which is satisfied for $\varphi = 0, \pm\pi, \pm 2\pi, \ldots$.

Because

$$\frac{\mathrm{d}^2 N}{\mathrm{d}\varphi^2} = -2mg\cos\varphi, \tag{2.34}$$

then

$$\left.\frac{\mathrm{d}^2 N}{\mathrm{d}\varphi^2}\right|_0 = -2mg, \qquad \left.\frac{\mathrm{d}^2 N}{\mathrm{d}\varphi^2}\right|_\pi = 2mg. \tag{2.35}$$

This means that in the lower (upper) pendulum position the thread tension achives its maximum (minimum).

The force minimum value is computed from (2.27):

$$N_{min} = \frac{mv_0^2}{l} + 2mg(-1 - \cos\varphi_0). \tag{2.36}$$

The thread will be stretched when $N_{min} \geq 0$, i.e., for

$$v_0 \geq \sqrt{2gl(1 + \cos\varphi_0)}. \tag{2.37}$$

At the end of this subsection we will study a *pendulum resultant motion*.

Let us assume that an oscillating mathematical pendulum undertakes a flat motion in plane Π, which rotates about a vertical axis crossing the pendulum clamping point (Fig. 2.3).

The equation of a relative pendulum motion expressed through the natural coordinates $\boldsymbol{\tau}, \mathbf{n}$ has the following form (projections of forces onto the tangent direction):

$$m(p_\tau^w + p_\tau^u + p_\tau^C) = -mg\sin\varphi, \tag{2.38}$$

and since the Coriolis acceleration $\mathbf{p}_\tau^C \perp \Pi$, then $p_\tau^C = 0$.

Projection of the translation acceleration onto a tangent to the particle trajectory is $p_\tau^u = (\omega^2 l \sin\varphi)\cos\varphi$, and finally (2.38) takes the form

$$ml\ddot{\varphi} = -mg\sin\varphi + m\omega^2 l \sin\varphi\cos\varphi \tag{2.39}$$

or

$$\ddot{\varphi} = (\omega^2\cos\varphi - \alpha^2)\sin\varphi. \tag{2.40}$$

Observe that now we have two sets of equilibrium positions yielded by the equation

$$(\omega^2\cos\varphi - \alpha^2)\sin\varphi = 0. \tag{2.41}$$

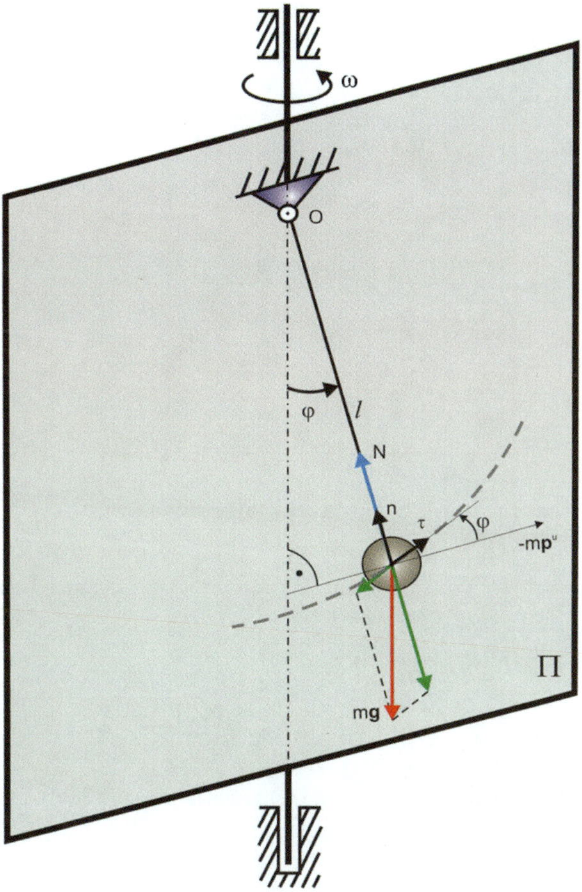

Fig. 2.3 Resultant mathematical pendulum motion oscillating in plane Π rotating with angular velocity ω

Besides the previously discussed set governed by the equation $\sin\varphi = 0$, i.e., $\varphi_n = n\pi, \quad n \in C$, the additional set of equilibrium positions is given by the following formula:

$$\varphi_n = \pm \arccos\frac{\alpha^2}{\omega^2} + 2\pi n, \quad n \in C \tag{2.42}$$

if $\omega \geq \alpha$.

Strongly non-linear equations of motion of the pendulum (2.2) can also be solved in an exact way. Following the introduction of non-dimesional time $\tau = \alpha t$ we obtain

$$\varphi'' + \sin\varphi = 0, \tag{2.43}$$

and multiplying by sides through φ' we have

$$\varphi'\varphi'' \equiv \frac{\mathrm{d}}{\mathrm{d}\tau}\left(\frac{\varphi'^2}{2}\right) = -\frac{\mathrm{d}\varphi}{\mathrm{d}\tau}\sin\varphi \tag{2.44}$$

or

$$\mathrm{d}\left(\frac{\varphi'^2}{2}\right) = -\sin\varphi\mathrm{d}\varphi. \tag{2.45}$$

Integration of (2.45) yields

$$\frac{\varphi'^2}{2} = -[(-\cos\varphi) + \cos\varphi_0], \tag{2.46}$$

that is,

$$\varphi' = \pm\sqrt{2}\sqrt{|\cos\varphi - \cos\varphi_0|}, \tag{2.47}$$

where $\cos\varphi_0$ is a constant of integration.

Because

$$\cos\varphi = 1 - 2\sin^2\left(\frac{\varphi}{2}\right), \tag{2.48}$$

(2.47) takes the form

$$\varphi' = -\sqrt{2}\sqrt{2\left|\sin^2\left(\frac{\varphi_0}{2}\right) - \sin^2\left(\frac{\varphi}{2}\right)\right|}, \tag{2.49}$$

and following separation of the variables and integration we have

$$-\int_{\varphi_0}^{\varphi} \frac{\mathrm{d}\left(\frac{\varphi}{2}\right)}{\sqrt{\left|\sin^2\left(\frac{\varphi_0}{2}\right) - \sin^2\left(\frac{\varphi}{2}\right)\right|}} = \int_0^t \mathrm{d}\tau. \tag{2.50}$$

The obtained integral cannot be expressed in terms of elementary functions and is called an elliptic integral because it also appears during calculation of the length of an elliptical curve.

For the purpose of its calculation we introduce two parameters – k, called the *elliptic modulus*, and u, called the *amplitude* – according to the following equations:

$$\sin\left(\frac{\varphi}{2}\right) = \sin\left(\frac{\varphi_0}{2}\right)\sin\theta = k\sin\theta, \qquad k = \sin\left(\frac{\varphi_0}{2}\right),$$

$$\cos\left(\frac{\varphi}{2}\right) = \sqrt{1 - \sin^2\left(\frac{\varphi}{2}\right)} = \sqrt{1 - k^2\sin^2\theta},$$

$$\mathrm{d}\left(\sin\left(\frac{\varphi}{2}\right)\right) = \cos\left(\frac{\varphi}{2}\right)\mathrm{d}\left(\frac{\varphi}{2}\right) = k\mathrm{d}(\sin\theta) = k\cos\theta\mathrm{d}\theta,$$

$$\mathrm{d}\left(\frac{\varphi}{2}\right) = \frac{k\cos\theta\mathrm{d}\theta}{\cos\left(\frac{\varphi}{2}\right)} = \frac{k\cos\theta\mathrm{d}\theta}{\sqrt{1 - k^2\sin^2\theta}}. \tag{2.51}$$

From the first equation of (2.51) we have

$$\sin^2\left(\frac{\varphi_0}{2}\right) = k^2, \tag{2.52}$$

and the integral from (2.50) takes the form

$$\int_{\varphi(t)}^{\varphi_0} \frac{k\cos\theta \mathrm{d}\theta}{\sqrt{1-k^2\sin^2\theta}\sqrt{k^2-k^2\sin^2\theta}} = \int_{\varphi(t)}^{\varphi_0} \frac{\mathrm{d}\theta}{\sqrt{1-k^2\sin^2\theta}}. \tag{2.53}$$

In the foregoing integral one should additionally alter the limits of integration. We have here a conservative system; let oscillations of the system be characterized by period T (we consider the case of a lack of rotation of the pendulum). After time $T/2$ starting from the initial condition φ_0 according to the first equation of (2.51) we have $\sin\theta = -1$, which implies $\theta(T/2) = 3\pi/2$. In turn, at the instant the motion began according to that equation we have $\sin\theta = 1$, that is, $\theta(0) = \pi/2$. Finally, (2.50) takes the form

$$\int_{\pi/2}^{3\pi/2} \frac{\mathrm{d}\theta}{\sqrt{1-k^2\sin^2\theta}} = 2\int_0^{\pi/2} \frac{\mathrm{d}\theta}{\sqrt{1-k^2\sin^2\theta}} = \frac{T}{2}. \tag{2.54}$$

The desired period of the pendulum oscillations is equal to

$$T = 4\int_0^{\pi/2} \frac{\mathrm{d}\theta}{\sqrt{1-k^2\sin^2\theta}}. \tag{2.55}$$

The integral

$$F(\theta^*, k) = \int_0^{\theta^*} \frac{\mathrm{d}\theta}{\sqrt{1-k^2\sin^2\theta}} \tag{2.56}$$

is called an *elliptic incomplete integral of the first kind*. Introducing variable

$$z = \sin\theta \tag{2.57}$$

we can represent integral (2.56) in the following equivalent form:

$$\int_0^{\theta^*} = \frac{\mathrm{d}z}{\cos\theta\sqrt{1-k^2z^2}} \int_0^{\sin\theta^*} \frac{\mathrm{d}z}{\sqrt{(1-z^2)(1-k^2z^2)}}. \tag{2.58}$$

A graph of dependency $T(k)$ on the basis of (2.55) is presented in Fig. 2.4.

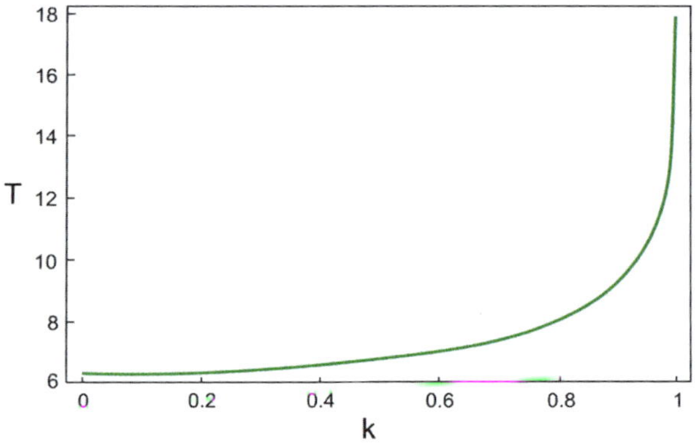

Fig. 2.4 Graph of dependency of period of pendulum oscillations on initial deflection $k = \sin(\varphi_0/2)$

2.2 The Physical Pendulum

In this section we will consider the plane motion of a material body suspended on the horizontal axis and allowed to rotate about it. It is the physical pendulum depicted in Fig. 2.5.

The pendulum is hung at point O, and the straight line passing through that point and the pendulum mass center C defines the axis OX_1. We assume that the angle of rotation of pendulum φ is positive and the sense of the axis perpendicular to the OX_1 axis is taken in such a way that the Cartesian coordinate system $OX_1X_2X_3$ is a right-handed one (the X_3 axis is perpendicular to the plane of the drawing).

Neglecting the resistance to motion in a radial bearing O and the resistance of the medium, the only force producing the motion is the component of gravity force tangent to a circle of radius s. The equation of moments of force about point O has the form

$$I_O\ddot{\varphi} = -msg\sin\varphi, \tag{2.59}$$

where I_O is the moment of inertia of the physical pendulum with respect to pivot point O, i.e., the axis X_3.

By analogy to the equation of motion of a mathematical pendulum (2.2), we will reduce (2.59) to the form

$$\ddot{\varphi} + \alpha_f^2\sin\varphi = 0, \tag{2.60}$$

where

$$\alpha_f^2 = \frac{mgs}{I_O} = \frac{g}{l_{\mathrm{red}}}. \tag{2.61}$$

From (2.61) it follows that the introduced *reduced length of a physical pendulum* equals $l_{\mathrm{red}} = \frac{I_C}{ms}$. All the calculations conducted in Sect. 2.1 hold also in this case

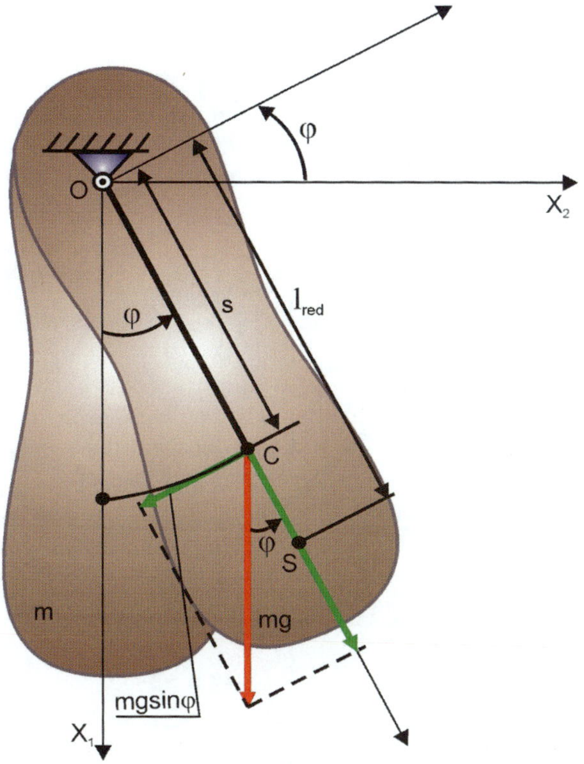

Fig. 2.5 Physical pendulum of mass m

owing to the same mathematical model. In particular, the period of small oscillations of the physical pendulum about the equilibrium position $\varphi = 0$ equals

$$T = \frac{2\pi}{\alpha_f} = 2\pi\sqrt{\frac{I_O}{mgs}} = 2\pi\sqrt{\frac{l_{\text{red}}}{g}}. \tag{2.62}$$

Let the moment of inertia with respect to the axis parallel to X_3 and passing through the mass center of the physical pendulum be I_C. According to the parallel axis theorem the moment of inertia I_O reads

$$I_O = I_C + ms^2 = m\left(\frac{I_O}{m} + s^2\right) = m\left(i_C^2 + s^2\right), \tag{2.63}$$

where i_C is a radius of gyration with respect to the axis passing through the mass center of the pendulum.

According to the previous calculations the reduced length of the physical pendulum is equal to

$$l_{\text{red}} = \frac{I_O}{ms} = s + \frac{i_C^2}{s}. \tag{2.64}$$

From the preceding equation it follows that the reduced length of a physical pendulum is a function of s. If $s \to 0$, then $l_{\mathrm{red}} \to \infty$, whereas if $s \to \infty$, then $l_{\mathrm{red}} \to \infty$. According to (2.62) the period of small oscillations of the pendulum $T \to \infty$ for $s \to 0$ and $T \to 0$ for $s \to \infty$. Our aim will be to determine such a value of s, i.e., the distance of the point of rotation of the pendulum from its mass center, for which the period is minimal. According to (2.64)

$$\frac{\mathrm{d}l_{\mathrm{red}}}{\mathrm{d}s} = 1 - \frac{i_C^2}{s^2} = 0, \tag{2.65}$$

$$\frac{\mathrm{d}^2 l_{\mathrm{red}}}{\mathrm{d}s^2} = \frac{2}{s^3} i_C^2 > 0. \tag{2.66}$$

From the two foregoing equations it follows that the function $l_{\mathrm{red}}(s)$ attains the minimum for the value $s = i_C$ because its second derivative for such s is positive.

In Fig. 2.5 point S is marked at the distance l_{red} from the axis of rotation. We will call that point the *center of swing corresponding to the pivot point O*. In other words, if we concentrate the total mass of the physical pendulum at point S, then we will obtain a mathematical pendulum of length l_{red}.

Let us suspend the pendulum at point S obtained in that way and determine the corresponding center of swing S^*. The length reduced to point S, according to (2.64), is equal to

$$l_{\mathrm{red}}^* = SS^* = CS + \frac{i_C^2}{CS}. \tag{2.67}$$

Because

$$CS = OS - OC = l_{\mathrm{red}} - s = \frac{i_C^2}{s}, \tag{2.68}$$

from (2.67) we have

$$l_{\mathrm{red}}^* = l_{\mathrm{red}} - s + s = l_{\mathrm{red}}. \tag{2.69}$$

From the foregoing calculations it follows that the pivot point of the pendulum and the corresponding center of swing play an identical role with respect to one another.

In Sect. 2.1 we mentioned the *first integral of motion*. Now, as distinct from that approach, we will determine the relationships between the velocities and the displacement of the pendulum based on the theorem of the conservation of mechanical energy. Let the mechanical energy of a physical pendulum be given by

$$T(t) + V(t) = C + mgs, \tag{2.70}$$

where C is a certain constant, i.e., the stored energy of the pendulum introduced by the initial conditions, and mgs denotes the potential energy of the system in the equilibrium position $\varphi = 0$.

Because we are dealing with a conservative system, the sum of kinetic energy T and potential energy V does not change in time and is constant for every time instant t in the considered case (Fig. 2.5):

$$T = \frac{1}{2} I_O \dot{\varphi}^2, \tag{2.71}$$

$$V = mgs(1 - \cos\varphi), \tag{2.72}$$

where V is the potential energy of the pendulum deflected through the angle φ. Substituting (2.71) and (2.72) into (2.70) we have

$$\frac{I_O \dot{\varphi}^2}{2} - mgs\cos\varphi = C, \tag{2.73}$$

and hence

$$\dot{\varphi}^2 = \frac{2}{I_O}(C + mgs\cos\varphi). \tag{2.74}$$

Let us note that (2.74) is analogous to the previously obtained equation for a mathematical pendulum (2.8) since we have

$$\dot{\varphi}^2 = 2C_f + 2\alpha_f^2 \cos\varphi, \tag{2.75}$$

where $C_f = C/I_O$ and α_f^2 is defined by (2.61).

If the initial conditions of the pendulum have the form $\varphi(0) = \varphi_0$, $\dot{\varphi}(0) = 0$, then from (2.73) we obtain

$$C = -mgs\cos\varphi_0. \tag{2.76}$$

This means that the initial energy is associated only with the potential energy, and taking into account (2.76) in (2.74) we have

$$\dot{\varphi}^2 = \frac{2mgs}{I_O}(\cos\varphi - \cos\varphi_0). \tag{2.77}$$

The last equation corresponds to (2.8).

2.3 Planar Dynamics of a Triple Physical Pendulum

2.3.1 *Equations of Motion*

Our goal is to introduce a mathematical model of a 2D triple physical pendulum. The *mathematical model* [3] is a description of the system dynamics with the aid of equations, in this case, ordinary differential equations. The mathematical model is a mathematical expression of physical laws valid in the considered system. In order to proceed with writing the equations, we must first have a physical model understood as a certain conception of physical phenomena present in the system. One should remember that the physical model to be presented below does not exist in reality but is an idea of a triple physical pendulum. If there existed a real object, which we would also call a *triple physical pendulum*, it would be able to correspond to our physical model approximately at best. By saying here that the real system and the model correspond to each other approximately, we mean that all physical phenomena taken into account in the model occur in the real object. Additionally, the influence of the physical phenomena occurring in the real system and not taken into account in the model (they can be treated as disturbances) on the observed (measured) quantities that are of interest to us is negligible. Those observed quantities in the triple pendulum can be, for instance, three angles describing the position of the pendulum at every time instant. Clearly, there exists a possibility of further development of this model so that it incorporates increasingly more physical phenomena occurring in a certain existing real system and, consequently, becomes closer to that real system. An absolute agreement, however, will not be attainable. On the other hand, one may also think about the opposite situation, in which to the theoretically created idea of a pendulum we try to match a real object, that is, the test stand. Then we build it in such a way that in the stand only the laws and physical phenomena assumed in the model are, to the best possible approximation, valid. The influence of other real phenomena on the quantities of interest should be negligible.

Our physical model of a triple pendulum (Fig. 2.6) consists of three ($i = 1, 2, 3$) absolutely rigid bodies moving in a vacuum in a uniform gravitational field of lines that are parallel and directed against the axis X_2 of the global coordinate system $O_1X_1X_2X_3$, connected to each other by means of revolute joints O_i and connected to an absolutely rigid base [5]. Those joints have axes perpendicular to the plane $O_1X_1X_2$ so that the whole system moves in planar motion. We assume that in the joints there exists viscous damping, that is, that the resistive moment counteracting the relative motion of two pendulums connected to each other is proportional (with a certain proportionality factor c_i) to their relative angular velocity. We also assume that mass centers of particular pendulums (C_i) lie in planes determined by the axes of joints by which the given pendulum is connected to the rest of the system (this does not apply to the third pendulum). This last assumption allows for a decrease in the number of model parameters – to be precise, the number of parameters establishing the positions of mass centers of particular pendulums. Each of the pendulums has its own local coordinate system $C_iX_1^{(i)}X_2^{(i)}X_3^{(i)}$ ($i = 1, 2, 3$) of the origin at the mass center of the given pendulum and the axis $X_3^{(i)}$ perpendicular to the plane of motion. Geometrical parameters that determine the positions of mass centers (e_i) and the distances between the joints (l_1, l_2) are indicated in the figure. Moreover, each of the pendulums possesses mass m_i and mass moment of inertia I_i with respect to the axis C_iX_3 passing through the mass center (centroidal axis) and perpendicular to the plane of motion. The first pendulum is acted upon by an external moment $M_e(t)$. The configuration of the system is uniquely described by three angles ψ_i, as shown in Fig. 2.6.

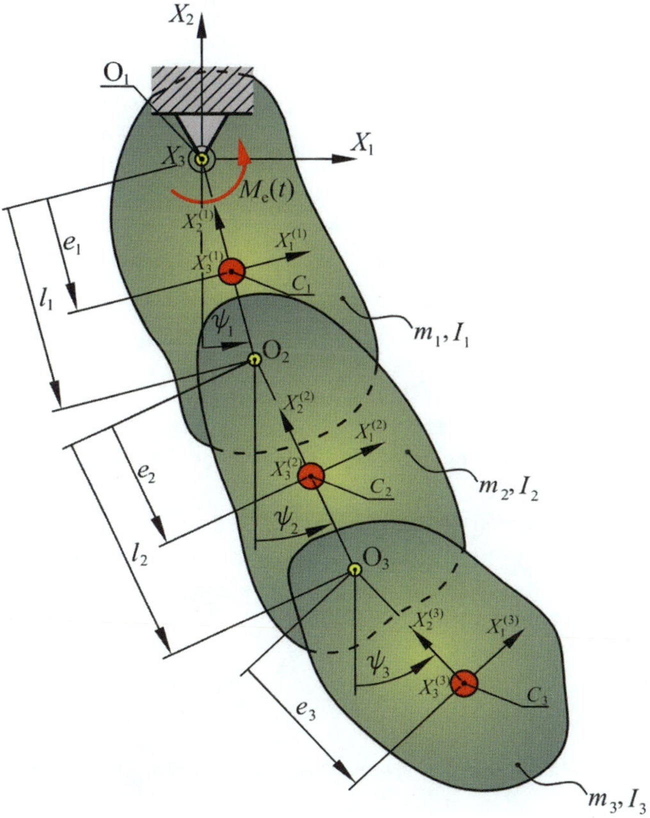

Fig. 2.6 A triple physical pendulum

For the derivation of equations of motion we will make use of Lagrange's equations of the second kind having the following form:

$$\frac{\mathrm{d}}{\mathrm{d}t}\left(\frac{\partial T}{\partial \dot{q}_n}\right) - \frac{\partial T}{\partial q_n} + \frac{\partial V}{\partial q_n} = Q_n, \quad n = 1, \ldots, N, \tag{2.78}$$

where N is the number of generalized coordinates, q_n the nth generalized coordinate, T the kinetic energy of the system, V the potential energy, and Q_n the nth generalized force. In the case of the considered triple pendulum one may choose three angles ψ_1, ψ_2, and ψ_3 as generalized coordinates (describing uniquely the system configuration). Then, (2.78) take the form

$$\frac{\mathrm{d}}{\mathrm{d}t}\left(\frac{\partial T}{\partial \dot{\psi}_n}\right) - \frac{\partial T}{\partial \psi_n} + \frac{\partial V}{\partial \psi_n} = Q_n. \tag{2.79}$$

Gravity forces can be included in two ways. Firstly, one may put them into equations as appropriate components of the generalized forces Q_n. Secondly, after taking into account that the gravity forces are conservative, one can put them into equations as the appropriate potential energy. The latter method is more convenient and will be applied here. The potential energy (of gravity forces) is as follows:

$$V = V_0 + \sum_{n=1}^{3} m_n g x_{2C_n}, \tag{2.80}$$

where V_0 is an arbitrary constant, g the acceleration of gravity, and x_{2Cn} the coordinate determining the position along the X_2 axis of the mass center of the nth pendulum.

The kinetic energy of the system is the sum of the kinetic energies of each of the bodies. In turn, the kinetic energy of a body is the sum of its kinetic energy for the translational motion (with velocity of the mass center) and for the motion about the mass center. That relative motion, generally, is the motion about a point (that is, the instantaneous rotational motion), whereas in our special case of planar motion it is the rotational motion of the nth pendulum (of the axis of rotation $C_i X_3^{(n)}$). Thus, the kinetic energy of the system of three connected pendulums is equal to

$$T = \frac{1}{2}\sum_{n=1}^{3} m_n \left(\dot{x}_{1C_n}^2 + \dot{x}_{2C_n}^2\right) + \frac{1}{2}\sum_{n=1}^{3} I_n \dot{\psi}_n^2. \tag{2.81}$$

The coordinates of mass centers occurring in expressions (2.80) and (2.81) are equal to

$$\begin{aligned}
x_{1C_1} &= e_1 \sin\psi_1, \\
x_{1C_2} &= l_1 \sin\psi_1 + e_2 \sin\psi_2, \\
x_{1C_3} &= l_1 \sin\psi_1 + l_2 \sin\psi_2 + e_3 \sin\psi_3, \\
x_{2C_1} &= -e_1 \cos\psi_1, \\
x_{2C_2} &= -l_1 \cos\psi_1 - e_2 \cos\psi_2, \\
x_{2C_3} &= -l_1 \cos\psi_1 - l_2 \cos\psi_2 - e_3 \cos\psi_3,
\end{aligned} \tag{2.82}$$

whereas their time derivatives

$$\begin{aligned}
\dot{x}_{1C_1} &= e_1 \dot{\psi}_1 \cos\psi_1, \\
\dot{x}_{1C_2} &= l_1 \dot{\psi}_1 \cos\psi_1 + e_2 \dot{\psi}_2 \cos\psi_2, \\
\dot{x}_{1C_3} &= l_1 \dot{\psi}_1 \cos\psi_1 + l_2 \dot{\psi}_2 \cos\psi_2 + e_3 \dot{\psi}_3 \cos\psi_3,
\end{aligned}$$

$$\begin{aligned}
\dot{x}_{2C_1} &= e_1\,\dot{\psi}_1 \sin\psi_1, \\
\dot{x}_{2C_2} &= l_1\,\dot{\psi}_1 \sin\psi_1 + e_2\,\dot{\psi}_2 \sin\psi_2, \\
\dot{x}_{2C_3} &= l_1\,\dot{\psi}_1 \sin\psi_1 + l_2\,\dot{\psi}_2 \sin\psi_2 + e_3\,\dot{\psi}_3 \sin\psi_3.
\end{aligned} \tag{2.83}$$

Inserting relationships (2.82) and (2.83) into expressions (2.80) and (2.81), applying suitable operations, using certain trigonometric identities, and grouping the terms, we obtain

$$V = -\sum_{n=1}^{3} M_n \cos\psi_n \tag{2.84}$$

and

$$T = \frac{1}{2}\sum_{n=1}^{3} B_n \dot{\psi}_n^2 + \sum_{n=1}^{2}\sum_{j=n+1}^{3} N_{nj}\dot{\psi}_n\dot{\psi}_j \cos\left(\psi_n - \psi_j\right), \tag{2.85}$$

where the following symbols were used:

$$\begin{aligned}
M_1 &= m_1 g e_1 + (m_2 + m_3)\, g l_1, \\
M_2 &= m_2 g e_2 + m_3 g l_2 \\
M_3 &= m_3 g e_3, \\
B_1 &= I_1 + {e_1}^2 m_1 + {l_1}^2 (m_2 + m_3), \\
B_2 &= I_2 + {e_2}^2 m_2 + {l_2}^2 m_3, \\
B_3 &= I_3 + {e_3}^2 m_3, \\
N_{12} &= m_2 e_2 l_1 + m_3 l_1 l_2, \\
N_{13} &= m_3 e_3 l_1, \\
N_{23} &= m_3 e_3 l_2.
\end{aligned} \tag{2.86}$$

Inserting relations (2.84) and (2.85) into the left-hand sides of (2.79) and differentiating, we obtain

$$\begin{aligned}
\frac{\mathrm{d}}{\mathrm{d}t}\left(\frac{\partial T}{\partial \dot{\psi}_1}\right) - \frac{\partial T}{\partial \psi_1} + \frac{\partial V}{\partial \psi_1} = {} & B_1 \ddot{\psi}_1 + N_{12} \cos(\psi_1 - \psi_2)\, \ddot{\psi}_2 \\
& + N_{13} \cos(\psi_1 - \psi_3)\, \ddot{\psi}_3 \\
& + N_{12} \sin(\psi_1 - \psi_2)\, \dot{\psi}_2^2 + N_{13} \sin(\psi_1 - \psi_3)\, \dot{\psi}_3^2 \\
& + M_1 \sin\psi_1,
\end{aligned}$$

$$\begin{aligned}
\frac{\mathrm{d}}{\mathrm{d}t}\left(\frac{\partial T}{\partial \dot{\psi}_2}\right) - \frac{\partial T}{\partial \psi_2} + \frac{\partial V}{\partial \psi_2} &= B_2\ddot{\psi}_2 + N_{12}\cos(\psi_1 - \psi_2)\,\ddot{\psi}_1 \\
&\quad + N_{23}\cos(\psi_2 - \psi_3)\,\ddot{\psi}_3 \\
&\quad - N_{12}\sin(\psi_1 - \psi_2)\,\dot{\psi}_1^2 + N_{23}\sin(\psi_2 - \psi_3)\,\dot{\psi}_3^2 \\
&\quad + M_2\sin\psi_2, \\
\frac{\mathrm{d}}{\mathrm{d}t}\left(\frac{\partial T}{\partial \dot{\psi}_3}\right) - \frac{\partial T}{\partial \psi_3} + \frac{\partial V}{\partial \psi_3} &= B_3\ddot{\psi}_3 + N_{13}\cos(\psi_1 - \psi_3)\,\ddot{\psi}_1 \\
&\quad + N_{23}\cos(\psi_2 - \psi_3)\,\ddot{\psi}_2 \\
&\quad - N_{13}\sin(\psi_1 - \psi_3)\,\dot{\psi}_1^2 - N_{23}\sin(\psi_2 - \psi_3)\,\dot{\psi}_2^2 \\
&\quad + M_3\sin\psi_3. \qquad (2.87)
\end{aligned}$$

Now we should determine the right-hand sides, that is, what kinds of generalized forces act along particular generalized coordinates. When using generalized forces Q_n one should take into account in the equations all non-conservative forces acting on the system. This will be the moment of force $M_e(t)$ acting on the first pendulum but also moments of resistive forces at the connections of the pendulums. Because the generalized coordinates are angles, the generalized forces must be moments of force. Moreover, generalized coordinates describe the absolute angular positions of individual pendulums; therefore, the generalized forces will be the moments of force acting on particular pendulums. Therefore, it is already known that the moment $M_e(t)$ will be the component of the first generalized force. Also, on particular pendulums additionally act the moments associated with viscous damping at particular revolute joints. We may write it in the following way:

$$\begin{aligned}
Q_1 &= M_e(t) + M_{01} + M_{21}, \\
Q_2 &= M_{12} + M_{32}, \\
Q_3 &= M_{23}, \qquad (2.88)
\end{aligned}$$

where M_{ij} is the moment of force with which the ith pendulum or the base ($i = 0$) acts on the jth pendulum by means of viscous damping at a joint connecting two pendulums. Positive directions of particular moments are consistent with the adopted positive direction common for all generalized coordinates. Clearly, according to Newton's third law (action and reaction principle), $M_{ij} = -M_{ji}$ must hold. For viscous damping (proportional to the relative velocity) we have

$$\begin{aligned}
M_{01} &= -c_1\dot{\psi}_1, \\
M_{12} &= -c_2\left(\dot{\psi}_2 - \dot{\psi}_1\right) = -M_{21}, \\
M_{23} &= -c_3\left(\dot{\psi}_3 - \dot{\psi}_2\right) = -M_{32}. \qquad (2.89)
\end{aligned}$$

While taking the appropriate signs in the preceding formulas we keep in mind that the moment of damping has to counteract the relative motion of the connected pendulums. Finally, the generalized forces read

$$\begin{aligned} Q_1 &= M_e(t) - c_1\dot{\psi}_1 + c_2\left(\dot{\psi}_2 - \dot{\psi}_1\right), \\ Q_2 &= -c_2\left(\dot{\psi}_2 - \dot{\psi}_1\right) + c_3\left(\dot{\psi}_3 - \dot{\psi}_2\right), \\ Q_3 &= -c_3\left(\dot{\psi}_3 - \dot{\psi}_2\right), \end{aligned} \tag{2.90}$$

where c_i is the viscous damping coefficient at joint O_i.

Equating formulas (2.90) to (2.87) and moving also the damping force to the left-hand side of every equation we obtain a system of three second-order ordinary differential equations (the mathematical model) describing the dynamics of the triple pendulum:

$$\begin{aligned} B_1\ddot{\psi}_1 &+ N_{12}\cos(\psi_1 - \psi_2)\,\ddot{\psi}_2 + N_{13}\cos(\psi_1 - \psi_3)\,\ddot{\psi}_3 \\ &+ N_{12}\sin(\psi_1 - \psi_2)\,\dot{\psi}_2^2 + N_{13}\sin(\psi_1 - \psi_3)\,\dot{\psi}_3^2 \\ &+ c_1\dot{\psi}_1 - c_2\left(\dot{\psi}_2 - \dot{\psi}_1\right) + M_1\sin\psi_1 = M_e(t), \\ B_2\ddot{\psi}_2 &+ N_{12}\cos(\psi_1 - \psi_2)\,\ddot{\psi}_1 + N_{23}\cos(\psi_2 - \psi_3)\,\ddot{\psi}_3 \\ &- N_{12}\sin(\psi_1 - \psi_2)\,\dot{\psi}_1^2 + N_{23}\sin(\psi_2 - \psi_3)\,\dot{\psi}_3^2 \\ &+ c_2\left(\dot{\psi}_2 - \dot{\psi}_1\right) - c_3\left(\dot{\psi}_3 - \dot{\psi}_2\right) + M_2\sin\psi_2 = 0, \\ B_3\ddot{\psi}_3 &+ N_{13}\cos(\psi_1 - \psi_3)\,\ddot{\psi}_1 + N_{23}\cos(\psi_2 - \psi_3)\,\ddot{\psi}_2 \\ &- N_{13}\sin(\psi_1 - \psi_3)\,\dot{\psi}_1^2 - N_{23}\sin(\psi_2 - \psi_3)\,\dot{\psi}_2^2 \\ &+ c_3\left(\dot{\psi}_3 - \dot{\psi}_2\right) + M_3\sin\psi_3 = 0. \end{aligned} \tag{2.91}$$

Equation (2.91) can also be represented in a more concise and clear form using matrix notation

$$\mathbf{M}(\boldsymbol{\psi})\,\ddot{\boldsymbol{\psi}} + \mathbf{N}(\boldsymbol{\psi})\,\dot{\boldsymbol{\psi}}^2 + \mathbf{C}\dot{\boldsymbol{\psi}} + \mathbf{p}(\boldsymbol{\psi}) = \mathbf{f}_e(t), \tag{2.92}$$

where

$$\mathbf{M}(\boldsymbol{\psi}) = \begin{bmatrix} B_1 & N_{12}\cos(\psi_1 - \psi_2) & N_{13}\cos(\psi_1 - \psi_3) \\ N_{12}\cos(\psi_1 - \psi_2) & B_2 & N_{23}\cos(\psi_2 - \psi_3) \\ N_{13}\cos(\psi_1 - \psi_3) & N_{23}\cos(\psi_2 - \psi_3) & B_3 \end{bmatrix},$$

$$\mathbf{N}(\boldsymbol{\psi}) = \begin{bmatrix} 0 & N_{12}\sin(\psi_1 - \psi_2) & N_{13}\sin(\psi_1 - \psi_3) \\ -N_{12}\sin(\psi_1 - \psi_2) & 0 & N_{23}\sin(\psi_2 - \psi_3) \\ -N_{13}\sin(\psi_1 - \psi_3) & -N_{23}\sin(\psi_2 - \psi_3) & 0 \end{bmatrix},$$

$$\mathbf{C} = \begin{bmatrix} c_1 + c_2 & -c_2 & 0 \\ -c_2 & c_2 + c_3 & -c_3 \\ 0 & -c_3 & c_3 \end{bmatrix}, \quad \mathbf{p}(\boldsymbol{\psi}) = \begin{bmatrix} M_1 \sin\psi_1 \\ M_2 \sin\psi_2 \\ M_3 \sin\psi_3 \end{bmatrix}, \quad \mathbf{f}_e(t) = \begin{bmatrix} M_e(t) \\ 0 \\ 0 \end{bmatrix},$$

$$\boldsymbol{\psi} = \begin{bmatrix} \psi_1 \\ \psi_2 \\ \psi_3 \end{bmatrix}, \quad \dot{\boldsymbol{\psi}} = \begin{bmatrix} \dot\psi_1 \\ \dot\psi_2 \\ \dot\psi_3 \end{bmatrix}, \quad \ddot{\boldsymbol{\psi}} = \begin{bmatrix} \ddot\psi_1 \\ \ddot\psi_2 \\ \ddot\psi_3 \end{bmatrix}, \quad \dot{\boldsymbol{\psi}}^2 = \begin{bmatrix} \dot\psi_1^2 \\ \dot\psi_2^2 \\ \dot\psi_3^2 \end{bmatrix}. \tag{2.93}$$

The unknowns of a system of differential equations describing the dynamics of a triple pendulum [in the form of (2.91) or (2.92)] are the functions $\psi_1(t)$, $\psi_2(t)$, and $\psi_3(t)$, which means that the solution of those equations describes the motion of the pendulum.

Let us also draw attention to the parameters of those equations. While there are 15 of the physical parameters of the pendulum (l_1, l_2, e_1, e_2, e_3, m_1, m_2, m_3, I_1, I_2, I_3, g, c_1, c_2, and c_3 – for now we omit parameters of excitation), in the equations there are actually 11 independent parameters (M_1, M_2, M_3, B_1, B_2, B_3, N_{13}, N_{23}, c_1, c_2, and c_3), because out of quantities (2.86) one is dependent on the remaining ones:

$$\frac{N_{12}}{N_{13}} = \frac{M_2}{M_3} = \frac{m_2 e_2}{m_3 e_3} + \frac{l_2}{e_3}, \tag{2.94}$$

and hence

$$N_{12} = N_{13}\frac{M_2}{M_3}. \tag{2.95}$$

From the fact that fewer parameters (11) occur in equations of motion than the number of physical parameters of a pendulum (15) it follows that the same pendulum in the sense of dynamics (i.e., behaving in the same way) one may build in an infinite number of ways.

2.3.2 *Numerical Simulations*

Differential equations (2.92) are strongly non-linear equations and do not have an exact analytical solution. For the investigation of their solutions, numerical and analytical approximate methods remain at our disposal. Here one of the most popular and effective numerical methods for the solution of ordinary differential equations – the fourth-order Runge–Kutta method [6] – was applied.

However, the mathematical model of a triple pendulum (2.91) or (2.92) has the form of a system of second-order equations. In order to be able to apply classical algorithms for integration of differential equations, we have to reduce these equations to the form of a system of first-order equations. Let us take a system state vector (the vector of state variables) as

$$\mathbf{x} = \begin{bmatrix} \psi_1 \\ \psi_2 \\ \psi_3 \\ \dot{\psi}_1 \\ \dot{\psi}_2 \\ \dot{\psi}_3 \end{bmatrix} = \begin{bmatrix} \boldsymbol{\psi} \\ \dot{\boldsymbol{\psi}} \end{bmatrix}, \tag{2.96}$$

then the system of six first-order differential equations has the following general form

$$\dot{\mathbf{x}} = \mathbf{f}(\mathbf{x}, t) = \begin{bmatrix} \dot{\boldsymbol{\psi}} \\ \ddot{\boldsymbol{\psi}} \end{bmatrix}. \tag{2.97}$$

Let us note that we need angular accelerations of the pendulums given explicitly as functions of system state $\mathbf{x}$ and time t. Then we have to solve (2.92) with respect to $\ddot{\psi}$, treating them as algebraic equations. Because it is a system of linear equations, we have

$$\ddot{\boldsymbol{\psi}} = [\mathbf{M}(\boldsymbol{\psi})]^{-1}\left[\mathbf{f}_e(t) - \mathbf{N}(\boldsymbol{\psi})\dot{\boldsymbol{\psi}}^2 - \mathbf{C}\dot{\boldsymbol{\psi}} - \mathbf{p}(\boldsymbol{\psi})\right]. \tag{2.98}$$

Inserting relation (2.98) into (2.97), we eventually obtain

$$\dot{\mathbf{x}} = \left[\begin{matrix} \dot{\boldsymbol{\psi}} \\ [\mathbf{M}(\boldsymbol{\psi})]^{-1}\left[\mathbf{f}_e(t) - \mathbf{N}(\boldsymbol{\psi})\dot{\boldsymbol{\psi}}^2 - \mathbf{C}\dot{\boldsymbol{\psi}} - \mathbf{p}(\boldsymbol{\psi})\right] \end{matrix}\right]\Bigg|_{\boldsymbol{\psi}=\begin{bmatrix} x_1 \\ x_2 \\ x_3 \end{bmatrix}, \dot{\boldsymbol{\psi}}=\begin{bmatrix} x_4 \\ x_5 \\ x_6 \end{bmatrix}, \dot{\boldsymbol{\psi}}^2=\begin{bmatrix} x_4^2 \\ x_5^2 \\ x_6^2 \end{bmatrix}}. \tag{2.99}$$

In every step of the integration we must perform an inversion of the matrix $\mathbf{M}(\boldsymbol{\psi})$. Because of its size, it is possible to use for this purpose an exact analytical expression. In the case of a slightly larger system, in practice there would remain only the possibility of using one of the existing numerical methods for inverting a matrix. If we investigate (integrate) differential equations (2.92) by means of a numerical method, in fact not only will the mathematical model consist of these equations, but also the method itself should be considered as an integral part of the model. Then a system with continuous time is approximated by a system with discrete time, and the differential equations themselves are approximated by difference equations.

In order to conduct an illustrative numerical simulation of a pendulum (i.e., to find the numerical solution of the model), we have to adopt some concrete values for model parameters and initial conditions. Let us assume that this special case of

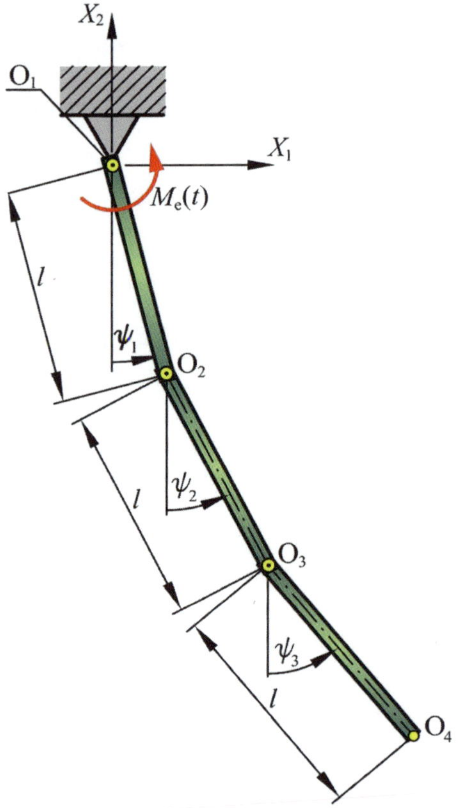

Fig. 2.7 Special case of a triple physical pendulum – a system of three identical pin-joined rods

a triple physical pendulum are three identical rods connected by means of joints located at their ends, as shown in Fig. 2.7.

Then we will have

$$l_1 = l_2 = l,$$

$$m_1 = m_2 = m_3 = m,$$

$$e_1 = e_2 = e_3 = \frac{l}{2},$$

$$I_1 = I_2 = I_3 = \frac{ml^2}{12}, \tag{2.100}$$

where l is the length of a single pendulum and m its mass, and expressions (2.86) will take the form

$$M_1 = \frac{5}{2}mgl, \qquad M_2 = \frac{3}{2}mgl, \qquad M_3 = \frac{1}{2}mgl,$$

$$B_1 = \frac{7}{3}ml^2, \qquad B_2 = \frac{4}{3}ml^2, \qquad B_3 = \frac{1}{3}ml^2,$$

$$N_{12} = \frac{3}{2}ml^2, \qquad N_{13} = \frac{1}{2}ml^2, \qquad N_{23} = \frac{1}{2}ml^2. \tag{2.101}$$

In turn, setting $g = 10\,\text{m/s}^2$, $m = 1$ kg and $l = 1$ m we obtain

$$\begin{aligned} M_1 &= 25\,\text{kg}\cdot\text{m}^2\cdot\text{s}^{-2}, & M_2 &= 15\,\text{kg}\cdot\text{m}^2\cdot\text{s}^{-2}, & M_3 &= 5\,\text{kg}\cdot\text{m}^2\cdot\text{s}^{-2}, \\ B_1 &= \frac{7}{3}\,\text{kg}\cdot\text{m}^2, & B_2 &= \frac{4}{3}\,\text{kg}\cdot\text{m}^2, & B_3 &= \frac{1}{3}\,\text{kg}\cdot\text{m}^2, \\ N_{12} &= \frac{3}{2}\,\text{kg}\cdot\text{m}^2, & N_{13} &= \frac{1}{2}\,\text{kg}\cdot\text{m}^2, & N_{23} &= \frac{1}{2}\,\text{kg}\cdot\text{m}^2. \end{aligned} \tag{2.102}$$

Viscous damping in the joints are taken as

$$c_1 = c_2 = c_3 = 1\,\text{N}\cdot\text{m}\cdot\text{s}. \tag{2.103}$$

We take the moment acting on the first pendulum as harmonically varying in time

$$M_e(t) = q\sin(\omega t), \tag{2.104}$$

where q is the amplitude and ω the angular frequency of pendulum excitation. These two parameters will vary for different simulation examples shown later, whereas the remaining parameters from (2.102) to (2.103) will be constant.

The free motion of a pendulum (the pendulum is not subjected to external excitation, i.e., $q = 0$) for initial conditions $\psi_1(0) = \psi_2(0) = \psi_3(0) = \dot{\psi}_1(0) = \dot{\psi}_2(0) = 0$ and $\dot{\psi}_3(0) = 1$ rad/s is shown in Fig. 2.8. Vibrations decay because the energy of the pendulum is dissipated through damping in the joints, and no new energy is simultaneously supplied (no excitation). Therefore, the solution tends to a stable equilibrium position, and the only stable equilibrium position in this system is $\psi_1 = \psi_2 = \psi_3 = 0$.

In Fig. 2.9, in turn, we present the excited transient motion of a pendulum ($q = 25\,\text{N}\cdot\text{m}$ and $\omega = 3\,\text{rad/s}$), which starts at the time instant $t = 0$ from zero initial conditions (bright) and tends to the stable periodical solution (by analogy to the stable equilibrium position) marked dark. The solution is marked bright for the time $t \in (0,\ 150\,\text{s})$, whereas for $t \in (150\,\text{s},\ 200\,\text{s})$ it is marked dark. Clearly, for the time $t = 150$ s only a pendulum with a good approximation moves on a periodic solution, whereas in reality it constantly approaches it and reaches it for $t \to \infty$.

The motion of a pendulum is presented in Figs. 2.9–2.11 as the motion of the tip of the third rod of the pendulum (point O_4 in Fig. 2.7) in the plane of motion of the pendulum (coordinates x_{1O_4} and x_{2O_4} describe the position of point O_4 in the coordinate system $O_1X_1X_2$). One should remember, however, that the space (plane) $x_{1O_4} - x_{2O_4}$ is a 2D subspace of the system phase space, which is actually 7D (apart from three angles ψ_1, ψ_2, and ψ_3, and three angular velocities $\dot{\psi}_1$, $\dot{\psi}_2$, and $\dot{\psi}_3$ we add here a phase of the periodic excitation $M_e(t)$). The graphs presented in the coordinates $x_{1O_4} - x_{2O_4}$ are projections of phase trajectories onto this subspace and

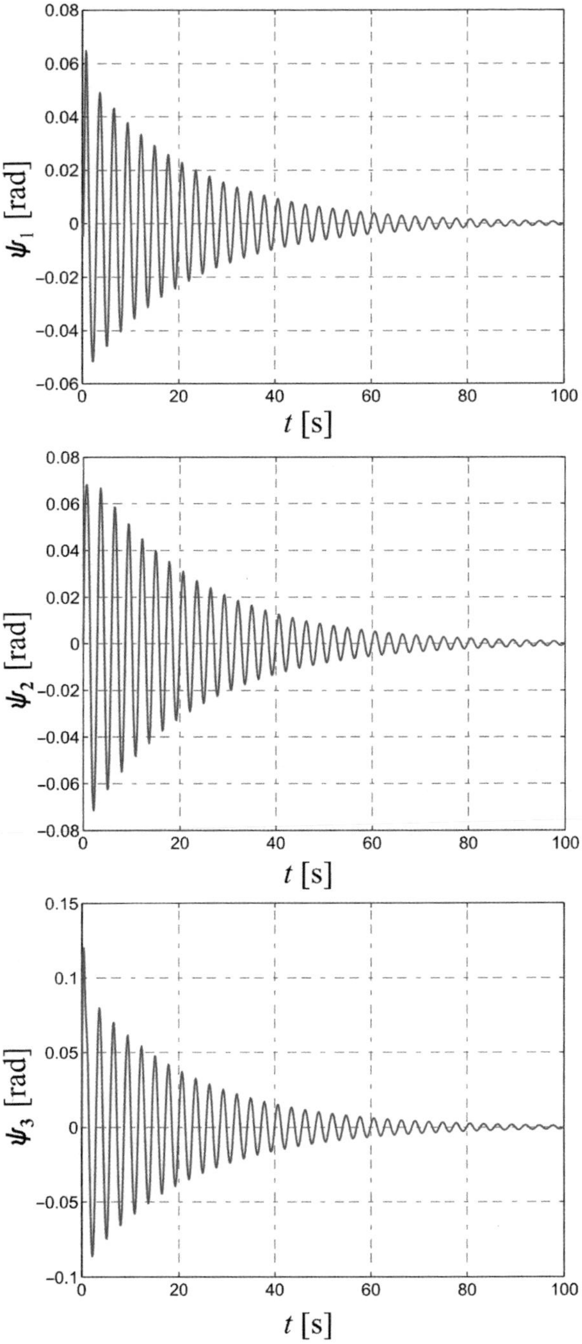

Fig. 2.8 Decaying motion of pendulum not subjected to external excitation ($q = 0$)

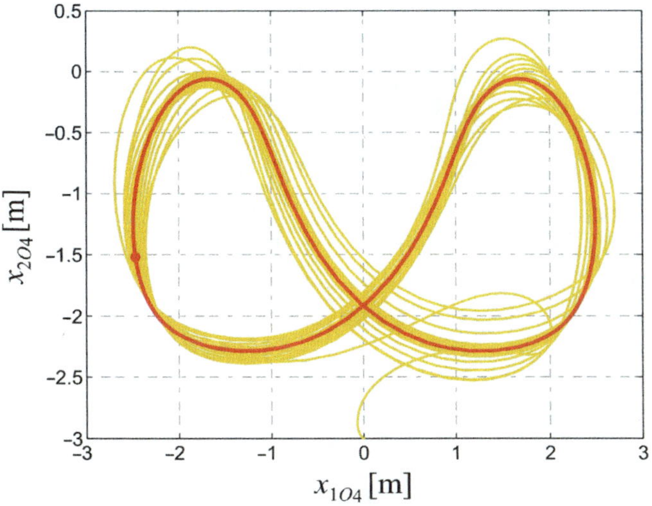

Fig. 2.9 Excited motion of pendulum tending to stable periodic solution (dark) for $q = 25\,\mathrm{N} \cdot \mathrm{m}$ and $\omega = 3\,\mathrm{rad/s}$

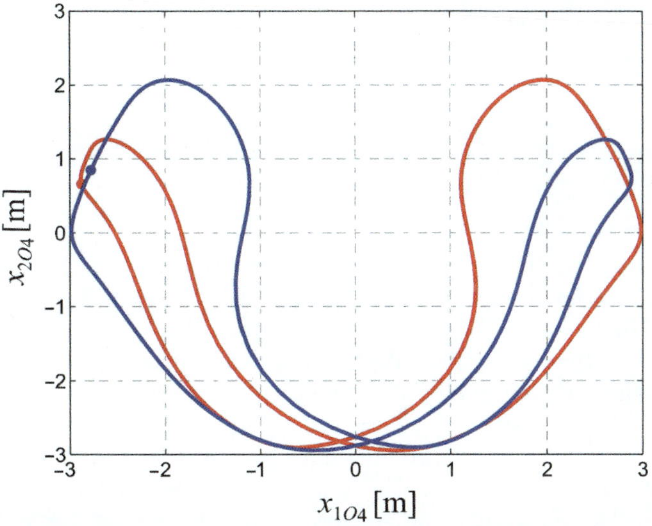

Fig. 2.10 Two coexisting periodic solutions for $q = 25\,\mathrm{N} \cdot \mathrm{m}$ and $\omega = 2.022\,\mathrm{rad/s}$ attained from different initial conditions

do not contain complete information. If we observe in Fig. 2.9 a periodic solution (dark) in the form of a closed line (i.e., the motion is repetitive), then the tip of the third rod moving on this line returns to its previous position (e.g., to the point marked with a dark circle), and the values of all state variables repeat themselves (positions and angular velocities and the phase of the periodic excitation).

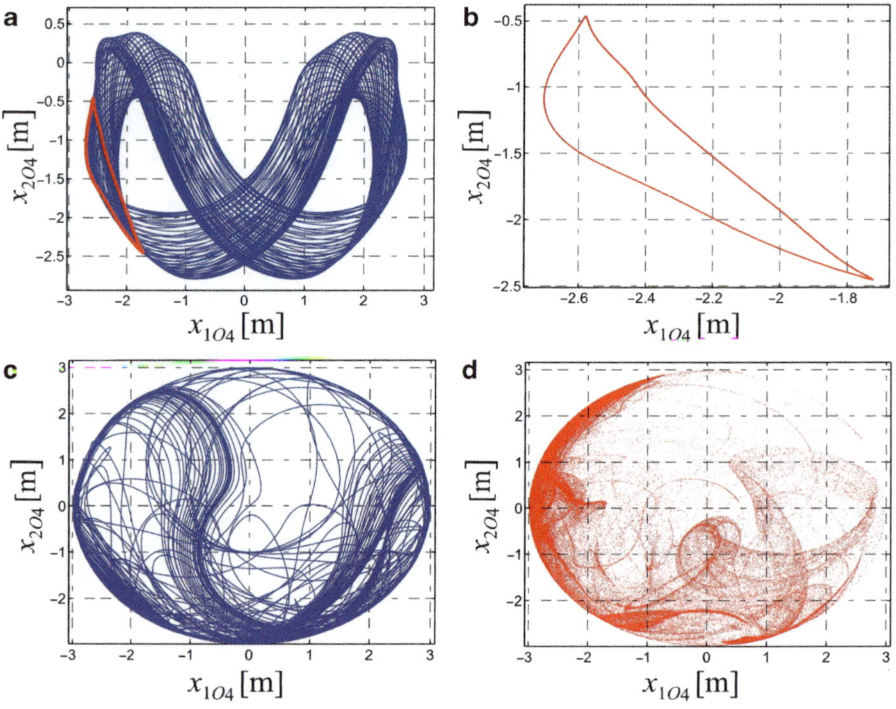

Fig. 2.11 Quasiperiodic solution for $\omega = 2.8\,\text{rad/s}$ [trajectory (**a**), Poincaré section (**b**] and chaotic solution for $\omega = 2\,\text{rad/s}$ [trajectory (**c**), Poincaré section (d)]; excitation amplitude $q = 25\,\text{N}\cdot\text{m}$

In non-linear systems the coexistence of many solutions (for the same parameters) is possible. An example is two periodic solutions for the excitation parameters $q = 25\,\text{N}\cdot\text{m}$ and $\omega = 2.022\,\text{rad/s}$ shown in Fig. 2.10. Each of the solutions is attained from different initial conditions: the solution marked in bright from zero initial conditions at the instant $t = 0$, and the solution marked in dark from initial conditions $\psi_1(0) = \psi_2(0) = \psi_3(0) = \dot{\psi}_3(0) = 0$ and $\dot{\psi}_1(0) = \dot{\psi}_2(0) = -1\,\text{rad/s}$. On the graph the initial transient motion for $t \in (0,\ 150\,\text{s})$ is neglected, and only the motion in the time interval $t \in (150,\ 200\,\text{s})$ is shown, when it has already taken place in a sufficient approximation on the appropriate periodic solution. Each solution is asymmetrical, whereas the system is symmetrical. In turn, the two solutions together form an object symmetrical with respect to the axis of symmetry of the pendulum. The symmetry of the system implies that asymmetrical solutions may appear only in such twin pairs.

From the previous examples we see that the typical behavior of a damped and periodically excited pendulum is that after waiting some time and neglecting certain initial transient motion, the pendulum starts to move periodically. This period is always a multiple of the period of excitation. However, it happens sometimes that the pendulum never starts its periodic motion regardless of the length of the transient period we would like to skip. An example is the solution shown in Fig. 2.11a for the excitation parameters $q = 25\,\mathrm{N \cdot m}$ and $\omega = 2.8\,\mathrm{rad/s}$. The pendulum starts at the time instant $t = 0$ from zero initial conditions. The transient motion for $t \in (0,\ 200\,\mathrm{s})$ was skipped and only the motion for $t \in (200,\ 350\,\mathrm{s})$ is shown on the graph. One may check that after skipping a time interval of transient motion of an arbitrary length, the pendulum still is going to behave qualitatively in the same way as shown in Fig. 2.11a.

For a more detailed analysis of aperiodic motions a tool called a *Poincaré section* (also a *Poincaré map*) is very useful. In the case of a system with periodic excitation, the simplest way to create such a section is by sampling the state of the system in the intervals equal to the period of excitation. A Poincaré section of the solution from Fig. 2.11a, obtained by sampling the position of the tip of the third rod of the pendulum at time instances $t_i = iT$ $(i = 1, 2, 3, \ldots)$, where $T = 2\pi/\omega$ is the period of excitation, is shown in Fig. 2.11b.

On this occasion we obviously skip an appropriate number of initial points in order to remove the transient motion. The Poincaré section shown in Fig. 2.11b contains 3,500 points. As can be seen, these points form a continuous line. This is a characteristic of *quasiperiodic motion*.

In Fig. 2.11c, d another case of aperiodic motion of the pendulum is shown for the excitation parameters $q = 25\,\mathrm{N{\cdot}m}$ and $\omega = 2\,\mathrm{rad/s}$. The pendulum starts its motion at the time instant $t = 0$ from zero initial conditions. The transient motion for $t \in (0,\ 200\,\mathrm{s})$ was skipped. The motion of the tip of the third rod for $t \in (200,\ 400\,\mathrm{s})$ is shown in Fig. 2.11c, whereas Fig. 2.11d shows the corresponding Poincaré section obtained in the same way as for the quasiperiodic solution, this time composed of 10^6 points. It is a typical section for *chaotic motion*, that is, the set of points approximating this motion is an infinite set.

In the end, we should add that a Poincaré section for a *periodic solution* of period nT (only those kinds of solutions are possible in a system with periodic excitation), where T is a period of excitation and n is an integer number, will form a set consisting of n separate points. An example would be the individual points plotted in Figs. 2.9 and 2.10.

2.3.3 *Dynamic Reactions in Bearings*

Lagrange's equations enable a relatively easy derivation of equations of motion of complex dynamical systems, since, for instance, they allow for avoiding the direct determination of dynamic reactions in a system. However, when these reactions have to be determined, it turns out that in equations of motion alone there is not enough information, and additional analysis of the physical system is required. That is precisely the case for a triple physical pendulum. If we want to determine the dynamic reactions in its three joints, we have to consider separately the motion

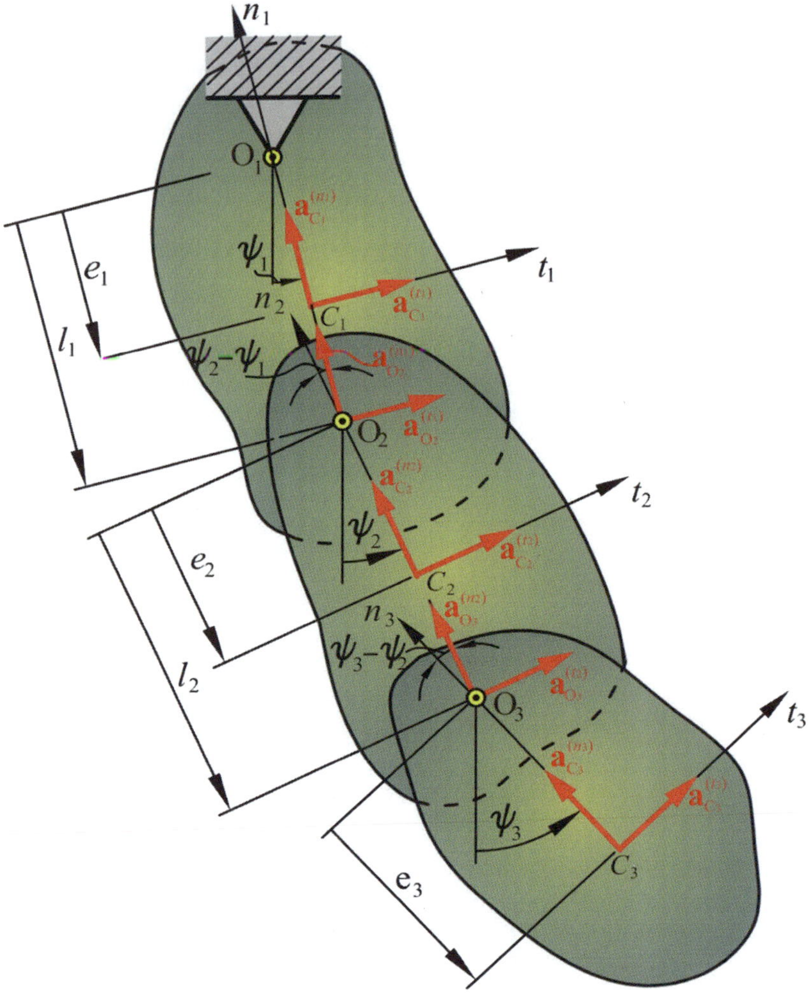

Fig. 2.12 Accelerations of characteristic points of a pendulum and their decomposition in local coordinate systems

of each of the bodies under the action of external forces. In order to obtain the relationships allowing us to determine the reactions, we will have to find the accelerations of the mass centers of particular bodies.

The accelerations of the mass center of the first pendulum C_1 and the joint O_2 (Fig. 2.12) can be expressed in terms of tangential and normal components:

$$\mathbf{a}_{C_1} = \mathbf{a}_{C_1}^{(t_1)} + \mathbf{a}_{C_1}^{(n_1)}, \qquad \mathbf{a}_{O_2} = \mathbf{a}_{O_2}^{(t_1)} + \mathbf{a}_{O_2}^{(n_1)}, \tag{2.105}$$

where

$$a_{C_1}^{(t_1)} = \varepsilon_1 e_1 = \ddot{\psi}_1 e_1, \qquad a_{C_1}^{(n_1)} = \omega_1^2 e_1 = \dot{\psi}_1^2 e_1, \tag{2.106a}$$

$$a_{O_2}^{(t_1)} = \varepsilon_1 l_1 = \ddot{\psi}_1 l_1, \qquad a_{O_2}^{(n_1)} = \omega_1^2 l_1 = \dot{\psi}_1^2 l_1. \tag{2.106b}$$

In turn, the acceleration of the mass center of the second pendulum C_2 can be represented as

$$\mathbf{a}_{C_2} = \mathbf{a}_{O_2} + \mathbf{a}_{C_2/O_2}^{(t_2)} + \mathbf{a}_{C_2/O_2}^{(n_2)} \tag{2.107}$$

or, taking into account relationship (2.105), as

$$\mathbf{a}_{C_2} = \mathbf{a}_{O_2}^{(t_1)} + \mathbf{a}_{O_2}^{(n_1)} + \mathbf{a}_{C_2/O_2}^{(t_2)} + \mathbf{a}_{C_2/O_2}^{(n_2)}, \tag{2.108}$$

where

$$a_{C_2/O_2}^{(t_2)} = \varepsilon_2 e_2 = \ddot{\psi}_2 e_2, \qquad a_{C_2/O_2}^{(n_2)} = \omega_2^2 e_2 = \dot{\psi}_2^2 e_2. \tag{2.109}$$

The total acceleration of point C_2 can also be decomposed into the following two components (Fig. 2.12):

$$\mathbf{a}_{C_2} = \mathbf{a}_{C_2}^{(t_2)} + \mathbf{a}_{C_2}^{(n_2)}, \tag{2.110}$$

and the best way to determine them is to project the right-hand side of (2.108) onto the directions t_2 and n_2:

$$\begin{aligned} a_{C_2}^{(t_2)} &= a_{O_2}^{(t_1)} \cos(\psi_2 - \psi_1) + a_{O_2}^{(n_1)} \sin(\psi_2 - \psi_1) + a_{C_2/O_2}^{(t_2)}, \\ a_{C_2}^{(n_2)} &= -a_{O_2}^{(t_1)} \sin(\psi_2 - \psi_1) + a_{O_2}^{(n_1)} \cos(\psi_2 - \psi_1) + a_{C_2/O_2}^{(n_2)}, \end{aligned} \tag{2.111}$$

and when we take into account relationships (2.106b) and (2.109), the acceleration components take the form

$$\begin{aligned} a_{C_2}^{(t_2)} &= \ddot{\psi}_1 l_1 \cos(\psi_2 - \psi_1) + \dot{\psi}_1^2 l_1 \sin(\psi_2 - \psi_1) + \ddot{\psi}_2 e_2, \\ a_{C_2}^{(n_2)} &= -\ddot{\psi}_1 l_1 \sin(\psi_2 - \psi_1) + \dot{\psi}_1^2 l_1 \cos(\psi_2 - \psi_1) + \dot{\psi}_2^2 e_2. \end{aligned} \tag{2.112}$$

In an analogous way we can proceed with the acceleration of point O_3:

$$\mathbf{a}_{O_3} = \mathbf{a}_{O_2} + \mathbf{a}_{O_3/O_2}^{(t_2)} + \mathbf{a}_{O_3/O_2}^{(n_2)}, \tag{2.113a}$$

$$\mathbf{a}_{O_3} = \mathbf{a}_{O_2}^{(t_1)} + \mathbf{a}_{O_2}^{(n_1)} + \mathbf{a}_{O_3/O_2}^{(t_2)} + \mathbf{a}_{O_3/O_2}^{(n_2)}, \tag{2.113b}$$

$$\mathbf{a}_{O_3} = \mathbf{a}_{O_3}^{(t_2)} + \mathbf{a}_{O_3}^{(n_2)}, \tag{2.113c}$$

where

$$a^{(t_2)}_{O_3/O_2} = \varepsilon_2 l_2 = \ddot{\psi}_2 l_2, \qquad a^{(n_2)}_{O_3/O_2} = \omega_2^2 l_2 = \dot{\psi}_2^2 l_2. \tag{2.114}$$

Projecting (2.113b) onto directions t_2 and n_2 we obtain

$$\begin{aligned} a^{(t_2)}_{O_3} &= a^{(t_1)}_{O_2}\cos(\psi_2 - \psi_1) + a^{(n_1)}_{O_2}\sin(\psi_2 - \psi_1) + a^{(t_2)}_{O_3/O_2}, \\ a^{(n_2)}_{O_3} &= -a^{(t_1)}_{O_2}\sin(\psi_2 - \psi_1) + a^{(n_1)}_{O_2}\cos(\psi_2 - \psi_1) + a^{(n_2)}_{O_3/O_2}, \end{aligned} \tag{2.115}$$

and after taking into account relations (2.106b) and (2.114) we obtain

$$\begin{aligned} a^{(t_2)}_{O_3} &= \ddot{\psi}_1 l_1 \cos(\psi_2 - \psi_1) + \dot{\psi}_1^2 l_1 \sin(\psi_2 - \psi_1) + \ddot{\psi}_2 l_2, \\ a^{(n_2)}_{O_3} &= -\ddot{\psi}_1 l_1 \sin(\psi_2 - \psi_1) + \dot{\psi}_1^2 l_1 \cos(\psi_2 - \psi_1) + \dot{\psi}_2^2 l_2. \end{aligned} \tag{2.116}$$

The acceleration of point C_3 can be represented as

$$\mathbf{a}_{C_3} = \mathbf{a}_{O_3} + \mathbf{a}^{(t_3)}_{C_3/O_3} + \mathbf{a}^{(n_3)}_{C_3/O_3} \tag{2.117}$$

or, after taking into account (2.113b), as

$$\mathbf{a}_{C_3} = \mathbf{a}^{(t_1)}_{O_2} + \mathbf{a}^{(n_1)}_{O_2} + \mathbf{a}^{(t_2)}_{O_3/O_2} + \mathbf{a}^{(n_2)}_{O_3/O_2} + \mathbf{a}^{(t_3)}_{C_3/O_3} + \mathbf{a}^{(n_3)}_{C_3/O_3}, \tag{2.118}$$

where

$$a^{(t_3)}_{C_3/O_3} = \varepsilon_3 e_3 = \ddot{\psi}_3 e_3, \qquad a^{(n_3)}_{C_3/O_3} = \omega_3^2 e_3 = \dot{\psi}_3^2 e_3. \tag{2.119}$$

The total acceleration of point C_3 we also decompose into the following two components (Fig. 2.12):

$$\mathbf{a}_{C_3} = \mathbf{a}^{(t_3)}_{C_3} + \mathbf{a}^{(n_3)}_{C_3}, \tag{2.120}$$

and projecting (2.117) onto directions t_3 and n_3 we obtain

$$\begin{aligned} a^{(t_3)}_{C_3} &= a^{(t_1)}_{O_2}\cos(\psi_3 - \psi_1) + a^{(n_1)}_{O_2}\sin(\psi_3 - \psi_1) \\ &\quad + a^{(t_2)}_{O_3/O_2}\cos(\psi_3 - \psi_2) + a^{(n_2)}_{O_3/O_2}\sin(\psi_3 - \psi_2) + a^{(t_3)}_{C_3/O_3}, \\ a^{(n_3)}_{C_3} &= -a^{(t_1)}_{O_2}\sin(\psi_3 - \psi_1) + a^{(n_1)}_{O_2}\cos(\psi_3 - \psi_1) \\ &\quad - a^{(t_2)}_{O_3/O_2}\sin(\psi_3 - \psi_2) + a^{(n_2)}_{O_3/O_2}\cos(\psi_3 - \psi_2) + a^{(n_3)}_{C_3/O_3}, \end{aligned} \tag{2.121}$$

and taking into account relationships (2.106b), (2.114), and (2.119) we obtain

$$\begin{aligned} a_{C_3}^{(t_3)} &= \ddot{\psi}_1 l_1 \cos(\psi_3 - \psi_1) + \dot{\psi}_1^2 l_1 \sin(\psi_3 - \psi_1) \\ &\quad + \ddot{\psi}_2 l_2 \cos(\psi_3 - \psi_2) + \dot{\psi}_2^2 l_2 \sin(\psi_3 - \psi_2) + \ddot{\psi}_3 e_3, \\ a_{C_3}^{(n_3)} &= -\ddot{\psi}_1 l_1 \sin(\psi_3 - \psi_1) + \dot{\psi}_1^2 l_1 \cos(\psi_3 - \psi_1) \\ &\quad - \ddot{\psi}_2 l_2 \sin(\psi_3 - \psi_2) + \dot{\psi}_2^2 l_2 \cos(\psi_3 - \psi_2) + \dot{\psi}_3^2 e_3. \end{aligned} \tag{2.122}$$

The dynamic reactions of the action of the links of a pendulum to one another and to the base can be represented as the sum of the following components (Fig. 2.13):

$$\mathbf{R}_{O_1} = \mathbf{R}_{O_1}^{(t_1)} + \mathbf{R}_{O_1}^{(n_1)}, \tag{2.123a}$$

$$\mathbf{R}_{O_2} = \mathbf{R}_{O_2}^{(t_2)} + \mathbf{R}_{O_2}^{(n_2)}, \tag{2.123b}$$

$$\mathbf{R}_{O_3} = \mathbf{R}_{O_3}^{(t_3)} + \mathbf{R}_{O_3}^{(n_3)}. \tag{2.123c}$$

Due to space limitations, Fig. 2.13 does not contain the moments of forces of the actions of links on one another and of the base action through joints, since in the following calculationswe do not use moment equations but force equations only.

For each link the equations expressing the acceleration of its mass center under the action of external forces have the following form:

$$\begin{aligned} m_1 \mathbf{a}_{C_1} &= \mathbf{R}_{O_1} + m_1 \mathbf{g} - \mathbf{R}_{O_2}, \\ m_2 \mathbf{a}_{C_2} &= \mathbf{R}_{O_2} + m_2 \mathbf{g} - \mathbf{R}_{O_3}, \\ m_3 \mathbf{a}_{C_3} &= \mathbf{R}_{O_3} + m_3 \mathbf{g}, \end{aligned} \tag{2.124}$$

and projecting these equations onto directions t_1, n_1, t_2, n_2, t_3, and n_3 we obtain

$$\begin{aligned} m_1 a_{C_1}^{(t_1)} &= R_{O_1}^{(t_1)} - m_1 g \sin(\psi_1) \\ &\quad - R_{O_2}^{(t_2)} \cos(\psi_2 - \psi_1) + R_{O_2}^{(n_2)} \sin(\psi_2 - \psi_1), \\ m_1 a_{C_1}^{(n_1)} &= R_{O_1}^{(n_1)} - m_1 g \cos(\psi_1) \\ &\quad - R_{O_2}^{(t_2)} \sin(\psi_2 - \psi_1) - R_{O_2}^{(n_2)} \cos(\psi_2 - \psi_1), \\ m_2 a_{C_2}^{(t_2)} &= R_{O_2}^{(t_2)} - m_2 g \sin(\psi_2) \\ &\quad - R_{O_3}^{(t_3)} \cos(\psi_3 - \psi_2) + R_{O_3}^{(n_3)} \sin(\psi_3 - \psi_2), \end{aligned}$$

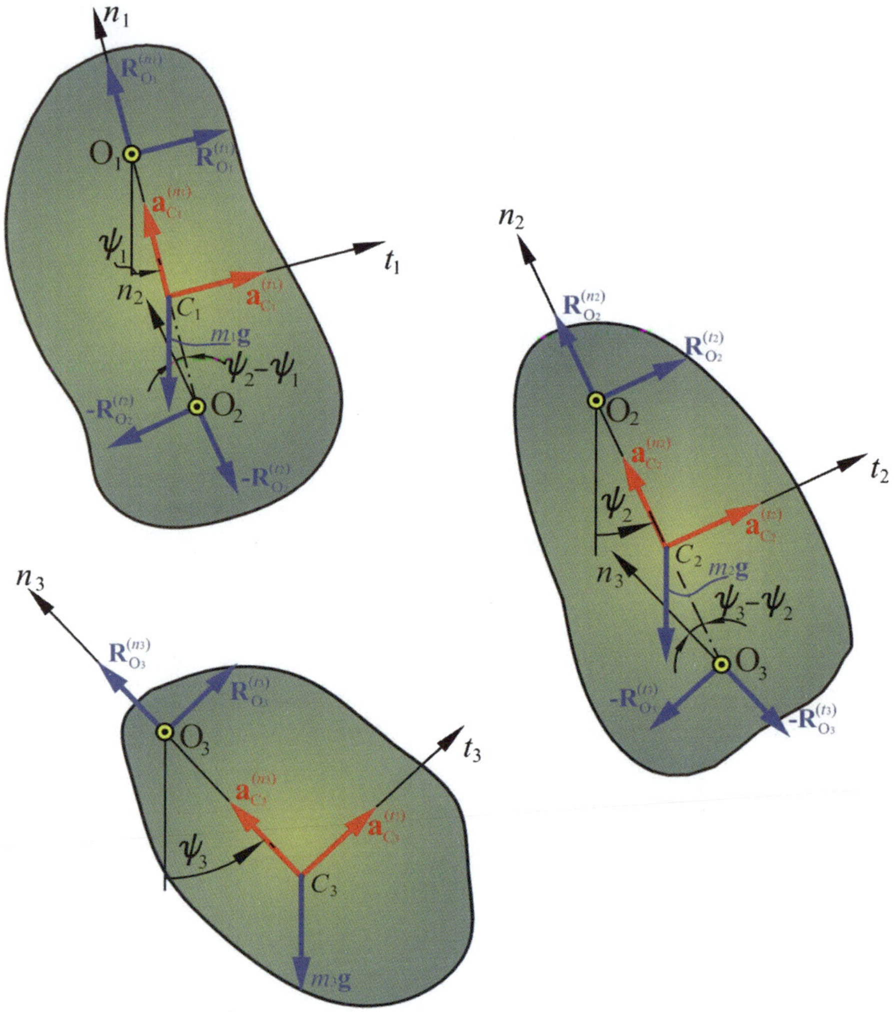

Fig. 2.13 External forces acting on particular links of a pendulum and the accelerations of the mass centers of the links (force couples acting at joints not shown)

$$
\begin{aligned}
m_2 a_{C_2}^{(n_2)} &= R_{O_2}^{(n_2)} - m_2 g \cos(\psi_2) \\
&\qquad - R_{O_3}^{(t_3)} \sin(\psi_3 - \psi_2) - R_{O_3}^{(n_3)} \cos(\psi_3 - \psi_2), \\
m_3 a_{C_3}^{(t_3)} &= R_{O_3}^{(t_3)} - m_3 g \sin(\psi_3), \\
m_3 a_{C_3}^{(n_3)} &= R_{O_3}^{(n_3)} - m_3 g \cos(\psi_3). \qquad (2.125)
\end{aligned}
$$

Those equations can be solved with respect to the components of the dynamic reactions, and taking into account relationships (2.106a), (2.112), and (2.122), and bearing in mind that $\sin(-x) = -\sin(x)$ and $\cos(-x) = \cos(x)$, we obtain

$$
\begin{aligned}
R_{O_3}^{(t_3)} &= m_3\left[g\sin\psi_3 + e_3\ddot{\psi}_3 + l_1\left(\ddot{\psi}_1\cos(\psi_1-\psi_3) - \dot{\psi}_1^2\sin(\psi_1-\psi_3)\right)\right.\\
&\qquad \left. + l_2\left(\ddot{\psi}_2\cos(\psi_2-\psi_3) - \dot{\psi}_2^2\sin(\psi_2-\psi_3)\right)\right],\\
R_{O_3}^{(n_3)} &= m_3\left[g\cos\psi_3 + e_3\dot{\psi}_3^2 + l_1\left(\ddot{\psi}_1\sin(\psi_1-\psi_3) + \dot{\psi}_1^2\cos(\psi_1-\psi_3)\right)\right.\\
&\qquad \left. + l_2\left(\ddot{\psi}_2\sin(\psi_2-\psi_3) + \dot{\psi}_2^2\cos(\psi_2-\psi_3)\right)\right],\\
R_{O_2}^{(t_2)} &= m_2\left[g\sin\psi_2 + e_2\ddot{\psi}_2 + l_1\left(\ddot{\psi}_1\cos(\psi_1-\psi_2) - \dot{\psi}_1^2\sin(\psi_1-\psi_2)\right)\right]\\
&\qquad + R_{O_3}^{(n_3)}\sin(\psi_2-\psi_3) + R_{O_3}^{(t_3)}\cos(\psi_2-\psi_3),\\
R_{O_2}^{(n_2)} &= m_2\left[g\cos\psi_2 + e_2\dot{\psi}_2^2 + l_1\left(\ddot{\psi}_1\sin(\psi_1-\psi_2) + \dot{\psi}_1^2\cos(\psi_1-\psi_2)\right)\right]\\
&\qquad + R_{O_3}^{(n_3)}\cos(\psi_2-\psi_3) - R_{O_3}^{(t_3)}\sin(\psi_2-\psi_3),\\
R_{O_1}^{(t_1)} &= m_1\left[g\sin\psi_1 + e_1\ddot{\psi}_1\right] + R_{O_2}^{(n_2)}\sin(\psi_1-\psi_2) + R_{O_2}^{(t_2)}\cos(\psi_1-\psi_2),\\
R_{O_1}^{(n_1)} &= m_1\left[g\cos\psi_1 + e_1\dot{\psi}_1^2\right] + R_{O_2}^{(n_2)}\cos(\psi_1-\psi_2) - R_{O_2}^{(t_2)}\sin(\psi_1-\psi_2).
\end{aligned}
\tag{2.126}
$$

Now the absolute values of total reactions can be calculated as

$$
\begin{aligned}
R_{O_1} &= \sqrt{\left(R_{O_1}^{(t_1)}\right)^2 + \left(R_{O_1}^{(n_1)}\right)^2},\\
R_{O_2} &= \sqrt{\left(R_{O_2}^{(t_2)}\right)^2 + \left(R_{O_2}^{(n_2)}\right)^2},\\
R_{O_3} &= \sqrt{\left(R_{O_3}^{(t_3)}\right)^2 + \left(R_{O_3}^{(n_3)}\right)^2}.
\end{aligned}
\tag{2.127}
$$

Some examples of time plots of dynamic reactions calculated from relations (2.126) and (2.127) are presented in Figs. 2.14–2.16.

The time plot shown in Fig. 2.14 corresponds to the solution shown in Fig. 2.8, that is, to the decaying motion of a pendulum without excitation. It can be seen that reactions decrease relatively quickly to a value close to a static reaction for a system at rest. In Fig. 2.15 we present the time plot of dynamic reactions in periodic motion of the pendulum shown in Fig. 2.9 ($q = 25\,\mathrm{N\cdot m}$ and $\omega = 3\,\mathrm{rad/s}$). Here greater values of reactions are visible. In turn, in Fig. 2.16 we present a certain select part of the time plot of dynamic reactions for the chaotic solution shown in

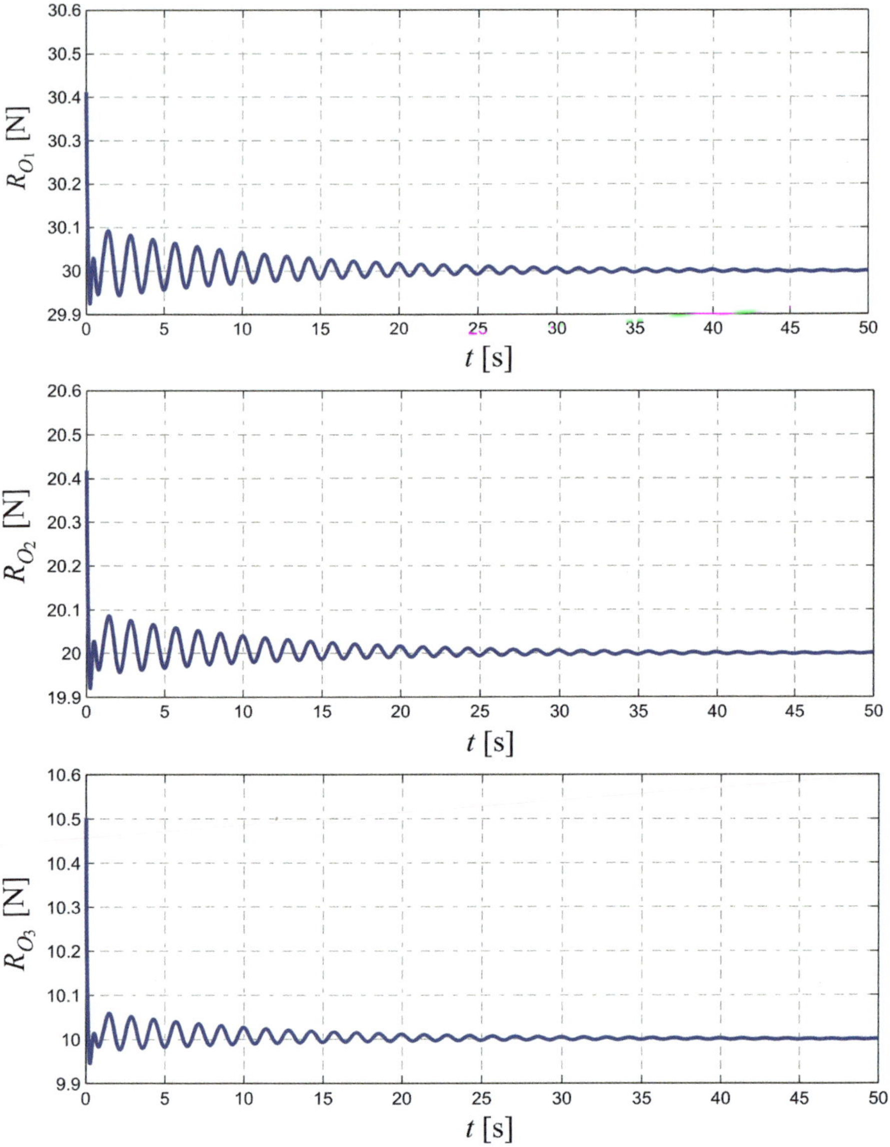

Fig. 2.14 Dynamic reactions in bearings for decaying motion of pendulum without external excitation ($q = 0$)

Fig. 2.11c, that is, for the parameters of excitation $q = 25\,\mathrm{N \cdot m}$ and $\omega = 2\,\mathrm{rad/s}$. The most rapid changes in the dynamic reactions of bearings are visible there. It should be emphasized, however, that it is only part of an irregular time plot, and the instantaneous values of reactions may be even greater.

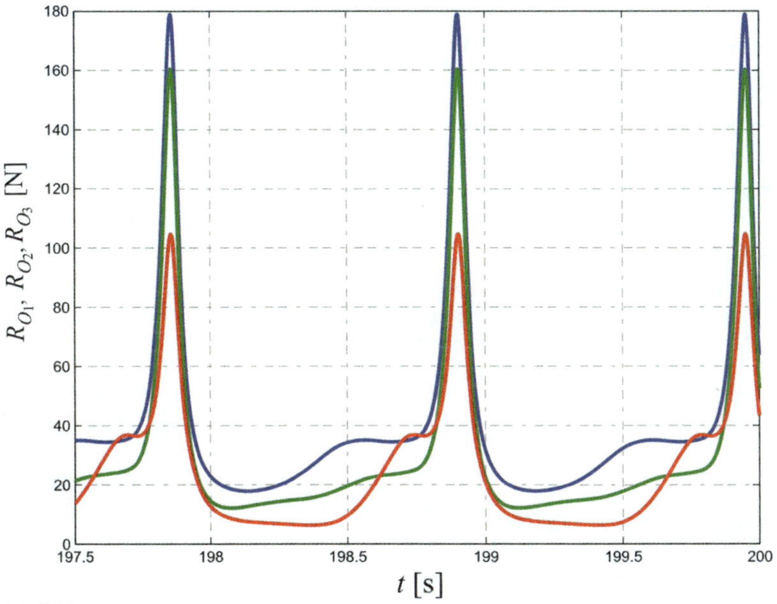

Fig. 2.15 Dynamic reactions in bearings for periodic motion of pendulum for $q = 25\,\mathrm{N \cdot m}$ and $\omega = 3\,\mathrm{rad/s}$; color codes: R_{O_1}—*1*, R_{O_2}—*2*, R_{O_3}—*3*

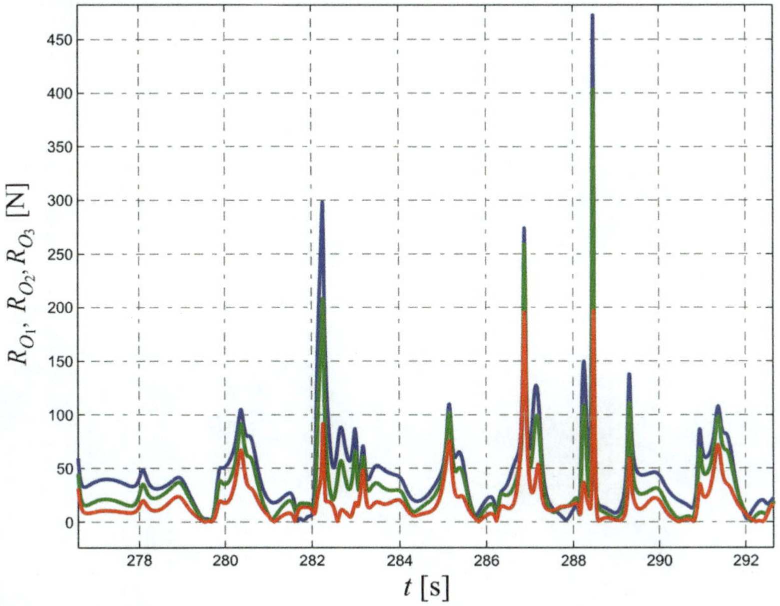

Fig. 2.16 Dynamic reactions in bearings for chaotic motion of pendulum for $q = 25\,\mathrm{N \cdot m}$ and $\omega = 2\,\mathrm{rad/s}$; color codes: R_{O_1}—*1*, R_{O_2}—*2*, R_{O_3}—*3*

Supplementary sources for the material in this chapter include [7–12]. In addition, numerous books are devoted to the periodic, quasiperiodic, and chaotic dynamics of lumped mechanical systems including [13–15].

References

1. J. Awrejcewicz, *Classical Mechanics: Statics and Kinematics* (Springer, New York, 2012)
2. J. Awrejcewicz, *Oscillations of Lumped Deterministic Systems* (WNT, Warsaw, 1996)
3. J. Awrejcewicz, *Mathematical Modelling of Systems* (WNT, Warsaw, 2006)
4. J. Leyko, *General Mechanics* (PWN, Warsaw, 2006)
5. G. Kudra, *Analysis of Bifurcation and Chaos in Three Coupled Physical Pendulums with Impacts* (Ph.D., Lodz, 2002)
6. J. Awrejcewicz (ed.), *Numerical Simulations of Physical and Engineering Processes* (InTech, Rijeka, 2011)
7. S.T. Thornton, J.B. Marion, *Classical Dynamics of Particles and Systems* (Saunders College Publishers, New York, 1995)
8. L.N. Hand, J.D. Finch, *Analytical Mechanics* (Cambridge University Press, Cambridge, 1998)
9. J.V. José, E.J. Saletan, *Classical Dynamics: A Contemporary Approach* (Cambridge University Press, Cambridge, 1998)
10. M.D. Ardema, *Newton–Euler Dynamics* (Springer, Berlin, 2005)
11. A.J. Brizard, *An Introduction to Lagrangian Mechanics* (World Scientific, Singapore, 2008)
12. P.W. Johnson, *Classical Mechanics with Applications* (World Scientific, Singapore, 2010)
13. J. Awrejcewicz, *Bifurcation and Chaos in Simple Dynamics Systems* (World Scientific, Singapore, 1989)
14. J. Awrejcewicz, *Bifurcation and Chaos in Coupled Oscillators* (World Scientific, Singapore, 1991)
15. J. Awrejcewicz, C.-H. Lamarque, *Bifurcation and Chaos in Nonsmooth Mechanical Systems* (World Scientific, Singapore, 2003)

Chapter 3
Statics and Dynamics in Generalized Coordinates

3.1 Constraints and Generalized Coordinates

We will consider a discrete (lumped) material system (DMS) in Euclidean space $\mathbf{E}^3$ composed of N particles of masses $m_1, m_2, \ldots, m_N$ (see [1]), presented in Fig. 3.1, which, as mentioned earlier, will be called a *discrete mechanical system*.

The position of every particle of a mechanical system is described by a radius vector $\mathbf{r}_n$ in the adopted right-handed Cartesian coordinate system.

Here it is worth emphasizing that, despite introducing the notion of a particle already at the very beginning of our book, the majority of interacting bodies cannot be described (even approximately) by such a notion. The problem is that a "physical point" always has a certain finite size, and additionally we are often faced with a body in a liquid or gaseous state.

Rigid bodies from a mechanical point of view are understood as non-deformable bodies and sufficiently large in comparison to a particle (although the local deformability of a rigid body is allowed, for instance, in the description of an impact phenomenon).

We will treat surfaces bounding rigid bodies as barriers that do not let in (out) other bodies including particles. The restriction of motion of those particles leads to the introduction of the notion of *constraints*.

According to the axioms of classical mechanics, masses are positive ($m_n > 0$) and time-independent ($\mathrm{d}m_n/\mathrm{d}t = 0$), and a system is additive (mass of the whole system $m = \sum_{n=1}^{N} m_n$). A state of the mechanical system presented in Fig. 3.1 is described by radius vectors $\mathbf{r}_n$ of particles $n = 1, \ldots, N$ and the velocities of these particles $\mathbf{v}_n$. A mechanical system can also be acted upon by certain forces. However, regardless of the action of these forces, certain restrictions called constraints can be imposed on vectors $\mathbf{r}_n$ and $\mathbf{v}_\nu$. If there appears at least one such restriction, we call the given mechanical system a *constrained system* or a *system with constraints*. If there are no restrictions, the mechanical system is called a *free system*.

J. Awrejcewicz, *Classical Mechanics: Dynamics*, Advances in Mechanics and Mathematics 29, DOI 10.1007/978-1-4614-3740-6_3,

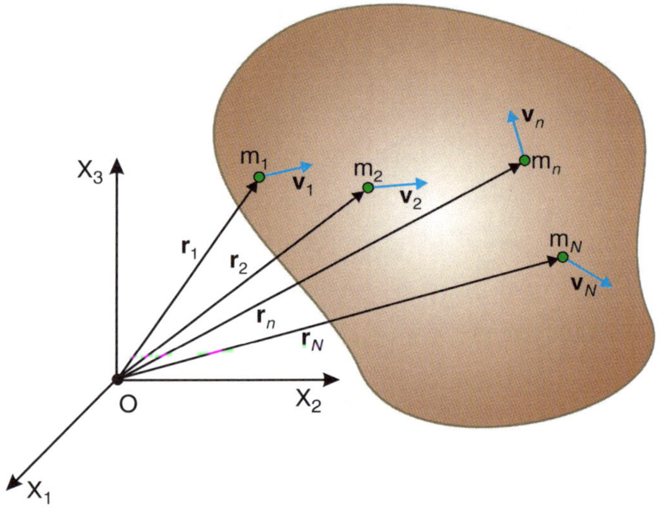

Fig. 3.1 A discrete mechanical system, $n = 1, \ldots, N$

Let a particle be moving on a certain plane with which we associate the coordinate system OX_1X_2. The equation of constraints for the particle has the form $x_3 = 0$, where x_3 denotes the coordinate of the axis OX_3 perpendicular to the plane. Let a particle be moving on a sphere of radius $r = r(t)$. If we take the coordinate system $OX_1X_2X_3$ at the center of the sphere, then the equation of constraints has the form $x_1^2+x_2^2+x_3^2-r^2(t) = 0$, where (x_1, x_2, x_3) are the coordinates of the particle.

We call a mechanical system a *system of a finite number of degrees of freedom* if it is possible to introduce a finite-dimensional space $\mathbf{R}^M$ and a set of points Ω in this space such that there exists a one-to-one relationship between the possible positions of particles of the mechanical system and all the points of the set $\Omega \subset \mathbf{R}^M$ [2].

We call the set Ω a *configuration manifold (configuration space)* of a mechanical system if the aforementioned relationship is differentiable following transition from one set of coordinates to another. In order to illustrate the introduced notions we present here two examples ([2]). The first example is presented in Fig. 3.2.

The motion of the end of the weightless rod with concentrated mass at this end takes place along a circle of radius l.

The second example involves the planar motion of a double mathematical pendulum (Fig. 3.3).

Choosing, in an arbitrary way, two mutually perpendicular torus sections it is possible to measure from them the angles φ_1 and φ_2; then the positions of the pendulum ends are represented by the points on the torus with a one-to-one correspondence.

In the example depicted in Fig. 3.2 it would seem that instead of a circle one might take a segment $[0, 2\pi]$ on the axis $O\varphi$. However, such a choice is improper because the one-to-one correspondence between the position of the pendulum

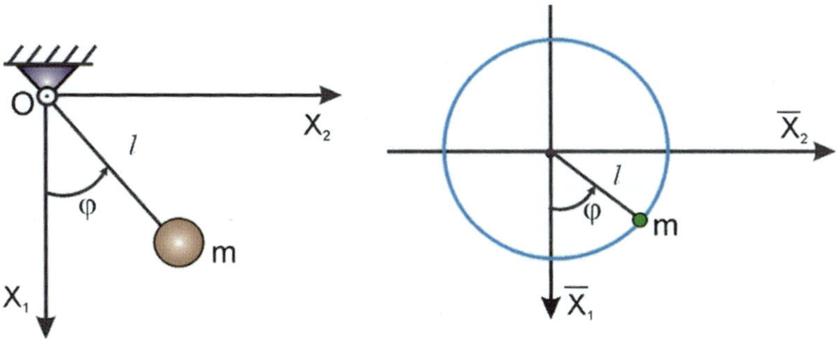

Fig. 3.2 A mathematical pendulum and its configuration manifold (*a circle*)

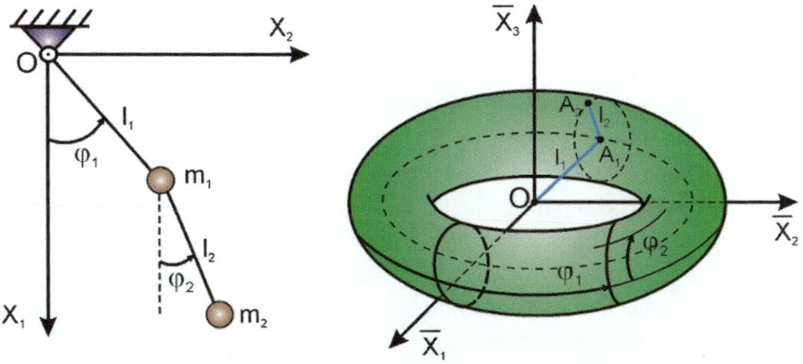

Fig. 3.3 A double mathematical pendulum and its configuration manifold (*a torus*)

$\varphi + 2n\pi$ $(n = \pm 1, \pm 2, \ldots)$ and the points φ belonging only to the interval $[0, 2\pi]$ is not preserved. The start of the segment (point O) should be connected with its end (point 2π), and then we obtain the circle. In turn, in the example from Fig. 3.3, one might take the square of side $[0, 2\pi]$ in the coordinate system $O\varphi_1\varphi_2$. In order to preserve the one-to-one correspondence between the position of the double pendulum described by the point (φ_1, φ_2) one should "glue" together the sides of a square, at first, for example, along φ_1 obtaining a cylinder, and then along φ_2, obtaining a torus.

Apart from the aforementioned three-dimensional Euclidean space $\mathbf{R}^3$ it is convenient to introduce a real space $\mathbf{R}^{3N}$ in which its *single* point $(x_1^1, x_2^1, x_3^1, \ldots, x_1^N, x_2^N, x_3^N)$ represents all N particles of the space $\mathbf{R}^3$. Let us note that in this case N trajectories, associated with every one of N points of the space $\mathbf{E}^3$, are represented by one trajectory in the space $\mathbf{R}^{3N}$. Such a space we call a *configuration space (configuration manifold).*

We call the minimum number of independent coordinates necessary to describe the motion of a mechanical system (they can be linear displacements or rotations) the *generalized coordinates* of this system [3–5]. It follows that the number of generalized coordinates is equal to the number of degrees of freedom. In both previousy mentioned cases we have already dealt with constraints that enabled the realization of the configuration manifold of the considered mechanical system (the base and the connection of links).

The system shown in Fig. 3.2 has constraints that in the general case can depend on displacement, velocity, and time.

Let us note that these restrictions are "external" because they do not follow from the motion of the investigated system. The relationships describing these restrictions (algebraic equations, algebraic inequalities, differential equations, or their combinations) are called *equations of constraints.* We call systems with imposed constraints *constrained systems* because their "freedom of motion" is constrained in a certain way.

Constraints that depend only on position are called *geometric constraints* (they do not depend on velocity), and a constraint equation is $f(\mathbf{r}_n, t) = 0$. *Kinematic (differential) constraints* additionally depend on the velocity, and the equation of the constraint has the form $f(\mathbf{r}_n, \mathbf{v}_n, t) = 0$. Constraints dependent on time are called *rheonomic (time-dependent) constraints*, whereas those not dependent on time are called *scleronomic (time-independent) constraints.*

The constraints listed above belong to the group of *bilateral constraints* described by algebraic equations, whereas *unilateral constraints* are described by algebraic inequalities. Constraints are called *holonomic constraints* if they are geometric or their equation can be integrated, i.e., it is possible to obtain their equation in the form of a function dependent on displacement and time. Kinematic constraints that cannot be reduced to the aforementioned form are called *non-holonomic (nonintegrable) constraints* [2–6].

The notion of a *non-holonomic system* was introduced by Hertz[1] (1894), but it had been considered earlier by Euler (1739). Euler is known as the first scientist to consider the small vibrations of a rigid body rolling without sliding on a horizontal surface. The fact that constraints are imposed on velocities fundamentally distinguishes non-holonomic systems from systems of Lagrange,[2] Routh,[3] or Hamilton.[4]

We will present now an example of non-holonomic constraints that will later be considered in more detail. Let an ice skate, modeled as a thin rod in contact with ice, be moving on ice (on the plane OX_1X_2). Let the velocity of point A of the rod $\mathbf{v}_A = [\dot{x}_1, \dot{x}_2]$ form an angle φ with the axis OX_1, so we have $\dot{x}_{1A} \equiv \mathbf{v}_A \circ \mathbf{E}_1 = v_A \cos\varphi$ and $\dot{x}_{2A} \equiv \mathbf{v}_A \circ \mathbf{E}_2 = v_A \sin\varphi$. In this case we are dealing with a hockey skate, where angle φ can change and is a coordinate.

---

[1]Heinrich Hertz (1857–1894), German physicist and mechanician working on contact problems and electromagnetic waves.

[2]Joseph Lagrange (1736–1813), distinguished French and Italian mathematician and astronomer, working also in Berlin.

[3]Edward Routh (1831–1907), English mathematician who played a significant role in the theory of control and stability.

[4]William Hamilton (1805–1865), Irish mathematician, physician, and astronomer.

Because the problem is planar, the constraints are described by the following equation:

$$\dot{x}_{2A} = \dot{x}_{1A} \tan \varphi. \tag{3.1}$$

We will show that (3.1) is not integrable. To this end, let us assume that the equation is integrable, i.e., there exists an analytical relationship between the quantities x_1, x_2, and φ of the form

$$f\,(x_1, x_2, \varphi, t) = 0. \tag{3.2}$$

Differentiating (3.2) we obtain

$$\dot{f} = \frac{\partial f}{\partial x_1}\dot{x}_1 + \frac{\partial f}{\partial x_2}\dot{x}_1 \tan \varphi + \frac{\partial f}{\partial \varphi}\dot{\varphi} + \frac{\partial f}{\partial t} = 0, \tag{3.3}$$

where relationship (3.1) was taken into account.

By assumption, $\dot{x}_1$ and $\dot{\varphi}$ are independent. Thus we have

$$\frac{\partial f}{\partial x_1} + \frac{\partial f}{\partial x_2} \tan \varphi = 0, \quad \frac{\partial f}{\partial \varphi} = 0, \quad \frac{\partial f}{\partial t} = 0. \tag{3.4}$$

Because angle φ is arbitrary, from equations (3.4) it follows that

$$\frac{\partial f}{\partial x_1} = 0, \quad \frac{\partial f}{\partial x_2} = 0, \quad \frac{\partial f}{\partial \varphi} = 0, \quad \frac{\partial f}{\partial t} = 0. \tag{3.5}$$

This means that the function f does not depend on x_1, x_2, and φ, which is in contradiction with the initial assumption. A complete solution of the problem of ice skate motion on ice will be presented in Example 3.4.

It is worth drawing the reader's attention to the difficulties associated with the analysis of non-holonomic mechanical systems because of their peculiar properties [6]:

1. They can be derived from the d'Alembert–Lagrange principle but not from Hamilton's principle.
2. The law of conservation of energy is valid, but the angular momentum of the system may be not conserved.
3. Non-holonomic systems are so-called almost Poisson systems but not Poisson systems.
4. The phase flow associated with non-holonomic systems may not be conserved in a phase space, which leads to the formulation of a new concept of asymptotic stability.

In [6] can be found the latest achievements and history regarding non-holonomic systems. Here we present only an example of a simple system of this type analyzed by L. Euler. That is, we will consider the geometry of a disk rolling without sliding on a horizontal plane. Here we can distinguish the "most" general motion of the disk, that is, the case where it is falling, marked in Fig. 3.4 by time instant t_1, when

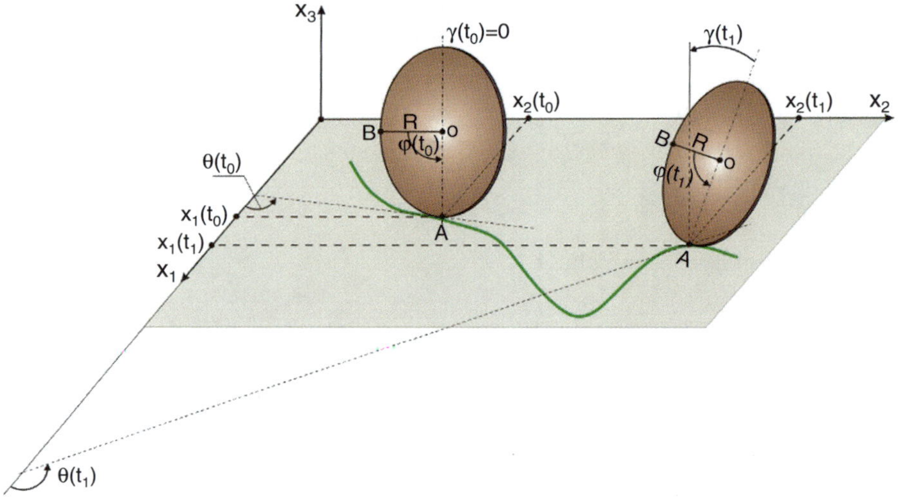

Fig. 3.4 General motion of a disk (coin) on a plane in the case where the axis of the disk is vertical (instant t_0) and where the disk is falling (instant t_1)

the disk has 5 degrees of freedom, i.e., $x_1(t)$, $x_2(t)$, $\Theta(t_1)$, $\gamma(t_1)$ and $\varphi(t_1)$. The coordinates of the point of contact between the disk and the plane are denoted by x_1 and x_2. The angle γ is formed between the plane of the disk and the vertical line, whereas the angle θ is the angle of rotation of the disk measured in the plane X_1X_2. The second of the distinguished cases corresponds to a situation where the plane of the disk is perpendicular to the horizontal plane ($\gamma = 0$). The third case (the simplest one) involves the motion of the disk when its plane is perpendicular to the horizontal plane, and the motion takes place along a straight line. Let us consider kinematic relationships at point A for the case $\gamma = 0$, that is, the relationships following from the process of rolling without sliding, of the form

$$v_A = R\frac{d\varphi}{dt}, \quad \frac{dx_1}{dt} = v_A \cos\Theta, \quad \frac{dx_2}{dt} = v_A \sin\Theta. \tag{3.6}$$

From the foregoing kinematic relationships we obtain equations of constraints in the configuration space, that is,

$$\begin{aligned} dx_1 &= R\cos\Theta d\varphi, \\ dx_2 &= R\sin\Theta d\varphi, \end{aligned} \tag{3.7}$$

but we cannot determine the curve along which the disk moves in the horizontal plane.

The notion of *constraints* describes the manner in which particles and rigid bodies interact when they begin to come into contact with each other.

If at the instant after the initial contact a particle "sticks" to the surface of the rigid body and rests there for the remaining time of the observation, then such constraints are called the *bilateral constraints* of the considered particle.

If following the initial contact the particle remains stuck for some time, and then it separates, then such constraints are called the *unilateral constraints* of the particle.

If the duration of contact between the particle and the body is very short, and a sudden change in the sense of the particle velocity occurs, then we are dealing with an impact and the relationship between the velocity before and after the impact is defined by the notion of a *coefficient of restitution* (Chap. 5).

In a material system, *unilateral constraints* may spontaneously change into *bilateral constraints*, and vice versa. For instance, the textbook [1] describes the self-excited vibrations of two particles (blocks) connected together elastically (by means of massless springs) lying on a non-deformable belt moving with a constant velocity. In such a system, stick-slip, slip-stick, and slip-slip motions can be observed.

Our planet Earth, in a certain approximation, can be treated as a material system with constraints. Houses, bridges, factories, etc., are solid bodies remaining on the ground, so for those "particles," Earth determines the *bilateral constraints*. Planes, balloons, rockets, and missiles are examples of bodies for which Earth determines the set of *unilateral constraints*.

Unilateral constraints and bilateral constraints can also have the following physical interpretation. If in the considered DMS we limit the independence of the positions and velocities of the system's particles by connecting them with rigid massless rods, then the constraints are bilateral constraints. If we connect the particles with a flexible inextensible thread that cannot break, then we are dealing with unilateral constraints.

Equations of bilateral constraints have the form

$$f_m(t, \mathbf{r}_1, \dot{\mathbf{r}}_1, \ldots, \mathbf{r}_N, \dot{\mathbf{r}}_N) = 0, \quad m = 1, \ldots, M. \tag{3.8}$$

If the equals sign in the equation above is replaced with the (≥ 0) or (≤ 0) sign, then the restrictions imposed on the motion of the particles of the DMS are called unilateral constraints.

From the foregoing discussion it follows that M relationships (3.8) are described by non-linear differential equations that depend on the time, position, and velocity of every particle of a DMS. Already here we encounter a very serious obstacle of a mathematical nature since, as a rule, the solution to a problem that involves (3.8) is impossible by means of an analytical method.

Example 3.1. Demonstrate that the connection of two particles by means of a rigid massless rod (inextensible thread) introduces bilateral constraints (unilateral constraints).

If we denote the distance between masses m_1 (particle A_1) and m_2 (particle A_2) in Fig. 3.1 by r_{12}, then in the adopted coordinate system we have

$$f_1 \equiv (\mathbf{r}_1 - \mathbf{r}_2)^2 = \mathbf{r}_{12}^2,$$

or in scalar form

$$(x_{1A_1} - x_{1A_2})^2 + (x_{2A_1} - x_{2A_2})^2 + (x_{3A_1} - x_{3A_3})^2 - r_{12}^2 = 0,$$

which, according to (3.8), classifies these constraints as bilateral constraints.

In the case of constraints imposed by means of an inextensible thread, we obtain

$$(x_{1A_1} - x_{1A_2})^2 + (x_{2A_1} - x_{2A_2})^2 + (x_{3A_1} - x_{3A_2})^2 - r_{12}^2 \le 0,$$

which classifies these constraints as unilateral constraints.

In this case neither particle A_1 nor particle A_2 moves, but during their motion their relative distance can only change from zero up to r_{12}. □

Let particle A_1 of mass m_1 (Fig. 3.1) move with a constant velocity $\dot{\mathbf{r}}_1 = \mathbf{v} \equiv$ *const.* We are going to determine the type of constraints imposed on this particle. We have

$$\dot{\mathbf{r}}_1 = \mathbf{v},$$

and after scalar multiplication by $\dot{\mathbf{r}}_1$ we obtain

$$f_1 \equiv \dot{\mathbf{r}}_1 \circ \dot{\mathbf{r}}_1 - \mathbf{v}^2 = 0,$$

or in scalar notation

$$\dot{x}_1^2 + \dot{x}_2^2 + \dot{x}_3^2 - v^2 = 0.$$

According to (3.8), here as well we are dealing with bilateral constraints, and the restrictions concern the velocity of particle A_1. Its components lie on a sphere of radius v.

Unilateral constraints in this case will appear when we assume that the velocity of particle A_1 cannot exceed velocity $\mathbf{v}$. Such a problem is equivalent to a restriction in the form of inequality

$$\dot{x}_1^2 + \dot{x}_2^2 + \dot{x}_3^2 - v^2 \le 0, \tag{3.9}$$

which defines this constraint as unilateral.

The transition from unilateral to bilateral constraints takes place through a change from inequality to equality, that is, when unilateral constraints undergo "tightening."

Let the considered mechanical system have M_1 holonomic constraints. Thus it has

$$K_1 = 3N - M_1 \tag{3.10}$$

degrees of freedom. There are $3N$ Cartesian coordinates describing the motion of a DMS, whereas there are K_1 generalized coordinates q_k $(k = 1, \ldots, K_1)$ (their number decreases because of the imposed constraints, and hence $K_1 < 3N$).

Cartesian coordinates can be expressed in terms of generalized coordinates in the following way:

$$\begin{aligned} x_{1n} &= x_{1n}\left(q_1,\ldots,q_k,\ldots,q_{K_1}\right), \\ x_{2n} &= x_{2n}\left(q_1,\ldots,q_k,\ldots,q_{K_1}\right), \\ x_{3n} &= x_{3n}\left(q_1,\ldots,q_k,\ldots,q_{K_1}\right), \end{aligned} \tag{3.11}$$

where $n = 1,\ldots,N$.

Relationships of type (3.11) are called *equations of constraints in the resolved form* [7–10].

The minimum number of parameters required for a complete description of any position of a DMS is called the *number of its independent generalized coordinates*.

As generalized coordinates one can take K_1 from the Cartesian coordinates x_{1n}, x_{2n}, x_{3n}, where $n = 1,\ldots,N$, so as to enable the solution of M_1 equations of constraints. In practice that approach is not always convenient. Completely different independent quantities $q_1,\ldots,q_{K_1}$ describing the configuration space of a system are often introduced. The generalized coordinates can be distances, angles, or surfaces, or they may not have any physical meaning. The functions (3.11) substituted into constraint equations have to turn these equations into identity relations, on the assumption that we are dealing with K_1 independent generalized coordinates. Because $K_1 \le 3N$, the rank of the rectangular matrix

$$\mathbf{A} = \begin{bmatrix} \dfrac{\partial x_{11}}{\partial q_1} & \cdots & \dfrac{\partial x_{11}}{\partial q_{K_1}} \\ \dfrac{\partial x_{12}}{\partial q_1} & \cdots & \dfrac{\partial x_{12}}{\partial q_{K_1}} \\ \dfrac{\partial x_{13}}{\partial q_1} & \cdots & \dfrac{\partial x_{13}}{\partial q_{K_1}} \\ \vdots & & \vdots \\ \dfrac{\partial x_{1N}}{\partial q_1} & \cdots & \dfrac{\partial x_{1N}}{\partial q_{K_1}} \\ \dfrac{\partial x_{2N}}{\partial q_1} & \cdots & \dfrac{\partial x_{2N}}{\partial q_{K_1}} \\ \dfrac{\partial x_{3N}}{\partial q_1} & \cdots & \dfrac{\partial x_{3N}}{\partial q_{K_1}} \end{bmatrix} \tag{3.12}$$

is equal to K_1.

If all positions of the system can be described by relations (3.11), the generalized coordinates are *global coordinates*. If (3.11) is satisfied only for certain configurations of the system, then the generalized coordinates are *local coordinates*, and it is necessary to choose different local coordinates to represent all possible system configurations. However, in most cases such problems do not emerge and the choice of generalized coordinates is suggested naturally by the considered mechanical system. Relationships inverse to (3.11) that describe the dependencies of K_1 coordinates q_k on $3N$ Cartesian coordinates of the DMS are usually very difficult to obtain in practice.

Holonomic constraints (also called *geometric* or *finite constraints*) are described by algebraic equations of the form

$$f_{m_1}(t, \mathbf{r}_1, \ldots, \mathbf{r}_N) = 0 \tag{3.13}$$

or

$$f_{m_1}(t, x_{11}, \ldots, x_{1N}, x_{21}, \ldots, x_{2N}, x_{31}, \ldots, x_{3N}) = 0, \quad m_1 = 1, \ldots, M_1.$$

Relationships of type (3.13) are called *equations of constraints in implicit form* [7–10]. With these constraints, restrictions are imposed on the positions of particles of a DMS (they can be unilateral or bilateral).

Kinematic (differential) constraints are described by a system of non-linear differential equations of the form

$$f_{m_2}(t, \mathbf{r}_1, \ldots, \mathbf{r}_N, \dot{\mathbf{r}}_1, \ldots, \dot{\mathbf{r}}_N) = 0 \tag{3.14}$$

or

$$f_{m_2}\{t, x_{11}, \ldots, x_{1N}, x_{21}, \ldots, x_{2N}, x_{31}, \ldots, x_{3N},$$
$$\dot{x}_{11}, \ldots, \dot{x}_{1N}, \dot{x}_{21}, \ldots, \dot{x}_{2N}, \dot{x}_{31}, \ldots, \dot{x}_{3N}\} = 0, \quad m_2 = 1, \ldots, M_2,$$

and they are indentical to constraints described by (3.8). Kinematic constraints can also be unilateral or bilateral constraints.

Further, we will deal with a case of kinematic constraints that boils down to an analysis of relationships that are linear with respect to generalized velocities of the form

$$\sum_{m_2=1}^{M_2} \mathbf{c}_{m_2 n}(t, \mathbf{r}_1, \ldots, \mathbf{r}_N)\,\mathrm{d}\mathbf{r}_n + \beta_{m_2}(t, \mathbf{r}_1, \ldots, \mathbf{r}_N)\,\mathrm{d}t = 0,$$
$$n = 1, \ldots, N, \quad m_2 = 1, \ldots, M_2. \tag{3.15}$$

The number of degrees of freedom of a system expressed by independent generalized coordinates is defined through (3.13) and is equal to $K_1 = 3N - M_1$ since kinematic constraints do not change the configuration manifold of a DMS.

Unilateral constraints can be *holonomic* or *non-holonomic*. Holonomic unilateral constraints are expressed analytically by restrictions imposed on the generalized coordinates in the form of the following inequalities:

$$f_m\left(t, x_{11}, \ldots, x_{1N}, x_{21}, \ldots, x_{2N}, x_{31}, \ldots, x_{3N}\right) \geqslant 0, \quad m = 1, \ldots, M. \tag{3.16}$$

The described holonomic constraints were *time-dependent constraints* because time occurred explicitly in the equations and inequalities describing them. If time does not occur in the aforementioned equations and inequalities, then the constraints are called *time-independent constraints*. The differential constraints (3.15) are called *time-independent constraints* (scleronomic constraints) if the vector functions $\mathbf{c}_{m_2n}$ do not depend explicitly on time, and additionally $\beta_{m_2} \equiv 0$.

The existence of constraints in a DMS significantly influences the dynamics of such a system. Constraints on every particle supply an additional force called the reaction of constraints. Moreover, every trajectory of motion in $\mathbf{R}^{3N}$ space lies on a constraint surface described by (3.13), (3.14) or (3.15) or by inequalities (3.16). The constrained motion of a system with constraints causes the imposition of restrictions on the displacements, velocities, and accelerations of particles of a DMS so as to satisfy the equation of constraints.

Equations of non-holonomic constraints (3.15) read as follows:

$$\sum_{n=1}^{N} \mathbf{c}_{m_2n}\dot{\mathbf{r}} + \beta_{m_2} = 0, \quad m_2 = 1, 2, \ldots, M_2. \tag{3.17}$$

In turn, differentiating equations of holonomic constraints (3.13) we obtain

$$\frac{\mathrm{d}f_{m_1}}{\mathrm{d}t} = \sum_{n=1}^{N} \frac{\partial f_{m_1}}{\partial \mathbf{r}_n}\dot{\mathbf{r}}_n + \frac{\partial f_{m_1}}{\partial t} = 0, \quad m_1 = 1, 2, \ldots, M_1, \tag{3.18}$$

and (3.18) have a form analogous to that of non-integrable kinematic constraints (3.17).

During the motion of a mechanical system, the system's radius vectors, velocities, and accelerations undergo change according to the following equations:

$$\begin{aligned}
\mathbf{r}_n &= \mathbf{r}_n\left(q_1, \ldots, q_K, t\right), \quad n = 1, \ldots, N, \\
\mathbf{v} \equiv \dot{\mathbf{r}}_n &= \frac{\partial \mathbf{r}_n}{\partial q_k}\dot{q}_k + \frac{\partial \mathbf{r}_n}{\partial t}, \quad k = 1, \ldots, K, \\
\mathbf{a}_n \equiv \ddot{\mathbf{r}}_n &= \frac{\partial \mathbf{r}_n}{\partial q_k}\ddot{q}_k + \frac{\partial^2 \mathbf{r}_n}{\partial q_k \partial q_m}\dot{q}_k\dot{q}_m + 2\frac{\partial^2 \mathbf{r}_n}{\partial q_k \partial t}\dot{q}_k + \frac{\partial^2 \mathbf{r}_n}{\partial t^2},
\end{aligned} \tag{3.19}$$

where K denotes the minimum number of independent generalized coordinates.

We aim to express non-holonomic constraints (3.17) in terms of generalized coordinates. Substituting the first two equations of (3.19) into (3.17) we obtain

$$c_{m_2 n}(q_1,\ldots,q_K,t)\left(\frac{\partial \mathbf{r}_n}{\partial q_k}\dot{q}_k+\frac{\partial \mathbf{r}_n}{\partial t}\right)+\beta_{m_2}(q_1,\ldots,q_K,t)=0,$$
$$n=1,\ldots,N,\quad m_2=1,\ldots,M_2, \tag{3.20}$$

which can be transformed into the following form:

$$B_{m_2 k}(q_1,\ldots,q_K,t)\,\dot{q}_k+b_{m_2}(q_1,\ldots,q_K,t)=0,\ m_2=1,\ldots,M_2, \tag{3.21}$$

where

$$B_{m_2 k}=\frac{\partial \mathbf{r}_n}{\partial q_k}c_{m_2 n},\quad b_{m_2}=\frac{\partial \mathbf{r}_n}{\partial t}c_{m_2 n}+\beta_{m_2}, \tag{3.22}$$

and the summation convention applies in the preceding equations (and further).

In the case of both holonomic and non-holonomic systems, generalized coordinates can assume arbitrary values. As far as generalized velocities $\dot{q}_k$ are concerned, they are arbitrary only in the case of a holonomic system. In the case of a non-holonomic system, generalized velocities satisfy system of equations (3.21).

Proceeding in a similar way, one may differentiate with respect to time equations (3.17) and (3.18) yielding constraints of accelerations of a DMS introduced respectively by holonomic and non-holonomic constraints (see, e.g., [11]).

Virtual displacements $\delta\mathbf{r}_n$ can be expressed through generalized virtual displacements δq_k in the following way:

$$\delta\mathbf{r}_n=\frac{\partial \mathbf{r}_n}{\partial q_k}\delta q_k,\quad n=1,\ldots,N,\quad k=1,\ldots,K. \tag{3.23}$$

In the case of a holonomic system, variations δq_k are arbitrary, whereas in the case of a non-holonomic system [see (3.21) and (3.22)], they are related to each other by the following equations:

$$\sum_{k=1}^{K_2} B_{m_2 k}\delta q_k=0,\quad m_2=1,\ldots,M_2. \tag{3.24}$$

From the preceding calculations it follows that for holonomic systems the number of degrees of freedom is equal to the number of independent generalized coordinates. In the case of non-holonomic systems, the number of degrees of freedom is equal to $K_2 = 3N - M_2$, where, according to the previous notation, K_2 denotes the number of generalized coordinates, whereas M_2 corresponds to the number of non-integrable constraints.

Let us present now the physical interpretation of the obtained (3.17) and (3.18) in the number of $M = M_1 + M_2$. These equations impose M constraints on

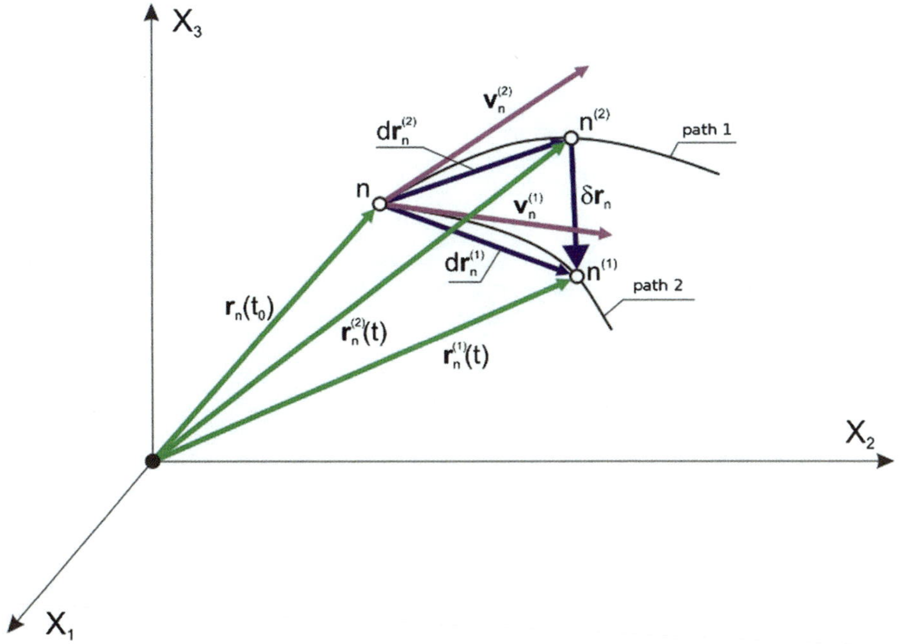

Fig. 3.5 A graphical illustration of *possible paths, displacements, and velocities* of particle n

the velocities $\dot{\mathbf{r}}_n$, $n = 1, \ldots, N$, of the considered system since at every time instant the velocities have to satisfy the aforementioned differential equations, and $M_1 + M_2 < 3N$.

If we considered only first-order ordinary differential equations (3.17) and (3.18) without the remaining $3N - M$ second-order differential equations, then, in general, there would be *infinitely many choices of velocities* $\dot{\mathbf{r}}_1, \dot{\mathbf{r}}_2, \ldots, \dot{\mathbf{r}}_N$ such that the differential equations of constraints would be satisfied [5]. Thus, for a fixed ("frozen") time instant there exist infinitely many possibilities from which to choose the set of displacements and velocities of all the particles of the DMS, and we call such displacements and velocities of the particles *possible velocities* and *possible displacements* [5].

This will be explained based on the notion of variation introduced in Sect. 1.4. Let us consider the *possible* motion of one of the particles of the DMS depicted in Fig. 3.5. In the figure are shown two possibilities of motion of particle n along possible paths 1 and 2.

According to Fig. 3.5, at the time instant $t = t_0 + \mathrm{d}t$ particle n can undergo a displacement $\mathrm{d}\mathbf{r}_n^{(1)}$ or $\mathrm{d}\mathbf{r}_n^{(2)}$ and end up in positions described by radius vectors $\mathbf{r}_n^{(1)}$ and $\mathbf{r}_n^{(2)}$. Let us note that

$$\delta\mathbf{r}_n = \mathrm{d}\mathbf{r}_n^{(1)} - \mathrm{d}\mathbf{r}_n^{(2)}. \tag{3.25}$$

Multiplying (3.17) and (3.18) by dt we obtain

$$\sum_{n=1}^{N} \frac{\partial f_m}{\partial \mathbf{r}_n}\mathbf{r}_n + \frac{\partial f_m}{\partial t}\mathrm{d}t = 0, \quad m_1 = 1, \ldots, M_1,$$

$$\sum_{n=1}^{N} \boldsymbol{c}_{m_2 n}\mathrm{d}\mathbf{r}_n + \beta_{m_2}\mathrm{d}t = 0, \quad m_2 = 1, \ldots, M_2, \tag{3.26}$$

and the preceding equations describe the increments of possible displacements $\mathrm{d}\mathbf{r}_n$ or possible displacements of vector $\mathbf{r}_n$.

Following substitution into (3.26) of possible displacements $\mathrm{d}\mathbf{r}_n^{(1)}$ and $\mathrm{d}\mathbf{r}_n^{(2)}$, and subtracting by sides the obtained equations for $m_1 = 1, \ldots, M_1$ and $m_2 = 1, \ldots, M_2$, and then taking into account (3.25), we obtain

$$\sum_{n=1}^{N} \frac{\partial f_m}{\partial \mathbf{r}_n}\delta\mathbf{r}_n = 0, \quad m_1 = 1, \ldots, M_1, \tag{3.27}$$

$$\sum_{n=1}^{N} \boldsymbol{c}_{m_2 n}\delta\mathbf{r}_n = 0, \quad m_2 = 1, \ldots, M_2. \tag{3.28}$$

The just obtained (3.27) and (3.28) do not contain terms with the time differential dt. Therefore, we will call variations $\delta\mathbf{r}_n$ *virtual displacements*. The latter will be coincident with *possible displacements* for the so-called "freezing" of constraints. The left-hand side of (3.27) can be represented as

$$\delta f_m = \sum_{n=1}^{N} \frac{\partial f_m}{\partial \mathbf{r}_n}\delta\mathbf{r}_n, \tag{3.29}$$

which is a variation of the function $f_m = f_m(\mathbf{r}_1, \mathbf{r}_2, \ldots, \mathbf{r}_n, t)$, but for the "frozen" time instant t, since according to the definition of variation δf_m denotes the deviation (variation) of the function f_m for the fixed ("frozen") time instant t. As was already mentioned, the imposition of constraints on a DMS has the effect that the particles of the DMS are not allowed to move arbitrarily in $\mathbf{R}^3$ space but move on certain surfaces conventionally called *constraint surfaces*.

If we are faced with scleronomic constraints, then the previously mentioned surface is fixed. In this case the differentials of vectors of possible displacements $\mathrm{d}\mathbf{r}_n^{(1)}$ and $\mathrm{d}\mathbf{r}_n^{(2)}$ and the vectors of virtual displacements $\delta\mathbf{r}_n$ are tangent to the surface of constraints.

Now, let us assume that we are dealing with rheonomic constraints. Then the previously mentioned vectors $\mathrm{d}\mathbf{r}_n^{(1)}$, $\mathrm{d}\mathbf{r}_n^{(2)}$, and $\delta\mathbf{r}_n$ are no longer tangent to the constraint surface, since this surface moves with a certain velocity, which we will denote by $\mathbf{v}$. In this case the vectors of possible velocities of particles $n^{(1)}$ and $n^{(2)}$ are as follows:

$$\dot{\mathbf{r}}_n^{(1)} = \mathbf{v} + \dot{\mathbf{r}}_{nt}^{(1)}, \quad \dot{\mathbf{r}}_n^{(2)} = \mathbf{v} + \dot{\mathbf{r}}_{nt}^{(2)}, \tag{3.30}$$

where $\dot{\mathbf{r}}_{nt}^{(1)}$ and $\dot{\mathbf{r}}_{nt}^{(2)}$ denote the velocities tangent to the surface moving with velocity $\mathbf{v}$.

According to (3.25) and (3.30), we have

$$\delta\mathbf{r}_n = \mathrm{d}\mathbf{r}_n^{(1)} - \mathrm{d}\mathbf{r}_n^{(2)} = \left(\dot{\mathbf{r}}_n^{(1)} - \dot{\mathbf{r}}_n^{(2)}\right)\mathrm{d}t = \left(\dot{\mathbf{r}}_{nt}^{(1)} - \dot{\mathbf{r}}_{nt}^{(2)}\right)\mathrm{d}t,$$

which means that vector $\delta\mathbf{r}_n$ would be coincident with *possible displacements* for *frozen constraints*, that is, for a non-moving surface. In this case one should prepare constraints in a certain way, that is, "freeze" them. Therefore, we also call the *virtual displacements "prepared displacements"* in mechanics.

If a particle moves on a smooth constraint surface, then we have

$$\mathbf{F}_n^R \circ \delta\mathbf{r}_n = 0, \tag{3.31}$$

where $\mathbf{F}_n^R$ is the reaction *normal* to the smooth (i.e., frictionless) surface at the position of particle n. In the case of a fixed constraint surface the total differential $\mathrm{d}\mathbf{r}_n = \delta\mathbf{r}_n$, whereas in the case of a movable constraint surface $\mathrm{d}\mathbf{r}_n \neq \delta\mathbf{r}_n$, but "freezing" of the constraints allows for the satisfaction of condition (3.31).

In mechanics a postulate of so-called *ideal constraints* is introduced. Adding (3.31) together for $n = 1, \ldots, N$ we get

$$\sum_{n=1}^{N} \mathbf{F}_n^R \circ \delta\mathbf{r}_n = 0, \tag{3.32}$$

which expresses this postulate in a proper way. The postulate means that in the case of ideal constraints the sum of works of all reaction forces $\mathbf{F}_n^R$ during virtual displacements $\delta\mathbf{r}_n$ (at any time instant) is equal to zero.

The introduction of such a postulate was necessary because of the need to determine unknown reactions $\mathbf{F}_n^R$. Let us then consider a constrained DMS. For each particle n the following equation is satisfied, which results from Newton's second law:

$$m_n\ddot{\mathbf{r}}_n = \mathbf{F}_n + \mathbf{F}_n^i + \mathbf{F}_n^R, \quad n = 1, \ldots, N, \tag{3.33}$$

and additionally M_1 equations of geometric constraints of the form

$$f_{m_1}(\mathbf{r}_1, \mathbf{r}_2, \ldots, \mathbf{r}_N, t), \quad m_1 = 1, \ldots, M_1 \tag{3.34}$$

and M_2 equations of kinematic constraints of the form

$$c_{m_2 n}(\mathbf{r}_1, \mathbf{r}_2, \ldots, \mathbf{r}_n, t) \circ \mathrm{d}\dot{\mathbf{r}}_n + \beta_{m_2}(\mathbf{r}_1, \mathbf{r}_2, \ldots, \mathbf{r}_N, t) = 0,$$
$$n = 1, \ldots, N, \quad m_2 = 1, \ldots, M_2. \tag{3.35}$$

In general, we have $6N + M_1 + M_2$ scalar first-order differential equations (3.33)–(3.35), $3N$ unknowns x_{1n}, x_{2n}, x_{3n}, and $3N$ unknowns $F^R_{nx_1}$, $F^R_{nx_2}$, $F^R_{nx_3}$, that is, $6N$ in total. This problem will be described more broadly and clarified during the derivation of Lagrange's equations.

Example 3.2. Determine the type of constraints of the systems shown in Figs. 3.2 and 3.3.

In the case of a simple pendulum, the coordinates of a particle of mass m satisfy the following restrictions:

$$f_1 \equiv x_3 = 0, \quad f_2 \equiv x_1^2 + x_2^2 + x_3^2 - l^2 = 0.$$

According to (3.13) they are geometric constraints, and from the foregoing system of equations we obtain an equation of the path of a particle of mass m, which is the circle

$$x_1^2 + x_2^2 = l^2.$$

In the case of a double mathematical pendulum, the restrictions imposed on the motion of particles of masses m_1 (particle A_1) and m_2 (particle A_2) have the form

$$\begin{aligned}
&f_1 \equiv x_{3,A_1} = 0,\\
&f_2 \equiv x_{1A_1}^2 + x_{2A_1}^2 + x_{3A_1}^2 - l_1^2 = 0,\\
&f_3 \equiv x_{3A_2} = 0,\\
&f_4 \equiv (x_{1A_2} - x_{1A_1})^2 + (x_{2A_2} - x_{2A_1})^2 + (x_{3A_2} - x_{3A_1})^2 - l^2 = 0.
\end{aligned}$$

The problem of determining the path of the point $A(x_{1A_1}, x_{2A_1}, x_{3A_1}, x_{1A_2}, x_{2A_2}, x_{3A_2})$, which represents a double mathematical pendulum, boils down to the solution of following two non-linear algebraic equations:

$$\begin{aligned}
&x_{1A_1}^2 + x_{2A_1}^2 = l_1^2,\\
&(x_{1A_2} - x_{1A_1})^2 + (x_{2A_2} - x_{2A_1})^2 = l_2^2.
\end{aligned}$$

In both cases we are dealing with *time-independent constraints*. However, if we had $l = l(t)$, $l_1 = l_1(t)$, $l_2 = l_2(t)$, where these time functions are given in advance, then the constraints would be *time-dependent constraints*.

According to (3.13), also here we have imposed geometric constraints. Those equations admit the following physical interpretation. Particle A_1 is always at the distance l_1 (in plane of the drawing) from point O. In turn, the other particle, A_2, lies with respect to A_1 on a circle of radius l_2 whose center is at point A_1. The surface on which point A lies is the surface of a torus, and on that surface two coordinates (φ_1, φ_2) suffice to describe the position of point A. □

Below we will consider an example of geometric and kinematic constraints imposed on the motion of a system of two particles.

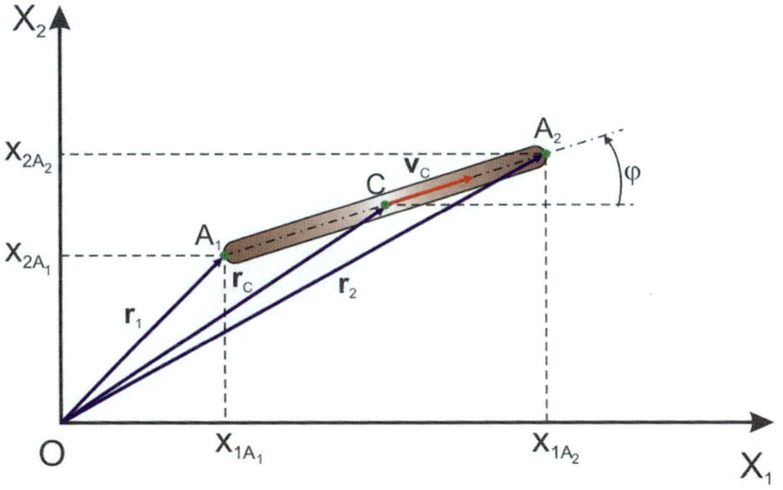

Fig. 3.6 Motion of a rod A_1A_2 on plane OX_1X_2

Example 3.3. Let two points A_1 and A_2 be connected by a massless rigid rod of length l $\left(\left|\overrightarrow{A_1A_2}\right| = l\right)$ and let them move on the plane OX_1X_2 (Fig. 3.6). On those two points are imposed constraints in such way that the velocity vector of the center of segment A_1A_2, denoted by $\mathbf{v}_C$, coincides with the rod axis (see also [5]).

A characteristic of such motion is that the velocity $\mathbf{v}_C$ of the center of segment A_1A_2 is always directed along vector $\overrightarrow{A_1A_2}$, that is,

$$\mathbf{v}_C = \lambda \overrightarrow{A_1A_2}.$$

Because

$$\mathbf{r}_2 = \mathbf{r}_1 + \overrightarrow{A_1A_2}$$

and

$$\mathbf{r}_C = \mathbf{r}_1 + \overrightarrow{A_1C},$$

$$\mathbf{r}_C = \mathbf{r}_2 - \overrightarrow{CA_2},$$

adding the preceding equations we have

$$2\mathbf{r}_C = \mathbf{r}_1 + \mathbf{r}_2 + \overrightarrow{A_1C} - \overrightarrow{CA_2},$$

that is,

$$\mathbf{r}_C = \frac{\mathbf{r}_1 + \mathbf{r}_2}{2},$$

and eventually, we obtain the vector equation

$$\mathbf{v}_C \equiv \frac{\dot{\mathbf{r}}_1 + \dot{\mathbf{r}}_2}{2} = \lambda(\mathbf{r}_2 - \mathbf{r}_1).$$

Multiplying the preceding equation in turn by unit vectors $\mathbf{E}_2$ and $\mathbf{E}_1$, and then dividing the first obtained scalar equation by the second, we arrive at

$$\frac{\dot{x}_{2A_1} + \dot{x}_{2A_2}}{\dot{x}_{1A_1} + \dot{x}_{1A_2}} = \frac{x_{2A_2} - x_{2A_1}}{x_{1A_2} - x_{1A_1}}.$$

Multiplying out the preceding equation we have

$$\dot{x}_{2A_1}(x_{1A_2} - x_{1A_1}) + \dot{x}_{2A_2}(x_{1A_2} - x_{1A_1}) + \dot{x}_{1A_1}(x_{2A_2} - x_{2A_1}) \\ + \dot{x}_{1A_2}(x_{2A_2} - x_{2A_1}) = 0.$$

The obtained equation describes *kinematic constraints*. It is not possible to integrate them directly to obtain geometric constraints, so they are non-holonomic constraints.

An additional equation of (geometric) constraints is (obviously the motion takes place in the plane OX_1X_2, and consequently $x_{3A_1} = x_{3A_2} = 0$)

$$\left(\overrightarrow{A_1A_2}\right)^2 = l^2,$$

that is,

$$(x_{1A_1} - x_{1A_2})^2 + (x_{2A_1} - x_{2A_2})^2 - l^2 = 0. \qquad \square$$

We will now explain in more detail the physical interpretation of the introduction of generalized coordinates $q_1, \ldots, q_{K_1}$ [see (3.11)] on the assumption that geometric constraints occur in the system.

Equations of geometric constraints (3.13) allow for the elimination of M_1 coordinates, and consequently the $K_1 = 3N - M_1$ coordinates remain in the equations. Following the introduction of generalized coordinates $q_1, q_2, \ldots, q_{3N-M_1}$, each of Cartesian coordinates $x_{1n}, x_{2n}, x_{3n}, n = 1, \ldots, N$, can be expressed through those generalized coordinates according to formula (3.11). Let us consider this statement based on Example 3.3. On the motion of particles A_1 and A_2 are imposed the following constraints:

$$\begin{aligned} x_{3A_1} &= 0, \\ x_{3A_2} &= 0, \\ (x_{1A_1} - x_{1A_2})^2 + (x_{2A_1} - x_{2A_2})^2 &= l^2, \end{aligned} \tag{3.36}$$

and we have then $N = 2,\ M_1 = 3$, that is, $K_1 = 3N - M_1 = 3$. It follows that the system demands that we introduce three generalized coordinates $q_1 = x_{1C}$, $q_2 = x_{2C}$, $q_3 = \varphi$. According to Fig. 3.6 we have

$$x_{1A_1} = q_1 - \frac{l}{2}\cos q_3, \quad x_{2A_1} = q_2 - \frac{l}{2}\sin q_3,$$
$$x_{1A_2} = q_1 + \frac{l}{2}\cos q_3, \quad x_{2A_2} = q_2 + \frac{l}{2}\sin q_3. \tag{3.37}$$

Substituting expressions (3.37) into the last equation of (3.36) we obtain

$$(l\sin q_3)^2 + (l\cos q_3)^2 = l^2, \tag{3.38}$$

which is identically satisfied.

From that follows the conclusion that the introduction of generalized coordinates q_1, q_2, and q_3 allowed constraints (3.36) to be identically satisfied. One may check that the equation describing kinematic constraints of the form

$$\dot{x}_{2A_1}(x_{1A_2} - x_{1A_1}) + \dot{x}_{2A_2}(x_{1A_2} - x_{1A_1})$$
$$+ \dot{x}_{1A_1}(x_{2A_2} - x_{2A_1}) + \dot{x}_{1A_2}(x_{2A_2} - x_{2A_1}) = 0 \tag{3.39}$$

will not be satisfied.

Based on earlier calculations [see relations (3.21) and (3.22)] we have

$$B_{m_2k} = \sum_{n=1}^{N}\left(\mathbf{c}_{m_2n}\frac{\partial \mathbf{r}_n}{\partial q_k}\right)$$
$$= \sum_{n=1}^{N}\left(c_{x_1m_2n}\frac{\partial x_{1n}}{\partial q_k} + c_{x_2m_2n}\frac{\partial x_{2n}}{\partial q_k} + c_{x_3m_2n}\frac{\partial x_{3n}}{\partial q_k}\right),$$
$$b_{m_2} = \sum_{n=1}^{N}\left(\mathbf{c}_{m_2n}\frac{\partial \mathbf{r}_n}{\partial t}\right) + \beta_{m_2}$$
$$= \sum_{n=1}^{N}\left(c_{x_1m_2n}\frac{\partial x_{1n}}{\partial t} + c_{x_2m_2n}\frac{\partial x_{2n}}{\partial t} + c_{x_3m_2n}\frac{\partial x_{3n}}{\partial t}\right) + \beta_{m_2}. \tag{3.40}$$

From (3.37) we obtain

$$\begin{aligned}
c_{x_1 11} &= x_{2A_1} - x_{2A_2} = -l \sin q_3, \\
c_{x_2 11} &= x_{1A_2} - x_{1A_1} = l \cos q_3, \\
c_{x_3 11} &= 0, \\
c_{x_1 12} &= x_{2A_1} - x_{2A_2} = -l \sin q_3, \\
c_{x_2 12} &= x_{1A_2} - x_{1A_1} = l \cos q_3, \\
c_{x_3 12} &= 0.
\end{aligned} \tag{3.41}$$

We have one equation of kinematic constraints, that is, $m_2 = M_2 = 1$.

From (3.40) it follows that $b_1 = 0$, because $\frac{\partial x_{1n}}{\partial t} = \frac{\partial x_{2n}}{\partial t} = \frac{\partial x_{3n}}{\partial t} = 0$ and $\beta_1 = 0$, and to determine coefficients B_{11}, B_{12}, and B_{13} the following partial derivatives are needed:

$$\begin{aligned}
&\frac{\partial x_{1A_1}}{\partial q_1} = 1, \frac{\partial x_{1A_1}}{\partial q_2} = 0, \frac{\partial x_{1A_1}}{\partial q_3} = \frac{l}{2} \sin q_3, \\
&\frac{\partial x_{2A_1}}{\partial q_1} = 0, \frac{\partial x_{2A_1}}{\partial q_2} = 1, \frac{\partial x_{2A_1}}{\partial q_3} = -\frac{l}{2} \cos q_3, \\
&\frac{\partial x_{1A_2}}{\partial q_1} = 1, \frac{\partial x_{1A_2}}{\partial q_2} = 0, \frac{\partial x_{1A_2}}{\partial q_3} = -\frac{l}{2} \sin q_3, \\
&\frac{\partial x_{2A_2}}{\partial q_1} = 0, \frac{\partial x_{2A_2}}{\partial q_2} = 1, \frac{\partial x_{2A_2}}{\partial q_3} = \frac{l}{2} \cos q_3.
\end{aligned} \tag{3.42}$$

The unknown coefficients are equal to

$$\begin{aligned}
B_{11} &= c_{x_1 11} \frac{\partial x_{1A_1}}{\partial q_1} + c_{x_1 12} \frac{\partial x_{1A_1}}{\partial q_1} = -2l \sin q_3, \\
B_{12} &= c_{x_2 11} \frac{\partial x_{2A_1}}{\partial q_2} + c_{x_1 12} \frac{\partial x_{2A_2}}{\partial q_2} = 2l \cos q_3, \\
B_{13} &= 0.
\end{aligned} \tag{3.43}$$

Finally, (3.21) takes the form

$$B_{11}\dot{q}_1 + B_{12}\dot{q}_2 = -2l\dot{q}_1 \sin q_3 + 2l\dot{q}_2 \cos q_3 = 0,$$

that is,

$$\dot{q}_2 = \dot{q}_1 \tan q_3 \tag{3.44}$$

Let us count the degrees of freedom of the analyzed system. We have $N = 2$, $M_1 = 3$, $M_2 = 1$, from which it follows that $W = 3N - (M_1 + M_2) = 2$, but we have three generalized coordinates (q_1, q_2, q_3). The number of generalized coordinates exceeds the number of degrees of freedom, but there is an additional equation (3.44) resulting from the existence of kinematic (non-holonomic) constraints.

In holonomic systems the number of degrees of freedom is always equal to the number of generalized coordinates.

If in Example 3.3 we substitute the rod by a skate, then we remove the possibility of both motion perpendicular to the blade and rotation about the vertical axis, so the skate will only be able to slide along the direction of the blade, and if it touches the ice surface, then it will move on that surface along the blade. If we relate this problem to the generalized coordinates of the blade in motion on the plane OX_1X_2, $\mathbf{q} = [x_{1C}, x_{2C}, \varphi]^{\mathrm{T}}$ (coordinates x_{1C}, x_{2C} and φ are dependent), then the equations of constraints take the form $\varphi - \varphi_0 = 0$ and $\dot{x}_{2C} = \dot{x}_{1C} \tan\varphi$, where the constant $\varphi_0 = \mathrm{const}$ is determined from the skate initial position on the ice. Following integration of the second equation we obtain $x_{2C} - x_{1C} \tan\varphi_0 + c = 0$, where constant c is also defined by initial conditions of the skate motion. The motion of point C takes place along a rectilinear path described by the initial conditions $x_{1C}(t_0)$, $x_{2C}(t_0)$, and φ_0. The skate has one degree of freedom, and its motion can be described by one independent variable.

If in Example 3.3 we replace the rod by a hockey skate (which is rounded and makes contact with the ice at point C) whose sliding is blocked along the direction transversal to the blade but can rotate about the vertical axis passing through point C, then we have three independent generalized coordinates x_{1C}, x_{2C}, and φ. The only non-holonomic constraints imposed on the motion of constraints are now the non-holonomic constraints $\dot{x}_{2C} = \dot{x}_{1C} \tan\varphi$.

On the surfaces described by the preceding equations, that is, on the surfaces of constraints, let us take an arbitrary point A whose radius vector is described by the equation $\mathbf{r}_A = (x_{1,1}, x_{2,1}, x_{3,1}, \ldots, x_{1,N}, x_{2,N}, x_{3,N})$. Through that point we draw an arbitrary smooth curve lying on the constraint surface, where the parameter s plays the role of a variable parameterizing this curve. The tip of the vector $\mathbf{r}(s) = (x_{1,1}(s), x_{2,1}(s), x_{3,1}(s), \ldots, x_{1,N}(s), x_{2,N}(s), x_{3,N}(s))$ determines the hodograph, that is, it lies on the curve. Let $\mathbf{r}(0) = \mathbf{r}_A$. The vector $\frac{\mathrm{d}\mathbf{r}(s)}{\mathrm{d}s}\big|_{s=0} \equiv \mathbf{r}'(0) = (\mathbf{r}'_1(0), \ldots, \mathbf{r}'_N(0))$ is tangent at point $\mathbf{r}_A$ to the curve lying on the constraint surface. There is a possible total of $K = 3N - M$ curves passing through point $\mathbf{r}_A$. All tangent vectors to these points form a K-dimensional vector space called a *tangent space* at point A.

The velocity vector $\dot{\mathbf{r}} = (\dot{\mathbf{r}}_1, \ldots, \dot{\mathbf{r}}_N)$ associated with every trajectory of motion lies in the tangent space.

Constraint surfaces are a certain part of the configuration space $\mathbf{R}^{3N}$. A particle (if it has constraints imposed on it) moves on a completely different surface than the space $\mathbf{R}^{3N}$, which is three-dimensional, infinite, and flat. If we consider a classic example of motion of a particle on a smooth spherical surface, then the configuration space is two-dimensional, finite, and curved.

Let one coordinate q_k change by a *virtual increment* δq_k. The radius vectors describing the positions of all the remaining points will change by

$$\delta \mathbf{r}_n = \delta \mathbf{r}_n (q_1, \ldots, q_K) = \frac{\partial \mathbf{r}_n}{\partial q_k} \delta q_k, \tag{3.45}$$

where in this case the summation convention does not apply.

The displacements of all the particles of the investigated mechanical system result from the action of external $\mathbf{F}_n$ and internal $\mathbf{F}_n^i$ forces, which perform certain *elementary works*. The sum of those elementary works is equal to

$$\sum_{n=1}^{N} \left(\mathbf{F}_n + \mathbf{F}_n^i\right) \circ \delta \mathbf{r}_n = \sum_{n=1}^{N} \left(\mathbf{F}_n + \mathbf{F}_n^i\right) \circ \frac{\partial \mathbf{r}_n}{\partial q_k} \delta q_k = Q_k \delta q_k = \delta W_k. \tag{3.46}$$

From the preceding equation it follows that the sum of virtual works of forces $(\mathbf{F}_n + \mathbf{F}_n^i)$, $n = 1, \ldots, N$, is equal to the work done by the so-called *generalized force Q_k during virtual displacement δq_k*.

Definition 3.1. *A physical quantity Q_k multiplied by a virtual increment δq_k of the generalized coordinate q_k is equal to the work done by the system of forces acting on the considered mechanical system during virtual displacements generated as a result of the increment of this generalized coordinate and is called the* generalized force.

From (3.46) it follows that

$$Q_k = \frac{\delta W_k}{\delta q_k} \tag{3.47}$$

and that Q_k corresponds to the generalized coordinates q_k.

The presented approach is often applied in practical calculations, where the mechanical system is intentionally subjected to the virtual displacement such that $\delta q_j = 0$ for $j \neq k$.

Let us now impose virtual increments δq_k on all generalized coordinates $q_1, \ldots, q_N$. They cause the virtual increments of each of the radius vectors $\mathbf{r}_n$ of the form

$$\delta \mathbf{r}_n = \frac{\partial \mathbf{r}_n}{\partial q_1} \delta q_1 + \frac{\partial \mathbf{r}_n}{\partial q_2} \delta q_2 + \cdots + \frac{\partial \mathbf{r}_n}{\partial q_K} \delta q_K = \frac{\partial \mathbf{r}_n}{\partial q_k} \delta q_k, \quad k = 1, \ldots, K. \tag{3.48}$$

There exists a certain arbitrariness of choice of generalized coordinates for each of the considered mechanical systems. Generalized coordinates allow for a complete geometrical description of the mechanical system being analyzed with respect to the introduced frame of reference. Sometimes, however, they do not have a straightforward physical interpretation. For instance, in Fig. 3.3 were introduced the so-called absolute angles φ_1 and φ_2 describing the motion of a double pendulum, but the so-called relative angles φ_1 and ψ_1 (where $\psi_1 = \varphi_2 - \varphi_1$), which are also generalized coordinates, might have been introduced as well.

3.2 Variational Principles of Jourdain and Gauss

Differentiation of (3.13) and (3.15) leads to the following equations:

$$\frac{\partial f_{m_1}}{\partial \mathbf{r}_n}\mathbf{a}_n + \frac{\partial^2 f_{m_1}}{\partial \mathbf{r}_n \partial \mathbf{r}_l}\mathbf{v}_l\mathbf{v}_n + 2\frac{\partial^2 f_{m_1}}{\partial t \partial \mathbf{r}_n}\mathbf{v}_n + \frac{\partial^2 f_{m_1}}{\partial t^2} = 0,$$

$$m_1 = 1, \ldots, M_1, \tag{3.49}$$

$$\boldsymbol{c}_{m_2 n}\mathbf{a}_n + \frac{\partial \boldsymbol{c}_{m_2 n}}{\partial \mathbf{r}_l}\mathbf{v}_l\mathbf{v}_n + \frac{\partial \boldsymbol{c}_{m_2 n}}{\partial t}\mathbf{v}_n + \frac{\partial \beta_{m_2}}{\partial \mathbf{r}_n}\mathbf{r}_n + \frac{\partial \beta_{m_2}}{\partial t} = 0\ ,$$

$$m_2 = 1, \ldots, M_2, \tag{3.50}$$

where the summation convention applies.

Because we assume the expression $3N - (M_1 + M_2)$ is positive (we are dealing with motion), for a given fixed time instant $t = t^*$ the system is in one position (configuration) described by the radius vectors $\mathbf{r}_n = \mathbf{r}_n^*$, but the sets of their velocities $\mathbf{v}_n = \mathbf{v}_n^*$ and accelerations $\mathbf{a}_n = \mathbf{a}_n^*$ can be chosen arbitrarily in infinitely many ways according to the imposed constraints.

The neighboring configuration of the system at the time instant $t^* + \Delta t$ is described by the radius vectors $\mathbf{r}_n^* + \Delta\mathbf{r}_n$, where $\Delta\mathbf{r}_n$ are *possible displacements* realized in time interval Δt. If Δt is small enough, we can determine displacements from the following equation:

$$\Delta\mathbf{r}_n = \mathbf{v}_n^*\Delta t + \frac{1}{2}\mathbf{a}_n^*(\Delta t)^2 + \cdots . \tag{3.51}$$

Because the sets $\mathbf{v}_n^*$ and $\mathbf{a}_n^*$ can be chosen arbitrarily in infinitely many ways (they are infinite sets), $\Delta\mathbf{r}_n$ is also an infinite set. Let us take two different virtual displacements (variations) corresponding to the same quantity Δt, which according to formula (3.51) are equal to

$$\Delta_1\mathbf{r}_n = \mathbf{v}_{n1}^*\Delta t + \frac{1}{2}\mathbf{a}_{n1}^*(\Delta t)^2 + \cdots ,$$

$$\Delta_2\mathbf{r}_n = \mathbf{v}_{n2}^*\Delta t + \frac{1}{2}\mathbf{a}_{n2}^*(\Delta t)^2 + \cdots . \tag{3.52}$$

Let us multiply (3.18) and (3.17) by Δt and use in turn (3.52). Next, after subtracting the corresponding equations from each other we obtain

$$\frac{\partial f_{m_1}}{\partial \mathbf{r}_n}\left(\mathbf{v}_{n1}^* - \mathbf{v}_{n2}^*\right)\Delta t = 0, \quad m_1 = 1, \ldots, M_1, \tag{3.53}$$

$$\boldsymbol{c}_{m_2 n}\left(\mathbf{v}_{n1}^* - \mathbf{v}_{n2}^*\right)\Delta t = 0, \quad m_2 = 1, \ldots, M_2. \tag{3.54}$$

In a similar way we proceed with (3.49) and (3.50), but now they are multiplied by $\frac{1}{2}(\Delta t)^2$, then subtracted from each other, to arrive at the equations

$$\frac{\partial f_{m_1}}{\partial \mathbf{r}_n}\left(\mathbf{a}_{n1}^* - \mathbf{a}_{n2}^*\right)\frac{(\Delta t)^2}{2} + \left[\left(\frac{\partial^2 f_{m_1}}{\partial \mathbf{r}_n \partial \mathbf{r}_l}\mathbf{v}_{l1}^*\right)\mathbf{v}_{n1}^* - \left(\frac{\partial^2 f_{m_1}}{\partial \mathbf{r}_n \partial \mathbf{r}_l}\mathbf{v}_{l2}^*\right)\mathbf{v}_{n2}^*\right]\frac{(\Delta t)^2}{2}$$
$$+2\frac{\partial^2 f_{m_1}}{\partial t\, \partial \mathbf{r}_n}\left(\mathbf{v}_{n1}^* - \mathbf{v}_{n2}^*\right) = 0, \quad m_1 = 1, \ldots, M_1, \tag{3.55}$$

$$\boldsymbol{c}_{m_2 n}\left(\mathbf{a}_{n1}^* - \mathbf{a}_{n2}^*\right)\frac{(\Delta t)^2}{2} + \left[\left(\frac{\partial \boldsymbol{c}_{m_2 n}}{\partial \mathbf{r}_l}\mathbf{v}_{l1}^*\right)\mathbf{v}_{n1}^* - \left(\frac{\partial \boldsymbol{c}_{m_2 n}}{\partial \mathbf{r}_l}\mathbf{v}_{l2}^*\right)\mathbf{v}_{n2}^*\right]\frac{(\Delta t)^2}{2}$$
$$+\frac{\partial \beta_{m_2}}{\partial \mathbf{r}_n}\left(\mathbf{r}_{n1}^* - \mathbf{r}_{n2}^*\right) = 0, \quad m_2 = 1, \ldots, M_2. \tag{3.56}$$

Subtracting by sides (3.52) we obtain

$$\delta \mathbf{r}_n \equiv \Delta_1 \mathbf{r}_n - \Delta_2 \mathbf{r}_n = \delta \mathbf{v}_n \Delta t + \delta \mathbf{a}_n \frac{(\Delta t)^2}{2} + \ldots, \tag{3.57}$$

where

$$\delta \mathbf{v}_n = \mathbf{v}_{n1}^* - \mathbf{v}_{n2}^*,$$
$$\delta \mathbf{a}_n = \mathbf{a}_{n1}^* - \mathbf{a}_{n2}^*, \quad n = 1, \ldots, N. \tag{3.58}$$

Historically, if in (3.57) we limit ourselves to the first approximation with respect to Δt, then we call the virtual displacement given by the equation

$$\delta \mathbf{r}_n = \delta \mathbf{v}_n \Delta t \tag{3.59}$$

the *Jourdain variational principle* [12, 13].[5]

If in (3.57) we assume $\delta \mathbf{v}_n = \mathbf{0}$ and $\delta \mathbf{a}_n \neq \mathbf{0}$, we obtain

$$\delta \mathbf{r}_n = \delta \mathbf{a}_n \frac{(\Delta t)^2}{2}, \tag{3.60}$$

which describes the *Gauss variational principle*.

[5]Philip E. B. Jourdain (1879–1919), English scientist.

The Jourdain variational principle, Gauss variational principle, and the more frequently and earlier applied d'Alembert principle are equivalent, and we use them as appropriate to describe the character of the solved problems. For example, the Jourdain variational principle gives quick results for determination of the selected reaction forces (reaction moments). In simple applications, it is commonly used by admitting the possibility of motion of the mechanical system along the direction of the reaction force that produces this motion, that is, we consider the system as "an instantaneous mechanism," which will be explained in more detail in Example 3.4.

The Gauss variational principle differs significantly from the d'Alembert–Lagrange principle and from the Jourdain variational principle as it is associated with the *notion of an extremum*. According to the previous calculations we assume that at a certain ("frozen") time instant all the particles of a material system will have positions $\mathbf{r}_n = \mathbf{r}_n^*$ and velocities $\mathbf{v}_n = \mathbf{v}_n^*$, $n = 1, \ldots, N$, equal to the actual positions and velocities of the system's particles. The accelerations in the system will undergo variations, and in this case they do not have to be infinitely small. We call such a variation a *Gauss variation*. Substituting (3.60) into the general equation of dynamics we obtain

$$(\mathbf{F}_n - m_n \mathbf{a}_n) \circ \delta \mathbf{a}_n = 0, \tag{3.61}$$

where the summation convention applies. Because forces $\mathbf{F}_n$ do not depend on accelerations $\mathbf{a}_n$ and $m_n = \text{const}$, it is easy to show that (3.61) can be obtained from the relationship for the extremum of function Γ of the form

$$\delta \Gamma = 0, \tag{3.62}$$

where

$$\Gamma = \frac{1}{2} m_n \left(\mathbf{a}_n - \frac{\mathbf{F}_n}{m_n} \right)^2. \tag{3.63}$$

The function $\Gamma = \Gamma(\mathbf{a}_n)$ is sometimes called the *system compulsory function*.

It has a stationary value for $\mathbf{a}_n = \mathbf{a}_n^*$, where $\mathbf{a}_n^*$ are actual accelerations of particles of a material system.

Let Γ^* be a system compulsory function, for actual accelerations. Let us consider the disturbance of actual accelerations of the form

$$\mathbf{a}_n = \mathbf{a}_n^* + \delta \mathbf{a}_n, \quad \delta \mathbf{a}_n > 0, \tag{3.64}$$

where now $\mathbf{a}_n$ denotes kinematically possible accelerations. Substituting formula (3.64) into (3.63) we have

$$\begin{aligned} \Gamma &= \frac{1}{2} m_n \left[(\mathbf{a}_n^* + \delta \mathbf{a}_n) - \frac{\mathbf{F}_n}{m_n} \right]^2 \\ &= \frac{1}{2} m_n \left[\mathbf{a}_n^{*2} + 2\mathbf{a}_n^* \delta \mathbf{a}_n + (\delta \mathbf{a}_n)^2 - 2\mathbf{a}_n^* \frac{\mathbf{F}_n}{m_n} - 2\delta \mathbf{a}_n \frac{\mathbf{F}_n}{m_n} + \left(\frac{\mathbf{F}_n}{m_n} \right)^2 \right], \end{aligned} \tag{3.65}$$

and in turn

$$\Gamma^* = \frac{1}{2}m_n\left(\mathbf{a}_n^* - \frac{\mathbf{F}_n}{m_n}\right)^2 = \frac{1}{2}m_n\left[\mathbf{a}_n^{*2} - 2\mathbf{a}_n^*\frac{\mathbf{F}_n}{m_n} + \left(\frac{\mathbf{F}_n}{m_n}\right)^2\right], \tag{3.66}$$

and from that we calculate

$$\Delta\Gamma = \Gamma - \Gamma^* = \left(m_n\mathbf{a}_n^* - \mathbf{F}_n\right)\delta\mathbf{a}_n + \frac{1}{2}m_n(\delta\mathbf{a}_n)^2. \tag{3.67}$$

From (3.67) it follows that $\Delta\Gamma$ attains its extremum (minimum) value for nonzero $\delta\mathbf{a}_n$ in the case of actual motion, since the first term on the right-hand side of (3.67) vanishes.

The variational principle of Jourdain can be represented in a slightly modified version convenient for direct application also to dynamic problems. We will begin the calculations with the general equation of dynamics

$$\sum_{n=1}^{N}(\mathbf{F}_n - m_n\dot{\mathbf{v}}_n)\circ\delta\mathbf{r}_n = 0. \tag{3.68}$$

Let us note that

$$\delta\mathbf{r}_n = \frac{\delta\mathbf{r}_n}{\delta q_k}\delta q_k = \frac{\delta\mathbf{v}_n}{\delta\dot{q}_k}\delta q_k, \tag{3.69}$$

and taking into account (3.69) in (3.68) we obtain

$$\left(\sum_{n=1}^{N}(\mathbf{F}_n - m_n\dot{\mathbf{v}}_n)\circ\frac{\delta\mathbf{v}_n}{\delta\dot{q}_k}\right)\delta q_k = 0. \tag{3.70}$$

Because δq_k are independent, from (3.70) we obtain the following system of equations:

$$\left(\sum_{n=1}^{N}(\mathbf{F}_n - m_n\dot{\mathbf{v}}_n)\circ\frac{\delta\mathbf{v}_n}{\delta\dot{q}_k}\right) = 0, \qquad k = 1,\ldots,K, \tag{3.71}$$

where K denotes the number of degrees of freedom of the considered DMS and q_k are generalized coordinates. (Note that $\mathbf{F}_n$ in (3.71) are forces that perform work.) If we further assume that $N = \infty$, then we can extend our calculations to the case of a rigid body. Let us introduce the system of local coordinates of origin at the mass center of the rigid body $O = C$ and let the position of material points n be represented by radius vectors $\boldsymbol{\rho}_n$. Let us note that if $N = \infty$, then $K = 6$. The second component of the scalar product in (3.71) can be represented in the form

$$\frac{\delta\mathbf{v}_n}{\delta\dot{q}_k} = \frac{\delta\mathbf{v}_C}{\delta\dot{q}_k} + \frac{\delta\boldsymbol{\omega}}{\delta\dot{q}_k}\times\boldsymbol{\rho}_n, \tag{3.72}$$

which follows from the König theorem.

The velocity of an arbitrary point n is equal to

$$\mathbf{v}_n = \mathbf{v}_C + \boldsymbol{\omega} \times \boldsymbol{\rho}_n, \tag{3.73}$$

and after differentiation with respect to time we obtain

$$\dot{\mathbf{v}}_n = \dot{\mathbf{v}}_C + \dot{\boldsymbol{\omega}} \times \boldsymbol{\rho}_n + \boldsymbol{\omega} \times \dot{\boldsymbol{\rho}}_n = \dot{\mathbf{v}}_C + \dot{\boldsymbol{\omega}} \times \boldsymbol{\rho}_n + \boldsymbol{\omega} \times (\boldsymbol{\omega} \times \boldsymbol{\rho}_n). \tag{3.74}$$

Substituting (3.72) and (3.74) into (3.71) we obtain

$$\begin{aligned}
&\sum_{n=1}^{N} \mathbf{F}_n \circ \frac{\delta \mathbf{v}_n}{\delta \dot{q}_k} - \sum_{n=1}^{N} m_n \dot{\mathbf{v}}_n \circ \frac{\delta \mathbf{v}_n}{\delta \dot{q}_k} = \sum_{n=1}^{N} \mathbf{F}_n \circ \frac{\delta \mathbf{v}_C}{\delta \dot{q}_k} \\
&\quad + \sum_{n=1}^{N} \mathbf{F}_n \circ \left(\frac{\delta \boldsymbol{\omega}}{\delta \dot{q}_k} \times \boldsymbol{\rho}_n \right) - \sum_{n=1}^{N} m_n (\dot{\mathbf{v}}_C + \dot{\boldsymbol{\omega}} \times \boldsymbol{\rho}_n \\
&\quad + \boldsymbol{\omega} \times (\boldsymbol{\omega} \times \boldsymbol{\rho}_n)) \circ \frac{\delta \mathbf{v}_C}{\delta \dot{q}_k} - \sum_{n=1}^{N} m_n \dot{\mathbf{v}}_C \circ \left(\frac{\delta \boldsymbol{\omega}}{\delta \dot{q}_k} \times \boldsymbol{\rho}_n \right) \\
&\quad - \sum_{n=1}^{N} m_n (\dot{\boldsymbol{\omega}} \times \boldsymbol{\rho}_n + \boldsymbol{\omega} \times (\boldsymbol{\omega} \times \boldsymbol{\rho}_n)) \circ \left(\frac{\delta \boldsymbol{\omega}}{\delta \dot{q}_k} \times \boldsymbol{\rho}_n \right) \\
&= (\mathbf{F} - m\dot{\mathbf{v}}_C) \circ \frac{\delta \mathbf{v}_C}{\delta \dot{q}_k} + \sum_{n=1}^{N} \frac{\delta \boldsymbol{\omega}}{\delta \dot{q}_k} \circ (\boldsymbol{\rho}_n \times \mathbf{F}_n) \\
&\quad - \frac{\delta \boldsymbol{\omega}}{\delta \dot{q}_k} \left(\sum_{n=1}^{N} m_n \boldsymbol{\rho}_n \times \dot{\mathbf{v}}_C \right) - \sum_{n=1}^{N} [m_n \boldsymbol{\rho}_n \times (\dot{\boldsymbol{\omega}} \times \boldsymbol{\rho}_n \\
&\quad + \boldsymbol{\omega} \times (\boldsymbol{\omega} \times \boldsymbol{\rho}_n))] \circ \frac{\delta \boldsymbol{\omega}}{\delta \dot{q}_k} = (\mathbf{F} - m\dot{\mathbf{v}}_C) \circ \frac{\delta \mathbf{v}_C}{\delta \dot{q}_k} \\
&\quad - \sum_{n=1}^{N} [m_n \boldsymbol{\rho}_n \times (\dot{\boldsymbol{\omega}} \times \boldsymbol{\rho}_n + \boldsymbol{\omega} \times (\boldsymbol{\omega} \times \boldsymbol{\rho}_n)) - \mathbf{M}_C] \circ \frac{\delta \boldsymbol{\omega}}{\delta \dot{q}_k} = 0,
\end{aligned} \tag{3.75}$$

where we used the relation $\mathbf{a} \circ (\mathbf{b} \times \mathbf{c}) = \mathbf{b}(\mathbf{c} \circ \mathbf{a})$ as well as the following relationships:

$$m = \sum_{n=1}^{N} m_n, \qquad \mathbf{M}_C = \sum_{n=1}^{N} \boldsymbol{\rho}_n \times \mathbf{F}_n, \qquad \sum_{n=1}^{N} m_n \boldsymbol{\rho}_n = \mathbf{0}. \tag{3.76}$$

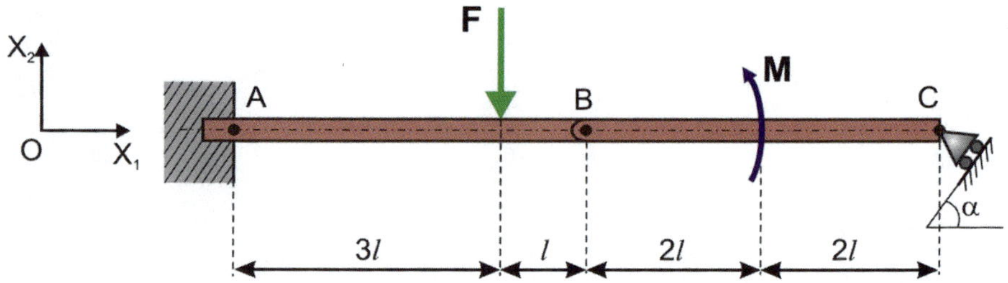

Fig. 3.7 Beam consisting of two pin-joined segments, with built-in support at point A and sliding support at point C situated at angle α with respect to the horizontal

The vector of angular momentum of a body with respect to its mass center equals

$$\begin{aligned}\mathbf{K}_C &= \sum_{n=1}^{N} \boldsymbol{\rho}_n \times m_n \mathbf{v}_n = \sum_{n=1}^{N} m_n \boldsymbol{\rho}_n \times (\mathbf{v}_C + \boldsymbol{\omega} \times \boldsymbol{\rho}_n) \\ &= \sum_{n=1}^{N} m_n \boldsymbol{\rho}_n \times \mathbf{v}_C + \sum_{n=1}^{N} [m_n \boldsymbol{\rho}_n \times (\boldsymbol{\omega} \times \boldsymbol{\rho}_n)] \\ &= \sum_{n=1}^{N} [m_n \boldsymbol{\rho}_n \times (\boldsymbol{\omega} \times \boldsymbol{\rho}_n)], \end{aligned} \tag{3.77}$$

where, during transformations, (3.73) was used.

The time derivative of angular momentum is equal to

$$\begin{aligned}\dot{\mathbf{K}}_C &= \sum_{n=1}^{N} \left[m_n \dot{\boldsymbol{\rho}}_n \times (\boldsymbol{\omega} \times \boldsymbol{\rho}_n) + m_n \boldsymbol{\rho}_n \times \frac{\mathrm{d}}{\mathrm{d}t} (\boldsymbol{\omega} \times \boldsymbol{\rho}_n) \right] \\ &= \sum_{n=1}^{N} [m_n \boldsymbol{\rho}_n \times (\dot{\boldsymbol{\omega}} \times \boldsymbol{\rho}_n) + \boldsymbol{\omega} \times (\boldsymbol{\omega} \times \boldsymbol{\rho}_n)]. \end{aligned} \tag{3.78}$$

Eventually, taking into account (3.78) in (3.75), we obtain

$$(\mathbf{F} - m\mathbf{v}_C) \circ \frac{\partial \mathbf{v}_C}{\partial \dot{q}_k} + (\mathbf{M}_C - \dot{\mathbf{K}}_C) \circ \frac{\partial \boldsymbol{\omega}_C}{\partial \dot{q}_k} = 0, \tag{3.79}$$

which also expresses the variational principle of Jourdain.

Example 3.4. Calculate the reactions of two beams AB and BC of lengths $4l$ connected by a pin joint at point B using the Jourdain variational principle. The geometry and loading of the beam are shown in Fig. 3.7 (see also [13]).

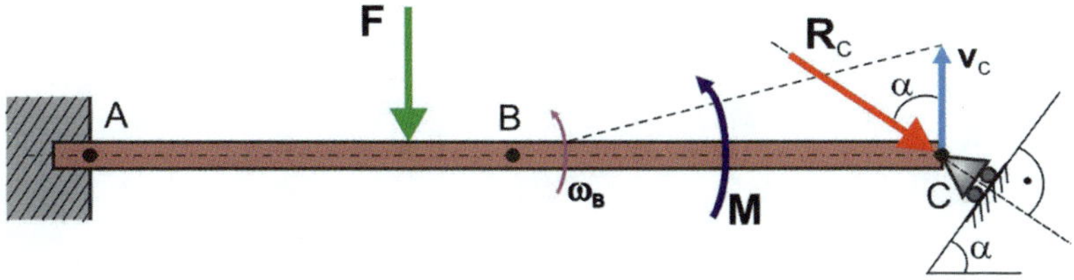

Fig. 3.8 Schematic showing how to determine the reaction $\mathbf{R}_{Ax_1}$

The Jourdain variational principle is associated with virtual velocities and for static cases has the following form:

$$\mathbf{F}_n \circ \delta \dot{\mathbf{r}}_n = 0,$$

and in scalar form, after dropping the variation symbol, we obtain

$$F_{x_{1n}} \dot{x}_{1n} + F_{x_{2n}} \dot{x}_{2n} + F_{x_{3n}} + \dot{x}_{3n} = 0,$$

where the summation convention applies.

The Jourdain variational principle, also known as the *principle of virtual power*, is valid for a material system with ideal, holonomic, and bilateral constraints. During application of this principle one should take into account the principle of independent action of forces (moments). Release from constraints takes place successively depending on the determined reactions. In the adopted coordinate system we admit a single possibility of motion (translation or rotation) and apply the reaction force (reaction moment) at the point associating the force with linear velocity (the moment with angular velocity), which will be presented subsequently.

We replace the actual supports with equivalent supports [13]. Let us calculate the horizontal reaction at point A, that is, $\mathbf{R}_{Ax_1}$. For this purpose we admit the possibility of horizontal motion of the beam, and we replace the action of the built-in support in the horizontal direction with a horizontal slider connected to the beam while at the same time imposing on it the virtual velocity $\mathbf{v}_{Ax_1}$ (Fig. 3.8).

We assume that the force $\mathbf{F}$ ($\mathbf{F} \perp \mathbf{R}_{Ax_1}$) does not influence the horizontal reaction and that the velocity $\mathbf{v}_B \parallel \mathbf{v}_{Ax_1}$. The principle of virtual power takes the form

$$R_{Ax_1} v_{Ax_1} + M\omega = 0.$$

The whole beam behaves like a rigid body of an instantaneous center of rotation at point S, and according to Fig. 3.8 we have $BS = 4l/\tan\alpha$. Substituting $v_{Ax_1} = v_B = \omega BS$ into the preceding equation we obtain

$$R_{Ax_1} = -\frac{M \tan\alpha}{4l}.$$

Let us now determine the horizontal reaction $\mathbf{R}_{Ax_2}$ (Fig. 3.9).

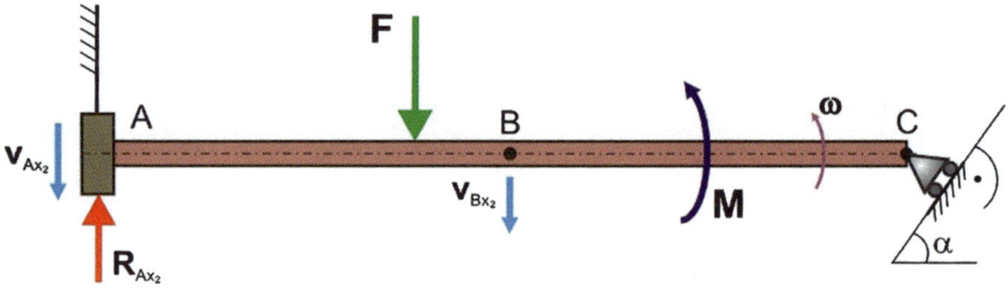

Fig. 3.9 Schematic showing how to determine the reaction $\mathbf{R}_{Ax_2}$

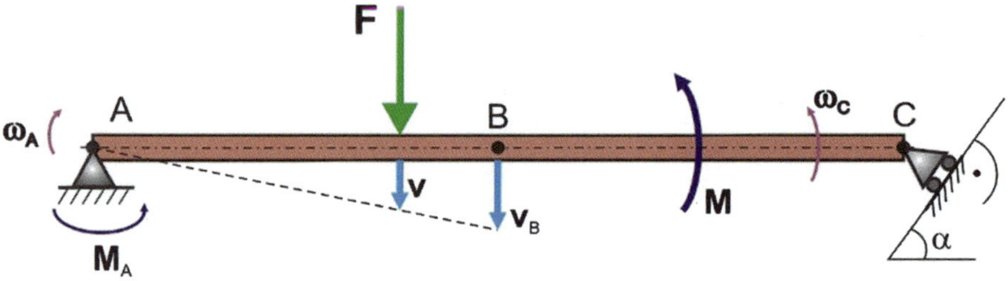

Fig. 3.10 Schematic showing how to determine a moment of reaction M_A

For this purpose we will treat the system of two beams as a rigid body ($\mathbf{v}_{Ax_2} = \mathbf{v}_{Bx_2}$) whose center of velocity is situated at point C. In this case the Jourdain variational principle has the form

$$(-R_{Ax_2} + F)\,v_{Ax_2} + M\omega = 0.$$

Now we determine ω from the relationship $v_{Bx_2} = 4l\omega$, which after substitution into the preceding equation yields

$$R_{Ax_2} = F + \frac{M}{4l}.$$

Let us determine now the reaction moment (of the built-in support) at point A, that is, M_A (Fig. 3.10).

In this case the principle of virtual power will take the form

$$-M_A\omega_A + Fv + M\omega_C = 0.$$

Because $v_B = 4l\omega = 4l\omega_C$, we have $\omega_A = \omega_C = \omega$, and in turn $v = 3l\omega$. Eventually from the preceding equation we obtain

$$M_A = M + 3lF.$$

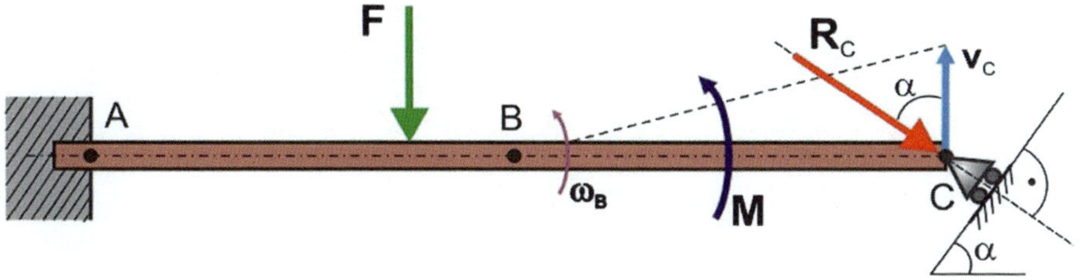

Fig. 3.11 Schematic showing how to determine a moment of reaction $\mathbf{R}_C$

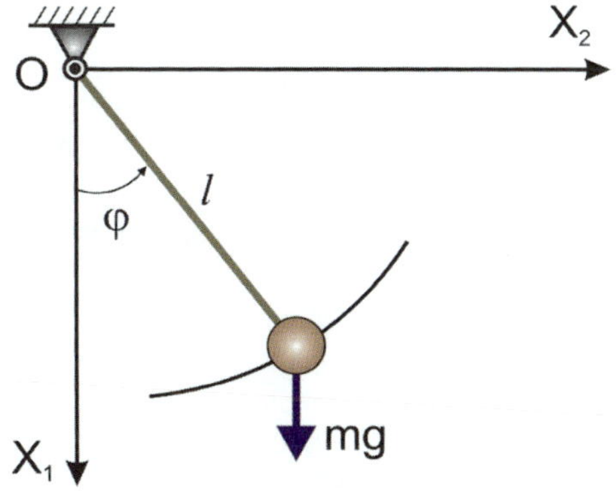

Fig. 3.12 Mathematical pendulum and the adopted coordinate system OX_1X_2

What is left to be determined is the reaction at point C, which is perpendicular to the roller support at this point (Fig. 3.11).

From the Jourdain variational principle we obtain

$$M\omega_B = R_C v_C \cos\alpha.$$

Because $v_c = 4l\omega_B$, finally we calulate

$$R_C = \frac{M}{4l\cos\alpha}.$$

Example 3.5. Derive the equation of motion of a mathematical pendulum (Fig. 3.12) using the Gauss variational principle.

According to (3.63) we have

$$\Gamma = \frac{1}{2}m\left[\left(\ddot{x}_1 - \frac{F_{x1}}{m}\right)^2 + \left(\ddot{x}_2 - \frac{F_{x2}}{m}\right)^2\right],$$

where according to Fig. 3.12

$$x_2 = l\sin\varphi, \quad x_1 = l\cos\varphi, \quad F_{x_1} = mg, \quad F_{x_2} = 0,$$

and we successively calculate

$$\dot{x}_1 = -l\dot{\varphi}\sin\varphi\,, \quad \dot{x}_2 = l\dot{\varphi}\cos\varphi$$

$$\ddot{x}_1 = -l\ddot{\varphi}\sin\varphi - l\dot{\varphi}^2\cos\varphi, \quad \ddot{x}_2 = l\ddot{\varphi}\cos\varphi - l\dot{\varphi}^2\sin\varphi.$$

Substituting $\ddot{x}_1$ and $\ddot{x}_2$ into the original equation we have

$$\begin{aligned}\Gamma &= \frac{1}{2}m\left[\left(-l\ddot{\varphi}\sin\varphi - l\dot{\varphi}^2\cos\varphi - g\right)^2 + \left(l\ddot{\varphi}\cos\varphi - l\dot{\varphi}^2\sin\varphi\right)^2\right]\\ &= \frac{1}{2}m(l^2\ddot{\varphi}^2\sin^2\varphi + l^2\dot{\varphi}^4\cos^2\varphi + g^2 + 2l^2\ddot{\varphi}\dot{\varphi}^2\sin\varphi\cos\varphi + 2lg\ddot{\varphi}\sin\varphi\\ &\qquad + 2lg\dot{\varphi}^2\cos\varphi + l^2\ddot{\varphi}^2\cos^2\varphi + l^2\dot{\varphi}^4\sin\varphi - 2l^2\ddot{\varphi}\dot{\varphi}^2\sin\varphi\cos\varphi)\\ &= \frac{1}{2}m\left(l^2\ddot{\varphi}^2 + 2lg\ddot{\varphi}\sin\varphi + l^2\dot{\varphi}^4 + 2lg\dot{\varphi}^2\cos\varphi + g^2\right).\end{aligned}$$

Condition (3.62) in this case takes the form

$$\frac{\partial\Gamma}{\partial\ddot{\varphi}} = m\left(l^2\ddot{\varphi} + lg\sin\varphi\right) = 0,$$

hence we get

$$\ddot{\varphi} + \frac{g}{l}\sin\varphi = 0. \qquad \square$$

Example 3.6. On an inclined plane of angle of inclination φ at point O is installed a light rod of length l and to its opposite end is mounted a disk of radius R and mass m that can roll on this inclined plane (Fig. 3.13). Determine the equations and the period of oscillations of the disk rolling without sliding for a small angle θ.

The vector of angular velocity of the disk is equal to

$$\boldsymbol{\omega} = \dot{\theta}\mathbf{e}_z - \dot{\psi}\mathbf{e}_r,$$

where the investigated mechanical system has one degree of freedom described by the generalized coordinate θ; the equation relating $\dot{\theta}$ and $\dot{\psi}$ follows from constraints of rolling of the form

$$\dot{\psi} = \frac{l}{R}\dot{\theta}.$$

The investigated system can be treated as a rigid body with a fixed point O. The angular momentum of the disk is equal to

$$\mathbf{K}_O = I_{X_3}\dot{\theta}\mathbf{e}_z - I_O\dot{\psi}\mathbf{e}_r,$$

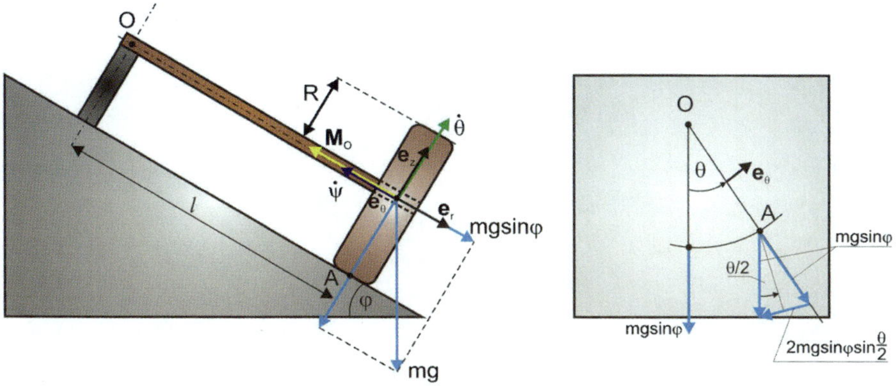

Fig. 3.13 Motion of a disk on an inclined plane

and its time derivative is

$$\dot{\mathbf{K}}_O = I_{X_3}\ddot{\theta}\mathbf{e}_z - I_O\ddot{\psi}\mathbf{e}_r + I_O\dot{\psi}\mathbf{e}_\theta.$$

The only external force performing work in the system is the gravity force, which produces a moment with respect to point O of magnitude

$$\mathbf{M}_O = -2mg\sin\varphi\sin\frac{\theta}{2}\mathbf{e}_z \cong -mg\theta\sin\varphi\mathbf{e}_z.$$

In the considered case (3.79) simplifies to

$$(\mathbf{M}_O - \dot{\mathbf{K}}_O) \circ \frac{\partial\boldsymbol{\omega}}{\partial\dot{q}_k} = 0, \qquad (*)$$

where now the magnitudes of moment and angular momentum are calculated with respect to the fixed pole O.

Because

$$\frac{\partial\boldsymbol{\omega}}{\partial\dot{\theta}} = \mathbf{e}_z - \frac{l}{R}\mathbf{e}_r,$$

following substitution of the values $\dot{\mathbf{K}}_O$, $\mathbf{M}_O$, and $\frac{\partial\omega}{\partial\dot\theta}$ calculated earlier, from equation (*) we obtain

$$(-mg\theta\sin\varphi\mathbf{e}_z - I_{X_3}\ddot{\theta}\mathbf{e}_z + I_O\ddot{\psi}\mathbf{e}_r - I_O\dot{\psi}\dot{\theta}\mathbf{e}_\theta) \circ \left(\mathbf{e}_z - \frac{l}{R}\mathbf{e}_r\right)$$
$$= -mg\theta\sin\varphi - I_{X_3}\ddot{\theta} - I_O\ddot{\psi}\frac{l}{R} = -mg\theta\sin\varphi - \left(I_{X_3} + I_O\left(\frac{l}{R}\right)^2\right)\ddot{\theta} = 0,$$

hence we eventually obtain an equation describing the small oscillations of the rolling disk in the form

$$\ddot{\theta} + \alpha^2 \theta = 0,$$

where

$$\alpha^2 = \frac{mg \sin \varphi}{I_{X_3} + I_O \left(\frac{l}{R}\right)^2}.$$

Because for the disk we have $I_O = \frac{mR^2}{2}$, $I_{X_3} = m(\frac{R^2}{4} + l^2)$, the period of oscillations is equal to $T = \frac{2\pi}{\alpha} = \pi \sqrt{\frac{R^2+6l^2}{gl \sin \varphi}}$. □

3.3 General Equation of Statics and Stability of Equilibrium Positions of Mechanical Systems in a Potential Force Field

The general equation of statics will be derived from the general equation of dynamics introduced in the following form in Chap. 1 of [14]:

$$(\mathbf{F}_n - m_n \mathbf{a}_n) \circ \delta \mathbf{r}_n = 0, \quad n = 1, \ldots, N, \tag{3.80}$$

which is valid for systems with ideal constraints.

Theorem 3.1. *From the many possible states of equilibrium (positions of equilibrium) of a mechanical system allowed for by ideal constraints, the actual state of equilibrium, valid in time interval $t_0 \le t \le t_*$, occurs when the elementary work done by all active forces during arbitrarily chosen virtual displacements is equal to zero, that is,*

$$\mathbf{F}_n \circ \delta \mathbf{r}_n = 0 \quad (t_0 \le t \le t_*). \tag{3.81}$$

Equation (3.81) is the necessary and sufficient condition of an actual equilibrium state and is called the *general equation of statics*.

Further we will consider equilibrium conditions of a system of particles (i) and a rigid body (ii).

Case (i). Let us consider a system of particles described by radius vectors $\mathbf{r}_n$. We aim at the formulation of conditions that must be satisfied by constraints imposed on the system in order for the system to remain in a state of equilibrium, represented by a certain equilibrium configuration (position) in time interval $t_0 \le t \le t_*$. If the equilibrium position corresponds to $\mathbf{r}_n = \mathbf{r}_n^*$, then according to (3.13) we have

$$f_{m_1}\left(\mathbf{r}_1^*, \ldots, \mathbf{r}_n^*, t\right) \equiv 0, \quad m_1 = 1, \ldots, M_1 \tag{3.82}$$

for $t \in [t_0, t^*]$. Velocities and accelerations corresponding to the positions $\mathbf{r}_n^*$ are equal to $\mathbf{v}_n = \mathbf{0}$, $\mathbf{a}_n = \mathbf{0}$ for $t \in [t_0, t^*]$. From the equation describing non-holonomic constraints (3.17) we obtain

$$\beta_{m_2}\left(\mathbf{r}_1^*, \ldots, \mathbf{r}_n^*, t\right) = 0, \quad m_2 = 1, \ldots, M_2 \tag{3.83}$$

for $t \in [t_0, t^*]$. The satisfaction of (3.82) and (3.83) means that the constraints imposed on the system enable the realization of the state of equilibrium of the mechanical system. However, the equilibrium of the mechanical system depends additionally on forces acting on this system, that is, whether or not the general equation of statics (3.81) is satisfied.

A proof of the necessity of condition (3.81) follows directly from the general equation of dynamics (3.80) after setting $\mathbf{a}_n = \mathbf{0}$. As shown in the monograph [11], a proof of the sufficiency of condition (3.81) is more complex. The complexity of this problem will be highlighted on the basis of the example below, which was excerpted from [11].

Example 3.7. Conduct an analysis of equilibrium conditions $x(0) = \dot{x}(0) = \ddot{x}(0)$ of a particle moving along the axis OX whose motion is described by the differential equation $\ddot{x} = Ax^a$, where $0 < a < 1$, $A > 0$.

Substituting $x = 0$ into the analyzed equation it can be seen that condition (3.81) is satisfied for any t. The condition of contraints allows for the existence of equilibrium position $x = 0$. Let us note, however, that despite the fact that this conservative, one-degree-of-freedom system is autonomous, the initial conditions $x(0) = \dot{x}(0) = 0$ do not guarantee that the particle will be kept in the equilibrium position $x = 0$. It follows from this observation that the analyzed equation, apart from the solution $x \equiv 0$, also has one more solution of the form

$$x(t) = A^* t^{a^*},$$

where

$$A^* = \left[\frac{A(1-a)^2}{2(1+a)}\right]^{\frac{1}{1-a}}, \quad a^* = \frac{2}{1-a}.$$

From the given relationship $a^* = 2(1-a)^{-1}$ it follows that $a^* - 2 = \epsilon > 0$. Differentiating this second non-trivial solution we have $\ddot{x}(t) = a^*(a^* - 1)A^* t^{\epsilon}$, hence we obtain $\ddot{x}(t) = 0$ for $t > 0$, although for $t = 0$ we have $\ddot{x}(0) = 0$.

The preceding example shows that despite the fact that the conditions of constraints and equality $x(0) = \dot{x}(0) = 0$ have been satisfied, it is not guaranteed that the particle will be in the equilibrium position $x = 0$.

The proof that condition (3.80) is a sufficient condition requires that the total uniqueness of the solution with respect to the initial conditions be taken into account.

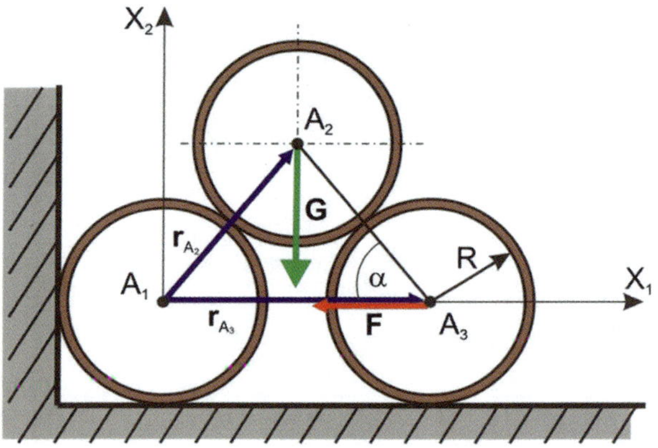

Fig. 3.14 Geometry of a system of three identical pipes with the forces performing virtual work marked

Example 3.8. Determine the magnitude of a horizontal force $\mathbf{F}$ that should be applied to a pipe of center A_3 in order for a mechanical system consisting of three identical pipes of radius R (Fig. 3.14) to remain in static equilibrium.

Virtual work is done only by two forces, that is, $\mathbf{G}_2 = \mathbf{G}$ and the desired force $\mathbf{F}$. From the isosceles triangle $\Delta A_1 A_2 A_3$ it follows that the radius vectors $\mathbf{r}_{A_2}$ and $\mathbf{r}_{A_3}$ are determined by the equations

$$\mathbf{r}_{A_2} = 2R\cos\alpha\mathbf{E}_1 + 2R\sin\alpha\mathbf{E}_2,$$
$$\mathbf{r}_{A_3} = 4R\cos\alpha\mathbf{E}_1,$$

and hence we obtain the variations

$$\delta\mathbf{r}_{A_2} = 2R\left(-\sin\alpha\mathbf{E}_1 + \cos\alpha\mathbf{E}_2\right)\delta\alpha,$$
$$\delta\mathbf{r}_{A_3} = -4R\sin\alpha\delta\alpha\mathbf{E}_1.$$

In turn,

$$\boldsymbol{G} = -G\mathbf{E}_2, \quad \mathbf{F} = -F\mathbf{E}_1,$$

so substituting it into (3.81) we obtain

$$\mathbf{F}\circ\delta\mathbf{r}_{A_3} + \boldsymbol{G}\circ\delta\mathbf{r}_{A_2} = 0.$$

From the preceding equation we have

$$(4RF\sin\alpha - 2RG\cos\alpha)\,\delta\alpha = 0,$$

hence we find that $F = G/(2\tan\alpha)$. □

The general equation of statics (3.81) can be expressed in terms of generalized coordinates in the following way:

$$\mathbf{F}_n \circ \delta\mathbf{r}_n = Q_m \delta q_m = 0,$$
$$n = 1, \ldots, N, \quad m = 1, \ldots, M. \tag{3.84}$$

In the case of holonomic constraints, $D^f = M$, where D^f defines the number of degrees of freedom of the DMS, and the quantities δq_m in (3.84) are independent of each other. In the case of non-holonomic constraints, they depend on each other.

From (3.84) it follows that in the equilibrium position $\mathbf{q} = \mathbf{q}_0 = \mathbf{0}$ for the case of holonomic constraints we obtain D^f equilibrium equations of the form

$$Q_m = 0, \quad m = 1, \ldots, M, \tag{3.85}$$

which allow for the determination of D^f desired coordinates of vector $\mathbf{q}_0[q_{10}, \ldots, q_{M0}]$.

In many problems of statics we encounter a special case where the only active forces are potential forces that follow from the gravitational force field.

In this case, the problem simplifies substantially since it boils down to the determination of the extremum of potential energy of the investigated mechanical system.

Let us recall that a force field is called a potential field if there exists a scalar function U such that it generates conservative forces acting on the system situated in that force field, that is,

$$\mathbf{F}_n = \frac{\partial U}{\partial \mathbf{r}_n}, \quad n = 1, \ldots, N, \tag{3.86}$$

where U is the *force (generating) function*. In turn, we call $V = -U$ the *potential* or *potential energy*, and forces $\mathbf{F}_n$ defined by (3.86) the *potential forces*.

The elementary work of potential forces in a *stationary potential field* is equal to

$$\begin{aligned}\delta W &= \mathbf{F}_n \circ \delta\mathbf{r}_n = \frac{\partial U}{\partial \mathbf{r}_n} \circ \delta\mathbf{r}_n \\ &= \frac{\partial U}{\partial x_{1n}}\delta x_{1n} + \frac{\partial U}{\partial x_{2n}}\delta x_{2n} + \frac{\partial U}{\partial x_{3n}}\delta x_{3n} = \delta U \\ &= -\delta V = -\frac{\partial V}{\partial q_m}\delta q_m = Q_m \delta q_m = 0.\end{aligned} \tag{3.87}$$

From (3.87) it follows that knowing the potential $V = V(\mathbf{r}_n)$ and for the case of holonomic constraints $D^f = M$, we have at our disposal D^f following algebraic

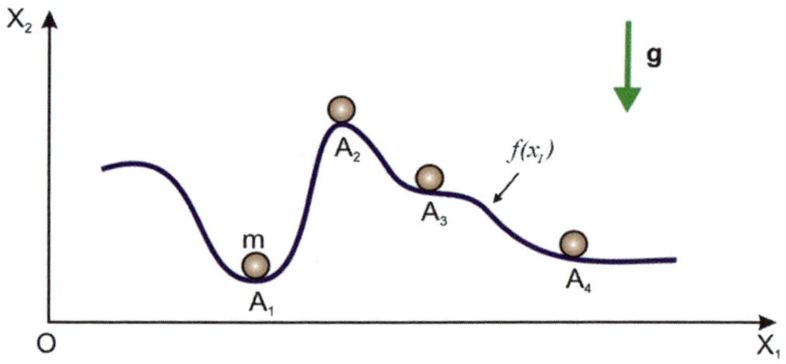

Fig. 3.15 Motion of a particle along curve $f(x_1)$ in a potential field

equations, which serve to determine the equilibrium position vector $\mathbf{r}_n = \mathbf{r}_n^0$, $n = 1, \ldots, N$:

$$Q_m = -\frac{\partial V}{\partial q_m} = 0, \quad m = 1, \ldots, M. \tag{3.88}$$

In a potential force field the satisfaction of condition (3.99) is equivalent to the condition

$$\mathrm{d}V = 0, \tag{3.89}$$

that is, the total differential of the potential energy is equal to zero.

Let a particle be moving in a potential force field in the vertical plane, where $x_2 \equiv h = f(x_1)$ denotes the elevation (position x_2) of this particle above the ground (Fig. 3.15).

The potential energy of a particle of mass m is equal to

$$V = mgf(x_1), \tag{3.90}$$

and condition (3.89) is satisfied at four points A_i of the curve. After a small displacement of the particle from the position A_1 (A_2) it always returns (does not return) to this position. In the case of an inflexion point A_3, the initial deflection of the material point from the equilibrium position results in permanent loss of this equilibrium position. In the case of a material point at position A_4, its motion takes place along the horizontal portion of the curve and in its vicinity there exist infinitely many new equilibrium positions determined by the initial conditions. We call point A_1 an *elliptic point*, point A_2 a *saddle point*, and point A_4 a *parabolic point*.

Let us emphasize, however, that conditions (3.88) are the necessary conditions for equilibrium, but they are not sufficient conditions. We skip the analysis of non-holonomic constraints, confining ourselves only to the commentary that in this case, select or all partial derivatives of the potential energy of the system may not be equal to zero.

Case (ii). We will now proceed to an analysis of equilibrium conditions in the second case, that is, in the case of a rigid body. From the previous considerations it follows that an arbitrary system of forces and moments acting on a rigid body can be replaced with the equivalent force system, and in the present case, with the main force vector $\mathbf{F}$ and the main moment of forces $\mathbf{M}_O$, where O is an arbitrary pole (point of reduction). If the body is a free body, the conditions $\mathbf{F} = \mathbf{0}$ and $\mathbf{M}_O = \mathbf{0}$ are the necessary and sufficient conditions of equilibrium of the body. If the body is constrained, we "mentally" release it from the constraints, introducing the reaction forces and reaction moments, and then we treat these forces as active forces. Also in this case, the equations $\mathbf{F} = \mathbf{0}$ and $\mathbf{M}_O = \mathbf{0}$ are valid, and now we determine from these equations the desired reaction forces and moments.

Theorem 3.2. *The necessary and sufficient condition for a body to remain in a state of equilibrium in the time interval $t_0 \leq t \leq t_*$ is a lack of motion of the body at the time instant t_0 and the satisfaction of the two equations*

$$\mathbf{F} = \mathbf{0}, \quad \mathbf{M}_O = \mathbf{0}, \tag{3.91}$$

that is, the main force vector and main moment of force (of the forces applied to the rigid body) with respect to any point O are equal to zero in the time interval $t_0 \leq t \leq t_*$.

Proof. A free rigid body is a scleronomic system, and its actual displacement realized during time $\mathrm{d}t$ is a virtual displacement. The elementary work of forces and moments of forces performed over a rigid body and related to its elementary displacement (it consists of the translation of pole O and body rotation with respect to this pole) is equal to

$$\delta W = \mathbf{F} \circ \mathbf{v}_O \mathrm{d}t + \mathbf{M}_O \circ \boldsymbol{\omega} \mathrm{d}t = 0, \tag{3.92}$$

where $\mathbf{v}_O$ denotes the velocity of pole O and $\boldsymbol{\omega}$ is the vector of angular velocity of the body. Let us impose on the body arbitrary instantaneous velocities $\mathbf{v}_O$ and $\boldsymbol{\omega}$ for time instants $t \in [t_0, t_*)$. From the general equation of statics in the form (3.92) equations (3.91) follow directly, which was to be demonstrated.

Let us return to the equilibrium conditions of a mechanical system in a potential force field that are described by equations (3.88). If we disturb the state of equilibrium $\mathbf{q} = \mathbf{q}_O$ of the system by introducing small displacements and initial velocities (disturbances), then the system might return to the position $\mathbf{q}_O$ (the disturbances tend to zero), or it might not return to this position (Fig. 3.15). We describe the equilibrium position $\mathbf{q}_O$ in the first case (point A_1 from Fig. 3.15) as stable and in the second case (points A_2, A_3, and A_4 in Fig. 3.15) as unstable.

Theorem 3.3. *If in the equilibrium position of a conservative mechanical system the energy has a local minimum, then this equilibrium position is stable.*

Although the proof of this theorem is omitted here, we will present some basic information regarding its interpretation.

Let $\mathbf{q}_0 = \mathbf{q}_0\left(q_1, q_2, \ldots, q_{D^f}\right) = 0$, where D^f denotes the number of degrees of freedom of the DMS, and let $V = V(0, \ldots, 0) = 0$. Here we make use of the following classical definition of stability (see, e.g., [1]). We call the equilibrium position $q_1 = \cdots = q_d = \cdots = q_{D^f} = 0$ a *stable position* if, for an arbitrary $\varepsilon > 0$, there exists $\delta = \delta(\epsilon)$ such that for all $t > t_0$ the following inequalities hold:

$$|q_d(t)| < \varepsilon, \quad |\dot{q}_d| < \varepsilon, \quad d = 1, \ldots, D^f, \tag{3.93}$$

at the initial conditions assumed earlier

$$|q_d(t_0)| < \delta, \quad |\dot{q}_d(t_0)| < \delta. \tag{3.94}$$

In other words, in the space $(\mathbf{q}, \dot{\mathbf{q}})$ the solution starting from a cube of edge 2δ will remain inside a cube of edge 2ϵ.

By assumption, the scalar function $V = V(q_1, \ldots, q_{D^f})$ has a local minimum, so there exists $\eta > 0$ such that in the neighborhood

$$|q_d| < \eta, \quad d = 1, \ldots, D^f, \tag{3.95}$$

the inequality

$$V(q_1, \ldots, q_{D^f}) > V(0, \ldots, 0) = 0 \tag{3.96}$$

is satisfied if at least one of the generalized coordinates q_d is not equal to zero. Because the kinetic energy in the proximity of the equilibrium position $(0, \ldots, 0)$ has the form

$$T = T_2 = \frac{1}{2} a_{ij}(q_1, \ldots, q_{D^f})\dot{q}_i \dot{q}_j > 0, \tag{3.97}$$

from formulas (3.96) and (3.97) it follows that the total energy of the system is equal to

$$E = V + T > 0 \tag{3.98}$$

if at least one of the generalized coordinates $q_d \neq 0$, $d = 1, \ldots, D^f$. In turn, because $V(0, \ldots, 0) = 0$ and $T(0, \ldots, 0) = 0$, we have $E(0, \ldots, 0) = 0$, and taking into account condition (3.94) we conclude that the function E attains the minimum at zero.

As we will present later in Sect. 3.4, knowledge of the kinetic energy and potential energy of a system of particles (rigid bodies) and of generalized forces allows for the derivation of the equations of motion of the system in the form of Lagrange equations of the second kind. The equations can be presented, for example, in the form

$$\frac{\mathrm{d}}{\mathrm{d}t}\frac{\partial T}{\partial \dot{q}_d} - \frac{\partial T}{\partial q_d} = Q_d, \quad d = 1, \ldots, D^f, \tag{3.99}$$

$$Q_d = Q_d^* - \frac{\partial D}{\partial \dot{q}_d} - \frac{\partial V}{\partial q_d}, \tag{3.100}$$

where T is the kinetic energy of the rigid system (or the rigid body), Q_d^* is the active force (non-potential and non-dissipative), D is the dissipation function, also known as the Rayleigh dissipation function describing energy dissipation, and V is the potential energy of the system.

In technical applications for the determination of equlilibrium positions we can use the four principles presented below.

(i) Lagrange's principle.
Inserting $T = 0$ (the state of equilibrium) in (3.99) we obtain $Q_d = 0,\ d = 1, \ldots, D$, which describes the equilibrium condition of an arbitrary system of forces with ideal holonomic constraints.

(ii) Principle regarding potential forces [described by (3.99)].
In the equilibrium position in a potential force field with ideal holonomic constraints the derivative of potential energy becomes zero with respect to all generalized coordinates, which indicates the existence of its extremum.

(iii) Dirichlet's[6] principle [follows also from (3.99)].
If a material system is in a potential force field, then the position determined by the minimum of the potential energy is the position of the stable equilibrium. The Dirichlet principle in the case of a mechanical system with one degree of freedom boils down to the satisfaction of two conditions:

$$\frac{\partial V}{\partial q_1} = 0, \quad \frac{\partial^2 V}{\partial q_1^2} > 0. \tag{3.101}$$

In the case of a system with two degrees of freedom $D^f = 2$, the system equilibrium position in a potential force field is stable when the following conditions are satisfied:

$$\begin{gathered} \frac{\partial V}{\partial q_1} = 0, \quad \frac{\partial^2 V}{\partial q_1^2} > 0, \\ \frac{\partial^2 V}{\partial q_1^2}\frac{\partial^2 V}{\partial q_2^2} - \left(\frac{\partial^2 V}{\partial q_1 \partial q_2}\right)^2 > 0, \\ \frac{\partial^2 V}{\partial q_1^2} > 0, \quad \frac{\partial^2 V}{\partial q_2^2} > 0. \end{gathered} \tag{3.102}$$

(iv) Torricelli's[7] principle.
This is a special case of Dirichlet's principle. If in a uniform gravitational field a constrained material system subjected to the action of ideal constraints reaches a minimum elevation (the equilibrium position) with respect to the chosen level, then such a position (configuration) of the system is a *static equilibrium position*.

---

[6]Johann Peter Dirichlet (1805–1859), German mathematician of French origin working in Wrocław, Göttingen, and Berlin.

[7]Evangelista Torricelli (1608–1647), Italian physicist and mathematician.

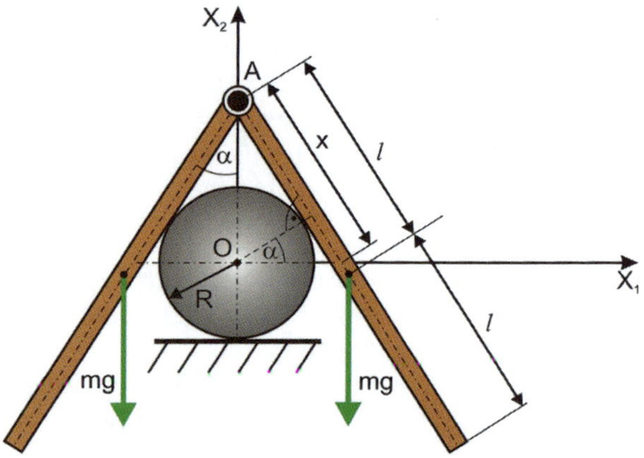

Fig. 3.16 Two heavy rods resting on a smooth cylinder in equilibrium position determined by angle 2α

Example 3.9. Two uniform rods of identical masses m and lengths l connected by a pin joint at point A rest on a smooth cylinder of radius R (Fig. 3.16). Determine the angle between the rods in the equilibrium position.

The potential energy of the rods with respect to the axis OX_1 has the form

$$V(\alpha) = -2mg(l - x)\cos\alpha,$$

and the desired value of x is equal to

$$x = R ctan\alpha + R tan\alpha = \frac{R}{\sin\alpha\cos\alpha}.$$

From the first equation we get

$$V(\alpha) = -2mg\left(l\cos\alpha - \frac{R}{\sin\alpha}\right),$$

and hence

$$V' \equiv \frac{\partial V(\alpha)}{\partial\alpha} = -2mg\left(-l\sin\alpha + \frac{R\cos\alpha}{\sin^2\alpha}\right) = 0.$$

From this equation we obtain

$$l\sin^3\alpha - R\cos\alpha = 0.$$

Dividing the preceding equation by $\cos\alpha$ we get

$$l\tan\alpha\sin^2\alpha - R\left(\sin^2\alpha + \cos^2\alpha\right) = 0,$$

which after dividing by $\cos^2\alpha$ leads to the equation

$$l\tan^3\alpha - R\tan^2\alpha - R = 0.$$

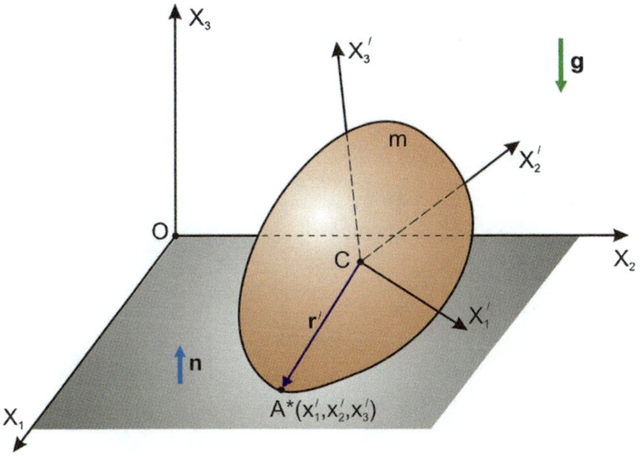

Fig. 3.17 *Convex* rigid body lying on plane OX_1X_2 and two coordinate systems (point of contact A^* is in proximity of point A)

Setting $\tan\alpha = y$, the desired root α_O can be determined using a graphical method shown by the equation

$$\frac{l}{R}y^3 = y^2 + 1.$$

The inequality condition associated with the preservation of the stability of equilibrium position takes the form

$$\frac{\partial^2 V}{\partial\alpha^2} = \frac{\partial V'}{\partial\alpha} = -2mg\left[-l\cos\alpha + R\left(\frac{-\sin^3\alpha - 2\sin\alpha\cos^2\alpha}{\sin^4\alpha}\right)\right]$$
$$= 2mg\left(l\cos\alpha + R\frac{\sin^2\alpha + 2\cos^2\alpha}{\sin^3\alpha}\right) > 0,$$

and following determination of the value of α_O it is possible to show that it is satisfied. □

Example 3.10. (See [11]) Investigate the stability of the equilibrium position of a rigid body lying on a perfectly smooth horizontal surface in a gravitational field. Let the surface Π be convex in the neighborhood of the point of contact with the body A on condition that the body' center of gravity lies on a vertical line passing through the body's center of mass C (Fig. 3.17).

With the body we associate the Cartesian coordinate system (body system) $CX_1'X_2'X_3'$ of origin at its mass center C and axes directed along its principal centroidal axes of inertia.

Let us describe by radius vector $\mathbf{r}'_A[x'_{1A}, x'_{2A}, x'_{3A}]$ the position of the point of contact with the plane, A, in the system $CX'_1X'_2X'_3$. Let the equation of the surface Π bounding the body be given by

$$f\left(x'_1, x'_2, x'_3\right) = 0.$$

We take the sign of the function in such a way that the unit vector $\mathbf{n}$ lying on a line perpendicular to the interior surface of the body at point A is described by the equation

$$\boldsymbol{n} = -\frac{\operatorname{grad} f}{|\operatorname{grad} f|}. \tag{$*$}$$

The equation of the surface in the body's system in the neighborhood of point A has (approximately) the following analytical form:

$$f \equiv -h - x'_{3A} + \frac{1}{2}\left(\frac{x'^2_{1A}}{R_1} + \frac{x'^2_{2A}}{R_2}\right) + \cdots = 0,$$

where point $A^* \neq A$ is now the point of contact of the surface Π of the investigated body with the horizontal plane after a disturbance (a small deflection) of the body from its equilibrium position associated with point A; $-h$ is the position of point C in the equilibrium position of the body (at this position $C = C(0, 0, -h)$); and R_1 and R_2 are the radii of curvature of the surface Π at point A^*.

As can be seen from Fig. 3.17, if the surface Π of the body is convex and is situated above the horizontal plane, then the radii of curvature $R_1 > 0$ and $R_2 > 0$.

A potential energy of the body is given by the equation

$$V = mgH, \quad H = -\left(\boldsymbol{n} \circ \overrightarrow{CA^*}\right),$$

where H denotes the distance between the mass center of the body and the horizontal surface of contact at point A^*. The normal vector, according to the preceding definition (*), is described by the formula

$$\boldsymbol{n} = -\frac{\nabla f}{|\nabla f|} = \frac{\left(\frac{\partial f}{\partial x'_1}\mathbf{E}'_1 + \frac{\partial f}{\partial x'_2}\mathbf{E}'_2 + \frac{\partial f}{\partial x'_3}\mathbf{E}'_3\right)}{|\nabla f|} \equiv n'_1\mathbf{E}'_1 + n'_2\mathbf{E}'_2 + n'_3\mathbf{E}'_3.$$

We calculate successively

$$n'_1 \equiv \frac{\partial f}{\partial x'_1} = \frac{x'_{1*}}{R_1}, \quad n'_2 \equiv \frac{\partial f}{\partial x'_2} = \frac{x'_{2*}}{R_2}.$$

Let us recall that $\sqrt{1-x} = 1 - \frac{1}{2}x + \frac{1}{8}x^2 + \ldots$, and because $(n'_1)^2 + (n'_2)^2 + (n'_3)^2 = 1$, we have

$$n'_3 = \sqrt{1 - \left[\left(\frac{x'_{1*}}{R_1}\right)^2 + \left(\frac{x'_{2*}}{R_2}\right)^2\right]} \cong 1 - \frac{1}{2}\left[\left(\frac{x'_{1*}}{R_1}\right)^2 + \left(\frac{x'_{1*}}{R_2}\right)^2\right].$$

In turn,

$$\begin{aligned} H &= -\left(\boldsymbol{n} \circ \overrightarrow{CA^*}\right) = -(n_1\mathbf{E}_1 + n_2\mathbf{E}_2 + n_3\mathbf{E}_3) \circ (x_1\mathbf{E}_1 + x_2\mathbf{E}_2 + x_3\mathbf{E}_3) \\ &= -(n_1x_1 + n_2x_2 + n_3x_3) = -\left\{ -\frac{x_1^2}{R_1} - \frac{x_2^2}{R_2} + \left[1 - \frac{1}{2}\left(\frac{x_1^2}{R_1} + \frac{x_2^2}{R_2}\right)\right] \right. \\ &\quad \left. \cdot \left[-h + \frac{1}{2}\left(\frac{x_1^2}{R_1} + \frac{x_2^2}{R_2}\right)\right] \right\} = \left\{ \frac{x_1^2}{R_1} + \frac{x_2^2}{R_2} - \left[-h + \frac{1}{2}\left(\frac{x_1^2}{R_1} + \frac{x_2^2}{R_2}\right) \right. \right. \\ &\quad \left. \left. + \frac{h}{2}\left(\frac{x_1^2}{R_1^2} + \frac{x_2^2}{R_2^2}\right) - \frac{1}{4}\left(\frac{x_1^2}{R_1^2} + \frac{x_2^2}{R_2^2}\right)\left(\frac{x_1^2}{R_1} + \frac{x_2^2}{R_2}\right) \right] \right\} \\ &\cong h + \frac{1}{2}\left(\frac{x_1^2}{R_1} + \frac{x_2^2}{R_2}\right) - \frac{h}{2}\left(\frac{x_1^2}{R_1^2} + \frac{x_2^2}{R_2^2}\right) \\ &\cong h + \frac{1}{2}\left(\frac{R_1 - h}{R_1^2}x_1^2 + \frac{R_2 - h}{R_2^2}x_2^2\right). \end{aligned}$$

The potential energy of a rigid body is equal to

$$V = mgH = mg\left[h + \frac{1}{2}\left(\frac{R_1 - h}{R_1^2}x_{1*}^2 + \frac{R_2 - h}{R_2^2}x_{2*}^2\right)\right].$$

We now make use of the conditions following from Dirichlet's principle, (3.102). We calculate successively

$$\frac{\partial V}{\partial x_{ix}} = 0, \quad i = 1, 2,$$

that is,

$$\frac{R_1 - h}{R_1^2}x_{1*} = 0, \quad \frac{R_2 - h}{R_2^2}x_{2*} = 0.$$

In turn,

$$\frac{\partial^2 V}{\partial x_1^{*2}} = \frac{R_1 - h}{R_1^2} > 0, \quad \frac{\partial^2 V}{\partial x_2^{*2}} = \frac{R_2 - h}{R_2^2} > 0, \quad \frac{\partial^2 V}{\partial x_1^* \partial x_2^*} = 0.$$

If $R_1 > h$ and $R_2 > h$, then the equilibrium position of the body is stable. For instance, in the case of a ball we have $R_1 = R_2 = R$ (and the stability condition boils down to the condition $R > h$) and for a cylinder we have $R_1 = R$, $R_2 = \infty$, and then also we obtain the stability condition described by the inequality $R > h$.

□

3.4 Lagrange's Equations of the First and Second Kind

Having defined the notion of generalized coordinates, we will now derive equations of motion of a discrete mechanical system with constraints in the generalized coordinates. According to Newton's second law, the motion of an arbitrary point of mass m_n is described by the equation

$$m_n \frac{\mathrm{d}\mathbf{v}_n}{\mathrm{d}t} \circ \delta \mathbf{r}_n = \left(\mathbf{F}_n + \mathbf{F}_n^i + \mathbf{F}_n^R\right) \circ \delta \mathbf{r}_n, \tag{3.103}$$

and using the earlier introduced summation convention this equation describes the motion of the whole material system.

Let us introduce the notion of *ideal constraints*. These are bilateral constraints such that the sum of the virtual works done by the reactions produced by those constraints during an arbitrary virtual displacement is equal to zero. This means that the virtual work

$$\delta W_n = \mathbf{F}_n^R \circ \delta \mathbf{r}_n = 0, \tag{3.104}$$

where the summation convention does not apply.

According to the calculations conducted earlier, the vector of reactions acting on particle A in the extended space $\mathbf{R}^{3N}$ has the form $\mathbf{F}_A^R = (\mathbf{F}_{1,A}^R, \ldots, \mathbf{F}_{N,A}^R)$, and the particle moves along the chosen smooth trajectory. Holonomic constraints (3.7) are called *ideal constraints* if for every trajectory vector $\mathbf{F}_A^R$ is perpendicular to the tangent space at this point, that is,

$$\mathbf{F}_A^R \circ \mathbf{r}'(0) = \mathbf{F}_{k,A}^R \circ \mathbf{r}_k'(0) = 0, \tag{3.105}$$

where, according to the previous remarks, the differentiation is carried out with respect to the parameter s and $\mathbf{r}_k'$ is the position vector.

Holonomic time-independent constraints (3.8) can be represented in the following way:

$$f_m\left(\mathbf{r}_{1,0}, \ldots, \mathbf{r}_{N,0}\right) = 0, \quad m = 1, \ldots, M. \tag{3.106}$$

Differentiating the preceding constraint equation along the trajectory of motion of particle A we obtain

$$\nabla_{\mathbf{r}_k} f_m\left(\mathbf{r}_{1,A}, \ldots, \mathbf{r}_{N,A}\right) \circ \mathbf{r}_k'(0), \quad k = 1, \ldots, N, \; m = 1, \ldots, M, \tag{3.107}$$

where the differential operator defined earlier in Chap. 4 of [14] was used. Let us recall that the operation ∇f, that is, the action of the operator ∇ on the scalar function f, yields a vector function called the gradient, that is, we obtain the vector

$$\nabla f_m \equiv (\nabla_{\mathbf{r}_1} f_m, \ldots, \nabla_{\mathbf{r}_N} f_m) \equiv \left(\frac{\partial f_m}{\partial \mathbf{r}_1}, \ldots, \frac{\partial f_m}{\partial \mathbf{r}_N}\right).$$

By assumption, the functions f_m, $m = 1, \ldots, M$ are independent, as are the vectors perpendicular to the tangents ∇f_m. The vectors ∇f_m, $m = 1, \ldots, M$ span complementary spaces (to tangent spaces) at every point of the constraint surface. The vectors of reactions must belong to the complementary spaces at every point of the space, that is, they should be expressed through vectors ∇f_m by the following relationship:

$$\mathbf{F}^R(\mathbf{r}_1, \ldots, \dot{\mathbf{r}}_N) = \lambda_m(\mathbf{r}_1, \ldots, \dot{\mathbf{r}}_N) \nabla f_m(\mathbf{r}_1, \ldots, \mathbf{r}_N), \quad m = 1, \ldots, M, \quad (3.108)$$

where λ_m are proportionality factors of the aforementioned vectors.

In the space $\mathbf{R}^3$, the preceding expression takes the form

$$\mathbf{F}^R(\mathbf{r}_k) = \lambda_m(\mathbf{r}_k, \dot{\mathbf{r}}_k) \nabla_{\mathbf{r}_k} f_m(\mathbf{r}_k), \quad m = 1, \ldots, M, \quad k = 1, \ldots, N. \quad (3.109)$$

The introduced functions λ_m, $m = 1, \ldots, M$ are called Lagrange multipliers (or undetermined Lagrange multipliers), and substituting them into (3.103) and dividing by $\delta \mathbf{r}_n$ we obtain the form of equations of motion called *Lagrange's equations of the first kind*. Those equations will be derived in the $\mathbf{E}^3$ space in the next section. The concept of the extended space $\mathbf{R}^{3N}$ was briefly discussed here because it is often used in so-called geometric mechanics.

The Lagrange multipliers enable us to determine reaction forces, which is required in many problems of mechanics.

Let us also note that in this approach we do not make use of the notions of work and virtual displacement. In the case of ideal constraints and problems of statics, that is, the determination of static equilibrium positions, d'Alembert's principle takes the form

$$\mathbf{F}_k(\mathbf{r}_1, \ldots, \mathbf{r}_N) \circ \mathbf{r}'_k = 0, \quad k = 1, \ldots, N, \quad (3.110)$$

where $\mathbf{F}_k$ is a resultant force acting on particle k.

Equations (3.108) and (3.110) allow for the determination of the desired equilibrium positions of a system with constraints. In the equilibrium positions, forces acting on a given particle of a DMS including reaction forces must vanish, and this condition, called a *necessary condition*, must be satisfied *simultaneously* for all particles.

It should be noted that our calculations concern ideally smooth constraints, which is rarely encountered in real systems. This approach does not allow for the introduction of friction. The friction force is tangent to the constraint surface and, additionally, proportional to the reaction force perpendicular to the surface of constraints.

Following introduction of the ideal constraints, (3.103) takes the form

$$m_n \frac{d\mathbf{v}_n}{dt} \circ \delta\mathbf{r}_n = \left(\mathbf{F}_n + \mathbf{F}_n^i\right) \circ \delta\mathbf{r}_n. \qquad (3.111)$$

Those equations will help in the derivation of Lagrange's equations of the second kind, which are characterized by the fact that they do not contain reactions of constraints but only forces of interaction of DMS characteristic points (i.e., the internal forces) and external forces independent of constraints.

Different authors derive Lagrange's equations of the first and the second kind in various ways. In this book, we rely on calculations excerpted from [5], and the versatility of this approach consists in the fact that it includes, during the derivation of Lagrange's equations of the second kind, both geometric (holonomic) constraints and non-holonomic (non-integrable) constraints.

Reaction forces of constraints are divided into two groups:

$$\mathbf{F}_n^R = \mathbf{F}_n^h + \mathbf{F}_n^n, \qquad (3.112)$$

where $\mathbf{F}_n^h$ ($\mathbf{F}_n^n$) are holonomic (non-holonomic) constraints. We relate the postulate of ideal constraints (3.32) at first only to holonomic constraints, and it takes the form

$$\sum_{n=1}^{N} \mathbf{F}_n^h \circ \delta\mathbf{r}_n = 0, \qquad (3.113)$$

and from (3.33), taking into account (3.113), we calculate

$$\mathbf{F}_n^h = m_n \ddot{\mathbf{r}}_n - \mathbf{F}_n^i - \mathbf{F}_n^n - \mathbf{F}_n, \quad n = 1, \ldots, N. \qquad (3.114)$$

Substituting (3.114) into relationship (3.113) we have (the forces $\mathbf{F}_n^i$ are omitted because they mutually cancel out)

$$\sum_{n=1}^{N} \left(\mathbf{F}_n^n + \mathbf{F}_n - m_n \ddot{\mathbf{r}}_n\right) \circ \delta\mathbf{r}_n = 0. \qquad (3.115)$$

Assume that in order to describe the investigated DMS we need q_k *independent generalized coordinates*, where $k = 1, 2, \ldots, K$. According to (3.48) we have

$$\delta\mathbf{r}_n = \sum_{k=1}^{K} \frac{\partial \mathbf{r}_n}{\partial q_k} \delta q_k. \qquad (3.116)$$

The sum of works of external forces $\mathbf{F}_n^e = \mathbf{F}_n$ during the virtual displacements $\delta\mathbf{r}_n$ is equal to

$$\delta W^e = \sum_{n=1}^{N} \mathbf{F}_n \circ \delta \mathbf{r}_n = \sum_{n=1}^{N} \mathbf{F}_n \circ \sum_{k=1}^{K} \frac{\partial \mathbf{r}_n}{\partial q_k} \delta q_k$$
$$= \sum_{k=1}^{K} \left(\sum_{n=1}^{N} \mathbf{F}_n \circ \frac{\partial \mathbf{r}_n}{\partial q_k} \right) \delta q_k = \sum_{k=1}^{K} Q_k \delta q_k, \quad (3.117)$$

where

$$Q_k(q_1, q_2, \ldots, q_K, t) = \sum_{n=1}^{N} \mathbf{F}_n \circ \frac{\partial \mathbf{r}_n}{\partial q_k}, \quad k = 1, \ldots, K. \quad (3.118)$$

Proceeding in a similar way we define the work of non-holonomic reactions during the virtual displacements $\delta \mathbf{r}_n$ of the form

$$\delta W^n = \sum_{n=1}^{N} \mathbf{F}_n^n \circ \delta \mathbf{r}_n = \sum_{k=1}^{K} \left(\sum_{n=1}^{N} \mathbf{F}_n^n \circ \frac{\partial \mathbf{r}_n}{\partial q_k} \right) \delta q_k$$
$$= \sum_{k=1}^{K} Q_k^n \delta q_k, \quad k = 1, \ldots, K. \quad (3.119)$$

In a similar way we determine also the work of d'Alembert's forces (the inertial forces) $\mathbf{F}_n^{in} = -m_n \ddot{\mathbf{r}}_n$ during virtual displacements $\delta \mathbf{r}_n$

$$\delta W^{in} = \sum_{n=1}^{N} \mathbf{F}_n^{in} \circ \delta \mathbf{r}_n = \sum_{k=1}^{K} \left(\sum_{n=1}^{N} \mathbf{F}_n^{in} \circ \frac{\partial \mathbf{r}_n}{\partial q_k} \right)$$
$$= \sum_{k=1}^{K} \left(-\sum_{n=1}^{N} m_n \frac{\partial \dot{\mathbf{r}}_n}{\partial t} \circ \frac{\partial \mathbf{r}_n}{\partial q_k} \right) \delta q_k = \sum_{k=1}^{K} Q_n^{in} \delta q_k. \quad (3.120)$$

The sum of all virtual works is

$$\sum_{k=1}^{K} \left(Q_k + Q_k^n + Q_k^{in} \right) \delta q_k = 0. \quad (3.121)$$

The component $\sum_{k=1}^{K} Q_k^{in} \delta q_k$ of the sum that follows from the action of the inertial forces are expressed through the kinetic energy of the DMS.

According to (3.111) we have

$$\left(m_n \frac{\mathrm{d}\mathbf{v}_n}{\mathrm{d}t} \right) \circ \left(\frac{\delta \mathbf{r}_n}{\delta q_k} \delta q_k \right) = m_n \frac{\mathrm{d}\mathbf{v}_n}{\mathrm{d}t} \circ \frac{\partial \mathbf{r}_n}{\partial q_k} \delta q_k,$$
$$n = 1, \ldots, N, \quad k = 1, \ldots, K. \quad (3.122)$$

Note that

$$\frac{\mathrm{d}}{\mathrm{d}t}\left(\mathbf{v}_n \circ \frac{\delta \mathbf{r}_n}{\delta q_i}\right) = \frac{\mathrm{d}\mathbf{v}_n}{\mathrm{d}t} \circ \frac{\partial \mathbf{r}_n}{\partial q_i} + \mathbf{v}_n \circ \frac{\mathrm{d}}{\mathrm{d}t}\left(\frac{\partial \mathbf{r}_n}{\partial q_i}\right). \tag{3.123}$$

Then, taking into account relationships (3.122) and (3.123), we obtain

$$\begin{aligned} m_n \frac{\mathrm{d}\mathbf{v}_n}{\mathrm{d}t} \circ \delta \mathbf{r}_n &= m_n \frac{\mathrm{d}\mathbf{v}_n}{\mathrm{d}t} \circ \frac{\partial \mathbf{r}_n}{\partial q_k}\delta q_k \\ &= m_n \left[\frac{\mathrm{d}}{\mathrm{d}t}\left(\mathbf{v}_n \circ \frac{\delta \mathbf{r}_n}{\delta q_k}\right) - \mathbf{v}_n \circ \frac{\mathrm{d}}{\mathrm{d}t}\left(\frac{\partial \mathbf{r}_n}{\partial q_k}\right)\right]\delta q_k, \end{aligned} \tag{3.124}$$

and hence

$$\begin{gathered} \left[\frac{\mathrm{d}}{\mathrm{d}t}\left(m_n \mathbf{v}_n \circ \frac{\delta \mathbf{r}_n}{\delta q_k}\right) - m_n \mathbf{v}_n \circ \frac{\mathrm{d}}{\mathrm{d}t}\left(\frac{\partial \mathbf{r}_n}{\partial q_k}\right)\right]\delta q_k = Q_k^{in}\delta q_k, \\ n = 1,\ldots,N, \quad k = 1,\ldots,K. \end{gathered} \tag{3.125}$$

Note that

$$\mathbf{v}_n = \frac{\mathrm{d}\mathbf{r}_n}{\mathrm{d}t} = \frac{\partial \mathbf{r}_n}{\partial q_1}\dot{q}_1 + \cdots + \frac{\partial \mathbf{r}_n}{\partial q_k}\dot{q}_k + \cdots + \frac{\partial \mathbf{r}_n}{\partial q_K}\dot{q}_K, \tag{3.126}$$

where $\dot{q}_k$ are *generalized velocities*.

The kinetic energy of the analyzed mechanical system can be expressed in terms of generalized coordinates and generalized velocities in the following way:

$$T = \frac{m_n v_n^2}{2} = \frac{m_n}{2}\left(\frac{\partial \mathbf{r}_n}{\partial q_k}\dot{q}_k\right)^2, \quad 1,\ldots,K, \quad n = 1,\ldots,N. \tag{3.127}$$

Let us calculate the derivatives of the kinetic energy with respect to the generalized velocity $\dot{q}_k$ and with respect to the generalized coordinate q_k

$$\frac{\partial T}{\partial \dot{q}_k} = m_n \left(\frac{\partial \mathbf{r}_n}{\partial q_k}\dot{q}_k\right) \circ \frac{\partial \mathbf{r}_n}{\partial q_k} = m_n \mathbf{v}_n \circ \frac{\partial \mathbf{r}_n}{\partial q_k}, \tag{3.128}$$

$$\frac{\partial T}{\partial \dot{q}_k} = m_n \left(\frac{\partial \mathbf{r}_n}{\partial q_k}\dot{q}_k\right) \circ \left(\frac{\partial^2 \mathbf{r}_n}{\partial q_i \partial q_k}\dot{q}_i\right) = m_n \mathbf{v}_n \circ \frac{\mathrm{d}}{\mathrm{d}t}\frac{\partial \mathbf{r}_n}{\partial q_k}, \tag{3.129}$$

where formula (3.126) was used.

Equations (3.125), after taking into account (3.128) and (3.129), take the form

$$\frac{\mathrm{d}}{\mathrm{d}t}\left(\frac{\partial T}{\partial \dot{q}_k}\right) - \frac{\partial T}{\partial q_k} = Q_k^{in}, \quad k = 1,\ldots,K. \tag{3.130}$$

Substituting (3.130) into (3.121) we obtain

$$\sum_{k=1}^{K}\left[Q_k + Q_k^n - \left(\frac{\mathrm{d}}{\mathrm{d}t}\frac{\partial T}{\partial \dot{q}_k} - \frac{\partial T}{\partial q_k}\right)\right]\delta q_k = 0. \tag{3.131}$$

We have already chosen the coordinates q_k to be independent, and hence from (3.131) we get

$$\frac{\mathrm{d}}{\mathrm{d}t}\frac{\partial T}{\partial \dot{q}_k} - \frac{\partial T}{\partial q_k} = Q_k + Q_k^n, \quad k = 1, \ldots, K. \tag{3.132}$$

In the preceding equations we have $2K$ unknowns, because the generalized coordinates $q_1, \ldots, q_k$ and the generalized non-holonomic reactions $Q_1^n, \ldots, Q_k^n$ are not known. We have at our disposal K (3.132) and M_2 equations of non-holonomic constraints (3.15), that is, in total $K + M_2 < 2K$, because $K > M_2$.

In order to determine any additional required equations we introduce the condition of ideal non-holonomic constraints of the form

$$\sum_{k=1}^{K} Q_k^n \delta q_k = 0. \tag{3.133}$$

The concept, described earlier, of constraints freezing allows for the replacement of non-holonomic constraints (3.15) with scleronomic constraints [this procedure is reduced to omitting in (3.15) terms β_{m_2} and additionally "freezing" time t in functions $\lambda_{m_2 n}$].

From (3.133) we have

$$Q_1^n \delta q_1 + Q_2^n \delta q_1 + \cdots + Q_K^n \delta q_K = 0, \tag{3.134}$$

and from (3.24) we obtain

$$\begin{aligned}
B_{11}\delta q_1 + B_{12}\delta q_2 + \cdots + B_{1K}\delta q_K &= 0,\\
B_{21}\delta q_1 + B_{22}\delta q_2 + \cdots + B_{2K}\delta q_K &= 0,\\
&\ \,\vdots\\
B_{M_2 1}\delta q_1 + B_{M_2 2}\delta q_2 + \cdots + B_{M_2 K}\delta q_K &= 0.
\end{aligned} \tag{3.135}$$

We have K Lagrange equations (3.132), and also M_2 additional equations (3.135) following from the non-holonomic constraints, whereas we need to determine K generalized coordinates and K non-holonomic reactions Q_k^n.

In order to solve this problem we make use of the method of auxiliary variables $\lambda_1, \lambda_2, \ldots, \lambda_M$ called the *Lagrange multiplers*. The desired K functions $Q_1^n, \ldots, Q_k^n$ are expressed in terms of M_2 Lagrange multipliers. Now the unknowns are $q_1, q_2, \ldots, q_k$, $\lambda_1, \lambda_2, \ldots, \lambda_{M_2}$ ($K + M_2$ in total), whereas we have at our disposal K Lagrange equations (3.132) and M_2 equations of non-holonomic constraints (3.21). To this end we multiply the first of equations (3.135) by λ_1, the second by λ_2, etc.

Next we add together (3.135) by ordering the terms at δq_k. Non-holonomic generalized forces Q_k^n, $k = 1, \ldots, K$ depend only on M_2 non-holonomic constraints, and then on the M_2 Lagrange multipliers λ_{m_2}.

By comparing (3.134) and (3.135) after the foregoing operation of multiplication through multipliers, adding by sides and ordering, and then by comparing the terms standing at the same coefficients δq_k we obtain

$$
\begin{aligned}
Q_1^n &= \lambda_1 B_{11} + \lambda_2 B_{21} +, \ldots, + \lambda_{M_2} B_{M_2 1}, \\
&\vdots \\
Q_k^n &= \lambda_1 B_{1k} + \lambda_2 B_{2k} +, \ldots, + \lambda_{M_2} B_{M_2 K},
\end{aligned}
\tag{3.136}
$$

or

$$
Q_k^n = \sum_{m_2=1}^{M_2} \lambda_{m_2} B_{m_2 k}, \quad k = 1, \ldots, K, \quad m_2 = 1, \ldots, M_2. \tag{3.137}
$$

Condition (3.133), taking into account (3.137), takes the following form:

$$
\sum_{k=1}^{K} \left(\sum_{m_2=1}^{M_2} \lambda_{m_2} B_{m_2 k} \right) \delta q_k = 0. \tag{3.138}
$$

Substituting (3.138) into Lagrange's equations and including the equation of constraints (3.21) we obtain

$$
\begin{aligned}
&\frac{\mathrm{d}}{\mathrm{d}t} \frac{\partial T}{\partial \dot{q}_k} - \frac{\partial T}{\partial q_k} = Q_k + \sum_{m_2=1}^{M_2} \lambda_{m_2} B_{m_2 k}, \quad k = 1, \ldots, K, \\
&\sum_{k=1}^{K} B_{m_2 k} \dot{q}_k + b_{m_2} = 0, \quad m_2 = 1, \ldots, M_2.
\end{aligned}
\tag{3.139}
$$

As is evident, we now have $K + M_2$ equations and $K + M_2$ unknowns, that is, $q_1, \ldots, q_k$, $\lambda_1, \ldots, \lambda_{M_2}$. Let us recall that according to our calculations, equations (3.139) describe the motion of N material points of the DMS under investigation with both holonomic and non-holonomic constraints. System of equations (3.139) is called a *system of Lagrange equations of the second kind with undetermined multipliers*.

Because a system is called non-holonomic if at least one of its constraints is non-holonomic, system (3.139) describes the dynamics of a non-holonomic DMS.

If $B_{m_2k} = 0$ and $b_{m_2} = 0$, then system of equations (3.139) reduces to the following equations:

$$\frac{\mathrm{d}}{\mathrm{d}t}\frac{\partial T}{\partial \dot{q}_k} - \frac{\partial T}{\partial q_k} = Q_k, \quad k = 1, \ldots, K. \tag{3.140}$$

System of equations (3.140) describes the motion of a non-holonomic DMS, and the number of independent generalized coordinates is equal to the number of degrees of freedom of the DMS, that is, $K = D^f$ (the motion of the DMS is restricted by introduction of the geometric constraints). If we release the DMS also from geometric constraints and the analyzed DMS becomes a free system, then Lagrange's equations of the second kind will not change and will be expressed through the same indices k as in the case of equations (3.140), where $K = D^f$ is understood as the number of degrees of freedom of the DMS with no constraints.

Let us now consider the case where on the DMS we have imposed M_1 geometric constraints (3.13), and additionally some generalized coordinates in the number of K_1 depend on each other, that is, there exist coordinates that do not satisfy equations of geometric constraints.

Following differentiation of holonomic constraints with respect to time we obtain relation (3.18), and taking into account the second equation of (3.19) we obtain

$$\begin{aligned}
&\sum_{n=1}^{N} \frac{\partial f_m}{\partial \mathbf{r}_n} \circ \left(\sum_{k=1}^{K_1} \frac{\partial \mathbf{r}_n}{\partial q_k} \dot{q}_k + \frac{\partial \mathbf{r}_n}{\partial t} \right) \\
&= \sum_{k=1}^{K_1} \left(\sum_{k=1}^{K_1} \frac{\partial f_m}{\partial \mathbf{r}} \circ \frac{\partial \mathbf{r}_n}{\partial q_k} \right) \dot{q}_k + \sum_{n=1}^{N} \frac{\partial f_m}{\partial \mathbf{r}_n} \circ \frac{\partial \mathbf{r}_n}{\partial t} \\
&= \sum_{k=1}^{K_1} B_{mk}^h \dot{q}_k + b_m^h.
\end{aligned} \tag{3.141}$$

The obtained result exhibits a similarity to the previously analyzed kinematic (non-holonomic) constraints. However, in this case the constraints were obtained by differentiation of the geometric constraints, that is, they are integrable (in contrast to the non-integrable non-holonomic constraints). The following procedure, analogous to the one described above, can be carried out also for the case of non-holonomic constraints, which boils down to omitting the index h in relation (3.141).

The general equation of dynamics for the considered case takes the following form:

$$\sum_{k=1}^{K_1} \left(\frac{\mathrm{d}}{\mathrm{d}t}\frac{\partial T}{\partial \dot{q}_k} - \frac{\partial T}{\partial q_k} - Q_k - Q_k^h \right) \delta q_k = 0, \tag{3.142}$$

but now only on some of the quantities δq_k are the restrictions of motion imposed.

From (3.13) it follows that

$$f_m(t, q_1, \ldots, q_{K_1}) = 0, \quad m = 1, \ldots, M_1. \tag{3.143}$$

Therefore, we have M_1 equations of geometric constraints, whereas we have introduced K_1 independent generalized coordinates. By proceeding in a way analogous to that described previously and concerning the analysis of the non-holonomic constraints using (3.141) we obtain the following system of equations describing the motion of a holonomic DMS of the form

$$\frac{\mathrm{d}}{\mathrm{d}t}\frac{\partial T}{\partial \dot{q}_k} - \frac{\partial T}{\partial q_k} = Q_k + \sum_{m=1}^{M_1} \lambda_m B_{mk}^h, \quad k = 1, \ldots, K_1,$$

$$\sum_{k=1}^{K_1} B_{mk}^h \dot{q}_k + b_m^h = 0, \quad m = 1, \ldots, M_1. \tag{3.144}$$

System (3.144) contains $K_1 + M_1$ equations with $K_1 + M_1$ unknowns, and M_1 undetermined Lagrange multipliers appear here; this system is called *Lagrange's equations of the first kind*. Reactions of the holonomic constraints are defined by $Q_k^h = \sum_{m=1}^{M_1} \lambda_m B_{mk}^h$.

Until now, we have considered a discrete mechanical system. If the number of masses tends to infinity, it is possible to pass to a continuous mechanical system. We will derive Lagrange's equations of the second kind for the system of a continuous mass distribution [2, 11].

We take a point of a continuous mechanical system described by a radius vector $\mathbf{r}(t)$ and separate in its proximity a small element of mass dm. Let dm be subjected to the action of force $\mathbf{f}$dm, where $\mathbf{f}$ is the force related to the unit of mass.

By Newton's second law we have

$$\mathrm{d}m \frac{\mathrm{d}^2\mathbf{r}(t)}{\mathrm{d}t^2} = \mathbf{f}\mathrm{d}m. \tag{3.145}$$

The virtual displacement $\delta\mathbf{r}$ is associated with the virtual work done by the force $\mathbf{f}$dm during that displacement, that is, $\mathbf{f}\mathrm{d}m\delta\mathbf{r}$. In view of that the virtual work done by the force acting on the continuous mechanical system of mass m is equal to

$$\delta W = \int_m \delta\mathbf{r} \circ \mathbf{f}\mathrm{d}m. \tag{3.146}$$

Using (3.23) we write

$$\delta W = \delta q_n \int_m \frac{\partial \mathbf{r}}{\partial q_n} \circ \mathbf{f}\mathrm{d}m = Q_n \delta q_n, \tag{3.147}$$

$$Q_n = \int_m \frac{\partial \mathbf{r}}{\partial q_n} \circ \mathbf{f}\mathrm{d}m, \tag{3.148}$$

where Q_n denote generalized forces or coefficients of a linear form of variation of the mechanical system in the configuration space, that is, we have N degrees of freedom of the system. If, as in the calculations concerning the discrete system, we assume that we have ideal constraints imposed on the system, then the virtual work of the ideal reactions during the virtual displacements is equal to zero. The remaining forces acting on the system are denoted by **f**, that is, just as previously.

From (3.145) and (3.23) we obtain

$$\delta q_n \int_m \frac{\partial \mathbf{r}}{\partial q_n} \circ (\ddot{\mathbf{r}} - \mathbf{f})\, \mathrm{d}m = 0, \quad n = 1, \ldots, N, \tag{3.149}$$

and taking into account (3.148) we have

$$\delta q_n \left(\int_m \frac{\partial \mathbf{r}}{\partial q_n} \circ \ddot{\mathbf{r}} \mathrm{d}m - Q_n \right) = 0. \tag{3.150}$$

For a holonomic system all δq_n are independent, and from formula (3.150) we obtain

$$\int_m \frac{\partial \mathbf{r}}{\partial q_n} \circ \ddot{\mathbf{r}} \mathrm{d}m = Q_n, \quad n = 1, \ldots, N. \tag{3.151}$$

Note that

$$\dot{\mathbf{r}} = \mathbf{v} = \frac{\partial \mathbf{r}}{\partial t} + \frac{\partial \mathbf{r}}{\partial q_n} \dot{q}_n, \tag{3.152}$$

and additionally

$$\frac{\partial \mathbf{v}}{\partial \dot{q}_n} = \frac{\partial \mathbf{r}}{\partial q_n}. \tag{3.153}$$

From equality (3.153) it follows that

$$\frac{\mathrm{d}}{\mathrm{d}t} \left(\frac{\partial \mathbf{r}}{\partial q_n} \right) = \frac{\partial^2 \mathbf{r}}{\partial t \partial q_n} + \frac{\partial^2 \mathbf{r}}{\partial q_i \partial q_n} \dot{q}_i, \tag{3.154}$$

and from expression (3.152) we have that

$$\frac{\partial \mathbf{v}}{\partial q_n} = \frac{\partial^2 \mathbf{r}}{\partial t \partial q_n} + \frac{\partial^2 \mathbf{r}}{\partial q_i \partial q_n} \dot{q}_i. \tag{3.155}$$

A comparison of (3.154) with (3.155) leads to the following conclusion:

$$\frac{\mathrm{d}}{\mathrm{d}t} \left(\frac{\partial \mathbf{r}}{\partial q_n} \right) = \frac{\partial \mathbf{v}}{\partial q_n}. \tag{3.156}$$

The problem boils down to the calculation of the integral occurring in (3.151), which takes the form

$$\int_m \frac{\partial \mathbf{r}}{\partial q_n} \circ \ddot{\mathbf{r}} \mathrm{d}m = \int_m \frac{\mathrm{d}}{\mathrm{d}t} \left(\frac{\partial \mathbf{r}}{\partial q_n} \circ \mathbf{v} \right) \mathrm{d}m - \int_m \frac{\mathrm{d}}{\mathrm{d}t} \left(\frac{\partial \mathbf{r}}{\partial q_n} \right) \circ \mathbf{v} \mathrm{d}m. \tag{3.157}$$

Using (3.153) and (3.155) in the preceding equation we obtain

$$\begin{aligned} \int_m \frac{\partial \mathbf{r}}{\partial q_n} \circ \ddot{\mathbf{r}} \mathrm{d}m &= \int_m \frac{\mathrm{d}}{\mathrm{d}t} \left(\frac{\partial \mathbf{v}}{\partial \dot{q}_n} \circ \mathbf{v} \right) \mathrm{d}m - \int_m \frac{\partial \mathbf{v}}{\partial q_n} \circ \mathbf{v} \mathrm{d}m \\ &= \frac{\mathrm{d}}{\mathrm{d}t} \left(\frac{\partial T}{\partial \dot{q}_n} \right) - \frac{\partial T}{\partial q_n}, \end{aligned} \tag{3.158}$$

where T denotes the kinetic energy of the mechanical system of mass m, that is,

$$T = \frac{1}{2} \int_m \mathbf{v} \circ \mathbf{v} \mathrm{d}m. \tag{3.159}$$

Eventually, from (3.151), taking into account (3.158), we obtain Lagrange's equations of the second kind for holonomic systems of the form (3.140).

Let us consider a potential force Q_n, that is, we assume the existence of a scalar function of time and generalized coordinates V such that

$$Q_n = -\frac{\partial V\,(t, q_1, \ldots, q_n)}{\partial q_n}, \quad n = 1, \ldots, N. \tag{3.160}$$

One may also introduce the notion of *generalized potential forces* Q_n of the form

$$\begin{aligned} Q_n = &-\frac{\partial V(t, q_1, \ldots, q_N, \dot{q}_1, \ldots, \dot{q}_N)}{\partial q_n} \\ &+ \frac{\mathrm{d}}{\mathrm{d}t} \left(\frac{\partial V(t, q_1, \ldots, q_N, \dot{q}_1, \ldots, \dot{q}_N)}{\partial \dot{q}_n} \right), \quad n = 1, \ldots, N. \end{aligned} \tag{3.161}$$

After the introduction of the Lagrangian function (also called Lagrangian or kinetic potential)

$$L = T - V, \tag{3.162}$$

Lagrange's equations of the second kind take the form

$$\frac{\mathrm{d}}{\mathrm{d}t} \left(\frac{\partial L}{\partial \dot{q}_n} \right) - \frac{\partial L}{\partial q_n} = 0, \quad n = 1, \ldots, N. \tag{3.163}$$

The obtained equations are valid for conservative systems.

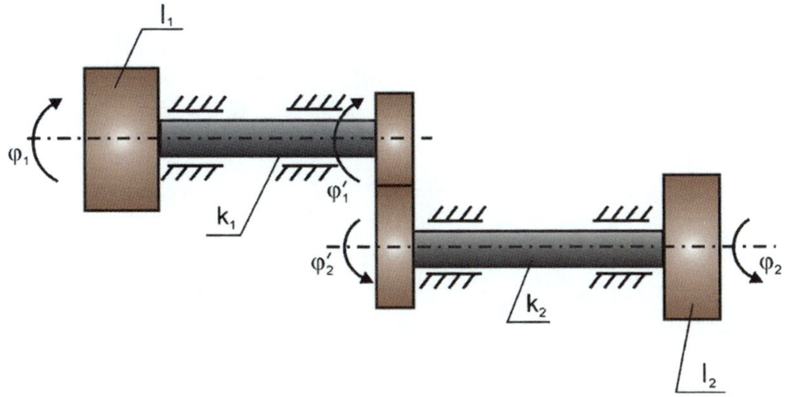

Fig. 3.18 Toothed gear transmission with disks of mass moments of inertia I_1 and I_2

Very often the introduced Lagrangian $L(t, q, \dot{q})$ is treated as an arbitrary function (a relationship of a function's independent variables). The class of Lagrangian dynamics includes a free and constrained DMS.

Finally, we note that the forces linearly dependent on the coordinates

$$Q_i = a_{ij} q_j, \quad a_{ij} = a_{ji}, \tag{3.164}$$

$$Q_i = c_{ij} \dot{q}_j, \quad c_{ij} = -c_{ji} \tag{3.165}$$

have respectively the potentials

$$V = \frac{1}{2} a_{ij} q_i q_j, \tag{3.166}$$

$$V = \frac{1}{2} c_{ij} q_i \dot{q}_j. \tag{3.167}$$

The forces described by (3.165) are called *gyroscopic forces*, and examples of them will be given later.

Example 3.11. Compose equations of motion of the mechanical system depicted in Fig. 3.18 using Lagrange's equations of the second kind. The following symbols are introduced in the figure: k_1, k_2 – torsional stiffnesses; z_1, z_2 – gear tooth numbers, and the shafts and gears are assumed to be massless; I_1, I_2 – moments of inertia of the rigid disks.

In the considered case, Lagrange's equations of the second kind take the form

$$\frac{\mathrm{d}}{\mathrm{d}t} \frac{\partial T}{\partial \dot{q}_n} + \frac{\partial V}{\partial q_n} = 0, \quad n = 1, 2.$$

The preceding form of Lagrange's equations of the second kind is valid only in the case where the system's kinetic energy does not depend on generalized coordinates. The kinetic and potential energies of the considered mechanical system have the form

$$T = \frac{I_1\dot{\varphi}_1^2}{2} + \frac{I_2\dot{\varphi}_2^2}{2},$$
$$V = \frac{1}{2}k_1\left(\varphi_1 - \varphi_1'\right)^2 + \frac{1}{2}k_2\left(\varphi_2 - \varphi_2'\right)^2.$$

The gear ratio of the toothed gear transmission is equal to

$$i = \left|\frac{\varphi_1'}{\varphi_2'}\right| = \frac{z_2}{z_1},$$

and if $z_2 > z_1$, then $\varphi_2' < \varphi_1'$. We calculate successively

$$\frac{\partial T}{\partial \dot{\varphi}_1} = I_1\dot{\varphi}_1, \quad \frac{\partial T}{\partial \dot{\varphi}_2} = I_2\dot{\varphi},$$
$$\frac{\partial V}{\partial \varphi_1} = k_1\left(\varphi_1 - \varphi_1'\right), \quad \frac{\partial V}{\partial \varphi_2} = k_2\left(\varphi_2 - \varphi_2'\right). \tag{$*$}$$

Let us introduce torsional moments as

$$M_i = k_i\left(\varphi_i - \varphi_i'\right), \quad i = 1, 2.$$

Assuming that power is conserved during the transmission of rotations we can write

$$M_1\omega_1 = M_2\omega_2,$$

where $\omega_1 = \dot{\varphi}_1'$, $\omega_2 = \dot{\varphi}_2'$, and hence we obtain

$$iM_1 = M_2.$$

We get further

$$ik_1\left(\varphi_1 - \varphi_1'\right) + k_2\left(\varphi_2 - \varphi_2'\right) = 0, \tag{$**$}$$

where

$$\varphi_1' = i\varphi_2'.$$

Substituting formulas $(*)$ into the Lagrange equations and taking into account equality we obtain the following two second-order differential equations:

$$I_1\ddot{\varphi}_1 + k_1\left(\varphi_1 - i\varphi_2'\right) = 0,$$
$$I_2\ddot{\varphi}_2 + k_2\left(\varphi_2 - \varphi_2'\right) = 0. \tag{$***$}$$

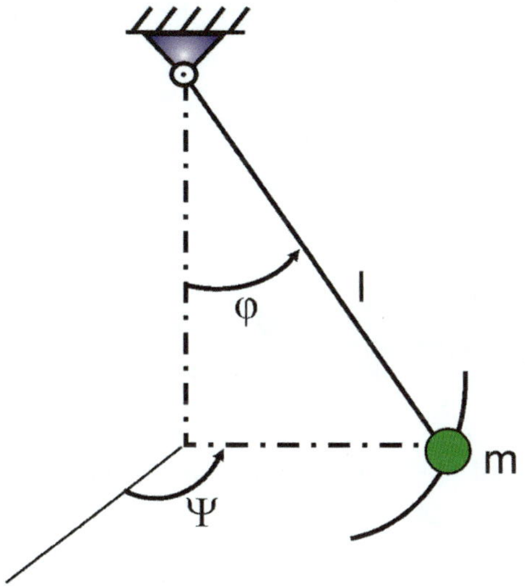

Fig. 3.19 Spherical pendulum and generalized coordinates φ and ψ

The angle φ_2' is expressed through relationship (∗∗), which takes the form

$$ik_1\left(\varphi_1 - i\varphi_2'\right) + k_2\left(\varphi_2' - \varphi_2\right) = 0,$$

hence we find

$$\varphi_2' = \frac{k_2}{k_2 + i^2k_1}\varphi_2 + \frac{ik_1}{k_2 + i^2k_1}\varphi_1.$$

Substituting the preceding expression into (∗∗∗) we obtain

$$I_1\ddot{\varphi}_1 + \frac{k_1k_2}{k_2 + i^2k_1}\varphi_1 - \frac{ik_1k_2}{k_2 + i^2k_1}\varphi_2 = 0,$$

$$I_2\ddot{\varphi}_2 - \frac{ik_1k_2}{k_2 + i^2k_1}\varphi_1 + \frac{i^2k_1k_2}{k_2 + i^2k_1}\varphi_2 = 0.$$

The obtained system of equations is linear and autonomous and can be solved analytically. □

Example 3.12. Derive equations of motion of the spherical pendulum shown in Fig. 3.19, where the mass m is attached at the end of a rigid weightless rod of length l.

Let us make use of Lagrange's equations of the form

$$\frac{\mathrm{d}}{\mathrm{d}t}\frac{\partial T}{\partial \dot{q}_n} - \frac{\partial L}{\partial q_n} = 0, \quad L = T - V.$$

The kinetic energy T and the potential energy V have the form

$$T = \frac{1}{2}m\left((l\dot{\varphi})^2 + \left(l\dot{\Psi}\sin\varphi\right)^2\right),$$
$$V = mgl\,(1-\cos\varphi)\,.$$

After differentiation we obtain

$$\frac{\partial T}{\partial\dot{\varphi}} = ml^2\dot{\varphi}, \qquad \frac{\partial T}{\partial\dot{\Psi}} = ml^2\dot{\Psi}\sin^2\varphi,$$
$$\frac{\partial L}{\partial\varphi} = \frac{1}{2}ml^2\dot{\Psi}^2\sin 2\varphi - mgl\sin\varphi, \qquad \frac{\partial L}{\partial\Psi} = 0,$$

and taking into account the Lagrange equations, we have

$$ml^2\ddot{\varphi} - \frac{1}{2}ml^2\dot{\Psi}^2\sin 2\varphi + mgl\sin\varphi = 0,$$
$$ml^2\ddot{\Psi}\sin^2\varphi + 2ml^2\dot{\Psi}\dot{\varphi}\sin\varphi\cos\varphi = 0. \qquad (*)$$

Following transformations, equations $(*)$ take the form

$$\ddot{\varphi} - \frac{1}{2}\dot{\Psi}^2\sin 2\varphi + \frac{g}{l}\sin\varphi = 0,$$
$$\ddot{\Psi}\sin^2\varphi + \dot{\Psi}\dot{\varphi}\sin 2\varphi = 0. \qquad (**)$$

The second equation of system $(**)$ can be written in the form

$$\frac{\mathrm{d}}{\mathrm{d}t}\left(\dot{\Psi}\sin^2\varphi\right) = 0,$$

which leads to its immediate integration, and eventually $(**)$ take the form

$$\ddot{\varphi} - \frac{1}{2}\dot{\Psi}^2\sin 2\varphi + \frac{g}{l}\sin\varphi = 0,$$
$$\dot{\Psi}\sin^2\varphi = C,$$

where C is a constant of integration.

If $\dot{\psi} = 0$, that is, we introduce the initial conditions of the form $\psi(0) = \dot{\psi}(0) = 0$, then from system of the preceding equations, we obtain

$$\ddot{\varphi} + \frac{g}{l}\sin\varphi = 0.$$

This means that the motion takes place in a plane and corresponds to the oscillations of the mathematical pendulum, already known to the reader. □

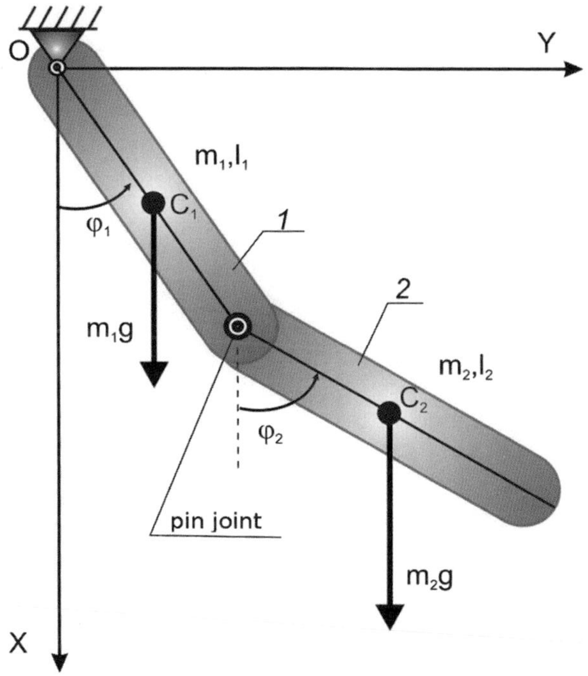

Fig. 3.20 Double compound pendulum (coordinates x_i, y_i, $i = 1, 2$ describe the positions of mass centers of particular pendulums)

Example 3.13. Derive equations of motion of the double compound pendulum (two thin pin-connected homogeneous rods) depicted in Fig. 3.20 using Lagrange's equations of the second kind and of the first kind.

The kinetic energy of the system has the form

$$T = \frac{1}{2}m_1\left(\dot{x}_1^2 + \dot{y}_1^2\right) + \frac{1}{2}m_2\left(\dot{x}_2^2 + \dot{y}_2^2\right) + \frac{1}{2}I_1\dot{\varphi}_1^2 + \frac{1}{2}I_2\dot{\varphi}_2^2,$$

and the constraint equations

$$\psi_1 \equiv x_1 - \frac{l_1}{2}\cos\varphi_1 = 0, \qquad \psi_2 \equiv y_1 - \frac{l_1}{2}\sin\varphi_1 = 0,$$

$$\psi_3 \equiv x_2 - l_1\cos\varphi_1 - \frac{l_2}{2}\cos\varphi_2 = 0, \quad \psi_4 \equiv y_2 - l_1\sin\varphi_1 - \frac{l_2}{2}\sin\varphi_2 = 0. \qquad (*)$$

The preceding equations are valid if the mass centers of both bodies are in the middle of segments l_1 and l_2.

Differentiating equations (∗) with respect to time we obtain

$$\dot{x}_1 = -\frac{l_1}{2}\dot{\varphi}_1 \sin\varphi_1, \qquad \dot{y}_1 = \frac{l_1}{2}\dot{\varphi}\cos\varphi_1,$$

$$\dot{x}_2 = -l_1\dot{\varphi}_1 \sin\varphi_1 - \frac{l_2}{2}\dot{\varphi}_2 \sin\varphi_2, \qquad \dot{y}_2 = l_1\dot{\varphi}_1\cos\varphi + \frac{l_2}{2}\dot{\varphi}_2\cos\varphi_2,$$

and hence

$$\begin{aligned}
\dot{x}_1^2 + \dot{y}_1^2 &= \frac{l_1^2}{4}\dot{\varphi}_1^2,\\
\dot{x}_2^2 + \dot{y}_2^2 &= l_1^2\dot{\varphi}_1^2\sin^2\varphi_1 + \frac{l_2^2}{4}\dot{\varphi}_2^2\sin^2\varphi_2 + l_1l_2\dot{\varphi}_1\dot{\varphi}_2\sin\varphi_1\sin\varphi_2\\
&\quad + l_1^2\dot{\varphi}_1^2\cos^2\varphi_1 + \frac{l_2^2}{4}\dot{\varphi}_2^2\cos^2\varphi_2 + l_1l_2\dot{\varphi}_1\dot{\varphi}_2\cos\varphi_1\cos\varphi_2\\
&= l_1^2\dot{\varphi}_1^2 + \frac{l_2^2}{4}\dot{\varphi}_2^2 + l_1l_2\dot{\varphi}_1\dot{\varphi}_2\cos(\varphi_1-\varphi_2).
\end{aligned}$$

Taking the preceding relationships into account in the formula for kinetic energy we obtain

$$\begin{aligned}
T &= \frac{1}{8}m_1l_1^2\dot{\varphi}_1^2 + \frac{1}{2}m_2l_1^2\dot{\varphi}_1^2 + \frac{1}{8}m_2l_2^2\dot{\varphi}_2^2\\
&\quad + \frac{1}{2}m_2l_1l_2\dot{\varphi}_1\dot{\varphi}_2\cos(\varphi_1-\varphi_2) + \frac{m_1l_1^2\dot{\varphi}_1^2}{24} + \frac{m_2l_2^2\dot{\varphi}_2^2}{24}\\
&= \frac{1}{6}m_1l_1^2\dot{\varphi}_1^2 + \frac{1}{2}m_2l_1^2\dot{\varphi}_1^2 + \frac{1}{6}m_2l_2^2\dot{\varphi}_2^2 + \frac{1}{2}m_2l_1l_2\dot{\varphi}_1\dot{\varphi}_2\cos(\varphi_1-\varphi_2).
\end{aligned}$$

Differentiating the kinetic energy T with respect to the generalized coordinates and generalized velocities we have

$$\begin{aligned}
\frac{\partial T}{\partial\dot{\varphi}_1} &= \frac{1}{3}m_1l_1^2\dot{\varphi}_1^2 + l_1^2m_2\dot{\varphi}_1^2 + \frac{1}{2}m_2l_1l_2\dot{\varphi}_2^2\cos(\varphi_1-\varphi_2),\\
\frac{\partial T}{\partial\dot{\varphi}_2} &= \frac{1}{3}m_2l_2^2\dot{\varphi}_2^2 + \frac{1}{2}m_2l_1l_2\dot{\varphi}_1\cos(\varphi_1-\varphi_2),\\
\frac{\mathrm{d}}{\mathrm{d}t}\left(\frac{\partial T}{\partial\dot{\varphi}_1}\right) &= \frac{1}{3}m_1l_1^2\ddot{\varphi}_1 + \frac{1}{2}m_2l_1l_2\ddot{\varphi}_2\cos(\varphi_1-\varphi_2)\\
&\quad - \frac{1}{2}m_2l_1l_2\dot{\varphi}_2(\dot{\varphi}_1-\dot{\varphi}_2)\sin(\varphi_1-\varphi_2) + l_1^2m_2\ddot{\varphi}_1,
\end{aligned}$$

$$\frac{\mathrm{d}}{\mathrm{d}t}\left(\frac{\partial T}{\partial \dot{\varphi}_2}\right) = \frac{1}{3}m_2 l_2^2 \ddot{\varphi}_2 + \frac{1}{2}m_2 l_1 l_2 \ddot{\varphi}_1 \cos(\varphi_1 - \varphi_2)$$

$$-\frac{1}{2}m_2 l_1 l_2 \dot{\varphi}_1(\dot{\varphi}_1 - \dot{\varphi}_2)\sin(\varphi_1 - \varphi_2),$$

$$\frac{\partial T}{\partial \varphi_1} = -\frac{1}{2}m_2 l_1 l_2 \dot{\varphi}_1 \dot{\varphi}_2 \sin(\varphi_1 - \varphi_2),$$

$$\frac{\partial T}{\partial \varphi_2} = \frac{1}{2}m_2 l_1 l_2 \dot{\varphi}_1 \dot{\varphi}_2 \sin(\varphi_1 - \varphi_2).$$

The potential energy of the system takes the form

$$V = C - m_1 g x_1 - m_2 g x_2,$$

where g denotes the acceleration of gravity and C is a certain constant. We have

$$V = C - \frac{1}{2}m_1 g l_1 \cos\varphi_1 - g m_2 \left(l_1 \cos\varphi_1 + \frac{l_2}{2}\cos\varphi_2\right),$$

and hence

$$\frac{\partial V}{\partial \varphi_1} = \frac{1}{2}m_1 g l_1 \sin\varphi_1 + m_2 g l_1 \sin\varphi_1,$$

$$\frac{\partial V}{\partial \varphi_2} = \frac{1}{2}m_2 g l_2 \sin\varphi_2.$$

Using Lagrange's equations of the form (3.163) we obtain

$$\left(\frac{1}{3}m_1 + m_2\right) l_1^2 \ddot{\varphi}_1 + \frac{1}{2}m_2 l_1 l_2 \ddot{\varphi}_2 \cos(\varphi_1 - \varphi_2) - \frac{1}{2}m_2 l_1 l_2 \dot{\varphi}_2(\dot{\varphi}_1 - \dot{\varphi}_2)\sin(\varphi_1 - \varphi_2)$$

$$+\frac{1}{2}m_2 l_1 l_2 \dot{\varphi}_1 \dot{\varphi}_2 \sin(\varphi_1 - \varphi_2) + g \sin\varphi_1 l_1 \left(\frac{m_1}{2} + m_2\right) = 0,$$

$$\frac{1}{3}m_2 l_2 \ddot{\varphi}_2 + \frac{1}{2}m_2 l_1 l_2 \ddot{\varphi}_1 \cos(\varphi_1 - \varphi_2) - \frac{1}{2}m_2 l_1 l_2 \dot{\varphi}_1(\dot{\varphi}_1 - \dot{\varphi}_2)\sin(\varphi_1 - \varphi_2)$$

$$+\frac{1}{2}m_2 l_1 \dot{\varphi}_1 \dot{\varphi}_2 \sin(\varphi_1 - \varphi_2) + \frac{1}{2} g m_2 l_2 \sin\varphi_2 = 0,$$

and after their transformations we obtain the following equations:

$$\left(\frac{1}{3}m_1 + m_2\right) l_1^2\ddot{\varphi}_1 + \frac{1}{2}m_2 l_1 l_2 \ddot{\varphi}_2 \cos(\varphi_1 - \varphi_2) + \frac{1}{2}m_2 l_1 l_2 \dot{\varphi}_2^2 \sin(\varphi_1 - \varphi_2)$$
$$+ g l_1 \left(\frac{m_1}{2} + m_2\right) \sin\varphi_1 = 0,$$
$$\frac{1}{2}m_2 l_1 l_2 \ddot{\varphi}_1 \cos(\varphi_1 - \varphi_2) - \frac{1}{2}m_2 l_1 l_2 \dot{\varphi}_1^2 \sin(\varphi_1 - \varphi_2)$$
$$+ \frac{1}{3}m_2 l_2^2 \ddot{\varphi}_2 + \frac{1}{2} g m_2 l_2 \sin\varphi_2 = 0.$$

The obtained system of second-order differential equations can be transformed into a system of second-order equations uncoupled with respect to angular accelerations $\ddot{\varphi}_n$, $n = 1, 2$.

An arbitrary plane motion of a rigid body of number n can be described by the motion of the mass center of this body $\{x_n, y_n\}$ and the angle of rotation φ_n of the form

$$m_n \ddot{x}_n = F_{nx} + \sum_{m=1}^{M} \lambda_m \frac{\partial \Psi_m}{\partial x_n},$$
$$m_n \ddot{y}_n = F_{ny} + \sum_{m=1}^{M} \lambda_m \frac{\partial \Psi_m}{\partial y_n}, \qquad n = 1, \ldots, N,$$
$$I_n \ddot{\varphi}_n = M_n + \sum_{m=1}^{M} \lambda_m \frac{\partial \Psi_m}{\partial \varphi_n}, \tag{$**$}$$

where M denotes the number of geometric constraints.

In the preceding equations, m_n denotes the mass of the body n, F_{nx} and F_{yn} denote the projections of forces acting on the mass center of the body n, and M_n are the moments of force reduced to the body centers. Functions ψ_m are constraint equations, and λ_m denote unknown Lagrange multipliers. We have then $3N$ equations $(**)$ with $3N + M$ unknowns. Additional M constraint equations are subject to the integration two times, and afterward they are included in the system of $3N$ differential equations.

We derive the Lagrange's equations of the first kind ($N = 2$, $M = 4$).

Let us list only non-zero derivatives of formulas (∗):

$$\frac{\partial \Psi_1}{\partial x_1} = 1, \qquad \frac{\partial \Psi_1}{\partial \varphi_1} = \frac{l_1}{2}\sin\varphi_1,$$

$$\frac{\partial \Psi_2}{\partial y_1} = 1, \qquad \frac{\partial \Psi_2}{\partial \varphi_1} = -\frac{l_1}{2}\cos\varphi_1,$$

$$\frac{\partial \Psi_3}{\partial x_2} = 1, \qquad \frac{\partial \Psi_3}{\partial \varphi_1} = l_1\sin\varphi_1,$$

$$\frac{\partial \Psi_3}{\partial \varphi_2} = \frac{l_2}{2}\sin\varphi_2, \qquad \frac{\partial \Psi_4}{\partial y_2} = 1,$$

$$\frac{\partial \Psi_4}{\partial \varphi_1} = -\,l_1\cos\varphi_1, \qquad \frac{\partial \Psi_4}{\partial \varphi_2} = -\frac{l_2}{2}\cos\varphi_2.$$

Eventually, from (∗∗) we obtain a system of differential equations with Lagrange multipliers:

$$\begin{aligned}
m_1\ddot{x}_1 &= m_1 g + \lambda_1,\\
m_1\ddot{y}_1 &= \lambda_2,\\
m_2\ddot{x}_2 &= m_2 g + \lambda_3,\\
m_2\ddot{y}_2 &= \lambda_4,\\
I_1\ddot{\varphi}_1 &= \lambda_1\frac{l_1}{2}\sin\varphi_1 - \lambda_2\frac{l_1}{2}\cos\varphi_1 + \lambda_3 l_1\sin\varphi_1 - \lambda_4 l_1\cos\varphi_1,\\
I_2\ddot{\varphi}_2 &= \lambda_3\frac{l_2}{2}\sin\varphi_2 - \lambda_4\frac{l_2}{2}\cos\varphi_2\,.
\end{aligned}$$

Additional four equations are yielded by differentiation of the constraint equations (∗) and they follow:

$$\begin{aligned}
&\ddot{x}_1 + \frac{l_1}{2}\ddot{\varphi}_1\sin\varphi_1 + \frac{l_1}{2}\dot{\varphi}_1^2\cos\varphi_1 = 0,\\
&\ddot{y}_1 - \frac{l_1}{2}\ddot{\varphi}_1\cos\varphi_1 + \frac{l_1}{2}\dot{\varphi}_1^2\sin\varphi_1 = 0,\\
&\ddot{x}_2 + l_1\ddot{\varphi}_1\sin\varphi_1 + \frac{l_2}{2}\ddot{\varphi}_2\sin\varphi_2 + l_1\dot{\varphi}_1^2\cos\varphi_1 + \frac{l_2}{2}\dot{\varphi}_2^2\cos\varphi_2 = 0,\\
&\ddot{y}_2 - l_1\ddot{\varphi}_1\cos\varphi_1 - \frac{l_2}{2}\ddot{\varphi}_2\cos\varphi_2 + l_1\dot{\varphi}_1^2\sin\varphi_1 + \frac{l_2}{2}\dot{\varphi}_2^2\sin\varphi_2 = 0,
\end{aligned}$$

□

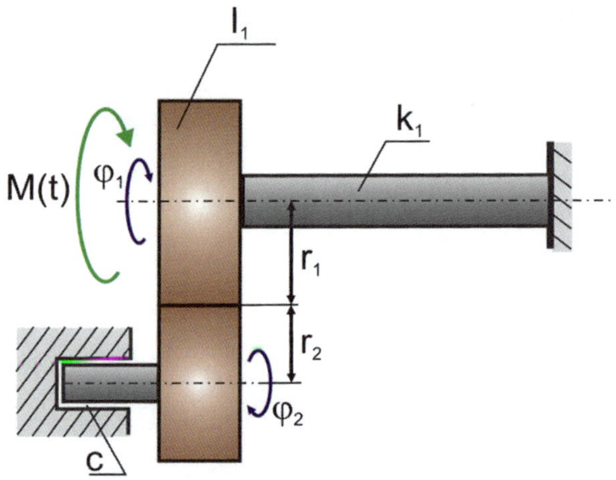

Fig. 3.21 Toothed gear transmission with viscous damping and driving torque $M(t)$

Example 3.14. Compose an equation of motion of the mechanical system shown in Fig. 3.21, and determine the intertooth force of a toothed gear transmission of mass moments of inertia I_1 and I_2. In the figure, k_1 denotes the torsional stiffness of the massless shaft, and additionally gear wheel (2) is viscously damped with the torque $M_t = -c\dot{\varphi}_2$, and gear wheel (1) is driven with the torque $M(t)$.

We will take the angles φ_1 and φ_2 as generalized coordinates that, as distinct from the case in Example 3.11, are now dependent on each other.

For the solution of the problem we will make use of Lagrange's equations of the first kind of the form

$$\frac{\mathrm{d}}{\mathrm{d}t}\frac{\partial T}{\partial \dot{\varphi}_n} + \frac{\partial V}{\partial \varphi_n} = Q_n + \lambda \frac{\partial f}{\partial \varphi_n}, \qquad n = 1, 2,$$

where Lagrange multiplier λ is the desired intertooth force, and additionally

$$Q_1 \delta\varphi_1 = M(t)\delta\varphi_1, \quad Q_2\delta\varphi_2 = -c\dot{\varphi}_2\delta\varphi_2,$$
$$f = r_1\varphi_1 - r_2\varphi_2 = 0, \quad T = \tfrac{I_1\dot{\varphi}_1^2}{2} + \tfrac{I_2\dot{\varphi}_2^2}{2}, \quad V = \tfrac{k_1\varphi_1^2}{2}.$$

Following differentiation and taking into account the geometric relation we have

$$\begin{aligned} I_1\ddot{\varphi}_1 + k_1\varphi_1 &= M(t) + \lambda r_1, \\ I_2\ddot{\varphi}_2 + c\dot{\varphi}_2 &= -\lambda r_2, \\ \varphi_2 &= i\varphi_1, \end{aligned} \qquad (*)$$

where $i = r_1/r_2$.

Equations $(*)$ make a system of three differential-algebraic equations with three unknowns: φ_1, φ_2, and λ.

However, the problem can be reduced to the analysis of one second-order non-homogeneous differential equation. Let us multiply the second equation of $(*)$ by i and then add the first and second equations, obtaining

$$I_1\ddot{\varphi}_1 + iI_2\ddot{\varphi}_2 + k_1\varphi_1 + ic\dot{\varphi}_2 = M(t),$$

and taking into account the third equation of $(*)$ we obtain

$$(I_1 + iI_2)\ddot{\varphi}_1 + i^2c\dot{\varphi}_1 + k_1\varphi_1 = M(t) \qquad (**)$$

or

$$(I_1 + i^2I_2)\ddot{\varphi}_2 + i^2c\dot{\varphi}_2 + k_1\varphi_2 = iM(t).$$

The reaction S_{12} of wheel (1) to wheel (2) is equal to

$$M_{R_1} = \lambda\frac{\partial f}{\partial \varphi_1} = \lambda r_1 \equiv S_{12}r_1,$$

and the reaction S_{21} of wheel (2) to wheel (1) equals

$$M_{R_2} = \lambda\frac{\partial f}{\partial \varphi_2} = -\lambda r_2 \equiv S_{12}r_2,$$

that is, $S_{12} = \lambda$ and $S_{21} = -\lambda$.

In order to determine the intertooth forces, we have to determine the Lagrange multiplier: from $(**)$ we determine $\varphi_1(t)$ in the known way, and then from the first equation of $(*)$ we determine $\lambda = \lambda(t)$. □

Example 3.15. Determine the equation of motion and reactions of constraints of the skate from Example 3.3 if the skate's mass is equal to m.

According to formula (3.44) we have

$$\dot{x}_{1C}\sin\varphi - \dot{x}_{2C}\cos\varphi = 0,$$

which defines the non-integrable non-holonomic constraints, and point C denotes the skate mass center.

We consider the stated problem without friction on the assumption that the position of the mass center of the skate is always at the same distance from the ice rink surface.

The kinetic energy of the skate equals

$$T = \frac{1}{2}m\left(\dot{x}_{1C}^2 + \dot{x}_{2C}^2\right) + \frac{1}{2}I_C\dot{\varphi}^2,$$

where I_C is the moment of inertia of the skate with respect to the axis perpendicular to the ice surface and passing through point C.

From Lagrange's equation (3.140) we obtain

$$\frac{\mathrm{d}}{\mathrm{d}t}\frac{\partial T}{\partial \dot{x}_{1C}} - \frac{\partial T}{\partial x_{1C}} = Q_{x_1} + \lambda \tan\varphi,$$
$$\frac{\mathrm{d}}{\mathrm{d}t}\frac{\partial T}{\partial \dot{x}_{2C}} - \frac{\partial T}{\partial x_{2C}} = Q_{x_2} - \lambda,$$
$$\frac{\mathrm{d}}{\mathrm{d}t}\frac{\partial T}{\partial \dot{\varphi}} - \frac{\partial T}{\partial \varphi} = Q_{\varphi}.$$

Let us take the following initial conditions of the skate motion: $x_{1C} = 0$, $x_{2C} = 0$, $\varphi = 0$. Furthermore, at the initial time instant let the skate mass center have linear velocity v_C and angular velocity ω_C, that is, $\dot{x}_{1_C}(0) = v_C$, $\dot{\omega} = \omega_C$ (from the equation of non-holonomic constraints we obtain $\dot{x}_{2C}(0) = 0$). Because the skate's potential energy during its motion does not change and there is no friction, we have $Q_{x_1} = 0$, Q_{x_2}, $Q_{\varphi} = 0$.

From Lagrange's equations of the first kind we obtain

$$m\ddot{x}_{1C} = \lambda\; tan\;\varphi, \quad m\ddot{x}_{2C} = -\lambda, \quad \ddot{\varphi} = 0.$$

Integrating twice the last equation we have $\varphi = \omega_C t$. This means that the skate during its motion rotates with a constant angular velocity ω_C about an axis perpendicular to the ice rink surface.

Combining the first two equations of motion gives

$$\ddot{x}_{1C} + \ddot{x}_{2C}\tan(\omega_C t) = 0. \tag{$*$}$$

Following differentiation of the equation of non-holonomic constraints we obtain

$$\ddot{x}_{1C}\tan\varphi + \dot{x}_{1C}\left(1 + \tan^2\varphi\right)\omega_C = \ddot{x}_{2C}. \tag{$**$}$$

Substituting ($**$) into ($*$) and following some transformations we obtain

$$\ddot{x}_{1C} + \omega_C \tan(\omega_C t)\,\dot{x}_{1C} = 0.$$

In order to solve the preceding equation we introduce the substitution

$$z = \dot{x}_{1C},$$

which allows us to decrease the order of the differential equation, and the problem boils down to the analysis

$$\dot{z} + \omega_C \tan(\omega_0 t)\, z = 0.$$

Separating the variables we obtain

$$\int \frac{\mathrm{d}z}{z} = -\int \tan\varphi \,\mathrm{d}\varphi,$$

and after integration we have

$$\ln|z| - \ln|\cos\varphi| = \ln C,$$

where C is a constant. Finally, we obtain

$$z = C\cos\varphi,$$

and from the initial condition $z(0) = v_0$ we have $C = v_C$. In view of that, we have

$$z \equiv \frac{\mathrm{d}x_{1C}}{\mathrm{d}t} = v_C \cos\omega_C t.$$

Following separation of the variables in the preceding equation and integration we obtain

$$x_{1C} = \frac{v_C}{\omega_C}\int\limits_0^{\varphi} \cos\varphi \,\mathrm{d}\varphi + C.$$

Because $x_{1C} = 0$, we have $C = 0$, and finally we obtain

$$x_{1C} = \frac{v_C}{\omega_C}\sin\omega_C t.$$

In order to determine $x_{2C}(t)$, using the constraints equation, we have

$$\dot{x}_{2C} = v_C \tan\varphi\cos\varphi = v_C \sin\varphi,$$

that is,

$$x_{2C} = \frac{v_C}{\omega_C}\int\limits_0^{\varphi} \sin\varphi \,\mathrm{d}\varphi + C,$$

which means that

$$x_{2C} = -\frac{v_C}{\omega_C}\left[\cos\varphi\right]_0^{\varphi} + C.$$

The constant C is determined from the following condition:

$$x_{2C}(0) \equiv 0 = -\frac{v_C}{\omega_C} + C,$$

that is, $C = v_C/\omega_C$. Finally,

$$x_{2C} = -\frac{v_C}{\omega_C}\left(1 - \cos\varphi\right).$$

In what follows we present a mechanical interpretation of the obtained solutions:

$$x_{1C} = \frac{v_C}{\omega_C}\sin\omega_C t, \quad x_{2C} = \frac{v_C}{\omega_C}(1-\cos\omega_C t).$$

Recall that the parametric equations of a circle of center $S(x_0, y_0)$ and radius r have the form

$$x = x_0 + r\cos\varphi,$$
$$y = y_0 + r\sin\varphi,$$

where φ is the arc measure of the angle.

In our case we have $r \equiv \frac{v_0}{\omega_C}$, $x_0 = 0$, $y_0 = \frac{v_0}{\omega_C}$. This means that during the motion of the skate, its center moves on a circle of radius v_0/ω_C whose center lies on the vertical axis of the adopted system of coordinates.

According to Lagrange's equations of the first kind the reactions of constraints read

$$R_{x_1} = \lambda\tan\omega_C t, \quad R_{x_2} = -\lambda. \tag{$***$}$$

The multiplier λ can be derived from the second equation of Lagrange's equations

$$\lambda = -m\ddot{x}_{2C} = -mv_C\omega_C\cos\omega_C t.$$

Finally, from equations $(***)$ we obtain

$$R_{x_1} = -mv_C\omega_C\sin\omega_C t, \quad R_{x_2} = mv_C\omega_C\cos\omega_C t.$$

The reaction of constraints $R = \sqrt{R_{x_1}^2 + R_{x_2}} = mv_C\omega_C$ has a constant value during motion of the skate. □

3.5 Properties of Lagrange's Equation

In a system with constraints, the motion of a DMS takes place in a certain subset Ω^K of the configuration space R^{3N}. The space Ω^K allows for the introduction of the so-called atlas of coordinates in which two different coordinate systems are defined simultaneously.

Note that, for example, in the case of particle motion on a spherical surface it is necessary to introduce two coordinate systems that would jointly cover completely the spherical surface. The dimension of the space Ω^K is equal to $K \le 3N$, and at every one of its points q it is possible to define the tangent space $T\Omega_q$, that is, the K-dimensional linear space of the vectors tangent to Ω^K at point q. The space Ω is a K-dimensional manifold. The tangent space $T\Omega_q$ is spanned by K tangent vectors $\gamma_1(1,0,\ldots,0),\ldots,\gamma_K(0,\ldots,0,1)$ and includes velocity vectors $\dot{q} = (\dot{q}_1,\ldots,\dot{q}_K)$ of every trajectory passing through point q.

Spaces Ω^K and $T\Omega_q$ can be merged into one $2K$-dimensional space $T\Omega^{2K}$, which we will call a tangent bundle of the manifold Ω^K. This space consists of pairs $(\mathbf{q}, \dot{\mathbf{q}})$ representing point q of manifold Ω^K and of every velocity vector $\dot{q}$ from the tangent space TQ_q.

Lagrange's equations (3.163) describe the motion on manifold Ω^K. The Lagrangian $L = (\mathbf{q}, \dot{\mathbf{q}}, t)$ is prescribed in the tangent bundle $T\Omega^{2K}$. If there exist constraints in a DMS, then the manifold Ω^K is described by constraint equations.

Lagrange's equation introduced on the basis of Newton's second law has many interesting properties. We will briefly describe them below [2].

1. *Covariance*. The methods introduced thus far of deriving Lagrange's equations of the second kind consisted in a description of kinetic and potential energies as functions q_n and $\dot{q}_n$ of generalized coordinates Q_n, where $n = 1,\ldots,N$. In the next step, differentiation of energies T and V (or of the Lagrangian L) yields N second-order differential equations. If we choose another set of generalized coordinates, the form of Lagrange's equations remains unaffected. If we subject the generalized coordinates to the transformation

$$q_i = q_i\left(t, \tilde{\mathbf{q}}\right), \tag{3.168}$$

then the form of Lagrange's equations will be the same. The transition from one system of generalized coordinates $\mathbf{q} = (q_1,\ldots,q_N)$ to the other $\tilde{\mathbf{q}} = (\tilde{q}_1,\ldots,\tilde{q}_N)$ is prescribed by the invertible and smooth transformation (3.168).

Since we have $L \to \tilde{L}$ and $Q \to \tilde{Q}$, the equations in new and old coordinates have the form

$$\begin{aligned}
\frac{\mathrm{d}}{\mathrm{d}t}\frac{\partial L}{\partial \dot{\mathbf{q}}} - \frac{\partial L}{\partial \mathbf{q}} &= \mathbf{Q},\\
\frac{\mathrm{d}}{\mathrm{d}t}\frac{\partial \tilde{L}}{\partial \dot{\tilde{\mathbf{q}}}} - \frac{\partial \tilde{L}}{\partial \tilde{\mathbf{q}}} &= \tilde{\mathbf{Q}},
\end{aligned} \tag{3.169}$$

because both Lagrangians $L = (\mathbf{q}, \dot{\mathbf{q}}, t)$ and $\tilde{L} = (\tilde{\mathbf{q}}, \dot{\tilde{\mathbf{q}}}, t)$ describe the same motions of a particle on manifold Ω^K. It follows that if $\mathbf{q}(t)$ is a trajectory of motion obtained from the equation generated by the Lagrangian $L = (\mathbf{q}, \dot{\mathbf{q}}, t)$, then the trajectory $\tilde{\mathbf{q}} = \tilde{\mathbf{q}}(t, \mathbf{q})$ obtained through transformation (3.168) also satisfies Lagrange's equation of the second kind but with Lagrangian $\tilde{L}$. The property of Lagrange's equation described thus far is called *covariance*.

Lagrange's equation of the form (3.169) points to the fact that in order to obtain differential equations of motion of a holonomic system in a potential field, one needs to know the form of only one function L.

2. *Calibration invariance*. If into Lagrange's equations (3.169) instead of the kinetic energy T we substitute

$$T(t, \mathbf{q}, \dot{\mathbf{q}}) = \frac{\mathrm{d}}{\mathrm{d}t} F(t, \mathbf{q}), \tag{3.170}$$

then they do not change, since we have

$$\frac{\mathrm{d}}{\mathrm{d}t}\left[\frac{\partial}{\partial \dot{\mathbf{q}}}\left(\frac{\mathrm{d}F}{\mathrm{d}t}\right)\right] - \frac{\partial}{\partial \mathbf{q}}\left(\frac{\mathrm{d}F}{\mathrm{d}t}\right) = \frac{\partial}{\partial \mathbf{q}}\left(\frac{\mathrm{d}F}{\mathrm{d}t}\right) - \frac{\partial}{\partial \mathbf{q}}\left(\frac{\mathrm{d}F}{\mathrm{d}t}\right) = 0. \tag{3.171}$$

If into the first of equations (3.169) instead of L we substitute $L + \frac{\mathrm{d}F}{\mathrm{d}t}$, then the equations will be not affected. Moreover, if in the Lagrangian $\tilde{L} = (\tilde{\mathbf{q}}, \dot{\tilde{\mathbf{q}}}, t)$ we change the variables $\tilde{\mathbf{q}}$ to $\mathbf{q}$ by means of transformation (3.168), then the new Lagrangian $\tilde{L} = (\mathbf{q}, \dot{\mathbf{q}}, t)$ determines the same trajectory as the Lagrangian $L = (\mathbf{q}, \dot{\mathbf{q}}, t)$. It follows that the difference $\tilde{L}(\mathbf{q}, \dot{\mathbf{q}}, t) - L(\mathbf{q}, \dot{\mathbf{q}}, t) = \frac{\mathrm{d}F(t,\mathbf{q})}{\mathrm{d}t}$ has to satisfy *identically* equation (3.163). We call the function $F(t, \mathbf{q})$ a *transformation function* from the Lagrangian $L = (\mathbf{q}, \dot{\mathbf{q}}, t)$ to the Lagrangian $\tilde{L} = (\tilde{\mathbf{q}}, \dot{\tilde{\mathbf{q}}}, t)$, and vice versa.
3. *Form of the kinetic energy*. During the derivation of Lagrange's equations it can be demonstrated that the kinetic energy has the form (Sect. 3.6)

$$T = T_2 + T_1 + T_0, \tag{3.172}$$

where

$$T_2 = \frac{1}{2} a_{ij}(t, \mathbf{q}) \dot{q}_i \dot{q}_j, \quad T_1 = b_i(t, \mathbf{q}) \dot{q}_i, \quad T_0 = T_0(t, \mathbf{q}). \tag{3.173}$$

4. *Non-singularity*. Substituting the energy forms (3.172) and (3.173) into Lagrange's equations we obtain

$$a_{nj} \ddot{q}_j + f_n(t, \mathbf{q}, \dot{\mathbf{q}}) = 0, \quad n = 1, \ldots, N. \tag{3.174}$$

Differential equations (3.174) can be solved easily with respect to accelerations, and then they can be represented in the form of normal (first-order) differential equations. It can be demonstrated that the square matrix $[a_{ij}]$ is a non-singular, symmetric, and positive-definite matrix. This means that Lagrange's equations satisfy the rule of determinism, that is, the Cauchy problem has a unique solution for the given initial conditions $\mathbf{q}_0 = \mathbf{q}(t_0)$, $\dot{\mathbf{q}}_0 = \dot{\mathbf{q}}(t_0)$.
5. *Hamilton's principle of least action*. By integrating the Lagrangian along the curve $\tilde{q}(t)$, where time is a parameter, we obtain the following number:

$$S = \int_{t_1}^{t_2} L\left(t, \tilde{\mathbf{q}}, \dot{\tilde{\mathbf{q}}}\right) \mathrm{d}t. \tag{3.175}$$

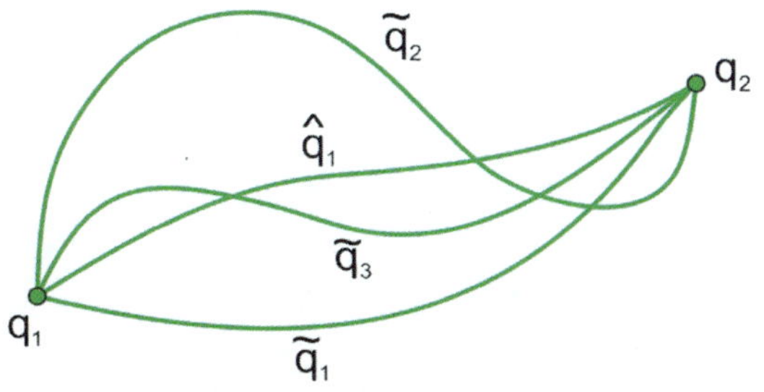

Fig. 3.22 Schematic illustrating Hamilton's principle of least action

Let us consider two arbitrary points 1 and 2 (Fig. 3.22) corresponding to time instants t_1 and t_2.

Point q_1 can be linked to point q_2 along various paths. Hamilton noticed that there exists a curve $\hat{q}$ joining those points such that it satisfies Lagrange's equation, that is,

$$\frac{\mathrm{d}}{\mathrm{d}t}\frac{\partial L}{\partial \dot{\hat{\mathbf{q}}}} - \frac{\partial L}{\partial \hat{\mathbf{q}}} = 0, \quad \hat{\mathbf{q}}(t_1) = q_1, \quad \hat{\mathbf{q}}(t_2) = q_2. \tag{3.176}$$

All curves joining two points can be parameterized through the introduction of a parameter α, where $\alpha = 0$ corresponds to the *actual* trajectory $\hat{q}$, which is the solution of boundary value problem (3.176). Because $\tilde{q} = \tilde{q}(\alpha, t)$, we have $S = S(\alpha)$, and we assume that the family of trajectories is differentiable with respect to the parameter α. Additionally, we have $\tilde{\mathbf{q}}(0, t) = \hat{\mathbf{q}}(t)$ and $S(\alpha) = \min$ for $\alpha = 0$. The condition of the minimum involves the satisfaction of the equation $\left.\frac{\mathrm{d}S}{\mathrm{d}\alpha}\right|_{\alpha=0} = 0$. We will formulate Hamilton's principle of least action on the basis of the following theorem.

Theorem 3.4. *The actual trajectory and only it is the extremum of the action according to the Hamilton.*

Proof. (see [2])
Differentiating (3.175) with respect to α we obtain

$$\frac{\mathrm{d}S}{\mathrm{d}\alpha} = \int_{t_1}^{t_2} \left(\frac{\partial L}{\partial \tilde{\mathbf{q}}}\frac{\partial \tilde{\mathbf{q}}}{\partial \alpha} + \frac{\partial L}{\partial \dot{\tilde{\mathbf{q}}}}\frac{\partial \dot{\tilde{\mathbf{q}}}}{\partial \alpha} \right) \mathrm{d}t. \tag{3.177}$$

Let us integrate by parts the second term of expression (3.177)

$$\int_{t_1}^{t_2} \frac{\partial L}{\partial \dot{\tilde{\mathbf{q}}}}\frac{\partial \dot{\tilde{\mathbf{q}}}}{\partial \alpha} dt = \left.\frac{\partial L}{\partial \dot{\tilde{\mathbf{q}}}}\right|_{t_1}^{t_2} \left.\frac{\partial \tilde{\mathbf{q}}}{\partial \alpha}\right|_{t_1}^{t_2} - \int_{t_1}^{t_2} \frac{\mathrm{d}}{\mathrm{d}t}\left(\frac{\partial L}{\partial \dot{\tilde{\mathbf{q}}}}\right)\frac{\partial \tilde{\mathbf{q}}}{\partial \alpha}\mathrm{d}t. \tag{3.178}$$

Because the curves pass through the points $\mathbf{q}_1 = \tilde{\mathbf{q}}_1$ and $\mathbf{q}_2 = \tilde{\mathbf{q}}_2$, we have $\left.\frac{\partial \tilde{\mathbf{q}}}{\partial \alpha}\right|_{t_1} = \left.\frac{\partial \tilde{\mathbf{q}}}{\partial \alpha}\right|_{t_2} = 0$, and after substituting (3.178) into formula (3.177), we obtain

$$\frac{\mathrm{d}S}{\mathrm{d}\alpha} = \int_{t_1}^{t_2} \frac{\partial \tilde{\mathbf{q}}}{\partial \alpha}\left(\frac{\partial L}{\partial \tilde{\mathbf{q}}} - \frac{\mathrm{d}}{\mathrm{d}t}\frac{\partial L}{\partial \dot{\tilde{\mathbf{q}}}}\right)\mathrm{d}t. \tag{3.179}$$

According to the definition of the trajectory as the *actual trajectory*, for $\alpha = 0$ we have $\frac{\partial L}{\partial \hat{\mathbf{q}}} - \frac{\mathrm{d}}{\mathrm{d}t}\frac{\partial L}{\partial \dot{\hat{\mathbf{q}}}} = 0$, which means that $\frac{\mathrm{d}S}{\mathrm{d}\alpha} = 0$. If, in turn, we take $\frac{\mathrm{d}S}{\mathrm{d}\alpha} = 0$ (the condition of minimum), then we obtain $\frac{\partial L}{\partial \hat{\mathbf{q}}} - \frac{\mathrm{d}}{\mathrm{d}t}\frac{\partial L}{\partial \dot{\hat{\mathbf{q}}}} = 0$, which was to be demonstrated.

In many textbooks this problem is considered based on the analysis of integral (3.175) understood as Hamilton's action and calculated along the trajectory $\tilde{q}(t)$. As was already mentioned, the actual (true) trajectory $\hat{q}$ satisfies Lagrange's equation (3.163).

Let us consider the difference of actions [i.e., the values of integrals (3.175)] calculated for the true trajectory $\hat{q}(t)$ and any trajectory $\tilde{q}(t)$ nearly coincident with the true one understood as

$$\delta q_i \equiv \tilde{q}_i(t) - \hat{q}_i(t), \quad t_1 \le t \le t_2, \quad i = 1, \ldots, K, \tag{3.180}$$

where δq_i, called *variations of trajectory*, are small.

According to the introduced definition of Hamilton's action we have [see (3.175)]

$$S = \int_{t_1}^{t_2} L(\tilde{\mathbf{q}}, \dot{\tilde{\mathbf{q}}}, t)\mathrm{d}t,$$

$$S_{\min} = \int_{t_1}^{t_2} L(\hat{\mathbf{q}}, \dot{\hat{\mathbf{q}}}, t)\mathrm{d}t, \tag{3.181}$$

that is,

$$\delta S = S - S_{\min} = \int_{t_1}^{t_2} \left[L(\tilde{\mathbf{q}}, \dot{\tilde{\mathbf{q}}}, t) - L(\hat{\mathbf{q}}, \dot{\hat{\mathbf{q}}}, t)\right]\mathrm{d}t. \tag{3.182}$$

In turn,

$$\begin{aligned} L(\tilde{\mathbf{q}}, \dot{\tilde{\mathbf{q}}}, t) &= L\left(\hat{\mathbf{q}} + \delta\mathbf{q}, \dot{\hat{\mathbf{q}}} + \delta\dot{\mathbf{q}}, t\right) - L\left(\hat{\mathbf{q}}, \dot{\hat{\mathbf{q}}}, t\right) \\ &\cong L\left(\hat{\mathbf{q}}, \dot{\hat{\mathbf{q}}}, t\right) + \left.\frac{\delta L}{\delta \mathbf{q}}\right|_{\hat{\mathbf{q}}, \dot{\hat{\mathbf{q}}}} \delta\mathbf{q} + \left.\frac{\delta L}{\delta \dot{\mathbf{q}}}\right|_{\hat{\mathbf{q}}, \dot{\hat{\mathbf{q}}}} \delta\dot{\mathbf{q}}, \end{aligned} \tag{3.183}$$

where in the preceding expansion into a series only the linear terms with respect to δq and $\delta \dot{q}$ were retained.

Substituting expression (3.183) into (3.182) we obtain

$$\delta S = \int_{t_1}^{t_2} \left(\left. \frac{\partial L}{\partial \mathbf{q}} \right|_{\hat{\mathbf{q}},\dot{\hat{\mathbf{q}}}} \delta \mathbf{q} + \left. \frac{\partial L}{\partial \dot{\mathbf{q}}} \right|_{\delta,\dot{\hat{\mathbf{q}}}} \right) \mathrm{d}t = 0. \tag{3.184}$$

The integral of the second equation on the right-hand side of (3.184) has the form

$$\int_{t_1}^{t_2} \left(\left. \frac{\partial L}{\partial \dot{\mathbf{q}}} \right|_{\hat{\mathbf{q}},\dot{\hat{\mathbf{q}}}} \delta \dot{\mathbf{q}} \right) \mathrm{d}t = \left[\frac{\partial L}{\partial \dot{\mathbf{q}}} \right]_{t_1}^{t_2} [\delta \mathbf{q}]_{t_1}^{t_2} - \int_{t_1}^{t_2} \frac{\mathrm{d}}{\mathrm{d}t} \left(\frac{\partial L}{\partial \dot{\mathbf{q}}} \right) \delta \mathbf{q} \mathrm{d}t, \tag{3.185}$$

where integration by parts was applied.

Because $\delta \mathbf{q}|_{t_1} = \delta \mathbf{q}|^{t_2} = 0$, after substituting formula (3.185) into equation (3.184) we obtain

$$\delta S = \delta \int_{t_1}^{t_2} L \left(\hat{\mathbf{q}}, \dot{\hat{\mathbf{q}}}, t \right) \mathrm{d}t = \int_{t_1}^{t_2} \delta \mathbf{q} \left[\frac{\delta L}{\delta \mathbf{q}} - \frac{\mathrm{d}}{\mathrm{d}t} \left(\frac{\delta L}{\delta \dot{\mathbf{q}}} \right) \right] \Bigg|_{\hat{\mathbf{q}},\dot{\hat{\mathbf{q}}}} \mathrm{d}t = 0. \tag{3.186}$$

The preceding integral we call a *first variation of (Hamilton's) action integral.*

In other words, attaining the stationary value (extremum value) on trajectory $\hat{q}(t)$ by integral (3.175) corresponds to the vanishing of the variation on the trajectory $\hat{q}(t)$ described by formula (3.176). And conversely, if the action described by formula (3.177) attains a stationary value on a certain trajectory $\hat{q}(t)$, it has to satisfy Lagrange's equation.

On the basis of the introduced Lagrangian function and the principle of least action, principles of Lagrangian mechanics are often formulated that unify the apparently distinct dynamic processes. □

3.6 First Integrals of Lagrange Systems

Let us consider an autonomous dynamic system of the form

$$\dot{x}_n = f_n(x_1, \ldots, x_N), \quad n = 1, \ldots, N. \tag{3.187}$$

Let system of differential equations (3.187) have general solutions of the form $x_n = \phi(t, C_1, \ldots, C_n)$, where now they are initial conditions for the aforementioned system of equations.

Let us consider the scalar function $F[x_1(t), x_2(t), \ldots, x_n(t)]$ determined in the same domain as the function f_i, where the following condition holds true:

$$F\,[x_1(t), x_2(t), \ldots, x_N(t)] \equiv \text{const.} \tag{3.188}$$

This means that along the trajectory $x_i(t)$ the function is constant (but is not identically equal to a constant). We call such a function F a *global first integral* or simply a *first integral*.

Differentiating formula (3.188) with respect to parameter t (time) we obtain

$$\frac{\mathrm{d}F}{\mathrm{d}t} = \frac{\partial F}{\partial x_n}\frac{\mathrm{d}x_n}{\mathrm{d}t} = \frac{\partial F}{\partial x_n} f_n(x_1, \ldots, x_N) \equiv 0. \tag{3.189}$$

If the scalar function F satisfies the condition of first integral, but only in a certain subdomain of the domain of the function $f_n(x_1, \ldots, x_N)$, then we call it a *local first integral*.

As was mentioned previously, if there exists such a global integral, then it allows for the reduction of a dimension of a phase space of the analyzed dynamical system.

Let us consider the conservative oscillator

$$\ddot{x} + x = 0. \tag{3.190}$$

We have then

$$\begin{aligned} \dot{x} &= v, \\ \dot{v} &= -x, \end{aligned} \tag{3.191}$$

and the global first integral has the form

$$F(x, v) = x^2 + v^2. \tag{3.192}$$

The solution of equality (3.191) reads

$$\begin{aligned} x &= A\sin t + B\cos t, \\ v &= A\cos t - B\sin t, \end{aligned} \tag{3.193}$$

that is,

$$\begin{aligned} F(x, v) &= A^2\sin^2 t + B^2\cos^2 t + A^2\cos^2 t + B^2\sin^2 t \\ &= A^2 + B^2. \end{aligned} \tag{3.194}$$

Thus it depends only on the initial conditions A and B, and for the given A and B it is a constant.

In a similar way we can determine the integral in the case of the non-autonomous system

$$\dot{x}_n = f_n(t, x_1, \ldots, x_N), \quad i = 1, \ldots, N. \tag{3.195}$$

Let us extend the phase space by one dimension by augmenting expression (3.189) with equation $\dot{t} = 1$. If a first integral is differentiable, it satisfies the condition

$$\frac{\partial F}{\partial t} + \frac{\partial F}{\partial x_n} f_n \equiv 0. \tag{3.196}$$

Now going back to the formalism introduced by Lagrange, the function $F(t, q, \dot{q})$, which assumes constant values during the motion of a system along a trajectory, is called the *first integral.*

We say that a mechanical system is *integrable* if it has a *global first integral.* According to this definition, as observed by Zhuravlev [2], a system with one degree of freedom with damping is not integrable.

We will now show how to find the first integral of Lagrange's equations when generalized forces have the potential. We multiply each kth Lagrange equation (3.163) (here we use index k instead of n) by $\dot{q}_k$, and then we add together those equations, obtaining

$$\dot{q}_k \frac{\mathrm{d}}{\mathrm{d}t} \frac{\partial L}{\partial \dot{q}_k} - \dot{q}_k \frac{\partial L}{\partial q_k} = 0. \tag{3.197}$$

Because

$$\frac{\mathrm{d}}{\mathrm{d}t}\left(\dot{q}_k \frac{\partial L}{\partial \dot{q}_k}\right) = \ddot{q}_k \frac{\partial L}{\partial \dot{q}_k} + \dot{q}_k \frac{\mathrm{d}}{\mathrm{d}t} \frac{\partial L}{\partial \dot{q}_k}, \tag{3.198}$$

$$\frac{\mathrm{d}}{\mathrm{d}t} L(t, \dot{\mathbf{q}}, \mathbf{q}) = \frac{\partial L}{\partial t} + \frac{\partial L}{\partial q_k}\dot{q}_k + \frac{\partial L}{\partial \dot{q}_k} = \ddot{q}_k, \tag{3.199}$$

and after taking into account (3.198) and (3.199) in (3.197) we obtain

$$\frac{\mathrm{d}}{\mathrm{d}t}\left(\dot{q}_k \frac{\partial L}{\partial \dot{q}_k}\right) - \frac{\mathrm{d}L}{\mathrm{d}t} + \frac{\partial L}{\partial t} = 0. \tag{3.200}$$

If the Lagrangian $L = L(\dot{q}, q)$, then we have $\frac{\partial L}{\partial t} = 0$, and from the preceding equation we obtain

$$\dot{q}_k \frac{\partial L}{\partial \dot{q}_k} - L = \text{const.} \tag{3.201}$$

The obtained first integral is called a *generalized integral of energy.*

Let us now examine the structure of the kinetic energy of a system. According to its definition and taking into account (3.19) we have

$$T = \frac{1}{2}\sum_{n=1}^{N} m_n \dot{\mathbf{r}}_n^2 = \frac{1}{2}\sum_{n=1}^{N} m_n \left(\sum_{k=1}^{K} \frac{\partial \mathbf{r}_n}{\partial q_k}\dot{q}_k + \frac{\partial \mathbf{r}_n}{\partial t}\right)^2$$
$$= \frac{1}{2}\sum_{k,i=1}^{K} a_{ki}\dot{q}_k\dot{q}_i + \sum_{k=1}^{K} a_k q_k + a_0, \tag{3.202}$$

where

$$a_{ki} = \sum_{n=1}^{N} m_n \frac{\partial \mathbf{r}}{\partial q_k}\frac{\partial \mathbf{r}}{\partial q_i}, \quad a_k = \sum_{n=1}^{N} m_n \frac{\partial \mathbf{r}_n}{\partial q_k}\frac{\partial \mathbf{r}_n}{\partial t},$$
$$a_0 = \frac{1}{2}\sum_{n=1}^{N} m_n \left(\frac{\partial \mathbf{r}_n}{\partial t}\right)^2. \tag{3.203}$$

One may conclude from (3.203) that the coefficients a_{ki}, a_k, and a_O are functions of $q_1, q_2, \ldots, q_K, t$, although they do not depend explicitly on time. According to (3.202) the kinetic energy of a DMS reads

$$T = T_2 + T_1 + T_0, \tag{3.204}$$

where

$$T_2 = \frac{1}{2}\sum_{k,i=1}^{K} a_{ki}\dot{q}_k\dot{q}_i,$$
$$T_1 = \sum_{k=1}^{K} a_k q_k,$$
$$T_0 = a_0. \tag{3.205}$$

One may show easily that the quadratic form $T_2 \geq 0$, and the determinant $\det [a_{ki}]_{k,i} \neq 0$.

Knowing the structure of the kinetic energy in a Lagrange system, one may determine the Lagrangian

$$L = T - V = L_2 + L_1 + L_0 = \frac{1}{2}a_{ki}\dot{q}_k\dot{q}_i + a_k\dot{q}_k + T_0 - V, \tag{3.206}$$

and hence the first integral (3.201) has the form

$$\frac{1}{2}a_{ki}\dot{q}_k\dot{q}_i - T_0 + V = T_2 - T_0 + V = \text{const}, \tag{3.207}$$

where $L_2 = T_2$, $L_1 = T_1$, and $L_0 = T - V$.

The total energy of the investigated system reads

$$E = T_2 + T_1 + T_0 + V, \tag{3.208}$$

where V denotes the potential energy.

From a comparison of (3.207) and (3.208) it follows that the total energy of the system is not conserved.

The next step will be the consideration of a conservative system. According to the definition given in [2], one is dealing with a conservative system if:

(1) $\frac{\partial \mathbf{r}}{\partial t} = 0$,
(2) $Q_k = \frac{\partial V}{\partial q_k}$,
(3) $\frac{\partial V}{\partial t} \equiv 0$.

Therefore, we have $T_1 = T_0 = 0$, and for a conservative system from (3.207) and (3.208) we obtain

$$E = T_2 + V = \text{const}, \tag{3.209}$$

which means that the sum of the kinetic and potential energy is constant at every time instant.

Until now, while defining the Lagrangian we assumed that the potential $V = V(q_n)$ and that we had $Q_n = -\frac{\partial V}{\partial q_n}$ (see, e.g., (3.160)).

However, one may also introduce the notion of the so-called *generalized potential* $V = V(q_i, \dot{q}_i, t)$, and then the generalized forces (3.160) take the form

$$Q_n = \frac{\mathrm{d}}{\mathrm{d}t}\frac{\partial V}{\partial \dot{q}_n} - \frac{\partial V}{\partial q_n}. \tag{3.210}$$

If the Lagrangian function $L(t, q_1, \ldots, q_N, \dot{q}_1, \ldots, \dot{q}_N)$ does not depend on certain coordinates, for instance, on $q_{K+1}, \ldots, q_N$, then we can immediately obtain $N - K$ first integrals of the considered mechanical system. Because Lagrange's equations in the field of potential forces and in the case of holonomic systems take the form

$$\frac{\mathrm{d}}{\mathrm{d}t}\frac{\partial L}{\partial \dot{q}_n} - \frac{\partial L}{\partial q_n} = 0, \quad n = 1, \ldots, K, \tag{3.211}$$

$$\frac{\mathrm{d}}{\mathrm{d}t}\frac{\partial L}{\partial \dot{q}_n} = 0, \quad n = K + 1, \ldots, N, \tag{3.212}$$

which means that

$$\frac{\partial L}{\partial \dot{q}_n} = C_n \equiv \text{const}, \quad n = K + 1, \ldots, N. \tag{3.213}$$

The coordinates $q_{K+1}, \ldots, q_N$ are called *cyclic coordinates*.

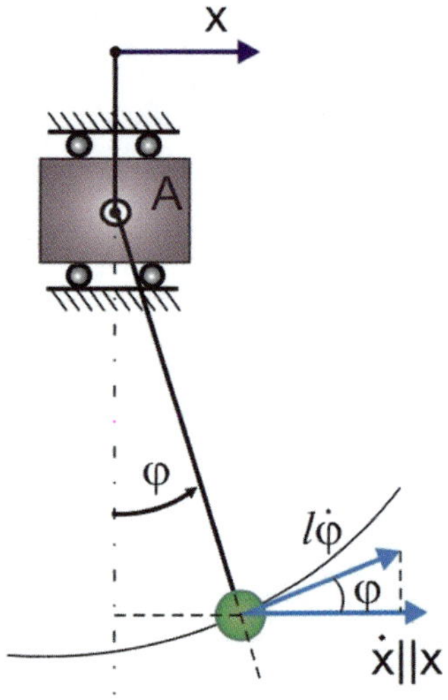

Fig. 3.23 Mathematical pendulum with a moveable pivot point

Definition 3.2. (of a *cyclic coordinate* [11]) *A coordinate q_C is called a cyclic coordinate if it does not occur in a Lagrangian function, that is, $\partial L/\partial q_C = 0$.*

Theorem 3.5. *If q_α is a cyclic coordinate, then according to* (3.213) *we obtain the Lagrange first integral. The remaining generalized coordinates of the analyzed system undergo changes in time defined by the system of $N - 1$ degrees of freedom, where the constant C_n plays the role of a parameter.*

Let us consider an example presented in [1] (Fig. 3.23).

The Lagrangian of the system is equal to

$$L = \frac{m}{2}\left[(\dot{x} + l\dot{\varphi}\cos\varphi)^2 + l^2\dot{\varphi}^2\sin^2\varphi\right] + mgl\cos\varphi. \tag{3.214}$$

The coordinate x is a cyclic coordinate because it does not occur in (3.214). According to (3.212) we have

$$\frac{\partial L}{\partial \dot{x}} = m(\dot{x} + l\dot{\varphi}\cos\varphi) = \text{const.} \tag{3.215}$$

If the system is conservative, then the total energy of the system is conserved, that is,

$$E = T + V = \frac{m}{2}\left(\dot{x}^2 + 2\dot{x}\dot{\varphi}l\cos\varphi + l^2\varphi^2\right) - mgl\cos\varphi. \tag{3.216}$$

Eliminating $\dot{x}$ from equation (3.215) and substituting into expression (3.216) we obtain first-order non-linear differential equation in $\varphi(t)$. As was mentioned, the order of the system is reduced by one.

The described method of determining first integrals leads to the reduction of the order of the analyzed equations and, consequaently, often to a simpler way of determining a solution. The analytical form expressed in terms of first integrals frequently describes important laws of mechanics connected with the conservation of certain physical quantities, which was already discussed in Chap. 1.

This last characteristic of first integrals is so important that very often the first integrals are determined despite the fact that the complete solution to the problem is already known. The first integrals enable the formulation of certain general laws of motion of an analyzed system and attribution to them of some physical meaning.

Below we present a method for making such conservation laws based on the schematic presented in Zhuravlev's monograph [2].

Let the system of differential equations (3.187) has a general solution of the form $\Phi_n(t, x_1, \ldots, x_N)$, where $x_1, \ldots, x_N$ are initial conditions. On that assumption we have

$$\frac{\mathrm{d}\phi_n}{\mathrm{d}t} = f_n\left[\phi_1(t, x_1, \ldots, x_N), \ldots, \phi_N(t, x_1, \ldots, x_N)\right]. \tag{3.217}$$

Theorem 3.6. *If there exists the integral*

$$F(x_1, \ldots, x_N) = \lim_{T \to \infty} \frac{1}{T} \int_0^T G\left[\phi_1(\tau, x_1, \ldots, x_N), \ldots, \phi_n(\tau, x_1, \ldots, x_N)\right] \mathrm{d}\tau, \tag{3.218}$$

which cannot be reduced to a constant, then the function $F(x_1, \ldots, x_N)$ is the first integral of the system of equations (3.187). It is worth noting that the choice of the function $G[\phi_1, \ldots, \phi_N]$ is arbitrary. It is chosen in such a way that the integral (3.218) can be calculated.

For the purpose of clarification of the formulated theorem, we present a slightly modified example excerpted from [2].

Example 3.16. Consider a conservative autonomous oscillator with one degree of freedom whose equations have the form

$$\begin{aligned} \dot{x}_1 &= x_2, \\ \dot{x}_2 &= -\alpha^2 x_1. \end{aligned} \tag{$*$}$$

It is known that in this case a general form of solutions is

$$\begin{aligned} x_1 &= \phi_1(t, C_1, C_2) = C_1 \cos \alpha t + C_2 \sin \alpha t, \\ x_2 &= \phi_2(t, C_1, C_2) = -\alpha C_1 \sin \alpha t + \alpha C_2 \cos \alpha t, \end{aligned} \tag{$**$}$$

and one may verify this by substitution of ϕ_1 and ϕ_2 into equations $(*)$.

According to Theorem 3.6 we aim to calculate the integral

$$F(C_1, C_2) = \lim_{k\to\infty} \frac{1}{kT_0} \int_0^{kT_0} \phi_1^2 \mathrm{d}t$$

because according to the freedom of choice of the function G we have taken it to be $G = \phi_1^2$, and the period $T_0 = 2\pi/\alpha$.

Let us observe that

$$\begin{aligned} F(C_1, C_2) &= \frac{k}{kT_0} \int_0^{T_0} \phi_1^2 \mathrm{d}t \\ &= \frac{\alpha}{2\pi} \int_0^{\frac{2\pi}{\alpha}} \left(C_1^2 \cos^2 \alpha t + C_2^2 \sin^2 \alpha t + C_1 C_2 \sin 2\alpha t\right) \mathrm{d}t \\ &= \frac{\alpha}{2\pi} \int_0^{\frac{2\pi}{\alpha}} \left[C_1^2 \left(\frac{1}{2} + \frac{1}{2}\cos 2\alpha t\right) + C_2^2 \left(\frac{1}{2} - \frac{1}{2}\cos 2\alpha t\right)\right. \\ &\quad \left. + C_1 C_2 \sin 2\alpha t\right] \mathrm{d}t = \frac{1}{2}\left(C_1^2 + C_2^2\right). \end{aligned}$$

From equations $(**)$ we obtain $x_1(0) = C_1$, $x_2(0) = \alpha C_2$, and finally $2F = x_1^2(0) + \frac{x_2^2(0)}{\alpha^2}$.

The obtained result means that while the system is in motion, its energy has a constant value determined by the introduced initial conditions. □

3.7 Routh's Equation

In some cases it is more convenient to pass to a higher class of problems, that is, from Lagrangian mechanics to Routhian mechanics. In order to do that we will at first become familiar with the so-called *Legendre transformation*, also known as a *potential transformation*.

Let us assume that from *old* variables x_k we would like to pass to *new* variables y_k, that is, $x_k \to y_k$, according to the relationships

$$
\begin{aligned}
y_1 &= f_1(x_1, \ldots, x_N), \\
&\vdots \\
y_n &= f_n(x_1, \ldots, x_N), \\
&\vdots \\
y_N &= f_N(x_1, \ldots, x_N).
\end{aligned} \tag{3.219}
$$

If functions f_n have a potential, that is, there exists a scalar function $V(x_1, \ldots, x_N)$ such that

$$
f_n = \frac{\partial V}{\partial x_n}, \quad n = 1, \ldots, N, \tag{3.220}
$$

then transformation (3.219) is called a *Legendre*[8] *transformation*. Moreover, if the determinant of the matrix (the Hessian[9]) $\det\left[\frac{\partial^2 V}{\partial x_k \partial x_l}\right] \neq 0$, then we call potential $V(x_1, \ldots, x_N)$ a *non-singular potential*.

If on the basis of (3.219) we can determine the inverse relationships in a one-to-one manner, that is, calculate the variables x_n according to the equalities

$$
\begin{aligned}
x_1 &= \varphi_1(y_1, \ldots, y_N), \\
&\vdots \\
x_n &= \varphi_n(y_1, \ldots, y_N), \\
&\vdots \\
x_N &= \varphi_N(y_1, \ldots, y_N),
\end{aligned} \tag{3.221}
$$

then we call the potential $V(x_1, \ldots, x_N)$ a *strongly non-singular potential*.

The following theorem is crucial to further calculations [2].

Theorem 3.7. *If transformation (3.219) is a potential transformation, and the corresponding potential $V(x_1, \ldots, x_N)$ is strongly non-singular, then the inverse transformation (3.221) is also a potential transformation, and the corresponding potential $V^*(y_1, \ldots, y_n)$ is also strongly non-singular, and the relationship between V and V^* reads*

[8] Adrien-Marie Legendre (1752–1833), French mathematician.

[9] Ludwig Otto Hesse (1811–1874), German mathematician mainly working on the problem of algebraic invariance.

$$V^*(y_1, \dots, y_N) = [x_i y_i - V(x_1, \dots, x_N)]_{x_n = \varphi_n(y_1,\dots,y_N)}, \tag{3.222}$$

where $x_i y_i = \sum_{i=1}^{N} x_i y_i$.

Proof. Let us perform the following differentiation:

$$\frac{\partial V^*}{\partial y_n} = \left[x_n + \frac{\partial x_i}{\partial y_n} y_i - \frac{\partial V}{\partial x_i} \frac{\partial x_i}{\partial y_n} \right]_{x_i = \varphi_i(y_1,\dots,y_N)}, \tag{3.223}$$

and we express the variables x_n through $y_1, \dots, y_N$ according to (3.221). According to the definitions introduced by (3.220) and (3.219) we have $\frac{\partial V}{\partial x_i} = y_i$, and from (3.223) we obtain

$$\frac{\partial V^*}{\partial y_n} = x_n \equiv \varphi_n(y_1, \dots, y_N). \tag{3.224}$$

This means that the transformation $x_n = \varphi_n(y_1, \dots, y_N)$ is a potential transformation. Potentials $V(x_1, \dots, x_N)$ and $V^*(x_1, \dots, x_N)$ are called *conjugate potentials*.

The presented information regarding the Legendre transformation will be used for the derivation of Routh's equations on the basis of Lagrange's equation and using the function $L = L(t, q_1, \dots, q_N, \dot{q}_1, \dots, \dot{q}_N)$.

Routh used the concepts of Lagrange and Hamilton to describe the state of a holonomic system.

In this case, part of the generalized velocities in the amount of $s = N - l$, that is, $\dot{q}_l, \dot{q}_{l+1}, \dots, \dot{q}_N$, will be subjected to the Legendre transformation. The role of the potential will be played by the kinetic energy, that is,

$$p_k = \frac{\partial L}{\partial \dot{q}_k}, \quad k = l + 1, \dots, N. \tag{3.225}$$

In other words, variables $q_1, \dots, q_N$, $\dot{q}_1, \dots, \dot{q}_l$ remain unchanged, and the introduced transformation has the form

$$(q_1, \dots, q_N, \dot{q}_1, \dots, \dot{q}_N) \mapsto (q_1, \dots, q_N, \dot{q}_1, \dots, \dot{q}_l, p_{l+1}, \dots, p_N). \tag{3.226}$$

We call the potential L^* (conjugate to L according to Theorem 3.6) a *Routhian function*, and it has the form

$$L^* \equiv R(t, q_1, \dots, q_N, \dot{q}_1, \dots, \dot{q}_l, p_{l+1}, \dots, p_N) = \dot{q}_k p_k - L, \tag{3.227}$$

where $k = l + 1, \dots, N$. The velocities $\dot{q}_k$ occurring on the right-hand side of (3.227) should be expressed through p_k determined from (3.225). Finally, according to (3.227), the Routhian function will depend on the new variables $q_1, \dots, q_N, \dot{q}_1, \dots, \dot{q}_l, p_{l+1}, \dots, p_N, t$.

Our aim is the application of the concept of Routh's potential to Lagrange's equations (3.163).

The total differential of the Routhian function [on the left-hand side of (3.227)] has the form

$$\mathrm{d}R = \sum_{i=1}^{l}\left(\frac{\partial R}{\partial q_i}\mathrm{d}q_i + \frac{\partial R}{\partial \dot{q}_i}\mathrm{d}\dot{q}_i\right) + \sum_{j=l+1}^{N}\left(\frac{\partial R}{\partial q_j}\mathrm{d}q_j + \frac{\partial R}{\partial p_j}\mathrm{d}p_j\right) + \frac{\partial R}{\partial t}\mathrm{d}t, \tag{3.228}$$

because $\dot{q}_j = p_j$ for $j = l+1,\ldots,N$.

The total differential of the right-hand side of (3.227) has the form

$$\mathrm{d}R = -\sum_{i=1}^{l}\left(\frac{\partial L}{\partial q_i}\mathrm{d}q_i + \frac{\partial L}{\partial \dot{q}_i}\mathrm{d}\dot{q}_i\right) + \sum_{j=l+1}^{N}\left(\dot{q}_j\mathrm{d}p_j + \underline{p_j\mathrm{d}\dot{q}_j} - \frac{\partial L}{\partial q_j}\mathrm{d}q_j - \underline{\frac{\partial L}{\partial \dot{q}_j}\mathrm{d}\dot{q}_j}\right) - \frac{\partial L}{\partial t}\mathrm{d}t, \tag{3.229}$$

where the underlined terms cancel out, because $p_j = \frac{\partial L}{\partial \dot{q}_j}$ [see relation (3.225)].

Comparing right-hand sides of (3.228) and (3.229) we obtain

$$\begin{aligned}
&\frac{\partial R}{\partial q_i} = -\frac{\partial L}{\partial q_i}, \quad \frac{\partial R}{\partial \dot{q}_i} = -\frac{\partial L}{\partial \dot{q}_i}, \quad i = 1,\ldots,l,\\
&\frac{\partial R}{\partial q_j} = -\frac{\partial L}{\partial q_j}, \quad \frac{\partial R}{\partial p_j} = \dot{q}_j, \quad j = l+1,\ldots,N,\\
&\frac{\partial R}{\partial t} = -\frac{\partial L}{\partial t}.
\end{aligned} \tag{3.230}$$

Because the analyzed system satisfies Lagrange's equations of the form

$$\frac{\mathrm{d}}{\mathrm{d}t}\left(\frac{\partial L}{\partial \dot{q}_i}\right) - \frac{\partial L}{\partial q_i} = 0, \quad i = 1,\ldots,N, \tag{3.231}$$

taking into account the first row of equations of system (3.230) we obtain

$$\frac{\mathrm{d}}{\mathrm{d}t}\left(\frac{\partial R}{\partial \dot{q}_i}\right) - \frac{\partial R}{\partial q_i} = 0, \quad i = 1,\ldots,l. \tag{3.232}$$

This means that the first l equations of Routh's system have the structure of Lagrange's equations, that is, the structure of second-order differential equations with respect to time.

The remaining equations of Routh's system are first-order differential equations of the form

$$\frac{dq_j}{dt} = \frac{\partial R}{\partial p_j},$$
$$\frac{dp_j}{dt} = \underline{\frac{d}{dt}\left(\frac{\partial L}{\partial \dot{q}_j}\right) = -\frac{d}{dt}\left(\frac{\partial R}{\partial \dot{q}_j}\right)} = -\frac{\partial R}{\partial q_j}, \quad j = l+1, \ldots, N, \tag{3.233}$$

where the underlined term can be neglected because it is useful only during the derivation of this relationship. □

3.8 Cyclic Coordinates

The usefulness of Routh's equations reveals in the case of the occurrence of the so-called *cyclic coordinates* (or *ignorable coordinates*), which were already mentioned in Sect. 3.7.

In a system of N degrees of freedom described by the generalized coordinates $q_1, q_2, \ldots, q_N$, let one of the coordinates q_j be a cyclic coordinate. Then $\frac{\partial L}{\partial q_j} = 0$.

Further we will consider a potential holonomic system described by Hamilton's equation (see Hamilton's mechanics in Chap. 4)

$$\frac{dq_i}{dt} = \frac{\partial H}{\partial p_i}, \quad \frac{dp_i}{dt} = -\frac{\partial H}{\partial q_i}, \quad i = 1, \ldots, N. \tag{3.234}$$

In the case of the cyclic coordinate we have

$$\frac{\partial L}{\partial q_j} = -\frac{\partial H}{\partial q_j} = -\frac{\partial R}{\partial q_j}. \tag{3.235}$$

Because $\frac{\partial H}{\partial q_j} = 0$, the function H does not depend on this coordinate, that is, $H = H(q_1, \ldots, q_{j-1}, q_{j+1}, \ldots, q_N, p_1, \ldots, p_{j-1}, p_j, p_{j+1}, \ldots, p_N, t)$. In light of that for such a coordinate we have [see the second equation of system (3.234)]

$$\frac{dp_j}{dt} = 0, \tag{3.236}$$

that is,

$$p_j = C_j \equiv \text{const.} \tag{3.237}$$

Because H depends on p_j, according to the first equation of system (3.234), we have

$$\frac{\mathrm{d}q_j}{\mathrm{d}t} = \frac{\partial H(q_1, \dots, q_{j-1}, q_{j+1}, \dots, q_N, p_1, \dots, p_{j-1}, p_j, p_{j+1}, \dots, p_N, t)}{\partial p_j}. \tag{3.238}$$

Let us assume that as a result of the integration of $2(N-1)$ equations (3.233) for $i \neq j$ we obtain

$$\begin{aligned} q_i &= q_i(C_j, C_1, \dots, C_{2(N-1)}, t), \\ p_i &= p_i(C_j, C_1, \dots, C_{2(N-1)}, t), \end{aligned} \tag{3.239}$$

where $C_1, C_2, \dots, C_{2(N-1)}$ are constants of integration following from the introduced initial conditions.

In light of that, (3.234) is integrable, and

$$q_j = \int_0^t \frac{\partial H(\tau)}{\partial p_j} \mathrm{d}\tau + C_{2N}, \tag{3.240}$$

where C_{2N} is the last constant subject to determination.

As can be seen, one cyclic coordinate reduced the number of equations subjected to integration by two (if we have k coordinates of this kind, the order of the system of equations subjected to integration is equal to $N - 2k$).

Let us introduce now the notion of cyclic coordinates in Routh's equations. Let us have, according to the previous agreement, $k = N - l$ cyclic coordinates. Then we obtain $k = N - l$ first integrals of a system of the form [see (3.237) and (3.234)]

$$p_j \equiv C_j = \frac{\partial L}{\partial \dot{q}_j}, \quad j = l + 1, \dots, N. \tag{3.241}$$

because from the second equation of system (3.233) we have

$$p_i = -\frac{\partial H}{\partial \dot{q}_i}.$$

According to (3.227), the Routhian function takes the form

$$R = \sum_{j=l+1}^{N} \dot{q}_j C_j - L, \tag{3.242}$$

where the functions $\dot{q}_j = \dot{q}_j(q_1, \dots, q_l, \dot{q}_1, \dots, \dot{q}_l, C_{l+1}, \dots, C_N)$ were obtained from (3.241) after solving them with respect to $\dot{q}_j$. It can be seen that the Routhian function does not depend on velocities $\dot{q}_{l+1}, \dots, \dot{q}_N$ corresponding to the cyclic coordinates $q_{l+1}, \dots, q_N$.

The problem boils down to the integration of l Lagrange equations (3.232) and to the determination of time changes of cyclic coordinates from differential equations of the following form: (see formula (3.233))

$$\frac{\mathrm{d}q_j}{\mathrm{d}t} = \frac{\partial R}{\partial C_j}, \quad \frac{\mathrm{d}p_j}{\mathrm{d}t} = 0, \quad j = l+1, \ldots, N. \tag{3.243}$$

In the next step, first we integrate the system of Lagrange's equations, and then we integrate equations (3.243).

Example 3.17. Let us consider the spherical pendulum from Example 3.12, where

$$T = \frac{1}{2}m\left[(l\dot{\varphi})^2 + \left(l\dot{\Psi}\sin\varphi\right)^2\right],$$
$$V = mgl\,(1-\cos\varphi),$$

and in light of that

$$L = \frac{1}{2}m\left[(l\dot{\varphi})^2 + \left(l\dot{\Psi}\sin\varphi\right)^2\right] - mgl(1-\cos\varphi).$$

Let us note that L does not depend on ψ, that is, ψ is the cyclic coordinate. According to (3.241) we have

$$p_\Psi \equiv C_1 = \frac{\partial L}{\partial \dot{\Psi}} = ml^2\dot{\Psi}\sin^2\varphi,$$

hence

$$\dot{\Psi} = \frac{C_1}{ml^2\sin^2\varphi}.$$

According to formula (3.242), the Routhian function has the following form:

$$\begin{aligned} R &= C_1\dot{\Psi} - \frac{1}{2}m\left[(l\dot{\varphi})^2 + \left(l\dot{\Psi}\sin\varphi\right)^2\right] + mgl(1-\cos\varphi) \\ &= -\frac{1}{2}ml^2\dot{\varphi}^2 + \frac{C_1^2}{ml^2\sin^2\varphi} - \frac{1}{2}ml^2\left[\frac{C_1^2}{(ml^2\sin^2\varphi)}^2\sin^2\varphi\right] + mgl(1-\cos\varphi) \\ &= -\frac{1}{2}ml^2\dot{\varphi}^2 + \frac{1}{2}\frac{C_1^2}{ml^2\sin^2\varphi} + mgl(1-\cos\varphi). \end{aligned}$$

In the preceding formula the so-called Routh's kinetic energy and Routh's potential energy are equal to

$$T^* = \frac{1}{2} m l^2 \dot{\varphi}^2,$$

$$V^* = \frac{1}{2} \frac{C_1^2}{m l^2 \sin^2 \varphi} + mgl(1 - \cos \varphi).$$

Because there is no dissipation of energy in the system, we have $T^* + V^* = \text{const} \equiv C_2$, that is,

$$\frac{1}{2} m l^2 \dot{\varphi}^2 + \frac{1}{2} \frac{C_1^2}{m l^2 \sin^2 \varphi} + mgl(1 - \cos \varphi) = C_2.$$

3.9 Kinetics of Systems of Rigid Bodies: A Three-Degree-of-Freedom Manipulator

3.9.1 Introduction

Now we will take up the kinetics of a system composed of four rigid bodies. During its analysis we will utilize the material introduced in previous chapters, especially Chap. 4 of [14] and this chapter.

Nowadays multibody systems are often used in robotics and in the construction of manipulators, where joints can be equipped with various mechatronic systems applied for the realization of a desired motion of links (bodies) [15–17].

The subject of this section is the analysis of the kinetics and controls of a three-degree-of-freedom manipulator with a SCARA configuration [18, 19].

3.9.2 A Physical and Mathematical Model

During modeling (e.g., see [20]) the following assumptions were made:

1. The manipulator consists of revolute and prismatic kinematic pairs (see Table 4.1 of book [14]).
2. The manipulator model is an open kinematic chain of connected rigid bodies, and each of the bodies is equipped with an independent driving motor.
3. Any backlashes in the kinematic pairs are neglected.

The analyzed three-degree-of-freedom manipulator with a SCARA configuration is presented in Fig. 3.24, where we introduced the coordinate systems of the bodies $O'X_1'X_2'X_3'$, $O''X_1''X_2''X_3''$, and $O'''X_1'''X_2'''X_3'''$, and after rotation of the system ($''$) we obtain the system ($'''$), where now on the axis X_1''' lies a body (particle) of mass m_4.

Let us apply the Denavit–Hartenberg notation (Chap. 4 of [14]), where masses, mass moments of inertia, and the positions of the centers of gravity of the bodies are presented in Fig. 3.24, and the degrees of freedom of the manipulator are denoted by h_1, θ_2, and θ_3.

The kinematic state equations of a manipulator will be derived based on Lagrange's equations of the form

$$\frac{\mathrm{d}}{\mathrm{d}t}\left(\frac{\partial L}{\partial \dot{\Theta}_i}\right) - \frac{\partial L}{\partial \Theta_i} = Q_i - R_i, \tag{3.244}$$

where $L = T - V$ is the Lagrangian function, T (V) denotes the kinetic energy (potential energy), θ_i is the joint variable associated with the ith kinematic pair, Q_i is the generalized force, and R_i is the generalized reaction force.

The kinetic energy of a system in translational motion is equal to

$$T_1 = \frac{1}{2}\left[\dot{h}_1^2(m_1 + m_2 + m_3) + m_4 v_{rm_4}^2\right], \tag{3.245}$$

where v_{rm_4} denotes the velocity of a body of mass m_4 in a space system, which according to the symbols introduced in Fig. 3.24 is equal to $v_{rm_4} = |\mathbf{v}_{rm_4}| = \frac{\mathrm{d}\mathbf{r}_{m_4}}{\mathrm{d}t} = \dot{\mathbf{r}}_{m_4}$.

The kinetic energy of rotational motion of bodies 2 and 3 is equal to

$$T_2 = \frac{1}{2}\left[\dot{\Theta}_3^2\left(I_{O3} + m_3 l_{O3}^2\right) + \dot{\Theta}_2^2\left(I_{O2} + m_2 l_{O2}^2 + I_{O3} + m_3\rho_{rO3}^2\right)\right],$$

$$\rho_{rO3} = \sqrt{x_{1rO3}^2 + x_{2rO3}^2}, \tag{3.246}$$

and x_i, $i = 1, 2$, are elements of vector $\mathbf{r}_{O3}$. The potential energy of a system of bodies is equal to

$$V = g h_1\left(m_1 + m_2 + m_3 + m_4\right), \tag{3.247}$$

where g denotes the gravitational acceleration. Vectors $\mathbf{r}_{O3}$ and $\mathbf{r}_{m_4}$ depend on the position (configuration) of the manipulator.

Following the introduction of the transformation matrices

$$\mathbf{A}_1 = \begin{bmatrix} 1 & 0 & 0 & 0 \\ 0 & 1 & 0 & 0 \\ 0 & 0 & 1 & h_1(t) \\ 0 & 0 & 0 & 1 \end{bmatrix},$$

$$\mathbf{A}_2 = \begin{bmatrix} \cos[\Theta_2(t)] & -\sin[\Theta_2(t)] & 0 & l_2\cos[\Theta_2(t)] \\ \sin[\Theta_2(t)] & \cos[\Theta_2(t)] & 0 & l_2\sin[\Theta_2(t)] \\ 0 & 0 & 1 & 0 \\ 0 & 0 & 0 & 1 \end{bmatrix},$$

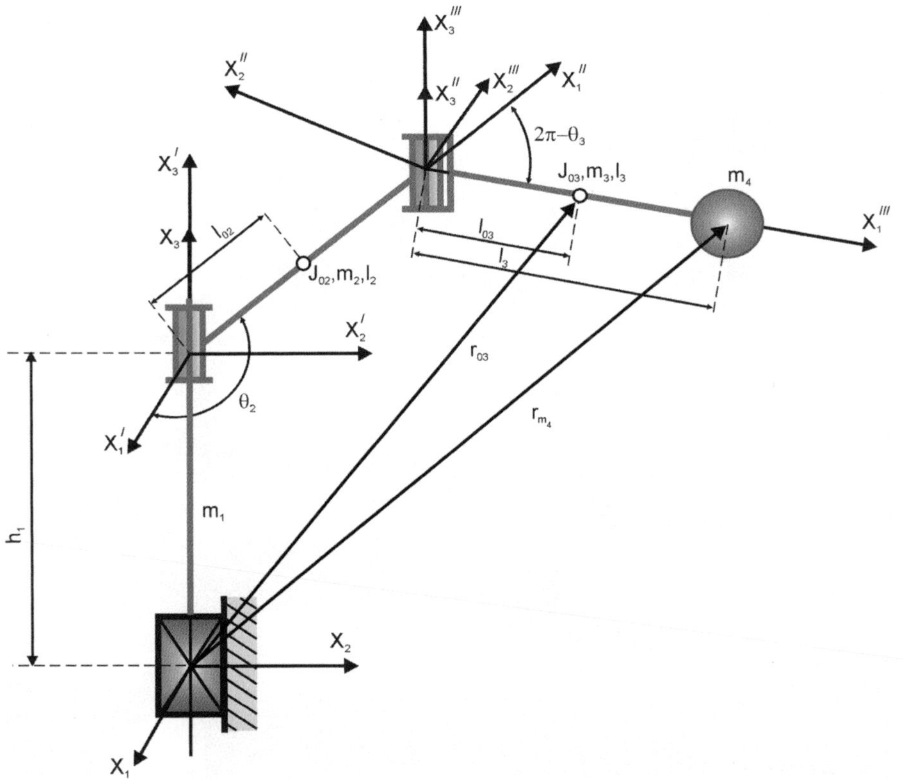

Fig. 3.24 A physical model of a three-degree-of-freedom manipulator with a SCARA configuration

$$\mathbf{A}_3 = \begin{bmatrix} \cos[\Theta_3(t)] & -\sin[\Theta_3(t)] & 0 & 0 \\ \sin[\Theta_3(t)] & \cos[\Theta_3(t)] & 0 & 0 \\ 0 & 0 & 1 & 0 \\ 0 & 0 & 0 & 1 \end{bmatrix}, \tag{3.248}$$

we obtain

$$\mathbf{T}_{O3} = \prod_{i=1}^{3} \mathbf{A}_i$$

$$= \begin{bmatrix} \cos[\Theta_2(t) + \Theta_3(t)] & -\sin[\Theta_2(t) + \Theta_3(t)] & 0 & l_2\cos[\Theta_2(t)] \\ \sin[\Theta_2(t) + \Theta_3(t)] & \cos[\Theta_2(t) + \Theta_3(t)] & 0 & l_2\sin[\Theta_2(t)] \\ 0 & 0 & 1 & h_1(t) \\ 0 & 0 & 0 & 1 \end{bmatrix}. \tag{3.249}$$

The vector

$$\mathbf{r}_{O3} = \mathbf{T}_{O3}\mathbf{l}_{O3} = \begin{bmatrix} l_2\cos[\Theta_2(t)] + l_{O3}\cos[\Theta_2(t)+\Theta_3(t)] \\ l_2\sin[\Theta_2(t)] + l_{O3}\sin[\Theta_2(t)+\Theta_3(t)] \\ h_1(t) \\ 1 \end{bmatrix}, \tag{3.250}$$

because we have $\mathbf{l}_{O3} = [l_{O3}, 0, 0]^{\mathrm{T}}$. Substituting matrix (3.250) into the second equation of (3.246) we obtain

$$\rho_{r_{O3}} = \Big[(l_2\cos[\Theta_2(t)] + l_3\cos[\Theta_2(t)+\Theta_3(t)])^2 + (l_2\sin[\Theta_2(t)] + l_{O3}\sin[\Theta_2(t)+\Theta_3(t)])^2\Big]^{\frac{1}{2}}. \tag{3.251}$$

In a similar way we find

$$\mathbf{r}_{m_4} = \begin{bmatrix} l_2\cos[\Theta_2(t)] + l_3\cos[\Theta_2(t)+\Theta_3(t)] \\ l_2\sin[\Theta_2(t)] + l_3\sin[\Theta_2(t)+\Theta_3(t)] \\ h_1(t) \\ 1 \end{bmatrix} \tag{3.252}$$

and

$$\mathbf{v}_{rm_4} = \begin{bmatrix} -l_2\sin[\Theta_2(t)]\dot{\Theta}_2(t) - l_3\sin[\Theta_2(t)+\Theta_3(t)]\left(\dot{\Theta}_2(t)+\dot{\Theta}_3(t)\right) \\ l_2\cos[\Theta_2(t)]\dot{\Theta}_2(t) + l_3\cos[\Theta_2(t)+\Theta_3(t)]\left(\dot{\Theta}_2(t)+\dot{\Theta}_3(t)\right) \\ \dot{h}_1(t) \\ 1 \end{bmatrix}, \tag{3.253}$$

$$v_{rm_4} = \Big\{\left(-l_2\sin[\Theta_2(t)]\dot{\Theta}_2(t) - l_3\sin[\Theta_2(t)+\Theta_3(t)]\left(\dot{\Theta}_2(t)+\dot{\Theta}_3(t)\right)\right)^2 + \left(l_2\cos[\Theta_2(t)]\dot{\Theta}_2(t) + l_3\cos[\Theta_2(t)+\Theta_3(t)]\left(\dot{\Theta}_2(t)+\dot{\Theta}_3(t)\right)\right)^2 + \left(\dot{h}_1(t)\right)^2\Big\}^{\frac{1}{2}}. \tag{3.254}$$

Eventually, the Lagrangian function takes the form

$$
\begin{aligned}
L = \frac{1}{2}\Bigg\{ & \left(\dot{h}_1(t)\right)^2 (m_1 + m_2 + m_3) + m_4 \Bigg[\big(-l_2 \sin[\Theta_2(t)]\dot{\Theta}_2(t) - l_3 \sin[\Theta_2(t)) \\
& + \Theta_3(t)] \left(\dot{\Theta}_2(t) + \dot{\Theta}_3(t)\right)^2 + \big(l_2 \cos[\Theta_2(t)]\dot{\Theta}_2(t) + l_3 \cos[\Theta_2(t)) \\
& + \Theta_3(t)] \left(\dot{\Theta}_2(t) + \dot{\Theta}_3(t)\right)^2 + \left(\dot{h}_1(t)\right)^2 \Bigg] \\
& + \dot{\Theta}_3(t)^2 \left(I_{O3} + m_3 l_{O3}^2\right) + \dot{\Theta}_2^2(t)^2 \left(I_{O2} + m_2 l_{O2}^2 + I_{O3}\right) \\
& + m_3 \Big((l_2 \cos[\Theta_2(t)] + l_{O3} \cos[\Theta_2(t) + \Theta_3(t)])^2 \\
& + \ (l_2 \sin[\Theta_2(t)] + l_{O3} \sin[\Theta_2(t) + \Theta_3(t)])^2 \Big) \Bigg\} \\
& - g h_1(t)(m_1 + m_2 + m_3 + m_4),
\end{aligned}
\tag{3.255}
$$

and after its substitution into (3.244) we obtain three second-order differential equations, which are given below in matrix form

$$
\begin{bmatrix} a_{11} & 0 & 0 \\ 0 & a_{22} & a_{23} \\ 0 & a_{32} & a_{33} \end{bmatrix}
\begin{bmatrix} \ddot{h}_1 \\ \ddot{\Theta}_2 \\ \ddot{\Theta}_3 \end{bmatrix}
+ \begin{bmatrix} 0 \\ c_2 \\ c_3 \end{bmatrix}
+ \begin{bmatrix} b_1 \\ 0 \\ 0 \end{bmatrix} g
= \begin{bmatrix} F_1 \\ M_2 \\ M_3 \end{bmatrix},
\tag{3.256}
$$

where

$$
\begin{aligned}
a_{11} &= m_1 + m_2 + m_3 + m_4, \\
a_{22} &= I_{O2} + I_{O3} + l_{O2}^2 m_2 \\
&\quad + m_3 \left(l_{O3}^2 + l_2^2 + 2 l_{O3} l_2 \cos[\Theta_3]\right) + m_4 \left(l_2^2 + l_3^2 + 2 l_2 l_3 \cos[\Theta_3]\right), \\
a_{23} &= m_4 l_3^2 + m_4 l_2 l_3 \cos[\Theta_3], \\
a_{32} &= a_{32}, \\
a_{33} &= I_{O3} + l_{O3}^2 m_3 + m_4 l_3^2, \\
c_2 &= -\left(l_{O3} + m_4 l_3\right) 2 l_2 \dot{\Theta}_2 \dot{\Theta}_3 \sin[\Theta_3] - \left(m_4 l_2 l_3 \sin[\Theta_3]\right) \dot{\Theta}_3^2, \\
c_3 &= \left(l_{O3} m_3 + m_4 l_3\right) l_2 \dot{\Theta}_2 \sin[\Theta_3], \\
b_1 &= a_{11}.
\end{aligned}
\tag{3.257}
$$

The obtained differential equations describe the kinematic state of a system without taking into account the resistance to motion. Although the first of

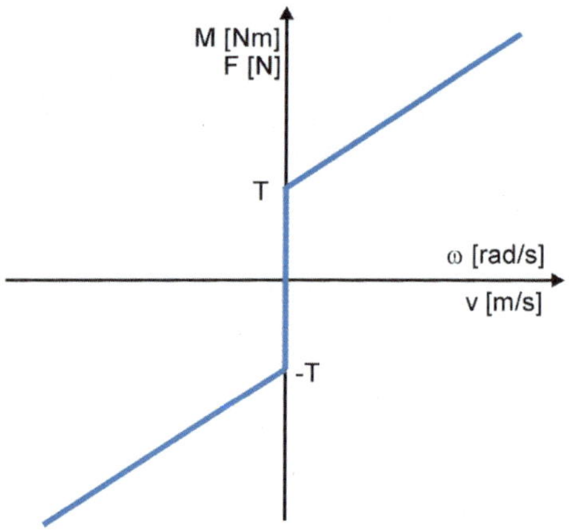

Fig. 3.25 Characteristic of force and moment of friction assumed for calculations

equations (3.256) is linear and independent of the others, the two remaining equations are strongly non-linear and coupled with each other. It was assumed that the force $F_1(t)$ came from the motor connected with link (body) 1.

Differential equations (3.256) were solved numerically, which was described in more detail in [19], where the adopted friction characteristic (Chap. 2 of [14]) was modeled as a linear function with discontinuity at zero (Fig. 3.25).

A real multibody object such as a manipulator is a mechatronic system (see also [20]). In the case of its control, a model of the object contains a driving system (actor), a multibody system (three links), and a measurement system (sensors). In the considered case during analysis using block diagrams, blocks simulating the friction and drive assemblies of particular manipulator links were also included. The drive block generates functions of driving moments for the given body, and thanks to the feedback on position and velocity, it is a primitive manipulator controller.

According to (3.256), the initial parameters of motion are the positions of particular manipulator links (h_1, θ_1, θ_2) in the corresponding body systems and the velocities of those links ($\dot{h}_1$, $\dot{\theta}_1$, $\dot{\theta}_2$) at the instant when the simulation (numerical calculations) starts.

The moments and forces of friction occurring at the ith manipulator body can be imposed through specification of two parameters: the magnitude of the static friction force and the ratio of increment of force to the increment of speed. The drive block consists of three elements: a controller, a regulator, and a drive unit. In this case the control is based on the method of linear functions with parabolic *connectors*. As a result, we obtain that the functions of velocity vs. time, the so-called *velocity profile*, are in the shape of a trapezium (Fig. 3.26). Because of that, constant accelerations during braking and startup are obtained.

At first, the controller forms a velocity profile in the shape of a trapezium. Then this signal is integrated, which leads to the determination of the displacement function. In the regulator, the difference between the actual and prescribed

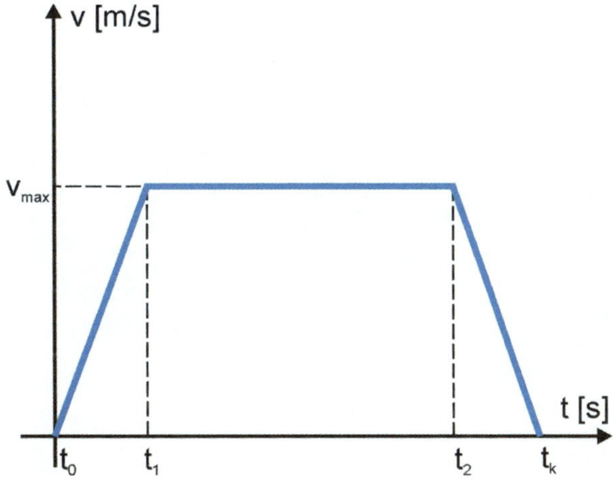

Fig. 3.26 A trapezium-shaped velocity profile

displacement generates a signal forcing the function of velocity, whereas the difference between the actual and prescribed velocity forms the function of the driving moment. The drive unit consists of a motor (with limited driving moment) and a transmission adjusting the velocity. The control of model movements takes place through specification of the velocity function in the form shown in Fig. 3.26. By adjusting the slopes of the trapezium's sides the required accelerations during startup and braking phases can be specified.

The settings of the PD controller are stored in additional blocks and are selected so as to guarantee the function of an aperiodic critical character. A maximum driving moment of the motor is selected separately for the positive and negative directions of motion. Additionally, a transmission ratio of the mechanical transmission is selected.

3.9.3 Results of Numerical Simulations

For the given three-parameter sets presented in [18, 19] a numerical simulation was conducted and plots of the position and velocity of particular links were obtained.

Below we will carry out the analysis of three cases.

(i) Correct movements of the manipulator.
We are dealing with correct movements of a manipulator when it reaches the prescribed positions and plots of the actual velocity coincide with those prescribed by the control system within the limits of a certain assumed tolerance. Let us consider, for instance, link 1, where the prescribed value for the position is equal to 0.75 m. As follows from Fig. 3.27a, the value of the end position of this link is close to the value 0.75 m. The velocity plot has a shape

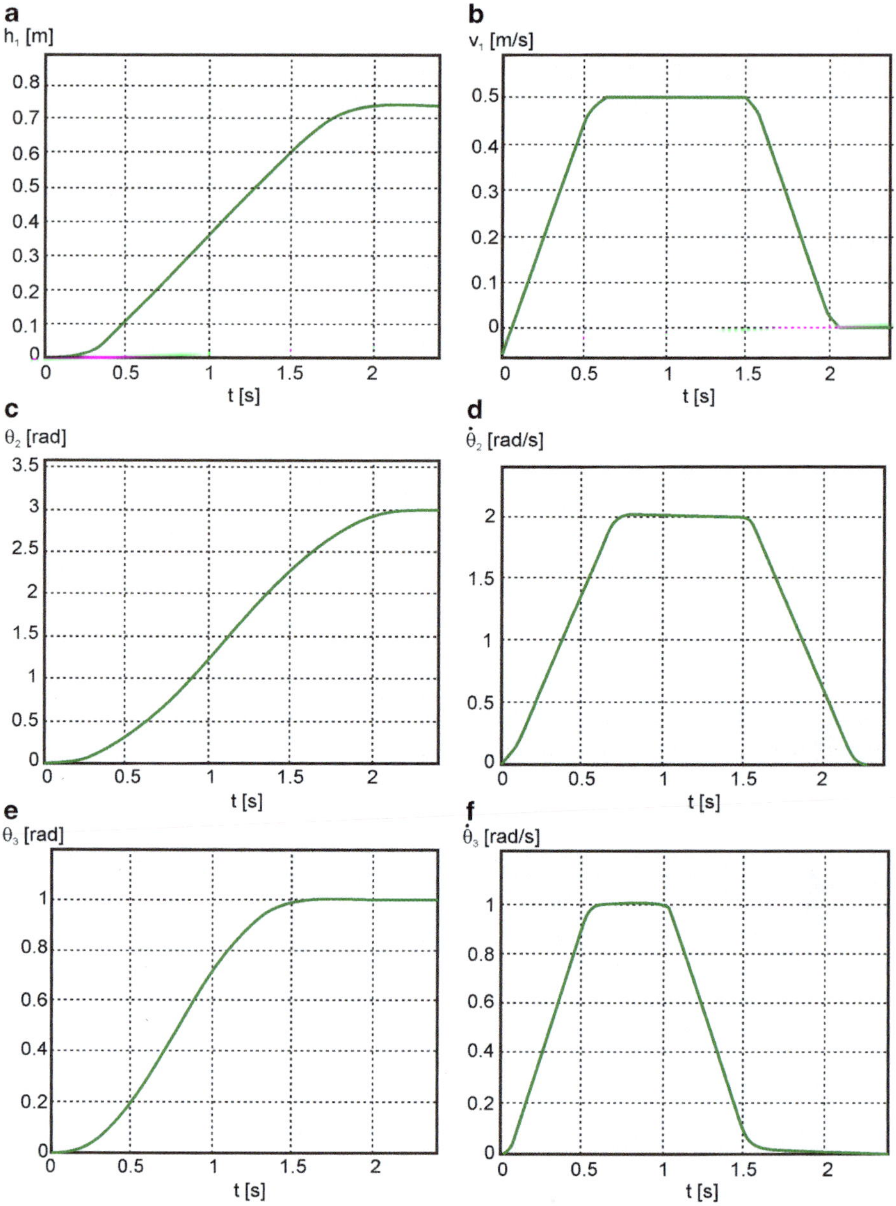

Fig. 3.27 Plots of positions and velocities of manipulator link no. 1 (**a**, **b**), 2 (**c**, **d**), and 3 (**e**, **f**) in the case of correct manipulator movements

imposed by the controller, and additionally all vertices of the trapezium are rounded. The visible shift of the actual velocity plot (Fig. 3.27b) with respect to the theoretical one points to the occurrence of error in the system. Similar observations concern the remaining links (see plots of $\theta_2(t)$, $\dot{\theta}_2(t)$ presented in Fig. 3.27c, d, and the plots of $\theta_3(t)$, $\dot{\theta}_3(t)$ presented in Fig. 3.27e, f).

(ii) A dynamic interaction.
In the case of the second set of parameters, manipulator links 2 and 3 move in the same direction. The plots presented in Fig. 3.28a–d demonstrate that the settings of the regulator and the powers of the motors are properly selected. The next four plots (Fig. 3.28e–h) present the results associated with the case where link number 3 moves with the same parameters as previously, but in the opposite direction. In Fig. 3.28h the deviation from the prescribed velocity function becomes evident. It occurs at the moment when braking of link 2 begins (about 1.5 s after the start). The driving moment of motor 3 is not sufficient to overcome the dynamic forces, which leads to the interaction between links 2 and 3. At that time instant the motion of the link 3 is not controlled and follows from the action of the link 2. In the case of 3.3 link 3 was under continuous control of the controller, and the maximum moment of the motor was sufficient for the realization of the input from the controller.

From the preceding calculations it follows that in the control systems commonly applied in industry today, changes in the direction of motion of only one link influence significantly the kinetic behavior of manipulators.

(iii) Kinetic effects associated with changes in the moment of inertia.
In the considered case (third set of parameters, Fig. 3.29a–f), the configuration of the manipulator is such that link 3 is rotated with respect to link 2 through the angle $\theta_3 = 1.5$ rad, and this angle remains fixed during the simulation (Fig. 3.29c). The velocity plot of link 2 (Fig. 3.29b) is correct.

After positioning of link 3 so that it is a straight line extension of link 2 ($\theta_3 = 0$), the velocity plot for link 2 shown in Fig. 3.29d was obtained. The form of this plot is the result of an overshoot. After the change in position of link 3, the moment of inertia of the assembly 2–3 is so large that motor 2 has insufficient power to follow the velocity input. At a certain time instant the driving moment of motor 2 no longer increases but settles at a constant level. This leads to an increase in the error of velocity and position, that is, to the overshoot of the manipulator.

From the discussed example it follows that the manipulator configuration has considerable influence on the process of its control. Although link 3 theoretically remains stationary, there exist small deviations of its position during motion caused by kinetic interactions (Fig. 3.29c).

Although the major source material for this chapter is in Polish and Russian, the reader may find complementary texts in [21–27].

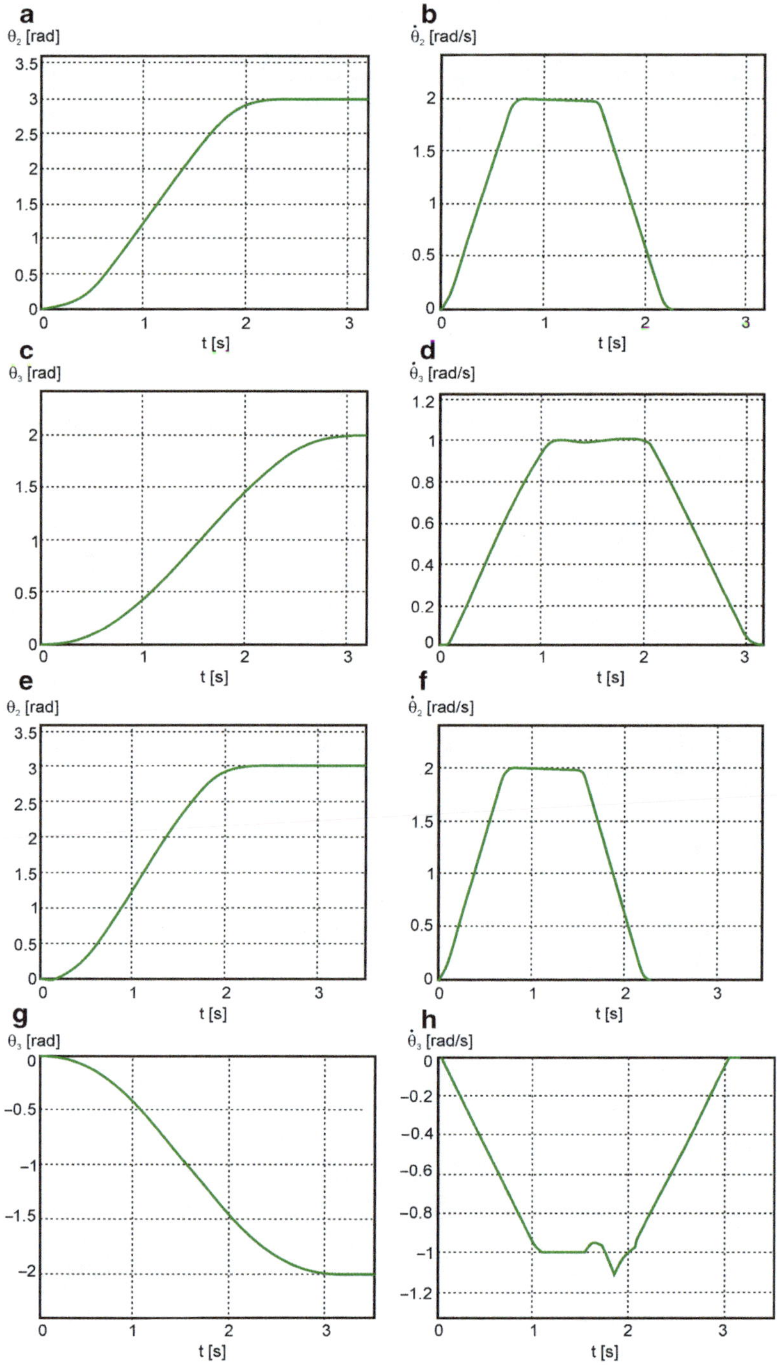

Fig. 3.28 Plots of positions and velocities of manipulator links of nos. 2 and 3 moving in the same direction (**a–d**) and in opposite directions (**e–h**)

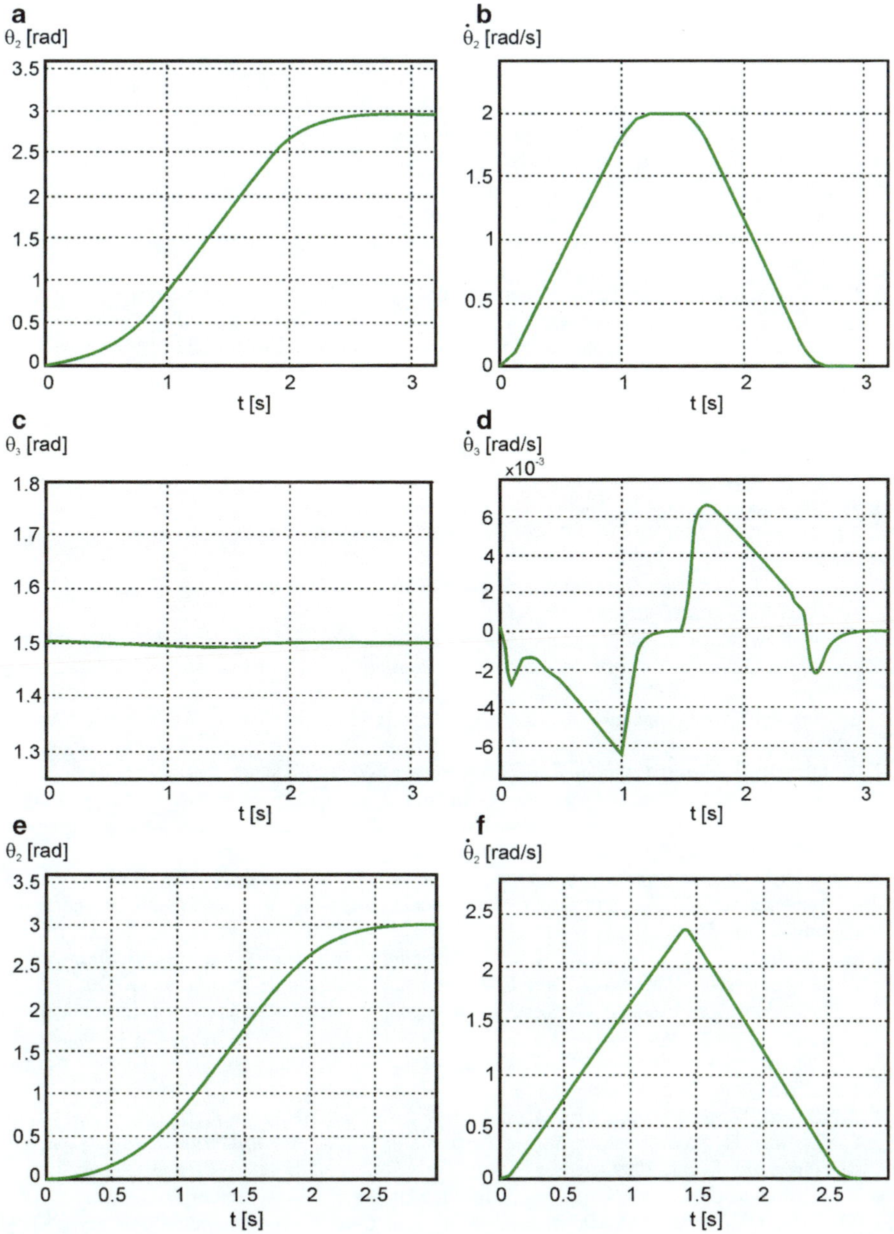

Fig. 3.29 Plots of positions and velocities of manipulator links of nos. 2 and 3 (case iii)

References

1. J. Awrejcewicz, *Oscillations of Lumped Deterministic Systems* (WNT, Warsaw, 1996)
2. V.E. Zhuravlev, *Fundamentals of Theoretical Mechanics* (Fizmatlit, Moscow, 2001)
3. R. Gutowski, *Analytical Mechanics* (PWN, Warsaw, 1971)
4. Z. Osinski, *General Mechanics* (PWN, Warsaw, 1994)
5. Z. Koruba, J.W. Osiecki, *Elements of Advanced Mechanics* (Kielce University of Technology, Kielce, 2007)
6. A.M. Bloch, J.E. Marsden, D.V. Zenkov, Nonholonomic dynamics. Notices AMS **52**(3), 324–333 (2005)
7. W. Blajer, A geometric unification of constrained system dynamics. Multibody Syst. Dyn. **1**(1), 3–21 (1997)
8. W. Schiehlen, Multibody system dynamics: roots and perspectives. Multibody Syst. Dyn. **1**(2), 149–188 (1997)
9. W. Blajer, *Methods of Multibody Systems Dynamics*. (Technical University of Radom Press, Radom, 1998)
10. W. Blajer, On the determination of joint reactions in multibody mechanisms. Trans. ASME J. Mech. Design **126**(2), 341–350 (2004)
11. A.D. Markeev, *Theoretical Mechanics* (Nauka, Moscow, 1990)
12. P.E.B. Jourdain, Note on an analogue of Gauss' principle of least constraint. Q. J. Pure Appl. Math. **XL**, 153–157, 1909
13. E. Wittbrodt, S. Sawiak, *General Mechanics: Theory and Problems* (Gdask Technical University, Gdask, Poland, 2005)
14. J. Awrejcewicz, *Classical Mechanics: Statics and Kinematics* (Springer, New York, 2012)
15. J.J. Craig, *Adaptive Control of Mechanical Manipulators* (Addison-Wesley, Reading, MA, 1987)
16. J. Knapczyk, P.A. Lebiediew, *Theory of Spatial Mechanisms and Manipulators* (WNT, Warsaw, 1990)
17. M. Olszewski, *Manipulators and Industrial Robots* (WNT, Warsaw, 1990)
18. J. Awrejcewicz, S. Pryk, in *Dynamical Model of a Three-Degree-of-Freedom Manipulator*, E. Onate, S.R. Idelsohn, eds. Computational Mechanics. New Trends and Applications, pp 1–110. (CMNE, Barcelona, Spain, 1998)
19. J. Awrejcewicz, *Mathematical Modelling of Systems* (WNT, Warsaw, 2007)
20. J. Awrejcewicz, Z. Koruba, *Classical Mechanics: Applied Mechanics and Mechatronics* (Springer, New York, 2012)
21. E.T. Whittaker, *A Treatise on the Analytical Dynamics of Particles and Rigid Bodies, with an Introduction to the Problem of Three Bodies* (Dover Publications, New York, 1937)
22. L.D. Landau, E.M. Lifshitz, *Mechanics Course of Theoretical Physics*, vol. 1. (Franklin Book Company, Boca Raton, FL, 1972)
23. S.T. Thornton, J.B. Marion, *Classical Dynamics of Particles and Systems* (Saunders College Publishers, New York, 1995)
24. L.N. Hand, J.D. Finch, *Analytical Mechanics* (Cambridge University Press, Cambridge, 1998)
25. C. Gignoux, B. Silvestre-Brac, *Solved Problems in Lagrangian and Hamiltonian Mechanics* (Springer, Dordrecht, 2009)
26. R.D. Gregory, *Classical Mechanics* (Cambridge University Press, Cambridge, 2011)
27. M. McCall, *Classical Mechanics: From Newton to Einstein: A Modern Introduction* (Wiley, Chichester, 2011)

Chapter 4
Classic Equations of Dynamics

4.1 Hamiltonian Mechanics

4.1.1 Hamilton's Equations

Let us return to Lagrange's equations of the second kind, which, for the system of N degrees of freedom loaded exclusively with potential forces, take the form

$$\frac{\mathrm{d}}{\mathrm{d}t}\left(\frac{\partial L}{\partial \dot{q}_n}\right) - \frac{\partial L}{\partial q_n} = 0, \quad n = 1, \ldots, N. \tag{4.1}$$

At the given time instant t the values of displacements $q_n(t)$ and velocities $\dot{q}_n(t)$ are known. The variables $q_n(t)$, $\dot{q}_n(t)$, t are called *Lagrangian variables* or *state variables*.

If we know the Lagrangian, instead of the generalized velocities $\dot{q}_n(t)$ we can choose quantities p_n, called *generalized impulses*, of the form

$$p_n = \frac{\partial L}{\partial \dot{q}_n}, \tag{4.2}$$

where the variables q_n, p_n, and t are called *Hamiltonian variables*.

Hamilton observed that after introducing the aforementioned variables, Lagrange's equations (4.1) take a form that exhibits a symmetry. The variables q_n and p_n will be called *canonically conjugate variables*, and the equations in those variables *Hamilton's canonical equations*.

In mechanics the quantities p_n are often called *generalized momenta*, that is, p_n has a dimension of momentum if the corresponding q_n ($\dot{q}_n$) has a dimension of displacement (linear velocity), or a dimension of angular momentum if q_n ($\dot{q}_n$) has a dimension of an angle (angular velocity). Also, applying the notation adopted in the present work, we have $\{p_n\} = [a_{mn}]\{\dot{q}_m\}$, which in tensor notation would have the

J. Awrejcewicz, *Classical Mechanics: Dynamics*, Advances in Mechanics and Mathematics 29, DOI 10.1007/978-1-4614-3740-6_4,

form $\dot{q}_n = a_{mn}\dot{q}^m$, where $[a_{mn}]$ is the matrix of generalized masses corresponding to q_n (i.e., the metric tensor of a covariant basis). In other words, p_n and $\dot{q}_n$ are covariant and contravariant components of one vector in a certain metric space, that is, $\mathbf{v} = \dot{q}_n\mathbf{E}_n = \dot{q}_n\mathbf{E}^n$.

The Legendre transformation (3.222), introduced earlier, now, for the function $L(q_n, \dot{q}_n, t)$ with respect to the coordinates $\dot{q}_n$, takes the form

$$H(q_n, p_n, t) = \sum_{n=1}^{N} p_n\dot{q}_n - L\left(q_1, \ldots, q_N, \dot{q}_1, \ldots, \dot{q}_N, t\right), \tag{4.3}$$

where the transformation is made with respect to all generalized velocities $n = 1, \ldots, N$. Let us note that $\dot{q}_n = \dot{q}_n(q_1, \ldots, q_N, p_1, \ldots, p_N, t)$, which follows from the solution of (4.2), i.e., using those equations we determine the generalized velocities in terms of the generalized momenta treating the generalized coordinates as parameters. The function $H(q_1, \ldots, q_N, p_1, \ldots, p_N, t)$ is called the *Hamiltonian function (the Hamiltonian).*

Proceeding similarly as in the case of calculations regarding the Routhian function let us calculate the total differential of the Hamiltonian function on the left-hand side of expression (4.3):

$$\mathrm{d}H = \sum_{n=1}^{N}\left(\frac{\partial H}{\partial q_n}\mathrm{d}q_n + \frac{\partial H}{\partial p_n}\mathrm{d}p_n\right) + \frac{\partial H}{\partial t}\mathrm{d}t. \tag{4.4}$$

In turn, the total differential of the right-hand side of (4.3) is equal to

$$\mathrm{d}H = \sum_{n=1}^{N}\left(\dot{q}_n\mathrm{d}p_n + \underline{p_n d\dot{q}_n - \frac{\partial L}{\partial \dot{q}_n}\mathrm{d}\dot{q}_n} - \frac{\partial L}{\partial q_n}\mathrm{d}q_n\right) - \frac{\partial L}{\partial t}\mathrm{d}t, \tag{4.5}$$

where the underlined terms cancel out.

Comparing (4.4) with (4.5) we obtain

$$\frac{\partial H}{\partial q_n} = -\frac{\partial L}{\partial q_n}, \quad \frac{\partial H}{\partial p_n} = \frac{\mathrm{d}q_n}{\mathrm{d}t}, \tag{4.6}$$

and

$$\frac{\partial H}{\partial t} = -\frac{\partial L}{\partial t}. \tag{4.7}$$

From (4.2) we get

$$\frac{\mathrm{d}p_n}{\mathrm{d}t} = \frac{\mathrm{d}}{\mathrm{d}t}\frac{\partial L}{\partial \dot{q}_n} = \frac{\partial L}{\partial q_n}. \tag{4.8}$$

Eventually, we obtain the equations

$$\frac{\mathrm{d}q_n}{\mathrm{d}t} = \frac{\partial H}{\partial p_n}, \qquad \frac{\mathrm{d}p_n}{\mathrm{d}t} = -\frac{\partial H}{\partial q_n}. \tag{4.9}$$

We call equations (4.9) *Hamilton's canonical equations.*

Reader interested in broadening their understanding of the problems discussed in this section may refer to [1–24].

Example 4.1. Determine the canonical form of Hamilton's equations for the spherical pendulum considered in Examples 3.13 and 3.18 of Chap. 3.

By definition (4.3) of the Hamiltonian we have

$$\begin{aligned}
p_\psi &= \frac{\partial L}{\partial \dot{\psi}} = ml^2\dot{\psi}\sin^2\varphi,\\
p_\varphi &= \frac{\partial L}{\partial \dot{\varphi}} = ml^2\dot{\varphi},\\
H &= \dot{\psi}p_\psi + \dot{\varphi}p_\varphi - L\\
&= \frac{p_\psi^2}{ml^2\sin^2\varphi} + \frac{p_\varphi^2}{ml^2} - \frac{1}{2}ml^2\frac{p_\varphi^2}{(ml^2)^2} - \frac{ml^2p_\psi^2\sin^2\varphi}{2\left(ml^2\sin^2\varphi\right)^2} - mgl(1-\cos\varphi)\\
&= \frac{1}{2}\frac{p_\varphi^2}{ml^2} + \frac{1}{2}\frac{p_\psi^2}{ml^2\sin^2\varphi} - mgl(1-\cos\varphi).
\end{aligned}$$

Hamilton's canonical equations (4.9) take the form

$$\begin{aligned}
\frac{\mathrm{d}\varphi}{\mathrm{d}t} &= \frac{\partial H}{\partial p_\varphi} = \frac{p_\varphi}{ml^2},\\
\frac{\mathrm{d}\psi}{\mathrm{d}t} &= \frac{\partial H}{\partial p_\psi} = \frac{p_\psi}{ml^2\sin^2\varphi},\\
\frac{\mathrm{d}p_\varphi}{\mathrm{d}t} &= -\frac{\partial H}{\partial \varphi} = \frac{1}{2}\frac{p_\psi^2}{ml^2}\frac{2\sin\varphi\cos\varphi}{\sin^4\varphi} - mgl\sin\varphi,\\
\frac{\mathrm{d}p_\psi}{\mathrm{d}t} &= -\frac{\partial H}{\partial \psi} = 0.
\end{aligned}$$

For $p_\psi = 0$ we have

$$\frac{\mathrm{d}\varphi}{\mathrm{d}t} = \frac{p_\varphi}{ml^2}, \qquad \frac{\mathrm{d}p_\varphi}{\mathrm{d}t} = -mgl\sin\varphi,$$

which are Hamilton's canonical equations for a mathematical pendulum. □

4.1.2 Jacobi–Poisson Theorem

In the literature on mechanics, the notion of the so-called Poisson bracket is used. The following expression is called the *Poisson bracket* of functions u and v:

$$[u,v] = \sum_{n=1}^{N} \frac{\partial u}{\partial q_n}\frac{\partial v}{\partial p_n} - \frac{\partial u}{\partial p_n}\frac{\partial v}{\partial q_n}, \tag{4.10}$$

where it is assumed that functions u and v are continuous and twice differentiable.

Let us cite here [9] certain properties of the Poisson brackets (the proofs are omitted):

$$\begin{aligned}
&\text{(i)}\ \ [u,v] = -[v,u],\\
&\text{(ii)}\ \ [Cu,v] = C[u,v], \quad C = \text{const},\\
&\text{(iii)}\ \ [u+v,w] = [u,w] + [v,w],\\
&\text{(iv)}\ \ \frac{\partial}{\partial t}[u,v] = \left[\frac{\partial u}{\partial t}, v\right] + \left[u, \frac{\partial v}{\partial t}\right],\\
&\text{(v)}\ \ [[u,v],w] = [[v,w],u] + [[w,u],v] = 0.
\end{aligned} \tag{4.11}$$

Let the variables p_n and q_n satisfy Hamilton's canonical equations (4.9), henceforth called Hamilton's equations. If the function $F(q_n(t),\ p_n(t),\ t) \equiv C =$ const, then it is a first integral of Hamilton's equations. Thus for any time instant t we have

$$\begin{aligned}
\frac{\mathrm{d}F}{\mathrm{d}t} &= \frac{\partial F}{\partial t} + \sum_{n=1}^{N}\left(\frac{\partial F}{\partial q_n}\frac{\mathrm{d}q_n}{\mathrm{d}t} + \frac{\partial F}{\partial p_n}\frac{\mathrm{d}p_n}{\mathrm{d}t}\right)\\
&= \frac{\partial F}{\partial t} + \sum_{n=1}^{N}\left(\frac{\partial F}{\partial q_n}\frac{\partial H}{\partial p_n} - \frac{\partial F}{\partial p_n}\frac{\partial H}{\partial q_n}\right) = \frac{\partial F}{\partial t} + [F,H] = 0.
\end{aligned} \tag{4.12}$$

It turns out that making use of the concept of the Poisson bracket enables a clear statement of the Jacobi–Poisson theorem, which allows for the construction of a first integral of a Hamiltonian system if at least two other of its first integrals F_1 and F_2 are known.

Theorem 4.1 (Jacobi–Poisson theorem). *If F_1 and F_2 are first integrals of Hamilton's equations (4.9), then their Poisson bracket $[F_1,\ F_2]$ is also a first integral of those Hamilton equations.*

Proof. According to property (iv) we have

$$\frac{\partial}{\partial t}[F_1,\ F_2] = \left[\frac{\partial F_1}{\partial t},\ F_2\right] + \left[F_1,\ \frac{\partial F_2}{\partial t}\right].$$

Because by assumption F_1 and F_2 are first integrals, we have

$$\frac{\partial F_i}{\partial t} + [F_i, H] = 0, \quad i = 1, 2,$$

which, after using the first equation of the proof and property (i), gives

$$\begin{aligned}\frac{\partial}{\partial t}[F_1, F_2] &= [-[F_1, H], F_2] + [F_1, -[F_2, H]] \\ &= [[H, F_1], F_2] - [-[F_2, H], F_1] \\ &= [[H, F_1], F_2] + [[F_2, H], F_1],\end{aligned}$$

where property (ii) for $C = -1$ was used as well. From property (v) for the functions F_1, F_2, and H we obtain

$$[[H, F_1], F_2] + [[F_1, F_2], H] + [[F_2, H], F_1] = 0,$$

that is,

$$[[H, F_1], F_2] + [[F_2, H], F_1] = -[[F_1, F_2], H],$$

and hence

$$\frac{\partial}{\partial t}[F_1, F_2] + [[F_1, F_2], H] = 0,$$

which we had set out to demonstrate. □

4.1.3 Canonical Transformations

With the aid of matrix notation Hamilton's equations (4.9) can be expressed in the form

$$\dot{\mathbf{x}} = \mathbf{I}\mathbf{H}', \tag{4.13}$$

where

$$\mathbf{x}_{2N} = \begin{bmatrix} q_1 \\ \vdots \\ q_N \\ p_1 \\ \vdots \\ p_N \end{bmatrix}, \quad \mathbf{I}_{2N} = \begin{bmatrix} \mathbf{0}_N & \mathbf{E}_N \\ -\mathbf{E}_N & \mathbf{0}_N \end{bmatrix}, \quad \mathbf{H}'_{1\times 2N} = \frac{\partial H}{\partial \mathbf{x}} = \begin{bmatrix} \frac{\partial H}{\partial q_1} \\ \vdots \\ \frac{\partial H}{\partial q_N} \\ \frac{\partial H}{\partial p_1} \\ \vdots \\ \frac{\partial H}{\partial p_N} \end{bmatrix}. \tag{4.14}$$

As can be seen, $\mathbf{x}$ is a column vector, matrix $\mathbf{I}$ is skew-symmetric, $\mathbf{H}'$ is a column matrix, and matrix $\mathbf{E}_N$ is a diagonal matrix composed of ones. It is easy to check that $\det \mathbf{I} = 1$, $\mathbf{I}^2 = -\mathbf{E}_{2N}$, $\mathbf{I}^{\mathrm{T}} = \mathbf{I}^{-1} = -\mathbf{I}$.

The use of canonical transformations is motivated by a desire to choose generalized coordinates (or their combination) such that the form of equations would enable us to find the first integrals of the considered problem. One such possibility has already been discussed on the example of Routh's equations and cyclic coordinates. Let us conduct the following transformation of coordinates $\mathbf{q}$ and $\mathbf{p}$:

$$Q_n = Q_n(\mathbf{q}, \mathbf{p}, t), \qquad P_n = P_n(\mathbf{q}, \mathbf{p}, t), \tag{4.15}$$

that is, we perform the transition to new coordinates $\mathbf{Q}$ and $\mathbf{P}$ in the following way: $\mathbf{q} \to \mathbf{Q}$, $\mathbf{p} \to \mathbf{P}$. We select the new coordinates Q_n and P_n $(n = 1, \ldots, N)$ so that the system of equations obtained after the transformation is a Hamiltonian system as well.

The Jacobian matrix of transformation (4.15) has the form

$$\mathbf{J} = \begin{bmatrix} \dfrac{\partial \mathbf{Q}}{\partial \mathbf{q}} & \dfrac{\partial \mathbf{Q}}{\partial \mathbf{p}} \\ \dfrac{\partial \mathbf{P}}{\partial \mathbf{q}} & \dfrac{\mathbf{P}}{\partial \mathbf{p}} \end{bmatrix} = \begin{bmatrix} \dfrac{\partial Q_1}{\partial q_1} & \cdots & \dfrac{\partial Q_1}{\partial q_N} & \dfrac{\partial Q_1}{\partial p_1} & \cdots & \dfrac{\partial Q_1}{\partial p_N} \\ \vdots & & \vdots & \vdots & & \vdots \\ \dfrac{\partial Q_N}{\partial q_1} & \cdots & \dfrac{\partial Q_N}{\partial q_N} & \dfrac{\partial Q_N}{\partial p_1} & \cdots & \dfrac{\partial Q_N}{\partial p_N} \\ \dfrac{\partial P_1}{\partial q_1} & \cdots & \dfrac{\partial P_1}{\partial q_N} & \dfrac{\partial P_1}{\partial p_1} & \cdots & \dfrac{\partial P_1}{\partial p_N} \\ \vdots & & \vdots & \vdots & & \vdots \\ \dfrac{\partial P_N}{\partial q_1} & \cdots & \dfrac{\partial P_N}{\partial q_N} & \dfrac{\partial P_N}{\partial p_1} & \cdots & \dfrac{\partial P_N}{\partial p_N} \end{bmatrix}. \tag{4.16}$$

Definition 4.1 (of the canonical transformation). If we choose matrix $\mathbf{J}$ by means of an appropriate selection of matrices $\mathbf{Q}$ and $\mathbf{P}$, so that the following relation holds true

$$\mathbf{J}^{\mathrm{T}}\mathbf{I}\mathbf{J} = C\mathbf{I}, \tag{4.17}$$

where C is a certain constant, and matrix $\mathbf{I}$ is defined by formula (4.14), then transformation (4.15) is called a canonical transformation.

Now we will check whether the identity transformations $Q_n = q_n$ and $P_n = p_n$ are canonical (to simplify the calculation we will take $N = 2$).

According to formula (4.16) we have

$$\mathbf{J} = \begin{bmatrix} 1 & 0 & 0 & 0 \\ 0 & 1 & 0 & 0 \\ 0 & 0 & 1 & 0 \\ 0 & 0 & 0 & 1 \end{bmatrix},$$

and substituting it into equality (4.17) we obtain

$$\begin{bmatrix}1&0&0&0\\0&1&0&0\\0&0&1&0\\0&0&0&1\end{bmatrix}\begin{bmatrix}0&0&1&0\\0&0&0&1\\-1&0&0&0\\0&-1&0&0\end{bmatrix}\begin{bmatrix}1&0&0&0\\0&1&0&0\\0&0&1&0\\0&0&0&1\end{bmatrix}=\begin{bmatrix}0&0&1&0\\0&0&0&1\\-1&0&0&0\\0&-1&0&0\end{bmatrix}\begin{bmatrix}1&0&0&0\\0&1&0&0\\0&0&1&0\\0&0&0&1\end{bmatrix}=\begin{bmatrix}0&0&1&0\\0&0&0&1\\-1&0&0&0\\0&-1&0&0\end{bmatrix},$$

which satisfies formula (4.16) for $C = 1$.

In the general case for $C = 1$ the transformation matrix $\mathbf{J}$ is called a *symplectic matrix*. If $C \neq 1$, then the transformation matrix $\mathbf{J}$ is called a *generalized symplectic matrix* of valence C. Knowing the valence of matrix $\mathbf{J}$ (i.e., the number C) we are able to determine the value of its determinant.

From (4.17) we obtain

$$\det\left(\mathbf{J}^{\mathrm{T}}\mathbf{I}\mathbf{J}\right) = \det\mathbf{J}^{\mathrm{T}}\det\mathbf{I}\det\mathbf{J} = \det(C\mathbf{I}). \tag{4.18}$$

It can be noticed easily (say, for $N = 2$) that

$$\det(C\mathbf{I}) = \begin{vmatrix}0&0&C&0\\0&0&0&C\\-C&0&0&0\\0&-C&0&0\end{vmatrix} = C(-1)^N\begin{vmatrix}0&0&C\\-C&0&0\\0&-C&0\end{vmatrix} = C^2\begin{vmatrix}-C&0\\0&-C\end{vmatrix} = C^{2N}.$$

Because $\det\mathbf{I} = 1$ and $\det\mathbf{J}^{\mathrm{T}} = \det\mathbf{J}$, from (4.18) we obtain

$$\det\mathbf{J} = \pm C^N. \tag{4.19}$$

Theorem 4.2. *If we take in the considered phase space the two successive canonical transformations* $\mathbf{y}_i = \mathbf{y}_i(\mathbf{x}, t)$ *with the valences* C_i, $i = 1, 2$, *respectively, then the resulting transformation* $\mathbf{y} = \mathbf{y}(\mathbf{x}, t) = \mathbf{y}_2[\mathbf{y}_1(\mathbf{x}, t), t]$ *is also a canonical one with the valence* $C = C_1C_2$.

Proof. Owing to the definition of the canonical transformation (4.17) we have

$$\mathbf{J}_i^{\mathrm{T}}\mathbf{I}\mathbf{J}_i = C_i\mathbf{I}, \qquad \mathbf{J}_1 = \frac{\partial\mathbf{y}_i}{\partial\mathbf{x}}, \qquad \mathbf{J}_2 = \frac{\partial\mathbf{y}_2}{\partial\mathbf{y}_1}, \qquad i = 1, 2.$$

Then

$$\mathbf{J} = \frac{\partial\mathbf{y}}{\partial\mathbf{x}} = \frac{\partial\mathbf{y}_2}{\partial\mathbf{y}_1}\frac{\partial\mathbf{y}_1}{\partial\mathbf{x}} = \mathbf{J}_2\mathbf{J}_1,$$

and consequently

$$\mathbf{J}^{\mathrm{T}}\mathbf{I}\mathbf{J} = (\mathbf{J}_2\mathbf{J}_1)^{\mathrm{T}}\mathbf{I}(\mathbf{J}_2\mathbf{J}_1) = \mathbf{J}_1^{\mathrm{T}}\mathbf{J}_2\mathbf{I}\mathbf{J}_2\mathbf{J}_1 = \mathbf{J}_1^{\mathrm{T}}C_2\mathbf{I}\mathbf{J}_1 = C_2\mathbf{J}_1^{\mathrm{T}}\mathbf{I}\mathbf{J}_1 = C_1C_2\mathbf{I}. \qquad \square$$

Theorem 4.3. *If we take the canonical transformation* $\mathbf{y} = \mathbf{y}(\mathbf{x}, t)$ *with valence* C, *then the inverse transformation* $\mathbf{x} = \mathbf{x}(\mathbf{y}, t)$ *is also a canonical one with valence* $1/C$.

Proof. Multiplying (4.17) by $(\mathbf{J}^{\mathrm{T}})^{-1}$ (left-hand sidedly) and by $\mathbf{J}^{-1}$ (right-hand sidedly) we obtain

$$(\mathbf{J}^{\mathrm{T}})^{-1}\mathbf{J}^{\mathrm{T}}\mathbf{I}\mathbf{J}\mathbf{J}^{-1} = C(\mathbf{J}^{\mathrm{T}})^{-1}\mathbf{I}\mathbf{J}^{-1}$$

or, equivalently,

$$\frac{1}{C}\mathbf{I} = (\mathbf{J}^{\mathrm{T}})^{-1}\mathbf{I}\mathbf{J}^{-1}.$$ □

Below we will present several theorems pertaining to canonical transformations (their proofs can be found in Markeev [9]).

Theorem 4.4. *A necessary and sufficient condition for transformation (4.15) to be canonical is the satisfaction of the following equations:*

$$[q_i, q_k] = 0, \quad [p_i, p_k] = 0, \quad [q_i, p_k] = C\delta_{ik}, \tag{4.20}$$

where the introduced Poisson brackets denote the valence defined by formula (4.17), and δ_{ik} *is the Kronecker delta (i.e.,* $\delta_{ik} = 1$ *for* $i = k$*).*

Theorem 4.5. *A necessary and sufficient condition for transformation (4.15) to be canonical is the satisfaction of the following equations:*

$$[Q_i, Q_k] = 0, \quad [P_i, P_k] = 0, \quad [Q_i, P_k] = C\delta_{ik}, \tag{4.21}$$

where $\mathbf{Q} = \mathbf{Q}(\mathbf{q}, \mathbf{p}, t)$, $\mathbf{P} = \mathbf{P}(\mathbf{q}, \mathbf{p}, t)$.

Theorem 4.6. *A necessary and sufficient condition for transformation (4.15) to be canonical is the existence of a constant* $C \neq 0$ *such that*

$$\delta F \equiv C\sum_{n=1}^{N} p_n\delta q_n - \sum_{n=1}^{N} P_n\delta Q_n \tag{4.22}$$

is the total differential of a certain function $F = F(\mathbf{q}, \mathbf{p}, t)$.

Now we will demonstrate that matrix $\mathbf{J}$ of the form (4.16), generated by the motion of a Hamiltonian system, satisfies canonical transformation (4.17) for $C = 1$. This means that the state of the system described at time instant $t = 0$ by the coordinates $\mathbf{q}_0$ and $\mathbf{p}_0$ is transformed into the coordinates $\mathbf{q}(t)$ and $\mathbf{p}(t)$ for time instant t, that is, $\mathbf{q}_0 \rightarrow \mathbf{q}$ and $\mathbf{p}_0 \rightarrow \mathbf{p}$.

Let us differentiate equation of motion (4.13) with respect to $(\mathbf{q}_0, \mathbf{p}_0)$ $(\mathbf{q}_0 = \mathbf{q}(0), \mathbf{p}_0 = \mathbf{p}(0))$; we obtain

$$\frac{\mathrm{d}}{\mathrm{d}t}\frac{\partial(\mathbf{q},\mathbf{p})}{\partial(\mathbf{q}_0,\mathbf{p}_0)} = \mathbf{I}\mathbf{H}''\frac{\partial(\mathbf{q},\mathbf{p})}{\partial(\mathbf{q}_0,\mathbf{p}_0)}, \tag{4.23}$$

where $\mathbf{q}^{\mathrm{T}} = \mathbf{q}^{\mathrm{T}}(\mathbf{q}_0, \mathbf{p}_0, t)$, $\mathbf{p}^{\mathrm{T}} = \mathbf{p}^{\mathrm{T}}(\mathbf{q}_0, \mathbf{p}_0, t)$.

In this case, matrix (4.16) has the form

$$\mathbf{J}_{2N\times 2N} = \left[\frac{\partial(\mathbf{q},\mathbf{p})}{\partial(\mathbf{q}_0,\mathbf{p}_0)}\right] = \begin{bmatrix} \frac{\partial q_1}{\partial q_{10}} \cdots \frac{\partial q_1}{\partial q_{N0}} & \frac{\partial q_1}{\partial p_{10}} \cdots \frac{\partial q_1}{\partial p_{N0}} \\ \vdots \quad\quad \vdots & \vdots \quad\quad \vdots \\ \frac{\partial q_N}{\partial q_{10}} \cdots \frac{\partial q_N}{\partial q_{N0}} & \frac{\partial q_N}{\partial p_{10}} \cdots \frac{\partial q_N}{\partial p_{N0}} \\ \frac{\partial p_1}{\partial q_{10}} \cdots \frac{\partial p_1}{\partial q_{N0}} & \frac{\partial p_1}{\partial p_{10}} \cdots \frac{\partial p_1}{\partial p_{N0}} \\ \vdots \quad\quad \vdots & \vdots \quad\quad \vdots \\ \frac{\partial p_N}{\partial q_{10}} \cdots \frac{\partial p_N}{\partial q_{N0}} & \frac{\partial p_N}{\partial p_{10}} \cdots \frac{\partial p_N}{\partial p_{N0}} \end{bmatrix}, \tag{4.24}$$

and matrix $\mathbf{H}''$ takes the form

$$\mathbf{H}''_{1\times 2N} = \left[\frac{\partial \mathbf{H}'}{\partial(\mathbf{q}_0,\mathbf{p}_0)}\right] = \begin{bmatrix} \frac{\partial^2 H}{\partial q_{10}^2} \cdots \frac{\partial^2 H}{\partial q_{10}\partial q_{N0}} & \frac{\partial^2 H}{\partial q_{10}\partial p_{10}} \cdots \frac{\partial^2 H}{\partial q_{10}\partial p_{N0}} \\ \vdots \quad\quad \vdots & \vdots \quad\quad \vdots \\ \frac{\partial^2 H}{\partial q_N\partial q_{10}} \cdots \frac{\partial^2 H}{\partial q_{N0}\partial q_{N0}} & \frac{\partial^2 H}{\partial q_{N0}\partial p_{10}} \cdots \frac{\partial^2 H}{\partial q_{N0}\partial p_{N0}} \\ \frac{\partial^2 H}{\partial p_{10}\partial q_{10}} \cdots \frac{\partial^2 H}{\partial p_{10}\partial q_{N0}} & \frac{\partial^2 H}{\partial p_{10}\partial p_{10}} \cdots \frac{\partial^2 H}{\partial p_{10}\partial p_{N0}} \\ \vdots \quad\quad \vdots & \vdots \quad\quad \vdots \\ \frac{\partial^2 H}{\partial p_{N0}\partial q_{10}} \cdots \frac{\partial^2 H}{\partial p_{N0}\partial q_{N0}} & \frac{\partial^2 H}{\partial p_N\partial p_{10}} \cdots \frac{\partial^2 H}{\partial p_{N0}\partial p_{N0}} \end{bmatrix}. \tag{4.25}$$

Equation (4.23) with matrices defined by (4.24) and (4.25) can be written in the form

$$\frac{\mathrm{d}}{\mathrm{d}t}\mathbf{J} = \mathbf{I}\mathbf{H}''\mathbf{J}. \tag{4.26}$$

Let us now transpose both sides of the preceding matrix equation

$$\frac{\mathrm{d}}{\mathrm{d}t}\mathbf{J}^{\mathrm{T}} = \mathbf{J}^{\mathrm{T}}\left(\mathbf{I}\mathbf{H}''\right)^{\mathrm{T}} = \mathbf{J}^{\mathrm{T}}\mathbf{H}''^{\mathrm{T}}\mathbf{I}^{\mathrm{T}} = -\mathbf{J}^{\mathrm{T}}\mathbf{H}''\mathbf{I}, \tag{4.27}$$

because it is easy to verify that $\mathbf{H}'' = \mathbf{H}''^{\mathrm{T}}$ [see matrix (4.25)] and $\mathbf{I}^{\mathrm{T}} = -\mathbf{I}$.

According to formula (4.17) for $C = 1$ we have

$$\mathbf{I} = \mathbf{J}^{\mathrm{T}}\mathbf{I}\mathbf{J}, \tag{4.28}$$

where $\mathbf{J}$ is described by (4.24).

Let us calculate the derivative [taking into account of (4.26) and (4.28)]

$$\begin{aligned}\frac{\mathrm{d}}{\mathrm{d}t}\left(\mathbf{J}^{\mathrm{T}}\mathbf{I}\mathbf{J}\right) &= \dot{\mathbf{J}}^{\mathrm{T}}\mathbf{I}\mathbf{J} + \mathbf{J}^{\mathrm{T}}\mathbf{I}\dot{\mathbf{J}} \\ &= -\mathbf{J}^{\mathrm{T}}\mathbf{H}''\mathbf{I}^2\mathbf{J} + \mathbf{J}^{\mathrm{T}}\mathbf{I}^2\mathbf{H}''\mathbf{J} \\ &= \mathbf{J}^{\mathrm{T}}\mathbf{H}''\mathbf{J} - \mathbf{J}^{\mathrm{T}}\mathbf{H}''\mathbf{J} \equiv \mathbf{0}\end{aligned} \tag{4.29}$$

because $\mathbf{I}^2 = -\mathbf{E}_{2N}$.

Thus we proved that the matrix $\mathbf{J}^{\mathrm{T}}\mathbf{I}\mathbf{J}$ is a constant matrix, and for time instant $t = 0$ it is equal to $\mathbf{I}$. This result can be stated in the form of the following theorem.

Theorem 4.7. *The transformation of a phase space* $(\mathbf{q}, \mathbf{p})$ *as a result of the motion of a Hamiltonian system is a canonical transformation of the valence* $C = 1$.

If we consider the motion of a point in a phase space $(\mathbf{q}, \mathbf{p})$ starting from the initial condition $\mathbf{q}(t_0) = \mathbf{q}_0$, $\mathbf{p}(t_0) = \mathbf{p}_0$, the trajectory of motion is determined by the equations $\mathbf{q} = \mathbf{q}(\mathbf{q}_0, \mathbf{p}_0, t)$ and $\mathbf{p} = \mathbf{p}(\mathbf{q}_0, \mathbf{p}_0, t)$.

Theorem 4.8 (Liouville's[1] theorem). *The volume of a phase space described by the formula*

$$V_t = \underbrace{\int \ldots \int}_{2N} \mathrm{d}q_1 \ldots \mathrm{d}q_N \mathrm{d}p_1 \ldots \mathrm{d}p_N \tag{4.30}$$

is conserved during the motion of a Hamiltonian system.

Proof. For the time instant $t = 0$ (4.30) has the form

$$V_0 = \underbrace{\int \ldots \int}_{2N} \mathrm{d}q_{10} \ldots \mathrm{d}q_N \mathrm{d}p_{10} \ldots \mathrm{d}p_{N0}.$$

Because matrix $\mathbf{J}$ is the Jacobian of transformation (4.15), using matrix (4.24) we have

$$\mathrm{d}\mathbf{x} = \mathbf{J}\mathrm{d}\mathbf{x}_0,$$

[1] Joseph Liouville (1809–1882), French mathematician who contributed greatly to number theory, complex analysis, differential geometry, and topology.

where

$$\mathrm{d}\mathbf{x} = \begin{bmatrix} \mathrm{d}q_1 \\ \vdots \\ \mathrm{d}q_N \\ \mathrm{d}p_1 \\ \vdots \\ \mathrm{d}p_N \end{bmatrix}, \quad \mathrm{d}\mathbf{x}_0 = \begin{bmatrix} \mathrm{d}q_{10} \\ \vdots \\ \mathrm{d}q_N \\ \mathrm{d}p_{10} \\ \vdots \\ \mathrm{d}p_N \end{bmatrix}.$$

Expressing variables $\mathrm{d}\mathbf{x}$ in terms of $\mathrm{d}\mathbf{x}_0$ occurring inside of the integral, we introduce the Jacobian of the transformation

$$V_t = \underbrace{\int \ldots \int}_{2N} |\det \mathbf{J}| \mathrm{d}q_{10} \ldots \mathrm{d}q_N \mathrm{d}p_{10} \ldots \mathrm{d}p_N .$$

According to equality (4.28) we have

$$\det \mathbf{I} = \det \mathbf{J}^{\mathrm{T}} \det \mathbf{I} \det \mathbf{J},$$

that is, $\det \mathbf{J} = \pm 1$.

In turn, for $t = 0$ we have $\mathbf{J} = \mathbf{E}_{2N}$, and because $\det \mathbf{E}_{2N} = 1$, we have $\det \mathbf{J} = 1$. According to the theorem proved earlier, the matrix $\mathbf{J}^{\mathrm{T}}\mathbf{I}\mathbf{J} = \mathbf{I}$ is a constant matrix for any time instant t. Also, for any time instant we have $\det |\mathbf{J}| = 1$, thus $V_t = V_0$, which we had set out to demonstrate. □

4.1.4 Non-Singular Canonical Transformations and Guiding Functions

If transformation (4.15) is canonical, and additionally we have

$$\det \frac{\partial \mathbf{Q}}{\partial \mathbf{p}} \neq 0, \tag{4.31}$$

then canonical transformation (4.15)

$$\mathbf{Q} = \mathbf{Q}(\mathbf{q}, \mathbf{p}, t), \qquad \mathbf{P} = \mathbf{P}(\mathbf{q}, \mathbf{p}, t), \tag{4.32}$$

is called a *non-singular canonical transformation*. According to the theorem associated with (4.22), if the transformation is canonical, then

$$\delta F(\mathbf{q}, \mathbf{p}(\mathbf{q}, \mathbf{Q}, t), t) \equiv C \sum_{n=1}^{N} p_n \delta q_n - \sum_{n=1}^{N} P_n \delta Q_n = \delta S(\mathbf{q}, \mathbf{Q}, t), \tag{4.33}$$

where the satisfaction of condition (4.31) allowed for the expression of $\mathbf{p}$ in the form of the function $\mathbf{p} = \mathbf{p}(\mathbf{q}, \mathbf{Q}, t)$. The function F in which such a change was made is denoted by S and is called the *guiding function of non-singular canonical transformation* (4.32). From (4.33) it follows that

$$\frac{\partial S(\mathbf{q}, \mathbf{Q}, t)}{\partial \mathbf{q}} = C\mathbf{p}, \quad \frac{\partial S(\mathbf{q}, \mathbf{Q}, t)}{\partial \mathbf{Q}} = -\mathbf{P}. \tag{4.34}$$

On the other hand, if the function $S = S(\mathbf{q}, \mathbf{Q}, t)$ is given such that

$$\det\left(\frac{\partial^2 S}{\partial \mathbf{q} \partial \mathbf{Q}}\right) \neq 0, \tag{4.35}$$

then equations (4.34) describe the non-singular canonical transformation of valence $C \neq 0$.

In turn, it is possible to pass from (4.34) to the form (4.32). From the first equation of (4.34) we can obtain $\mathbf{Q} = \mathbf{Q}(\mathbf{q}, \mathbf{p}, t)$. Subsequently, substituting $\mathbf{Q}$ into the second equation of (4.34) we obtain $\mathbf{P} = \mathbf{P}(\mathbf{q}, \mathbf{p}, t)$.

Knowing the guiding function S allows for an easy transition from one form of the Hamiltonian function [corresponding to the old coordinates $(\mathbf{q}, \mathbf{p})$] to another form [corresponding to the new coordinates $(\mathbf{Q}, \mathbf{P})$]. In this case (see [9]) we have

$$\hat{H} = CH + \frac{\partial S}{\partial t}, \tag{4.36}$$

where $\hat{H}$ is the new Hamiltonian function.

A great advantage of this approach consists in the fact that in order to change the coordinates, that is, to pass from the old to the new coordinates (which is required for the determination of first integrals), there is no need to carry out the transformations of, often, many functions (variables) $2N$, but it is enough to know the functions H and S.

4.1.5 Jacobi's Method and Hamilton–Jacobi Equations

Let us now make an attempt, using the notions introduced earlier, to integrate Hamilton's canonical equations of the form

$$\dot{\mathbf{q}} = \frac{\partial H}{\partial \mathbf{p}}, \quad \dot{\mathbf{p}} = -\frac{\partial H}{\partial \mathbf{q}}. \tag{4.37}$$

For $C = 1$, according to (4.34), we have

$$\frac{\partial S(\mathbf{q}, \mathbf{Q}, t)}{\partial \mathbf{q}} = \mathbf{p}, \quad \frac{\partial S(\mathbf{q}, \mathbf{Q}, t)}{\partial \mathbf{Q}} = -\mathbf{P}. \tag{4.38}$$

Hamilton's equations in the new coordinate system $(\mathbf{P}, \mathbf{Q})$ take the following form [equivalent to (4.37)]:

$$\dot{\mathbf{Q}} = \frac{\partial \hat{H}}{\partial \mathbf{P}}, \quad \dot{\mathbf{P}} = -\frac{\partial \hat{H}}{\partial \mathbf{Q}}, \tag{4.39}$$

where, according to function (4.36), we have

$$\hat{H}(\mathbf{Q}, \mathbf{P}, t) = H(\mathbf{q}, \mathbf{p}, t) + \frac{\partial S(\mathbf{q}, \mathbf{Q}, t)}{\partial t}, \tag{4.40}$$

where $\frac{\partial S}{\partial t}$ can be expressed in terms of functions $\mathbf{Q}$, $\mathbf{P}$ using (4.38).

The value of the new Hamiltonian function $\hat{H}$ depends on the function S. If we take it such that $\hat{H} \equiv 0$, then from (4.39) we obtain

$$\dot{\mathbf{Q}} = \mathbf{0}, \quad \dot{\mathbf{P}} = \mathbf{0}, \tag{4.41}$$

and after their integration

$$\mathbf{Q} = \mathbf{C}_1, \quad \mathbf{P} = \mathbf{C}_2, \tag{4.42}$$

where $\mathbf{C}_1$ and $\mathbf{C}_2$ are vectors of N constants, that is, $\mathbf{C}_i^{\mathrm{T}} = (C_1^{(i)}, \ldots, C_N^{(i)})^{\mathrm{T}}$, $i = 1, 2$.

Substituting equality (4.42) into the second equation of (4.38) we have

$$\mathbf{q} = \mathbf{q}(\mathbf{C}_1, \mathbf{C}_2, t), \tag{4.43}$$

and then from the first equation of (4.38) we get

$$\mathbf{p} = \mathbf{p}(\mathbf{C}_1, \mathbf{C}_2, t). \tag{4.44}$$

Substituting expressions (4.43) and (4.44) into equation (4.40) and taking into account (4.38) we obtain

$$\frac{\partial S(\mathbf{q}, \mathbf{Q}, t)}{\partial t} + H\left(\mathbf{q}, \frac{\partial S(\mathbf{q}, \mathbf{Q}, t)}{\partial \mathbf{q}}, t\right) = 0. \tag{4.45}$$

This is a partial differential equation called the *Hamilton–Jacobi equation*; it serves to determine the guiding function S dependent on $q_1(t), q_2(t), \ldots, q_N(t)$ and t, where the quantities $\mathbf{Q} = \mathbf{C}_1$ are treated as parameters (because they are constant).

Theorem 4.9 (Jacobi's theorem). *If $S = S(\mathbf{q}, \mathbf{C}_1, t)$ is a complete integral of Hamilton–Jacobi equation (4.45), that is, S depends on N constants $(C_1^1, \ldots, C_N^1)$, and we have $\left|\frac{\partial^2 S}{\partial \mathbf{q} \partial \mathbf{C}_1}\right| \neq 0$, then we find the solutions (4.43) and (4.44) of (4.37)* *from (4.38), i.e.*

$$\frac{\partial S}{\partial \mathbf{q}} = \mathbf{p}, \quad \frac{\partial S}{\partial \mathbf{C}_1} = -\mathbf{C}_2, \tag{4.46}$$

where the vector $\mathbf{C}_2$ *is composed of* N *arbitrary constants.*

It should be emphasized that in general there is no *recipe* for the integration of partial differential equation (4.45), and it is not always easier to solve this equation than ordinary differential equations (4.37). However, in many cases where the Hamiltonian function appears in technical applications, either function S can be directly determined or the process by which it is determined is significantly simplified.

4.1.6 Forms of the Hamilton–Jacobi Equations in the Case of Cyclic Coordinates and Conservative Systems

Let us consider the case where we have $N - K$ cyclic coordinates, that is, $q_{K+1}, \ldots, q_N$ are cyclic coordinates. The Hamiltonian function in this case takes the form

$$H = H(q_1, \ldots, q_K, p_1, \ldots, p_K, p_{K+1}, \ldots, p_N, t), \tag{4.47}$$

and, according to the previous calculations, because cyclic coordinates are integrable, the complete integral is equal to

$$S = C_{K+1}^{(1)} q_{K+1}, \ldots, C_N^{(1)} q_N + \hat{S}(q_1, \ldots, q_K, C_1^{(1)}, \ldots, C_N^{(1)}, t), \tag{4.48}$$

and after its substitution into Hamilton–Jacobi equation (4.45) we have

$$\frac{\partial \hat{S}}{\partial t} + H\left(q_1, \ldots, q_K, \frac{\partial \hat{S}}{\partial q_1}, \ldots, \frac{\partial \hat{S}}{\partial q_K}, C_{K+1}^{(1)}, \ldots, C_N^{(1)}, t\right) = 0. \tag{4.49}$$

The function $\hat{S}$ subjected to integration according to partial differential equation (4.49) is significantly simplified since it depends on $(N + 1 - K)$ variables.

In the case of conservative systems, the full energy of the system is conserved, and the Hamiltonian function assumes a constant value, that is,

$$H(q_1, \ldots, q_N, p_1, \ldots, p_N) = H^*. \tag{4.50}$$

This means that the Hamiltonian H^* does not depend explicity on t, and H^* is an arbitrary constant.

Substituting (4.50) into Hamilton–Jacobi equation (4.45) we obtain

$$\frac{\partial S}{\partial t} + H^* = 0, \tag{4.51}$$

that is, after integration

$$S = -H^* t + V, \tag{4.52}$$

where now V does not depend explicitly on time. Because $S = S(\mathbf{q}, \mathbf{Q})$, in our case we have $S = S(\mathbf{q}, \mathbf{C}_1)$.

From (4.50) and (4.52) we get

$$H^* = H\left(q_1, \ldots, q_N, \frac{\partial V}{\partial q_1}, \ldots, \frac{\partial V}{\partial q_N}\right), \tag{4.53}$$

which is known as the Hamilton–Jacobi equation for conservative systems. From (4.53) we compute $V = V(q_1, \ldots, q_N, C_1^{(1)}, \ldots, C_{N-1}^{(1)}, H^*)$. Equation (4.52) now takes the form

$$S = -H^* t + V\left(q_1, \ldots, q_N, C_1^{(1)}, \ldots, C_{N-1}^{(1)}, H^*\right), \tag{4.54}$$

which determines the complete integral of the Hamilton–Jacobi equation.

According to Jacobi's theorem [see formulas (4.46)] we get

$$\frac{\partial V}{\partial q_n} = p_n, \quad (n = 1, \ldots, N), \quad \frac{\partial V}{\partial C_n^{(1)}} = -C_n^{(2)}, \quad (n = 1, \ldots, N-1),$$
$$\frac{\partial V}{\partial H^*} = t - C_N^{(2)}, \tag{4.55}$$

where $C_1^{(2)}, \ldots, C_N^{(2)}$ are arbitrary constants.

The equation (4.55) can be interpreted in the following way. First, part of N equations of (4.55) define the impulses p_n, $n = 1, \ldots, N$; second, part of the $N - 1$ equations describe trajectories of the N-dimensional coordinate space of $q_1, \ldots, q_N$; all of them govern the dynamics of the conservative system.

Various extensions of the described approach (despite a huge number of monographs) can also be found in [25, 26] with an emphasis on mechanical applications.

4.2 Solution Methods for Euler–Lagrange Equations

4.2.1 Introduction

Other methods of analysis of second-order (or higher) non-linear differential equations obtained with the use of variation and called Euler or Euler–Lagrange equations (Chap. 3), which are relatively new and alternative as compared to asymptotic methods, are the so-called *Bogomolny decomposition* [6, 13] and *Bäcklund transformation* [1, 7, 17].

In the mechanics of deformable bodies, we commonly face the problem of static or dynamic deflections of beams, plates, and shells, which are described by non-linear partial differential equations (PDEs). This concerns both hyperbolic and elliptic PDEs. The mentioned variational methods were successfully applied in certain branches of physics for the analysis of non-linear equations of a field theory and, in particular, of the so-called *soliton equations*, which include the Korteweg–de Vries equation,[2] the sine-Gordon[3] equation, and the non-linear Schrödinger[4] equation. These methods are based on a theory of the so-called *strong necessary conditions* and, subsequently, on its modification (extension) called *semistrong necessary conditions*.

This approach made it possible to find the Bäcklund[5] transformation for a wide class of non-linear PDEs, including the aforementioned soliton equations. This, in turn, opened up the possibility of finding particular solutions to such equations and then to generate a whole "lattice" of such solutions (by application of the so-called *Bianchi*[6] *permutability theorem*).

4.2.2 Euler's Theorem and Euler–Lagrange Equations

Let us consider the functional

$$\phi[y] = \int_a^b F(x, y(x), y'(x))\mathrm{d}x, \tag{4.56}$$

where $x \in [a, b]$, $x \to y(x) \in \mathbf{R}$, and F has continuous partial derivatives.

---

[2]This equation is a mathematical model of waves on shallow water surfaces and is named after Dutch mathematicians D. Korteweg (1848–1941) and G. de Vries (1866–1934).

[3]Walter Gordon (1893–1939), twentieth-century English physicist.

[4]Erwin Schrödinger (1887–1961), an Austrian/Irish mathematician awarded the Nobel prize in 1933.

[5]Arthur Bäcklund (1845–1922), a Swedish mathematician and physicist.

[6]Luigi Bianchi (1856–1928), Italian mathematician who made important contributions to differential geometry.

Definition 4.2 (of a weak minimum). Functional (4.56) attains a weak minimum for $y = y^*$ if there exists $\varepsilon > 0$ such that $\phi[y] \geq \phi[y^*]$ for all $y \in K^1(y^*, \varepsilon)$, where

$$K^1(y^*, \varepsilon) = \left(\max|y(x) - y^*(x)| + \max\left|y'(x) - y^*(x)\right| < \varepsilon\right). \tag{4.57}$$

Definition 4.3 (of a strong minimum). Functional (4.56) attains a strong minimum for $y = y^*$, if there exists $\varepsilon > 0$ such that $\phi[y] \geq \phi[y^*]$ for all $y \in K^0(y^*, \varepsilon)$, where

$$K^0(y^*, \varepsilon) = \left(\max|y(x) - y^*(x)| < \varepsilon\right). \tag{4.58}$$

It can be demonstrated that, if the functional $\phi[y]$ attains a strong extremum for $y = y^*$, then it also attains a weak extremum for the function y^*, but not vice versa.

Theorem 4.10 (Euler's theorem). *If functional (4.56) attains an extremum for $y = y^*(x)$, then*

$$\frac{\mathrm{d}F}{\mathrm{d}y} - \frac{\mathrm{d}}{\mathrm{d}x}\left(\frac{\partial F}{\partial y'}\right) = 0 \tag{4.59}$$

for $y = y^*$, $y' = y'^*$.

Equation (4.59) is called *Lagrange's equation* or the *Euler–Lagrange equation*, but it is also known as *Euler's equation* or the *Euler–Poisson equation.*

The problem of determining the extremum of functional (4.56) boils down to the solution of (4.59). However, there exist no general methods of integration of non-linear equations of this kind, and here the previously introduced theories of the Bäcklund transformation and Bogomolny decomposition come to our aid.

The presented concepts of the solution of non-linear PDEs can be exploited to solve a wide class of equations that appear in various fields of physics by treating them as equations of the type (4.59), further called *Lagrange's systems*. Even if we are unable to construct functional (4.56), very often it is possible to reduce these equations to the Lagrange class by introducing certain transformations of independent variables of these equations.

A classic variational approach is based on the fact that the actual trajectory in a configuration space satisfies *Hamilton's principle of least action*. Since, if we consider an arbitrary trajectory $q = q(t)$ admissible by constraints, then the principle of least action is reduced to the following condition for the vanishing of variation:

$$\delta\phi[y] = \int_a^b \left(\frac{\mathrm{d}F}{\mathrm{d}y}\delta y + \frac{\partial F}{\partial y'}\delta y'\right)\mathrm{d}x = 0 \tag{4.60}$$

for the physical (actual) trajectory $q^* = q^*(t)$. Integrating by parts we obtain the Euler–Lagrange equation. From variational equation (4.60) it follows that

$$\frac{\mathrm{d}F}{\mathrm{d}y} = 0, \quad \frac{\mathrm{d}F}{\mathrm{d}y'} = 0. \tag{4.61}$$

Any solution of the preceding system of equations is simultaneously the solution of Euler–Lagrange equation (4.59). The solutions of equations (4.61) are called *strong solutions*, whereas solutions of equation (4.59) that are different from strong solutions are called *weak solutions*.

Let us note that the order of strong equations (4.61) is smaller than the order of original equations (4.56). It turns out that strong equations allow for the generation of many solutions by taking advantage of an *internal symmetry* of Hamilton's action integral.

For a four-dimensional Euclidean space (X_1, X_2, X_3, t) there exists a Lie[7] group of transformations having ten generators, and with every generator of this group is associated an integral of motion, as stated by the so-called *Noether*[8] *theorem* [14]. This theorem makes it possible to establish certain associations between symmetries of a system and certain properties of motion such as momentum, angular momentum, and energy.

Hamilton's variational principle applied to the Lagrangians L and L^* of the form

$$L^* = L + I \tag{4.62}$$

leads to the action principle of the form

$$\delta \int_a^b [L + I]\mathrm{d}t = 0. \tag{4.63}$$

Because for both Lagrangians L and L^* from Hamilton's principle we obtain identical Euler–Lagrange equations, we have

$$\delta \mathrm{I} = \delta \int_a^b I dt \equiv 0. \tag{4.64}$$

We call a functional I having the preceding property a *topological invariant*, and $I = L^* - L$ a *density of a topological invariant.*

According to the construction of functional (4.56), now we have

$$\int_a^b \left[\left(\frac{\mathrm{d}F}{\mathrm{d}y} + \frac{\mathrm{d}I}{\mathrm{d}y}\right)\delta y + \left(\frac{\partial F}{\partial y'} + \frac{\partial I}{\partial y'}\right)\delta y'\right]\mathrm{d}x = 0, \tag{4.65}$$

---

[7] Marius Sophus Lie (1842–1899), Norwegian mathematician.

[8] Emmy Noether (1882–1935), German mathematician known for her contributions to abstract algebra and theoretical physics.

hence we obtain the following system of strong equations:

$$\frac{\partial F}{\partial y} + \frac{\partial I}{\partial y} = 0, \qquad \frac{\partial F}{\partial y'} + \frac{\partial I}{\partial y'} = 0. \tag{4.66}$$

Any solution of the system of strong equations is simultaneously the solution of Euler–Lagrange equation (4.59).

The Euler–Lagrange equations are invariant with respect to calibrating a transformation, that is, they remain identical for both ϕ and $\phi^* = \phi + I$.

As it turns out, this property makes it possible to find many new analytical solutions for numerous equations of the Euler–Lagrange kind.

4.2.3 Bogomolny Equation and Decomposition

Both notions will be introduced using a non-variational approach. Let us consider a one-dimensional model of a scalar field whose free energy functional has the form

$$H[y] = \int \left[\frac{1}{2}\left(\frac{\mathrm{d}y(x)}{\mathrm{d}x}\right)^2 + V(y(x)) \right] \mathrm{d}x. \tag{4.67}$$

According to Euler's theorem, trajectory $y^* = y^*(x)$, on which functional (4.67) attains its minimum, satisfies the following Euler–Lagrange equation:

$$\frac{\mathrm{d}^2 y}{\mathrm{d}x^2} = -\frac{\partial V[y(x)]}{\partial y}. \tag{4.68}$$

Bogomolny proposed the following decomposition of (4.67):

$$\begin{aligned} H[y] &= \frac{1}{2}\int \left[\frac{\mathrm{d}y(x)}{\mathrm{d}x} \pm \sqrt{2[V(y) - C]}\right]^2 \mathrm{d}x \\ &\quad \mp \int \frac{\mathrm{d}y}{\mathrm{d}x}\sqrt{2(V - C)}\mathrm{d}x + \int C\,\mathrm{d}x \\ &= \int \left[\frac{1}{2}\left(\frac{\mathrm{d}y}{\mathrm{d}x}\right)^2 \pm \frac{\mathrm{d}y}{\mathrm{d}x}\sqrt{2(V - C)} + (V - C)\right]\mathrm{d}x \\ &= \int \frac{\mathrm{d}y}{\mathrm{d}x}\sqrt{2(V - C)}\mathrm{d}x = \int \left[\frac{1}{2}\left(\frac{\mathrm{d}y}{\mathrm{d}x}\right)^2 + V\right]\mathrm{d}x, \end{aligned} \tag{4.69}$$

where C is a constant that describes the initial energy and $\left|\int C\,\mathrm{d}x\right| < \infty$.

The Bogomolny equation has the form

$$\frac{\mathrm{d}y}{\mathrm{d}x} \pm \sqrt{2[V(y) - C]} = 0. \tag{4.70}$$

4.2.4 Bäcklund Transformation

Nowadays Bäcklund transformations are introduced using both an algebraic variational and a geometric approach.

Definition 4.4 (of a Bäcklund transformation). Let the following two decoupled differential equations be given:

$$\begin{aligned} E(u, u_x, u_t, u_{xx}, \ldots, x, t) &= 0, \\ F(v, v_x, v_t, v_{xx}, \ldots, x, t) &= 0, \end{aligned} \tag{4.71}$$

where $u = u(x, t)$ and $v = v(x, t)$ are certain fields and variables t and x are independent and real.

The following two differential relations

$$R_i(u, u_x, u_t, v, v_x, v_t, \ldots, x, t) = 0, \quad i = 1, 2, \tag{4.72}$$

are called a Bäcklund transformation for (4.71) if:

(i) For u^0 satisfying the equation $E(u^0, u^0_x, u^0_t, u^0_{xx}, \ldots, x, t) = 0$, the solution of system of equations (4.72) of the form

$$\begin{aligned} R_1\left(u^0, u^0_x, u^0_t, v, v_x, v_t, \ldots, x, t\right) &= 0, \\ R_2\left(u^0, u^0_x, u^0_t, v, v_x, v_t, \ldots, x, t\right) &= 0 \end{aligned} \tag{4.73}$$

makes it possible to find $v(x, t)$, which at the same time is the solution of the second equation of (4.71);

(ii) For v^0 satisfying the equation $F(v^0, v^0_x, v^0_t, v^0_{xx}, \ldots, x, t) = 0$, the solution of system of equations (4.72) of the form

$$\begin{aligned} R_1\left(u, u_x, u_t, v^0, v^0_x, v^0_t, \ldots, x, t\right) &= 0, \\ R_2\left(u, u_x, u_t, v^0, v^0_x, v^0_t, \ldots, x, t\right) &= 0 \end{aligned} \tag{4.74}$$

makes it possible to determine $u = u(x, t)$, which at the same time is the solution of the first equation of (4.71).

If in (4.71) we have $E = F$, then (4.72) is called an *auto-Bäcklund transformation.*

Definition 4.5 (auto-Bäcklund transformation). Relations (4.72) are called strong Bäcklund (auto-Bäcklund) transformations if for every pair

$$R_1\left(u^0, u_x^0, u_t^0, v^0, v_x^0, v_t^0, \ldots, x, t\right) = 0,$$
$$R_2\left(u^0, u_x^0, u_t^0, v^0, v_x^0, v_t^0, \ldots, x, t\right) = 0 \tag{4.75}$$

the following equations are simultaneously satisfied:

$$E(u^0, u_x^0, u_t^0, u_{xx}^0, \ldots, x, t) = 0,$$
$$F(v^0, v_x^0, v_t^0, v_{xx}^0, \ldots, x, t) = 0. \tag{4.76}$$

Let us note that if the order of coupled equations (4.72) is equal to or greater than the order of the original equations (4.71), then the advantages of this approach are questionable.

Example 4.2. Let us consider an auto-Bäcklund transformation for the system of sine-Gordon equations of the form

$$\frac{\partial^2 u}{\partial x \partial t} = \sin u, \qquad \frac{\partial^2 v}{\partial x \partial t} = \sin v.$$

According to [17], the transformation has the form

$$\frac{1}{2}\left(\frac{\partial u}{\partial x} + \frac{\partial v}{\partial x}\right) = \beta \sin\left(\frac{u-v}{2}\right),$$

$$\frac{1}{2}\left(\frac{\partial u}{\partial t} - \frac{\partial v}{\partial t}\right) = \frac{1}{\beta}\sin\left(\frac{u+v}{2}\right).$$

Differentiating the first of the preceding equations with respect to t and the second with respect to x we obtain

$$\frac{1}{2}\left(\frac{\partial^2 u}{\partial x \partial t} + \frac{\partial^2 v}{\partial x \partial t}\right) = \frac{\beta}{2}\left(\frac{\partial u}{\partial t} - \frac{\partial v}{\partial t}\right)\cos\left(\frac{u-v}{2}\right)$$
$$= \sin\left(\frac{u+v}{2}\right)\cos\left(\frac{u-v}{2}\right),$$

$$\frac{1}{2}\left(\frac{\partial^2 u}{\partial x \partial t} - \frac{\partial^2 v}{\partial x \partial t}\right) = \frac{1}{2\beta}\left(\frac{\partial u}{\partial x} + \frac{\partial v}{\partial x}\right)\cos\left(\frac{u+v}{2}\right)$$
$$= \sin\left(\frac{u-v}{2}\right)\cos\left(\frac{u+v}{2}\right).$$

Adding and subtracting the preceding equations by sides we obtain the original equation.

Let us substitute a trivial solution $v = 0$ into the obtained Bäcklund relations. This leads to the following differential equations:

$$\frac{\partial u}{\partial x} = 2\beta \sin\left(\frac{u}{2}\right), \quad \frac{\partial u}{\partial t} = \frac{2}{\beta} \sin\left(\frac{u}{2}\right),$$

and after their integration we obtain the following particular sine-Gordon equation:

$$u(x,t) = 4 \arctan\left[C e^{\mp\left(\frac{1}{\beta}t + \beta x\right)}\right]. \qquad \square$$

Example 4.3. Derive the Bogomolny equation based on strong necessary conditions.

It can be demonstrated that the only non-trivial invariant of (4.64) is

$$I = \int G(y) y' \mathrm{d}x.$$

Substituting the foregoing relation into Euler's equation we obtain

$$\frac{\partial [G(y) y']}{\partial y} - \frac{\mathrm{d}}{\mathrm{d}x}\left[\frac{\partial [G(y) y']}{\partial y'}\right] \equiv 0.$$

Thus we have

$$F(x, y(x), y'(x)) = \frac{1}{2}\left(\frac{\mathrm{d}y}{\mathrm{d}x}\right)^2 + V[y(x)],$$
$$I(y) = G(y) y'.$$

According to formulas (4.66) a system of strong equations has the following form:

$$\frac{\partial V}{\partial y} + \frac{\partial G}{\partial y}\frac{\partial y}{\partial x} = 0, \quad \frac{\partial y}{\partial x} + G(y) = 0.$$

From the second equation we have

$$\frac{\partial y}{\partial x} = -G,$$

and after substitution of this equality into the first equation we obtain

$$\frac{\partial V}{\partial y} = \frac{\partial G}{\partial y} G.$$

Integrating the preceding equation we obtain

$$V[y(x)] = \frac{1}{2}G^2 + C,$$

where C is an integration constant (a real number). From the preceding equation we find

$$G = \pm\sqrt{2[V(y) - C]}.$$ □

4.3 Whittaker's Equations

Let a Hamiltonian function not depend explicitly on time, that is,

$$H = H(q_1, \ldots, q_N, p_1, \ldots, p_N) = H_0 \equiv \text{const}, \tag{4.77}$$

where $H_0 = H(q_{10}, \ldots, q_{N0}, p_{10}, \ldots, p_{N0})$, i.e., it is defined by the introduced initial conditions. Because $q_n = q_n(t)$ and $p_n = p_n(t)$, the motion of a Hamiltonian system can be understood as the motion of a point on a hyperplane (4.77). From (4.77) we determine

$$p_1 = -H^W(q_1, \ldots, q_N, p_2, \ldots, p_N, H_0). \tag{4.78}$$

Separating out variables q_1 and p_1 in Hamilton's canonical equations (4.9) we obtain

$$\frac{\mathrm{d}q_1}{\mathrm{d}t} = \frac{\partial H}{\partial p_1}, \quad \frac{\mathrm{d}p_1}{\mathrm{d}t} = -\frac{\partial H}{\partial q_1}, \tag{4.79}$$

$$\frac{\mathrm{d}q_n}{\mathrm{d}t} = \frac{\partial H}{\partial p_n}, \quad \frac{\mathrm{d}p_n}{\mathrm{d}t} = -\frac{\partial H}{\partial q_n}, \quad n = 2, \ldots, N. \tag{4.80}$$

Dividing both sides of (4.80) by the first equation of (4.79) we have

$$\frac{\mathrm{d}q_n}{\mathrm{d}q_1} = \left(\frac{\partial H}{\partial p_n}\right) \Big/ \left(\frac{\partial H}{\partial p_1}\right),$$

$$\frac{\mathrm{d}p_n}{\mathrm{d}q_1} = -\left(\frac{\partial H}{\partial p_n}\right) \Big/ \left(\frac{\partial H}{\partial p_1}\right), \quad n = 2, 3, \ldots, N. \tag{4.81}$$

Let us substitute p_1 defined by (4.78) into (4.77) and as a result obtain

$$H(q_1, \ldots, q_N, -H^W, p_2, \ldots, p_N) = H_0. \tag{4.82}$$

Differentiating the preceding equation with respect to q_n we obtain

$$\frac{\partial H}{\partial q_n} + \frac{\partial H}{\partial p_1}\frac{\partial p_1}{\partial q_n} = \frac{\partial H}{\partial q_n} - \frac{\partial H}{\partial p_1}\frac{\partial H^W}{\partial q_n} = 0, \qquad n = 2, \ldots, N. \tag{4.83}$$

Proceeding in an analogous way we get

$$\frac{\partial H}{\partial p_n} - \frac{\partial H}{\partial p_1}\frac{\partial H^W}{\partial p_n} = 0, \qquad n = 2, \ldots, N. \tag{4.84}$$

Taking into account (4.83) and (4.84) in the right-hand sides of equations (4.81) we obtain

$$\frac{dq_n}{dq_1} = \frac{\partial H^W}{\partial p_n}, \qquad \frac{dp_n}{\partial q_1} = -\frac{\partial H^W}{\partial q_n}. \tag{4.85}$$

If in the obtained equations we exchange q_1 with t, we arrive at the canonical form of Hamilton's equations (4.9). The role of a Hamiltonian function H is played by a Whittaker function H^W. Equation (4.85) are called *Whittaker's equations.*[9]

Integration of Whittaker's equations leads to the determination of velocities and momenta

$$\begin{aligned} q_n &= q_n(q_1, H_0, C_1, \ldots, C_{2N-2}), \\ p_n &= p_n(q_1, H_0, C_1, \ldots, C_{2N-2}), \qquad n = 2, 3, \ldots, N, \end{aligned} \tag{4.86}$$

where $C_1, \ldots, C_{2N-2}$ are arbitrary constants.

Substituting the quantities thus obtained into (4.78) we obtain

$$p_1 = f_1(q_1, H_0, C_1, \ldots, C_{2N-2}). \tag{4.87}$$

Equations (4.86) and (4.87) define the motion of a point of coordinates $p_1(q_1)$, $p_2(q_1), \ldots, p_N(q_1)$, $q_2(p_1), \ldots, q_N(p_1)$ on a phase hyperplane $H = H_0$.

In order to determine the dependency of coordinates on time one should make use of the first equation of system (4.79), which will take the form

$$\frac{\mathrm{d}q_1}{\mathrm{d}t} = f_2(q_1, H_0, C_1, \ldots, C_{2N-2}), \tag{4.88}$$

[9]Edmund Taylor Whittaker (1873–1956), English mathematician who contributed widely to applied mathematics and mathematical physics.

and separating the variables and integrating we eventually get

$$t = \int \frac{\mathrm{d}q_1}{f_2} + C_{2N-1}. \tag{4.89}$$

Solving the preceding algebraic equation with respect to q_1 we obtain

$$q_1 = q_1(t, H_0, C_1, \ldots, C_{2N-1}). \tag{4.90}$$

In [14] it is shown that using the Legendre transformation, Whittaker's equations can be reduced to Lagrange's equations of the form

$$\frac{\mathrm{d}}{\mathrm{d}q_1}\frac{\partial P}{\partial q_n'} - \frac{\partial P}{\partial q_n} = 0, \quad n = 2, 3, \ldots, N, \tag{4.91}$$

where

$$P = P(q_2, \ldots, q_N, q_2', \ldots, q_N', q_1, H_0) = \sum_{n=2}^{N} q_n' p_n - H^W,$$

$$q_n' = \frac{\mathrm{d}q_n}{\mathrm{d}q_1}, \tag{4.92}$$

and quantities p_n occurring in a Legendre function P are expressed by $q_2', \ldots, q_N'$ from equations

$$q_n' = \frac{\partial H^W}{\partial p_n}, \quad n = 2, 3, \ldots, N. \tag{4.93}$$

Equations in the form of (4.91) are called *Jacobi equations*.

4.4 Voronets and Chaplygin Equations

In many cases equations describing the motion of a DMS with Lagrange multipliers are burdensome in practical applications, especially if we are not interested in determining the reactions of constraints. P.V. Voronets[10] [3] proposed equations that do not have the aforementioned disadvantages. We will derive them on the basis of equations taken from [9], where we will be considering a scleronomic nonholonomic DMS.

[10]Peter Vasilevich Voronets (1871–1923), Russian scientist widely known for his research in the field of analytical mechanics.

In the considered case, (3.139) takes the form

$$\frac{\mathrm{d}}{\mathrm{d}t}\frac{\partial T}{\partial \dot{q}_w} - \frac{\partial T}{\partial q_w} = Q_w - \sum_{m=1}^{M} \lambda_m \alpha_{mw}, \quad w = 1, \ldots, W,$$

$$\frac{\mathrm{d}}{\mathrm{d}t}\frac{\partial T}{\partial \dot{q}_{W+m}} - \frac{\partial T}{\partial q_{W+m}} = Q_{W+m} + \lambda_m, \quad m = 1, \ldots, M,$$

$$\dot{q}_{W+m} = \sum_{w=1}^{W} \alpha_{mw}\dot{q}_w. \tag{4.94}$$

In the preceding equations the system has $W+M$ generalized coordinates, where M denotes the number of scleronomic non-holonomic constraints.

The third of equations (4.94) results from the equation of constraints described by the second equation of system (3.139), where coefficients B_{mk} do not depend explicitly on time and we have $b_m = 0$. That equation is obtained after imposing the condition that from K generalized velocities we have W independent generalized velocities, and the number of degrees of freedom of the DMS is equal to $W = K - M$. From this equation it follows that the dependent generalized velocities in the amount of M can be expressed by W independent generalized velocities.

It can be demonstrated [9] that the coefficients

$$A_{wk}^{(m)} = \left(\frac{\partial \alpha_{mw}}{\partial q_k} + \sum_{u=1}^{M} \frac{\partial \alpha_{mw}}{\partial q_{W+u}}\alpha_{uk}\right) - \left(\frac{\partial \alpha_{mk}}{\partial q_w} + \sum_{u=1}^{M} \frac{\partial \alpha_{mk}}{\partial q_{W+u}}\alpha_{uw}\right) \tag{4.95}$$

are not identically equal to zero for an arbitrary instant of motion of the analyzed DMS.

The third equation of system (4.94) allows for the following representation of the kinetic energy of a DMS:

$$T(q_1, \ldots, q_K, \dot{q}_1, \ldots, \dot{q}_K, t) = \bar{\Theta}(q_1, \ldots, q_K, \dot{q}_1, \ldots, \dot{q}_W, t). \tag{4.96}$$

Differentiating (4.96) with respect to $\dot{q}_1, \ldots, \dot{q}_W$ we obtain

$$\frac{\partial \bar{\Theta}}{\partial \dot{q}_w} = \frac{\partial T}{\partial \dot{q}_w} + \sum_{m=1}^{M} \frac{\partial T}{\partial \dot{q}_{W+m}}\alpha_{mw}, \quad w = 1, 2, \ldots, W, \tag{4.97}$$

where the third equation of system (4.94) was used in the transformations.

Differentiation of (4.97) with respect to time leads to the following equation:

$$\frac{\mathrm{d}}{\mathrm{d}t}\frac{\partial \bar{\Theta}}{\partial \dot{q}_w} = \underline{\frac{\mathrm{d}}{\mathrm{d}t}\frac{\partial T}{\partial \dot{q}_w}} + \sum_{m=1}^{M} \alpha_{mw}\underline{\frac{\mathrm{d}}{\mathrm{d}t}\frac{\partial T}{\partial \dot{q}_{W+m}}} + \sum_{m=1}^{M} \frac{\mathrm{d}\alpha_{mw}}{\mathrm{d}t}\frac{\partial T}{\partial \dot{q}_{W+m}}. \tag{4.98}$$

The terms underlined in the preceding equation will be calculated from (4.94). As a result of the transformations, the terms containing Lagrange multipliers λ_m are reduced, and (4.98) takes the form

$$\frac{\mathrm{d}}{\mathrm{d}t}\frac{\partial \bar{\Theta}}{\partial \dot{q}_w} = \frac{\partial T}{\partial q_w} + Q_w + \sum_{m=1}^{M} \alpha_{mw} \frac{\partial T}{\partial q_{W+m}} + \sum_{m=1}^{M} \alpha_{mw} Q_{W+m} + \sum_{m=1}^{M} \frac{\mathrm{d}\alpha_{mw}}{\mathrm{d}t} \frac{\partial T}{\partial \dot{q}_{W+m}}. \tag{4.99}$$

Since, according to the third equation of system (4.94), we obtain

$$\frac{\partial \Theta}{\partial q_l} = \frac{\partial T}{\partial q_l} + \sum_{m=1}^{M} \frac{\partial T}{\partial \dot{q}_{W+m}} \left(\sum_{k=1}^{W} \frac{\partial \alpha_{mk}}{\partial q_l} \dot{q}_k \right), \quad l = 1, 2, \ldots, K, \tag{4.100}$$

taking into account this result in (4.99) we have

$$\begin{aligned} \frac{\mathrm{d}}{\mathrm{d}t}\frac{\partial \bar{\Theta}}{\partial \dot{q}_w} = {} & \frac{\partial \bar{\Theta}}{\partial q_w} - \sum_{m=1}^{M} \frac{\partial T}{\partial \dot{q}_{W+m}} \left(\sum_{k=1}^{W} \frac{\partial \alpha_{mk}}{\partial q_w} \dot{q}_k \right) \\ & + Q_w + \sum_{m=1}^{M} \alpha_{mw} \frac{\partial \bar{\Theta}}{\partial q_{W+m}} \\ & - \sum_{u=1}^{k} \alpha_{uw} \left(\sum_{m=1}^{M} \frac{\partial T}{\partial \dot{q}_{W+m}} \left(\sum_{k=1}^{W} \frac{\partial \alpha_{mk}}{\partial q_{W+u}} \dot{q}_k \right) \right) \\ & + \sum_{m=1}^{M} \alpha_{mw} Q_{W+m} + \sum_{m=1}^{M} \frac{\mathrm{d}\alpha_{mw}}{\mathrm{d}t} \frac{\partial T}{\partial \dot{q}_{W+m}}. \end{aligned} \tag{4.101}$$

Transforming the preceding equations we eventually arrive at

$$\begin{aligned} \frac{\mathrm{d}}{\mathrm{d}t}\frac{\partial \bar{\Theta}}{\partial \dot{q}_w} - \frac{\partial \bar{\Theta}}{\partial q_w} = {} & Q_w + \sum_{m=1}^{M} \alpha_{mw} \left(Q_{W+m} + \frac{\partial \bar{\Theta}}{\partial q_{W+m}} \right) \\ & + \sum_{m=1}^{M} \frac{\partial T}{\partial \dot{q}_{W+m}} \left[\underline{\frac{d\alpha_{mw}}{dt} - \sum_{k=1}^{W} \left(\frac{\partial \alpha_{mk}}{\partial q_w} + \sum_{u=1}^{M} \frac{\partial \alpha_{mk}}{\partial q_{W+u}} \alpha_{uw} \right) \dot{q}_k} \right] \\ = {} & Q_w + \sum_{m=1}^{M} \alpha_{mw} \left(Q_{W+m} + \frac{\partial \bar{\Theta}}{\partial q_{W+m}} \right) + \sum_{m=1}^{M} \Theta_w \left(\sum_{k=1}^{W} A_{wk}^{(m)} \dot{q}_k \right), \\ \Theta_w = {} & \frac{\partial T}{\partial \dot{q}_{w+m}}, \qquad w = 1, 2, \ldots, W, \quad m = 1, 2, \ldots, M, \end{aligned} \tag{4.102}$$

where in the preceding equation the underlined term was replaced by the previously introduced coefficients $A_{wk}^{(m)}$ according to (4.96), and the notion of generalized momentum Θ_w was introduced.

The obtained equations (4.102) in the amount of W are called *Voronets equations* [3].

We have $K = W + M$ generalized coordinates, and in order to determine them one should solve the following system of $M + W$ equations:

$$\frac{\mathrm{d}}{\mathrm{d}t}\frac{\partial\bar{\Theta}}{\partial\dot{q}_w} - \frac{\partial\bar{\Theta}}{\partial q_w} = \Theta_w + \sum_{m=1}^{M}\alpha_{mw}\left(Q_{W+m} + \frac{\partial\bar{\Theta}}{\partial q_{W+m}}\right) + \sum_{m=1}^{M}\Theta_w\left(\sum_{k=1}^{W}A_{wk}^{(m)}\dot{q}_k\right),$$

$$\dot{q}_{W+m} = \sum_{w=1}^{W}\alpha_{mw}\dot{q}_w, \quad w = 1, 2, \ldots, W, \quad m = 1, 2, \ldots, M, \tag{4.103}$$

in which there are no longer any Lagrange multipliers.

If the kinetic energy T, coefficients α_{mw}, and generalized forces Q_l ($l = 1, \ldots, K$) do not depend on the generalized coordinates q_{W+m} ($m = 1, 2, \ldots, M$), then the Voronets equations take the following simpler form:

$$\frac{\mathrm{d}}{\mathrm{d}t}\frac{\partial\bar{\Theta}}{\partial\dot{q}_w} - \frac{\partial\bar{\Theta}}{\partial q_w} = Q_w + \sum_{m=1}^{M}\alpha_{mw}Q_{W+m} + \sum_{m=1}^{M}\left(\sum_{k=1}^{W}A_{wk}^{(m)}\dot{q}_k\right), \quad w = 1, \ldots, W, \tag{4.104}$$

where coefficients $A_{wk}^{(m)}$ are now described by the simple relationships

$$A_{wk}^{(m)} = \frac{\partial\alpha_{mw}}{\partial q_k} - \frac{\partial\alpha_{mk}}{\partial q_w}, \quad w, k = 1, \ldots, W, \quad m = 1, \ldots, M. \tag{4.105}$$

In this case we can further simplify the expressions for generalized forces Q_l ($l = 1, \ldots, K$) and for generalized momenta Θ_m ($m = 1, 2, \ldots, M$). If we express generalized velocities $\dot{q}_{w+m}$ ($m = 1, \ldots, M$) by independent generalized velocities $\dot{q}_1, \ldots, \dot{q}_W$ according to the third equation of system (4.94), then we obtain the system of equations dependent only on the generalized coordinates $q_1, \ldots, q_W$, which can be solved independently of an equation of constraints, that is, the third equation of system (4.94). Following application of the described algorithm to (4.104) we obtain *Chaplygin's equations.*[11]

[11] Sergey Alexeyevich Chaplygin (1869–1942), Russian mechanician and mathematician.

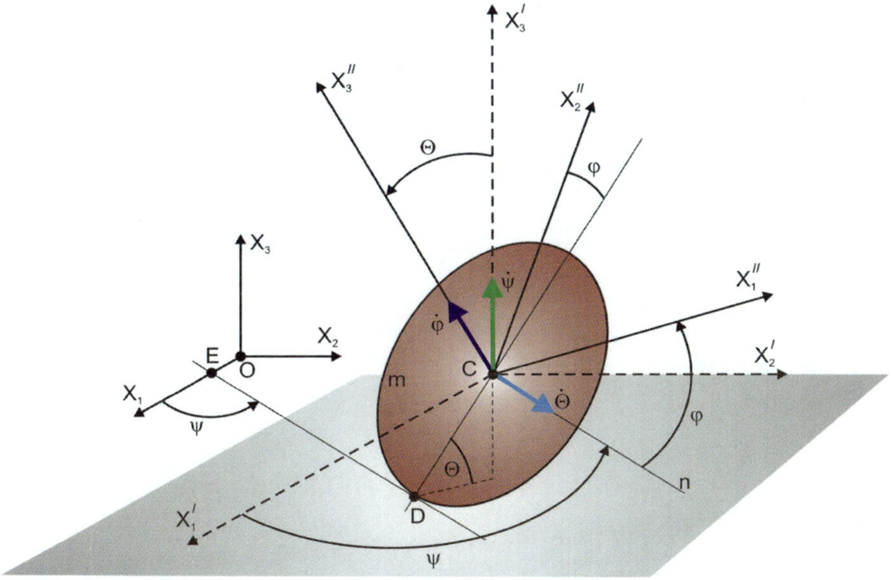

Fig. 4.1 Rolling without slip of a disk on a horizontal plane [9]

Chaplygin's equations simplify even more if generalized forces are potential forces and potential V does not depend on the generalized coordinates q_{W+m}. Then Chaplygin's equations take the form

$$\frac{\mathrm{d}}{\mathrm{d}t}\frac{\partial\bar{\Theta}}{\partial\dot{q}_w} - \frac{\partial\bar{\Theta}}{\partial q_w} = -\frac{\partial V}{\partial q_w} + \sum_{m=1}^{M}\Theta_m\left(\sum_{k=1}^{W}A_{wk}^{(m)}\dot{q}_k\right), \quad w = 1, \ldots, W. \tag{4.106}$$

Example 4.4. Let a homogeneous disk of mass m roll on a horizontal plane without slipping while in contact with the plane at a single point (Fig. 4.1). Derive the equation of motion of the disk.

In Fig. 4.1 the absolute coordinate system $OX_1X_2X_3$ is introduced in such a way that the axis OX_3 is perpendicular to the horizontal plane, on which the disk moves. With the disk is associated the body system $CX_1''X_2''X_3''$, where C is the mass center of the disk and the axis CX_3'' is perpendicular to the disk plane.

The position of the mass center of the disk in the absolute system $C = C(x_1, x_2, r\sin\Theta)$, where r is the radius of the disk and Θ is one of the introduced Euler angles.

The kinetic energy and potential energy of the disk are equal to

$$T = \frac{1}{2}m\left(\dot{x}_1^2 + \dot{x}_2^2 + \left(r\dot{\Theta}\cos\Theta\right)^2\right) + \frac{1}{2}\left(I_1\omega_1{}''^2 + I_2\omega_2''^2 + I_3\omega_3''^2\right),$$
$$V = mgr\sin\Theta,$$

where g is the acceleration of gravity, the vector of angular velocity of the disk is given by $\boldsymbol{\omega} = \omega_1''\mathbf{E}_1'' + \omega_2''\mathbf{E}_2'' + \omega_3''\mathbf{E}_3''$, and I_i are moments of inertia of the disk with respect to the principal centroidal axes of inertia CX_i'', $i = 1, 2, 3$ ($I_1 = I_2 = \frac{1}{4}mr^2$, $I_3 = mr^2$).

According to the calculations of Sects., 5.5.4 and 5.5.5 of [24] Euler's kinematic equations take the form

$$\omega_1'' = \dot{\psi}\sin\Theta\sin\varphi + \dot{\Theta}\cos\varphi,$$
$$\omega_2'' = \dot{\psi}\sin\Theta\cos\varphi - \dot{\Theta}\sin\varphi,$$
$$\omega_3'' = \dot{\psi}\cos\Theta + \dot{\varphi};$$

using these equations in the expression for the kinetic energy T we obtain

$$T = \frac{1}{2}m\left(\dot{x}_1^2 + \dot{x}_2^2\right) + \frac{1}{8}mr^2\left(1 + 4\cos^2\Theta\right)\dot{\Theta}^2$$
$$+ \frac{1}{8}mr^2\dot{\psi}^2\sin^2\Theta + \frac{1}{4}mr^2\left(\dot{\psi}\cos\Theta + \dot{\varphi}\right)^2.$$

The velocity of point D of contact of the disk with the horizontal plane is equal to

$$\mathbf{v}_D = \mathbf{v}_C + \boldsymbol{\omega} \times \overrightarrow{CD} = \mathbf{0}. \qquad (*)$$

According to Fig. 4.1, line DE is parallel to the line of nodes n. In turn, line $DC \perp DE$ and lies on the plane determined by the axes CX_3' and CX_3''. There is an angle φ between line DC and axis CX_2''.

In the absolute system we have

$$\mathbf{v}_C = \mathbf{E}_1\dot{x}_1 + \mathbf{E}_2\dot{x}_2 + \mathbf{E}_3 r\dot{\Theta}\cos\Theta,$$
$$\overrightarrow{CD} = \mathbf{E}_1 r\cos\Theta\sin\psi - \mathbf{E}_2 r\cos\Theta\cos\psi - \mathbf{E}_3 r\sin\Theta,$$
$$\boldsymbol{\omega} = \mathbf{E}_1\left(\dot{\Theta}\cos\psi + \dot{\varphi}\sin\psi\sin\Theta\right) + \mathbf{E}_2\left(\dot{\Theta}\sin\psi - \dot{\varphi}\cos\psi\sin\Theta\right)$$
$$+ \mathbf{E}_3\left(\dot{\psi} + \dot{\varphi}\cos\Theta\right).$$

We successively calculate

$$\omega\times\overrightarrow{CD} = \begin{vmatrix} \mathbf{E}_1 & \mathbf{E}_2 & \mathbf{E}_3 \\ \dot{\Theta}\cos\psi + \dot{\varphi}\sin\psi\sin\Theta & \dot{\Theta}\sin\psi - \dot{\varphi}\cos\psi\sin\Theta & \dot{\psi} + \dot{\varphi}\cos\Theta \\ r\cos\Theta\sin\psi & -r\cos\Theta\cos\psi & -r\sin\Theta \end{vmatrix}$$

$$= \mathbf{E}_1\left[-r\sin\Theta\left(\dot{\Theta}\sin\psi - \dot{\varphi}\cos\psi\sin\Theta\right) + r\cos\Theta\cos\psi\left(\dot{\psi} + \dot{\varphi}\cos\Theta\right)\right]$$

$$+ \mathbf{E}_2\left[r\sin\Theta\left(\dot{\Theta}\cos\psi + \dot{\varphi}\sin\psi\sin\Theta\right) + r\cos\Theta\sin\psi\left(\dot{\psi} + \dot{\varphi}\cos\Theta\right)\right]$$

$$- \mathbf{E}_3\left[r\cos\Theta\cos\psi\left(\dot{\Theta}\cos\psi + \dot{\varphi}\sin\psi\sin\Theta\right)\right.$$

$$\left.+ r\cos\Theta\sin\psi\left(\dot{\Theta}\sin\psi - \dot{\varphi}\cos\psi\sin\Theta\right)\right],$$

and from $(*)$ we obtain

$$\dot{x}_1 = r\sin\Theta\left(\dot{\Theta}\sin\psi - \dot{\varphi}\cos\psi\sin\theta\right)$$

$$- r\cos\Theta\cos\psi\left(\dot{\psi} + \dot{\varphi}\cos\Theta\right) = r\dot{\Theta}\sin\Theta\sin\psi$$

$$- r\dot{\varphi}\cos\psi\sin^2\Theta - r\dot{\varphi}\cos^2\Theta\cos\psi - r\dot{\psi}\cos\Theta\cos\psi$$

$$= r\left[\dot{\Theta}\sin\psi\sin\Theta - \left(\dot{\psi}\cos\Theta + \dot{\varphi}\right)\cos\psi\right],$$

$$\dot{x}_2 = -r\sin\Theta\left(\dot{\Theta}\cos\psi + \dot{\varphi}\sin\psi\sin\Theta\right)$$

$$- r\cos\Theta\sin\psi\left(\dot{\psi} + \dot{\varphi}\cos\Theta\right) = -r\dot{\Theta}\sin\Theta\cos\psi$$

$$- r\dot{\varphi}\sin^2\Theta\sin\psi - r\dot{\psi}\cos\Theta\sin\psi - r\dot{\varphi}\cos^2\Theta\sin\psi$$

$$= -r\left[\dot{\Theta}\sin\Theta\cos\psi + \left(\dot{\varphi} + \dot{\psi}\cos\Theta\right)\sin\psi\right],$$

$$\dot{x}_3 = -r\dot{\Theta}\cos\Theta\cos^2\psi + r\dot{\varphi}\sin\psi\cos\psi\sin\Theta\cos\Theta$$

$$+ r\dot{\Theta}\cos\Theta\sin^2\psi - r\dot{\varphi}\cos\Theta\cos\psi\sin\psi\sin\Theta$$

$$= r\dot{\Theta}\cos\Theta - r\dot{\Theta}\cos\Theta \equiv 0. \qquad (**)$$

Let us note that in the solutions describing the kinetic T and potential energy of the disk, and in equations of constraints $(**)$, the coordinates x_1 and x_2 do not occur. Equations of motion of the disk will be derived on the basis of Chaplygin's equations (4.106). Let us denote the generalized coordinates as

$$q_1 = \Theta, \quad q_2 = \varphi, \quad q_3 = \psi, \quad q_4 = x_1, \quad q_5 = x_2.$$

In our case [see (4.94)] we have $W = 3$ and $m = 2$. According to (4.94) we have to determine $\dot{q}_4 = \dot{x}_1$ and $\dot{q}_5 = \dot{x}_2$ from equations of constraints $(**)$, which with the new notation takes the form

$$\dot{q}_4 = r\,[\dot{q}_1 \sin q_3 \sin q_1 - (\dot{q}_3 \cos q_1 + \dot{q}_2) \cos q_3] = \alpha_{11}\dot{q}_1 + \alpha_{12}\dot{q}_2 + \alpha_{13}\dot{q}_3,$$

$$\dot{q}_5 = -r\,[\dot{q}_1 \sin q_1 \cos q_3 + (\dot{q}_2 + \dot{q}_3 \cos q_1) \sin q_3] = \alpha_{21}\dot{q}_1 + \alpha_{22}\dot{q}_2 + \alpha_{33}\dot{q}_3.$$

From the preceding equations we determine the desired coefficients α_{mw}, which are equal to

$$\alpha_{11} = r \sin q_1 \sin q_3, \qquad \alpha_{12} = -r \cos q_3, \quad \alpha_{13} = -r \cos q_1 \cos q_3,$$

$$\alpha_{21} = -r \sin q_1 \cos q_3, \quad \alpha_{22} = -r \sin q_3, \quad \alpha_{23} = -r \cos q_1 \sin q_3.$$

From (4.105) we calculate the coefficients $A^{(m)}_{wk}$. Non-zero values of those coefficients are equal to

$$A^{(1)}_{23} = -A^{(1)}_{23} = r \sin q_3, \qquad A^{(2)}_{23} = -A^{(2)}_{32} = -r \cos q_3.$$

Generalized momenta according to the second equation of (4.102) are equal to

$$\begin{aligned}
\Theta_1 &= \frac{\partial T}{\partial \dot{q}_4} = m\dot{x}_1 \\
&= mr\,(\dot{q}_1 \sin q_1 \sin q_3 - \dot{q}_2 \cos q_3 - \dot{q}_3 \cos q_4 \cos q_3)\,, \\
\Theta_2 &= \frac{\partial T}{\partial \dot{q}_5} = m\dot{x}_2 \\
&= mr\,(\dot{q}_1 \sin q_1 \cos q_3 + \dot{q}_2 \sin q_3 + \dot{q}_3 \cos q_1 \sin q_3)\,.
\end{aligned}$$

Chaplygin's equations (4.106) take the form

$$\frac{\mathrm{d}}{\mathrm{d}t}\frac{\partial \bar{\Theta}}{\partial \dot{\Theta}} - \frac{\partial \bar{\Theta}}{\partial \Theta} = -mgr\cos\Theta,$$

$$\frac{\mathrm{d}}{\mathrm{d}t}\frac{\partial \bar{\Theta}}{\partial \dot{\varphi}} - \frac{\partial \bar{\Theta}}{\partial \varphi} = mr^2\dot{\Theta}\dot{\psi},$$

$$\frac{\mathrm{d}}{\mathrm{d}t}\frac{\partial \bar{\Theta}}{\partial \dot{\psi}} - \frac{\partial \bar{\Theta}}{\partial \psi} = -mr^2\dot{\Theta}\dot{\varphi}\sin\Theta.$$

In the preceding equation the energy

$$
\begin{aligned}
T = \bar{\Theta} &= \frac{1}{2} m r^2 \big[\dot{\Theta} \sin\psi \sin\Theta - \big(\dot{\psi}\cos\Theta + \dot{\varphi}\big)\cos\psi\big]^2 \\
&+ \frac{1}{2} m r^2 \big[\dot{\Theta} \sin\Theta \cos\psi + \big(\dot{\varphi} + \dot{\psi}\cos\Theta\big)\sin\psi\big]^2 \\
&+ \frac{1}{8} m r^2 \left(1 + 4\cos^2\Theta\right)\dot{\Theta}^2 + \frac{1}{8} m r^2 \dot{\psi}^2 \sin^2\Theta + \frac{1}{4} m r^2 \big(\dot{\psi}\cos\Theta + \dot{\varphi}\big)^2 \\
&= \frac{5}{8} m r^2 \dot{\Theta}^2 + \frac{1}{8} m r^2 \dot{\psi}^2 \sin^2\Theta + \frac{3}{4} m r^2 \big(\dot{\psi}\cos\Theta + \dot{\varphi}\big)^2 .
\end{aligned}
$$

Substituting the preceding equation into Chaplygin's equations we get

$$
\begin{aligned}
&\dot{\Theta} + \dot{\psi}^2 \sin\Theta \cos\Theta + \frac{6}{5}\dot{\varphi}\dot{\psi}\sin\Theta + \frac{4}{5}\frac{g}{r}\cos\Theta = 0, \\
&\frac{\mathrm{d}}{\mathrm{d}t}\left(\dot{\varphi} + \dot{\psi}\cos\Theta\right) = \frac{2}{3}\dot{\Theta}\dot{\psi}\sin\Theta, \\
&\frac{\mathrm{d}}{\mathrm{d}t}\left[\left(\dot{\varphi} + \dot{\psi}\cos\Theta\right)\cos\Theta + \frac{1}{6}\dot{\psi}\sin^2\Theta\right] = -\frac{2}{3}\dot{\varphi}\,\dot{\Theta}\sin\Theta.
\end{aligned}
$$

Integrating the preceding equations (e.g., numerically) we find $\Theta = \Theta(t)$, and consequently we determine the third coordinate of the position of point C ($C = C(x_1(t), x_2(t), r\sin\Theta(t))$), where $x_1(t)$ and $x_2(t)$ are determined by integration of equations (**). □

4.5 Appell's Equations

In order to derive Appell's[12] equations, first we will introduce certain preliminary concepts regarding *pseudo-coordinates* useful for the analysis of non-holonomic systems [2]. Let a DMS with non-holonomic constraints in the amount of K have W degrees of freedom. We will call the additional introduced quantities π_w, $\dot{\pi}_w$, $\ddot{\pi}_w$, $w = 1, \ldots, W$ respectively *pseudo-displacements*, *pseudo-velocities*, and *pseudo-accelerations*. Let us construct linear combinations of generalized coordinates $\dot{q}_1, \ldots, \dot{q}_K$ of the form

$$
\dot{\pi}_w = \sum_{k=1}^{K} a_{wk}\left(q_1, q_2, \ldots, q_K, t\right)\dot{q}_k, \qquad w = 1, \ldots, W. \tag{4.107}
$$

---

[12]Paul Émile Appell (1855–1930), French mathematician and rector of the University of Paris.

The system of linear equations (4.107) determines $\dot{\pi}_1, \dot{\pi}_2, \ldots, \dot{\pi}_W$ pseudo-velocities, which may not have any physical meaning, that is, they may not be generalized velocities. A similar observation concerns pseudo-displacements and pseudo-accelerations. Generalized velocities satisfy the equation of non-holonomic constraints (3.24), and henceforth we will use the index $m_2 = m$.

Generalized velocities $\dot{q}_1, \dot{q}_2, \ldots, \dot{q}_K$ satisfy M (3.21), and expressing them in terms of pseudo-velocities according to relation (4.107), equation (3.21) will take the form

$$\dot{q}_k = \sum_{k=1}^{K} d_{wk}\dot{\pi}_w + g_k, \qquad k = 1, \ldots, K, \tag{4.108}$$

where now $d_{wk} = d_{wk}(q_1, q_2, \ldots, q_K, t)$ and $g_k = g_k(q_1, q_2, \ldots, q_K, t)$ and imposed pseudo-velocities $\dot{\pi}_w$ can have arbitrary values.

From (4.107) we obtain

$$\delta\pi_w = \sum_{k=1}^{K} a_{wk}\delta q_k, \qquad w = 1, \ldots, W. \tag{4.109}$$

This means that the variation of a pseudo-displacement $\delta\pi_w$ is equal to the sum of variations of generalized displacements δq_k.

Let us note that in the case of non-holonomic constraints, generalized displacements satisfy (3.24) of the form

$$\sum_{k=1}^{K} B_{mk}\delta q_k = 0, \qquad m = 1, \ldots, M. \tag{4.110}$$

From systems of equations (4.109) and (4.110) we express δq_k in terms of pseudo-displacements, obtaining

$$\delta q_k = \sum_{w}^{W} d_{wk}\delta\pi_w, \qquad k = 1, 2, \ldots, K. \tag{4.111}$$

Substituting relations (4.111) into (3.23) we obtain

$$\begin{aligned}\delta\mathbf{r}_n &= \sum_{k=1}^{K} \frac{\partial\mathbf{r}_n}{\partial q_k}\delta q_k = \sum_{k=1}^{K} \frac{\partial\mathbf{r}_n}{\partial q_k}\sum_{w}^{W} d_{wk}\partial\pi_w \\ &= \sum_{w=1}^{W} \mathbf{e}_{nw}\delta\pi_w, \quad n = 1, 2, \ldots, N,\end{aligned} \tag{4.112}$$

where

$$\mathbf{e}_{nw} = \sum_{k=1}^{K} \frac{\partial \mathbf{r}_n}{\partial q_k} d_{wk}, \qquad n = 1, 2, \ldots, N, \quad w = 1, 2, \ldots, W. \tag{4.113}$$

Our next task is the expression of $\mathbf{e}_{nw}$ in terms of generalized accelerations and pseudo-accelerations. To this end we differentiate (4.107) with respect to time, and we substitute the obtained generalized accelerations $\ddot{q}_k$, $k = 1, 2, \ldots, K$ into the third equation of (3.19) obtaining

$$\mathbf{a}_n = \sum_{w=1}^{W} \mathbf{e}_{nw} \ddot{\pi}_w + \mathbf{h}_n, \qquad n = 1, \ldots, N, \tag{4.114}$$

where now $\mathbf{h}_n = \mathbf{h}_n(\pi_1, \ldots, \pi_W, \dot{\pi}_1, \ldots, \dot{\pi}_W, t)$. Differentiating the preceding equation with respect to $\ddot{\pi}_w$, $w = 1, \ldots, W$ we obtain

$$\frac{\partial \mathbf{a}_n}{\partial \ddot{\pi}_w} = \mathbf{e}_{nw}, \qquad n = 1, \ldots, N, \quad w = 1, \ldots, W. \tag{4.115}$$

Eventually, substituting (4.115) into (4.112) we obtain

$$\delta \mathbf{r}_n = \sum_{w=1}^{W} \frac{\partial \mathbf{a}_n}{\partial \ddot{\pi}_w} \partial \pi_w, \qquad n = 1, \ldots, N. \tag{4.116}$$

Appell's equations, like the majority of equations discussed in this book, will be derived from the general equation of dynamics of the form

$$\sum_{n=1}^{N} (\mathbf{F}_n - m_n \mathbf{a}_n) \circ \delta \mathbf{r}_n = 0 \tag{4.117}$$

and by making use of relation (3.117) of the form

$$\sum_{n=1}^{N} \mathbf{F}_n \circ \delta \mathbf{r}_n = \sum_{k=1}^{K} Q_k \delta q_k. \tag{4.118}$$

The elementary work of active forces acting on a given DMS is equal to

$$\delta W = \sum_{n=1}^{N} \mathbf{F}_n \circ \delta \mathbf{r}_n = \sum_{k=1}^{K} \mathbf{Q}_k \circ \sum_{w=1}^{W} \frac{\partial \mathbf{a}_n}{\partial \ddot{\pi}_w} \delta \pi_w = \sum_{w=1}^{W} \pi_w \delta \pi_w, \tag{4.119}$$

where

$$\pi_w = \pi_w (q_1, q_k, \dot{\pi}_1, \ldots, \dot{\pi}_w, t) = \sum_{k=1}^{K} d_{wk} Q_k, \tag{4.120}$$

and in the course of the preceding calculations, transformations (4.116) were used. The quantities π_w will be called *generalized forces corresponding to pseudo-displacements* π_w, $w = 1, \ldots, W$.

Next we calculate the elementary work done by inertia forces

$$\begin{aligned}\delta W &= \sum_{n=1}^{N} m_n \mathbf{a}_n \circ \delta \mathbf{r}_n = \sum_{n=1}^{N} m_n \mathbf{a}_n \circ \sum_{w=1}^{W} \mathbf{e}_{nw} \delta \pi_w \\ &= \sum_{w=1}^{W} \left(\sum_{n=1}^{N} m_n \mathbf{a}_n \circ \frac{\partial \mathbf{a}_n}{\partial \ddot{\pi}_w} \right) \delta \pi_w \\ &= \sum_{w=1}^{W} \frac{\partial}{\partial \ddot{\pi}_w} \left(\frac{1}{2} \sum_{n=1}^{N} m_n \mathbf{a}_n^2 \right) \delta \pi_w = \sum_{w=1}^{W} \frac{\partial S}{\partial \ddot{\pi}_w} \delta \pi_w. \end{aligned} \tag{4.121}$$

Equations (4.112) and (4.115) were used during transformations, and the introduced function

$$S = \frac{1}{2} \sum_{n=1}^{N} m_n \mathbf{a}_n^2 \tag{4.122}$$

is called the *energy of accelerations* of a DMS, and $S = S(q_1, \ldots, q_K, \dot{\pi}_1, \ldots, \dot{\pi}_W, \ddot{\pi}_1, \ldots, \ddot{\pi}_W, t)$ according to the general equation of dynamics (4.117), and taking into account (4.119) and (4.122), we obtain *Appell equations*

$$\frac{\partial S}{\partial \ddot{\pi}_w} = \pi_w, \qquad w = 1, \ldots, W. \tag{4.123}$$

In order to solve the problem of the dynamics of a non-holonomic DMS, equations (4.123) should be solved simultaneously with M equations of non-holonomic constraints (3.21) and with W equations (4.107) defining pseudo-velocities. It can be shown that Appell's equations can be solved with respect to pseudo-accelerations $\ddot{\pi}_w$, $w = 1, \ldots, W$. Moreover, equations of constraints (3.21) [see (4.108)] and (4.107) can be solved with respect to generalized velocities $\dot{q}_k$, $k = 1, \ldots, K$. Eventually, we have at our disposal $K + W$ equations with unknowns $q_1, \ldots, q_K, \dot{\pi}_1, \ldots, \dot{\pi}_W$.

Imposing initial conditions $q_{10}, \ldots, q_{k0}, \dot{\pi}, \ldots, \dot{\pi}_{w0}$, from (4.108) we determine uniquely the initial conditions for generalized velocities $\dot{q}_{10}, \ldots, \dot{q}_{K0}$ according to the conditions of non-holonomic constraints. Thus, we have initial conditions $q_{10}, \ldots, q_{K0}, \dot{q}_{10}, \ldots, \dot{q}_{K0}$, which at once define a *Cauchy problem* for a given DMS.

If as $\dot{\pi}_w$ we take generalized velocities $\dot{q}_w$, $w = 1, \ldots, W$, then generalized forces π_w correspond to Q'_w, where

$$Q'_w = Q_w + \sum_{p=1}^{K-W} \alpha_{pw} Q_{w+p}, \qquad w = 1, \ldots, W. \tag{4.124}$$

For such a choice, the acceleration energy $S = S(q_1, \ldots, q_K, \dot{q}_1, \ldots, \dot{q}_W, \ddot{q}_1, \ldots, \ddot{q}_W, t)$, and Appell's equations take the form

$$\frac{\partial S}{\partial \ddot{q}_w} = Q'_w, \qquad w = 1, \ldots, W. \tag{4.125}$$

The preceding W Appell equations (also called the Gibbs–Appell equations [9]), augmented with M equations of non-holonomic constraints, describe the motion of the analyzed non-holonomic DMS. The number of those equations $M + W = K$ is equal to the number of generalized coordinates. If we are dealing with a holonomic DMS, then $K = W$, since $M = 0$ and $Q'_w = Q_w$, and (4.125) represent the form of Lagrange's equations of the second kind. Despite many advantages, the major drawback of Appell's equations is connected with the difficulties associated with determining the acceleration energy S essential for their formulation. It seems that in engineering practice it is easier to exploit Voronets's and Chaplygin's equations in which it is necessary to determine the kinetic energy T of the DMS under investigation.

Now we will demonstrate a method to determine the acceleration energy on an example of the motion of a rigid body with one point fixed. We introduce the absolute system of coordinates $OX_1X_2X_3$ and the body system $O''X''_1X''_2X''_3$. The point $O = O''$ is a fixed point of the body, and the axes of the body system coincide with its principal axes of inertia related to point O.

According to the definition of acceleration energy given by (4.122), in a body system we have

$$S = \frac{1}{2} \sum_{n=1}^{N} m_n \left(\mathrm{a}^2_{nx''_1} + \mathrm{a}^2_{nx''_2} + \mathrm{a}^2_{nx''_3} \right), \tag{4.126}$$

where N denotes the number of particles approximating the rigid body.

In Chap. 5 of [24] we already determined the acceleration of a particle n of a rigid body, which reads

$$\mathbf{a}_n = \boldsymbol{\varepsilon} \times \mathbf{r}_n + \boldsymbol{\omega} \left(\boldsymbol{\omega} \circ \mathbf{r}_n \right) - \omega^2 \mathbf{r}_n. \tag{4.127}$$

From the preceding vector equation we determine the coordinates of projections of vector $\mathbf{a}_n$ on the system axes $O''X''_1X''_2X''_3$, which are equal to

$$\mathrm{a}_{nx_1''} = -x_{1n}\left(\omega_2''^2 + \omega_3''^2\right) + x_{2n}\left(\omega_2''\omega_1'' - \dot{\omega}_3''\right) + x_{3n}\left(\omega_1''\omega_3'' + \dot{\omega}_2''\right),$$
$$\mathrm{a}_{nx_2''} = -x_{2n}\left(\omega_3''^2 + \omega_1''^2\right) + x_{3n}\left(\omega_3''\omega_2'' - \dot{\omega}_1''\right) + x_{1n}\left(\omega_2''\omega_1'' + \dot{\omega}_3''\right),$$
$$\mathrm{a}_{nx_3''} = -x_{3n}\left(\omega_1''^2 + \omega_2''^2\right) + x_{1n}\left(\omega_1''\omega_3'' - \dot{\omega}_2''\right) + x_{2n}\left(\omega_3''\omega_2'' + \dot{\omega}_1''\right). \quad (4.128)$$

Substituting (4.128) into (4.126), while taking into account $I_{X_1''X_2''} = 0$, $I_{X_1''X_3''} = 0$, $I_{X_2''X_3''} = 0$ (see Chap. 3 in [24]) and additionally neglecting terms independent of $\dot{\omega}_1''$, $\dot{\omega}_2''$, and $\dot{\omega}_3''$, we obtain

$$\begin{aligned} S &= \frac{1}{2}\left(\sum_{n=1}^{N} m_n x_{1n}^2\right)\left(\dot{\omega}_3''^2 + 2\omega_2''\omega_1''\dot{\omega}_3'' + \dot{\omega}_2''^2 - 2\omega_1''\omega_3''\dot{\omega}_2''\right) \\ &+ \frac{1}{2}\left(\sum_{n=1}^{N} m_n x_{2n}^2\right)\left(\dot{\omega}_1''^2 + 2\omega_3''^2\omega_2''\dot{\omega}_1'' + \dot{\omega}_3''^2 - 2\omega_2''\omega_1''\dot{\omega}_3''\right) \\ &+ \frac{1}{2}\left(\sum_{n=1}^{N} m_n x_{3n}^2\right)\left(\dot{\omega}_2''^2 + 2\omega_1''^2\omega_3''\dot{\omega}_2'' + \dot{\omega}_1''^2 - 2\omega_3''\omega_2''\dot{\omega}_1''\right) \\ &= \frac{1}{2}\left(I_{X_1''}\dot{\omega}_1''^2 + I_{X_2''}\dot{\omega}_2''^2 + I_{X_3''}\dot{\omega}_3''^2\right) + \left(I_{X_3''} - I_{X_2''}\right)\omega_2''\omega_3''\dot{\omega}_1'' \\ &+ \left(I_{X_1''} - I_{X_3''}\right)\omega_3''\omega_1''\dot{\omega}_2'' + \left(I_{X_2''} - I_{X_1''}\right)\omega_1''\omega_2''\dot{\omega}_3'', \end{aligned} \quad (4.129)$$

where $I_{X_1''}$, $I_{X_2''}$, and $I_{X_3''}$ are respectively the moments of inertia about axes $O''X_1''$, $O''X_2''$, and $O''X_3''$.

Let point C be the mass center of a given DMS moving in an arbitrary manner. We will determine the acceleration energy of the DMS taking the pole at point C. The acceleration of point m of the number n ($n = 1, \ldots, N$) is equal to

$$\mathbf{a}_n = \mathbf{a}_C + \mathbf{a}_n^w. \quad (4.130)$$

Substituting (4.130) into (4.122) we obtain

$$\begin{aligned} S &= \frac{1}{2}\left(\sum_{n=1}^{N} m_n\right)\mathbf{a}_C^2 + \left(\sum_{n=1}^{N} m_n \mathbf{a}_n^w\right) \circ \mathbf{a}_C + \frac{1}{2}\sum_{n=1}^{N} m_n\left(\mathbf{a}_n^w\right)^2 \\ &= \frac{1}{2}m\mathbf{a}_C^2 + \frac{1}{2}\sum_{n=1}^{N} m_n\left(\mathbf{a}_n^w\right)^2 \end{aligned} \quad (4.131)$$

because

$$\sum_{n=1}^{N} m_n = m, \qquad \sum_{n=1}^{N} m_n \mathbf{a}_n^w = M\mathbf{a}_C^w = \mathbf{0}. \quad (4.132)$$

The obtained result can be formulated in the form of a theorem analogous to König's theorem on the kinetic energy of a DMS.

Theorem 4.11. *The acceleration energy of a DMS is equal to the sum of the acceleration energy of a particle located at the mass center of the DMS and having a mass equal to the mass of the DMS and the acceleration energy resulting from the motion of the DMS relative to its mass center.*

Example 4.5. (see [9]) Let a homogeneous ball of radius r be moving without slip on a stationary horizontal surface. Demonstrate that the angular velocity vector of the ball is conserved during its motion.

We will introduce two Cartesian coordinate systems. The absolute system $OX_1X_2X_3$ has its origin at the point of contact of the ball with the horizontal surface of axis OX_3 directed against the acceleration of gravity. The body system $CX_1''X_2''X_3''$ has its origin at the mass center of the ball. Let $C = C(x_1, x_2, r)$ be the coordinates of the mass center of the ball in the $OX_1X_2X_3$ system, and the vector of angular velocity $\boldsymbol{\omega} = \Sigma \mathbf{E}_i \omega_i = \Sigma \mathbf{E}_i'' \omega_i''$.

From the condition of the absence of slip we obtain

$$\dot{x}_1 = \omega_2 r, \qquad \dot{x}_2 = -\omega_1 r. \tag{$*$}$$

Because for a ball $I_{X_1''} = I_{X_2''} = I_{X_3''} = \frac{2}{5}mr^2$, according to (4.129) we obtain

$$S_1 = \frac{1}{2}mr^2\left(\dot{\omega}_1''^2 + \dot{\omega}_2''^2 + \dot{\omega}_3''^2\right),$$

and S_1 is the acceleration energy of the ball resulting from its rotational motion relative to the mass center C.

According to (4.131) and taking into account S_1, the total acceleration energy of the ball equals

$$S = \frac{1}{2}m\left(\ddot{x}_1^2 + \ddot{x}_2^2\right) + \frac{1}{5}mr^2\left(\dot{\omega}_1''^2 + \dot{\omega}_2''^2 + \dot{\omega}_3''^2\right). \tag{$**$}$$

Differentiating ($*$) with respect to time and introducing pseudo-velocities $\dot{\pi}_i = \omega_i$, $i = 1, 2, 3$ we obtain

$$\ddot{x}_1 = r\ddot{\pi}_2, \qquad \ddot{x}_2 = -r\ddot{\pi}_1.$$

The angular acceleration of the ball is given by the equation

$$\varepsilon^2 = \dot{\omega}_1''^2 + \dot{\omega}_2''^2 + \dot{\omega}_3''^2 = \dot{\omega}_1^2 + \dot{\omega}_2^2 + \dot{\omega}_3^2 = \ddot{\pi}_1^2 + \ddot{\pi}_2^2 + \ddot{\pi}_3^2.$$

Taking into account the last two equations in $(**)$ we obtain

$$S = \frac{1}{2}mr^2\left(\ddot{\pi}_1^2 + \ddot{\pi}_2^2\right) + \frac{1}{5}mr^2\left(\ddot{\pi}_1^2 + \ddot{\pi}_2^2 + \ddot{\pi}_3^2\right)$$
$$= \frac{1}{10}mr^2\left[7\left(\ddot{\pi}_1^2 + \ddot{\pi}_2^2\right) + 2\ddot{\pi}_3\right].$$

Because the generalized forces π_w in Appell's equations (4.123) are equal to zero, the problem boils down to the integration of three equations

$$\frac{\partial S}{\partial \ddot{\pi}_i} = 0, \quad i = 1, 2, 3.$$

We successively get

$$\frac{\partial S}{\partial \ddot{\pi}_1} = \frac{7}{5}mr^2\ddot{\pi}_1 = 0, \quad \frac{\partial S}{\partial \ddot{\pi}_2} = \frac{7}{5}mr^2\ddot{\pi}_2 = 0, \quad \frac{\partial S}{\partial \ddot{\pi}_3} = \frac{2}{5}mr^2\ddot{\pi}_3 = 0.$$

This means that $\dot{\pi}_1 = \omega_1 = \text{const}$, $\dot{\pi}_2 = \omega_2 = \text{const}$, $\dot{\pi}_3 = \omega_3 = \text{const}$, that is, that the vector $\boldsymbol{\omega} = \mathbf{E}_1\omega_1 + \mathbf{E}_2\omega_2 + \mathbf{E}_3\omega_3$ is constant during the motion of the ball on the horizontal plane. □

References

1. A.V. Bäcklund, *Einiges über Kurven- and Flächen-transformationen*, vol 10 (Lund Universitets Arsskrift, 1875) (in German)
2. P. Appell, *Traité de Mécanique Rationnelle*, vol 4 (Gauthier-Villars, Paris, 1893–1896)
3. P.V. Voronets, On the equations of motion for non-holonomic systems. Mat. Sb. **22**, 659–668 (1901)
4. H. Goldstein, *Classical Mechanics*. (Addison-Wesley, Cambridge, MA, 1950)
5. L.D. Landau, E.M. Lifshitz, *Mechanics Course of Theoretical Physics*, vol 1. (Franklin Book Company, New York, 1972)
6. E.B. Bogomolny, Sov. J. Nucl. Phys. **24**, 449–454 (1976)
7. R.M. Miura, *Bäcklund Transformation*. (Springer, Berlin, 1976)
8. V.I. Arnold, *Mathematical Methods of Classical Mechanics*. (Springer, Berlin, 1989)
9. A.D. Markeev, *Theoretical Mechanics*. (Nauka, Moscow, 1990)
10. N.V. Butenin, N.A. Fufayev, *Introduction to Analytical Mechanics* (Nauka, Moscow, 1991)
11. S.T. Thornton, J.B. Marion, *Classical Dynamics of Particles and Systems* (Saunders College Publishers, New York, 1995)
12. L.N. Hand, *Analytical Mechanics* (Cambridge University Press, Cambridge, 1998)
13. J. Benoit, T. Saxena, T. Lookman, *Bogomol'nyi decomposition for vesicles of arbitrary genius*. J. Phys. A Math. Gen. **34**, 9417 (2001)
14. A.V. Borisov, I.S. Mamaev, *Dynamics of the Rigid Body*. (R&C Dynamics, Izhevsk, 2001)
15. G.J. Sussman, *Structure and Interpretation of Classical Mechanics* (MIT, MA, 2001)
16. L. Greiner, *Classical Mechanics: Systems of Particles and Hamiltonian Dynamics* (Springer, Berlin, 2002)

17. C. Rogers, W.K. Schief, *Bäcklund and Darboux Transformations* (Cambridge Texts in Applied Mathematics, Cambridge, 2003)
18. J.B. Tatum, *Classical Mechanics* (University of Victoria, Canada, 2003)
19. J. Awrejcewicz, V.A. Krysko, *Introduction to Asymptotic Methods* (Chapman and Hall/CRC Press, New York, 2006)
20. R. Abraham, J.E. Marsden, *Foundations of Mechanics* (AMS Chelsea Publishing, Chelsea, 2008)
21. A.J. Brizard, *An Introduction to Lagrangian Mechanics* (World Scientific, Singapore, 2008)
22. C. Gignoux, B. Silvestre-Brac, *Solved Problems in Lagrangian and Hamiltonian Mechanics* (Springer, Dordrecht, 2009)
23. A. Deriglazov, *Classical Mechanics: Hamiltonian and Lagrangian Formalism* (Springer, Berlin, 2010)
24. J. Awrejcewicz, *Classical Mechanics: Statics and Kinematics* (Springer, New York, 2012)
25. J. Awrejcewicz, A.G. Petrov, On the normal forms of Hamiltonian systems. Nonlinear Dyn. **48**, 185–197 (2007)
26. J. Awrejcewicz, A.G. Petrov, Nonlinear oscillations of an elastic two degrees-of-freedom pendulum. Nonlinear Dyn. **53**(1–2), 19–30 (2008)

Chapter 5
Theory of Impact

5.1 Basic Concepts

So far we have considered problems concerning the statics and dynamics of discrete (lumped) and continuous material systems when forces of action and reaction act upon these systems in a continuous fashion for the entire duration of a process. On the other hand, it is known that changes in system momentum leading to changes in velocity are associated with the action of a force or moment of force during a finite and often very short time interval. *A phenomenon is called an impact if we observe a sudden (instantaneous) change in the velocity of a particle caused by the action of instantaneous forces.* Despite the passage of several years, the notions of instantaneous changes in velocities and forces of an infinitely short duration time are intuitive, and to date they have not found an adequate mathematical description. If two bodies collide and the time of the collision process is very short, then we observe a continuous change in the velocity of the body, and because the collision usually lasts for a very short time, it is associated with the generation of relatively large forces. However, it should be emphasized that the notions of "small" and "large" quantities are relative and subjective.

If other forces also act on the analyzed system and the impact is associated with the vibrations of the system, which are characterized by a period of free vibrations with or without damping, then the duration of the collision process is short in comparison, for example, with the previously mentioned period of oscillations, whereas the instantaneous forces are large compared to the other forces present in the analyzed system. However, if there are no such reference quantities in the system, then the introduced idealized notions must be related, for example, to the duration of observation of the phenomenon.

According to Newton's second law, the motion of a particle whose position is determined by a radius vector $\mathbf{r}(t)$ is described by a second-order differential equation of the form

$$m\frac{\mathrm{d}^2\mathbf{r}}{\mathrm{d}t^2} = \mathbf{F}, \tag{5.1}$$

J. Awrejcewicz, *Classical Mechanics: Dynamics*, Advances in Mechanics and Mathematics 29, DOI 10.1007/978-1-4614-3740-6_5,

where $\mathbf{F}$ is the force acting on a particle of mass m. From this equation we obtain a variable quantity of motion of the particle in the time interval from t_0 to $t_0 + \tau$, since after integrating (5.1) we have

$$m\left(\dot{\mathbf{r}}(t) - \dot{\mathbf{r}}_0\right) = \int_{t_0}^{t_0+\tau} \mathbf{F} \mathrm{d}\tau. \tag{5.2}$$

Let us introduce the following notation: $\dot{\mathbf{r}}(t) = \mathbf{v}(t)$, $\dot{\mathbf{r}}_0 = \mathbf{v}_0$, and

$$\mathbf{J} = \int_{t_0}^{t_0+\tau} \mathbf{F} \mathrm{d}\tau. \tag{5.3}$$

In the case of an arbitrary duration of action of the force $\mathbf{F}$, that is, the quantity τ, (5.3) defines an *impulse of a force*. If the duration of a force action tends to zero, that is, $\tau \to 0$, and despite the fact that the limit

$$\mathbf{J} = \lim_{\tau \to 0} \int_{t_0}^{t_0+\tau} \mathbf{F} \mathrm{d}\tau \tag{5.4}$$

still exists, the quantity $\mathbf{J}$ is called an *impulse of impact*, and the force $\mathbf{F}$ itself is called an *impact force* or *instantaneous force*. If the value of the integral occurring in (5.4) is averaged out, then from this equation we obtain

$$\mathbf{F}_{\mathrm{av}} = \frac{\mathbf{J}}{\tau}. \tag{5.5}$$

Note that $\mathbf{F}_{\mathrm{av}} \to \infty$ for $\tau \to 0$, that is, the impact force reaches large magnitudes during the collision process. According to formula (5.2) for $\tau \to 0$ we obtain

$$m(\mathbf{v} - \mathbf{v}_0) = m(\Delta \mathbf{v}) = \Delta(m\mathbf{v}) = \mathbf{J}. \tag{5.6}$$

From the obtained equation it follows that the *impulse of impact* leads to a step change in the velocity of a particle, where $\mathbf{v}_0$ is the velocity before the impact and $\mathbf{v}$ is the velocity following impact. This is the first characteristic of the impact phenomenon.

We obtain the second feature of the impact phenomenon after the integration of (5.2). That is, we have

$$m(\mathbf{r}(t) - \mathbf{r}_0) = m\dot{\mathbf{r}}_0\tau + \int_{t_0}^{t_0+\tau} \left(\int_{t_0}^{t_0+\tau} \mathbf{F} \mathrm{d}\tau \right) \mathrm{d}\tau. \tag{5.7}$$

From the preceding equation it is easy to notice that for $\tau \to 0$ we have

$$\lim_{\tau \to 0} \int_{t_0}^{t_0+\tau} \mathbf{F} \mathrm{d}\tau = \mathbf{J}, \quad \lim_{\tau \to 0} \int_{t_0}^{t_0+\tau} \mathbf{J} \mathrm{d}\tau = 0, \tag{5.8}$$

and hence from (5.7) we get

$$\lim_{\tau \to 0} (\mathbf{r}(t_0 + \tau) - \mathbf{r}(t_0)) = \mathbf{0}, \tag{5.9}$$

which means that the particle position just before and just after impact is the same.

5.2 Fundamental Laws of a Theory of Impact

In Chap. 2 of [1] and Chap. 1 of this book, we presented the fundamental laws and theorems connected with the statics and dynamics of material systems during the action of constant forces, independent of time or acting during a finite and "long" time interval. Here they will be extended and adapted to the requirements of the analysis of the phenomenon of impact.

5.2.1 The Law of Conservation of Momentum During Impact

The law of conservation of momentum for a particle was discussed in Chap. 1. In the case of a discrete material system consisting of N separate particles, the law of conservation of momentum is expressed by the following equation [see the case of a particle described by (1.147)]:

$$\frac{\mathrm{d}\mathbf{P}}{\mathrm{d}t} = \frac{\mathrm{d}}{\mathrm{d}t}\left(\sum_{n=1}^{N} m_n \mathbf{v}_n\right) = \sum_{n=1}^{N} \mathbf{F}_n, \tag{5.10}$$

where on the left-hand side occurs a derivative of the quantity of motion of a system of particles and on the right-hand side the sum of forces acting on the system.

Integrating the preceding equation we obtain

$$\mathbf{P} - \mathbf{P}_0 = \int_{t_0}^{t_0+\tau} \sum_{n=1}^{N} \mathbf{F}_n \mathrm{d}\tau. \tag{5.11}$$

Taking into account the occurrence of *instantaneous forces (impact forces)*, we can write

$$\Delta\mathbf{P} = \mathbf{P} - \mathbf{P}_0 = \sum_{n=1}^{N} \lim_{\tau \to 0} \int_{t_0}^{t_0+\tau} \mathbf{F}_n \mathrm{d}\tau = \sum_{n=1}^{N} \mathbf{J}_n, \tag{5.12}$$

where $\mathbf{J}_n$ are *impulses of impact* coming from the impact forces $\mathbf{F}_n$. Based on the preceding equations it is possible to formulate the following theorem.

Theorem 5.1. *The change of momentum of a system of particles at the instant of impact is equal to the sum of external impulses of impact caused by all instantaneous forces acting on the system at the considered instant.*

5.2.2 The Law of Conservation of Angular Momentum During Impact

If external $\mathbf{F}^{\mathrm{e}}$ and internal forces $\mathbf{F}^{\mathrm{i}}$ act on every particle of a material system (a position of the particle is described by a radius vector $\mathbf{r}(t)$), then according to Newton's second law we have

$$m\ddot{\mathbf{r}} = \mathbf{F}^{\mathrm{e}} + \mathbf{F}^{\mathrm{i}}. \tag{5.13}$$

The calculations will be conducted with respect to a certain point O not associated with the adopted stationary coordinate system $O'X_1'X_2'X_3'$. Premultiplying (5.13) by the vector $(\mathbf{r} - \mathbf{r}_O)$ and integrating over the whole mass of the considered system we obtain

$$\int_m (\mathbf{r} - \mathbf{r}_O) \times \frac{\mathrm{d}^2\mathbf{r}}{\mathrm{d}t^2}\mathrm{d}m = \int_m (\mathbf{r} - \mathbf{r}_O) \times \mathbf{F}^{\mathrm{e}}\mathrm{d}m \tag{5.14}$$

because according to Newton's second law the internal forces cancel out.

Note that

$$\frac{\mathrm{d}}{\mathrm{d}t}\int_m (\mathbf{r} - \mathbf{r}_O) \times \frac{\mathrm{d}\mathbf{r}}{\mathrm{d}t}\mathrm{d}m = \int_m (\mathbf{r} - \mathbf{r}_O) \times \frac{\mathrm{d}^2\mathbf{r}}{\mathrm{d}t^2}\mathrm{d}m + \int_m \frac{\mathrm{d}}{\mathrm{d}t}(\mathbf{r} - \mathbf{r}_O) \times \frac{\mathrm{d}\mathbf{r}}{\mathrm{d}t}\mathrm{d}m, \tag{5.15}$$

and additionally the second term on the right-hand side of the preceding equation is equal to

$$\begin{aligned}
\int_m \frac{\mathrm{d}}{\mathrm{d}t}(\mathbf{r} - \mathbf{r}_O) \times \frac{\mathrm{d}\mathbf{r}}{\mathrm{d}t}\mathrm{d}m &= \int_m \frac{\mathrm{d}\mathbf{r}}{\mathrm{d}t} \times \frac{\mathrm{d}\mathbf{r}}{\mathrm{d}t}\mathrm{d}m - \int_m \frac{\mathrm{d}\mathbf{r}_O}{\mathrm{d}t} \times \frac{\mathrm{d}\mathbf{r}}{\mathrm{d}t}\mathrm{d}m \\
&= -\int_m \frac{\mathrm{d}\mathbf{r}_O}{\mathrm{d}t} \times \frac{\mathrm{d}\mathbf{r}}{\mathrm{d}t}\mathrm{d}m = -\frac{\mathrm{d}\mathbf{r}_O}{\mathrm{d}t} \times \int_m \frac{\mathrm{d}\mathbf{r}}{\mathrm{d}t}\mathrm{d}m \\
&= -\frac{\mathrm{d}\mathbf{r}_O}{\mathrm{d}t} \times \frac{\mathrm{d}}{\mathrm{d}t}\int_m \mathbf{r}\mathrm{d}m = -\frac{\mathrm{d}\mathbf{r}_O}{\mathrm{d}t} \times m\frac{\mathrm{d}\mathbf{r}_C}{\mathrm{d}t},
\end{aligned} \tag{5.16}$$

where $\mathbf{r}_C$ describes the position of the center of mass of the system (Fig. 5.1).

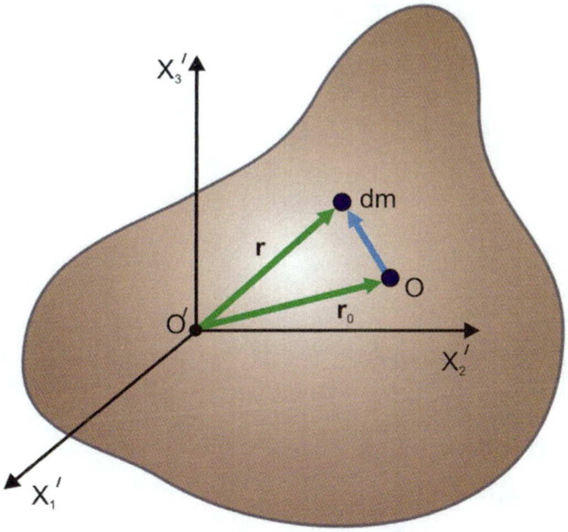

Fig. 5.1 Mechanical system with point O shown; about this point the angular momentum of the system is determined

If the analyzed mechanical system is continuous, then the system's angular momentum is described by the formula

$$\mathbf{K} = \int\limits_m (\mathbf{r} - \mathbf{r}_O) \times \mathbf{v} \mathrm{d}m, \tag{5.17}$$

and if it is discrete, then its angular momentum is equal to

$$\mathbf{K} = \sum_{n=1}^{N} m_n (\mathbf{r}_n - \mathbf{r}_O) \times \mathbf{v}_n, \tag{5.18}$$

where $\sum_{n=1}^{N} m_n = m$. For a continuous system we have

$$\frac{\mathrm{d}\mathbf{K}}{\mathrm{d}t} = \int\limits_m \frac{\mathrm{d}}{\mathrm{d}t}(\mathbf{r} - \mathbf{r}_O) \times \mathbf{v} \mathrm{d}m + \int\limits_m (\mathbf{r} - \mathbf{r}_O) \times \frac{\mathrm{d}^2\mathbf{r}}{\mathrm{d}t^2} \mathrm{d}m, \tag{5.19}$$

and taking into account formulas (5.14) and (5.16) we obtain

$$\frac{\mathrm{d}\mathbf{K}}{\mathrm{d}t} = -\frac{\mathrm{d}\mathbf{r}_O}{\mathrm{d}t} \times m \frac{\mathrm{d}\mathbf{r}_C}{\mathrm{d}t} + \int\limits_m (\mathbf{r} - \mathbf{r}_O) \times \mathbf{F}^{\mathrm{e}} \mathrm{d}m. \tag{5.20}$$

Integrating the preceding equation with respect to time we obtain

$$\Delta \mathbf{K} = \mathbf{K} - \mathbf{K}_O = \int_{t_0}^{t_0+\tau} \int_m (\mathbf{r} - \mathbf{r}_O) \times \mathbf{F}^{\mathrm{e}} \mathrm{d}m \mathrm{d}\tau - [\mathbf{r}_O \times m \mathbf{r}_{\mathrm{C}}]_{t_0}^{t_0+\tau}$$

$$= \int_m (\mathbf{r} - \mathbf{r}_O) \times \left(\int_{t_0}^{t_0+\tau} \mathbf{F}^{\mathrm{e}} \mathrm{d}\tau \right) \mathrm{d}m = \int_m (\mathbf{r} - \mathbf{r}_O) \times \mathbf{J} \mathrm{d}m \tag{5.21}$$

for a continuous mechanical system. Similar calculations conducted for a discrete mechanical system [see (5.17)] lead to the following result:

$$\Delta \mathbf{K} = \mathbf{K} - \mathbf{K}_O = \sum_{n=1}^{N} \mathbf{r}_n \times \mathbf{J}_n, \tag{5.22}$$

where $\mathbf{J}_n$ is the impulse of impact on particle n.

The results of those calculations can be summarized in the form of the following theorem.

Theorem 5.2. *A change in the angular momentum of any mechanical system with respect to an arbitrary pole O at the instant of impact is equal to the moment of external impulses of impact acting on the system.*

It can be demonstrated that the preceding theorem is valid in the case of an arbitrary pole, either moving or fixed [2].

We will derive a *general equation of the theory of impact*. The general equation of motion for a discrete material system has the form

$$\sum_{n=1}^{N} (m_n \ddot{\mathbf{r}}_n - \mathbf{F}_n) \circ \delta \mathbf{r}_n = 0. \tag{5.23}$$

Integrating the equations resulting from Newton's second law with respect to (time of) impact duration we obtain

$$\lim_{\tau \to 0} \sum_{n=1}^{N} \int_{t_0}^{t_0+\tau} (m_n \ddot{\mathbf{r}}_n - \mathbf{F}_n) \, \mathrm{d}\tau = \sum_{n=1}^{N} m_n \Delta \dot{\mathbf{r}}_n - \lim_{\tau \to 0} \int_{t_0}^{t_0+\tau} \mathbf{F}_n \mathrm{d}\tau. \tag{5.24}$$

Forces $\mathbf{F}_n$ are resolved into impact forces $\mathbf{F}_n^{\mathrm{i}}$ associated with impact and "non-impact" (i.e., ordinary) forces $\mathbf{F}_n^{ni}$; then from (5.23), and taking into account expression (5.24), we obtain a *general equation of impact*:

$$\sum_{n=1}^{N} [m_n (\dot{\mathbf{r}} - \dot{\mathbf{r}}_n(0)) - \mathbf{J}_n] \circ \delta \mathbf{r}_n = 0, \tag{5.25}$$

where:

$$\lim_{\tau\to 0}\int\limits_{t_0}^{t_0+\tau}\mathbf{F}_n^{i}\mathrm{d}\tau=\mathbf{J},\quad \lim_{\tau\to 0}\int\limits_{t_0}^{t_0+\tau}\mathbf{F}_n^{ni}\mathrm{d}\tau=0. \tag{5.26}$$

5.3 Particle Impact Against an Obstacle

Let a particle of mass m hit the surface of an obstacle whose mass is so large that the impact does not cause any change in its position (Fig. 5.2).

In the introduced coordinate system the angle of incidence (angle of rebound) is denoted α (β). We assume that at the instant of contact of the mass m with the obstacle the only instantaneous force (impact force) is the normal force $\mathbf{N}$ (the tangent force is equal to zero because we assume that the constraints are ideal). The velocity $\mathbf{v}_1$ denotes the velocity of the mass m just before impact and $\mathbf{v}_2$ – right after impact. The law of conservation of momentum for a particle in vector notation $m(\mathbf{v}_2-\mathbf{v}_1)=\mathbf{J}$ has the following representation in the system OX_1X_2 (the equation is successively multiplied by $\mathbf{E}_1$ and $\mathbf{E}_2$ yielding $v_{2x_1}-v_{1x_1}=0, m(v_{2x_2}-v_{1x_2})=\mathrm{J}$):

$$\begin{aligned} mv_2\cos(90^\circ-\beta)-mv_1\cos(90^\circ-\alpha)&=0,\\ mv_2\cos\beta-mv_1\cos(180^\circ-\alpha)&=\mathrm{J}, \end{aligned} \tag{5.27}$$

that is,

$$\begin{aligned} m(v_2\sin\beta-v_1\sin\alpha)&=0,\\ m(v_2\cos\beta+v_1\cos\alpha)&=\mathrm{J}, \end{aligned} \tag{5.28}$$

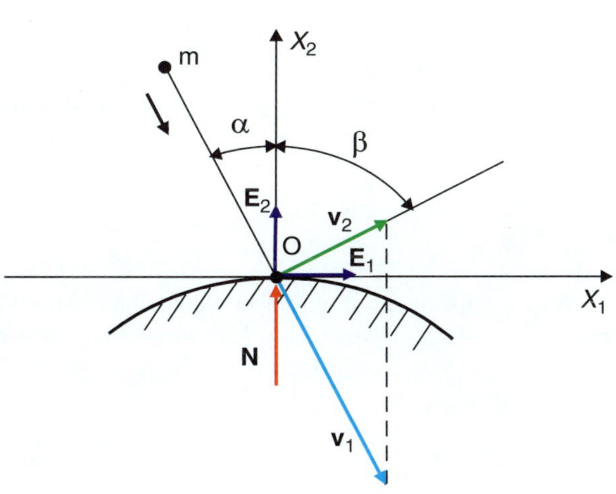

Fig. 5.2 Impact of mass m against a stationary obstacle at its point O

because vector **J** has the same sense as vector **N**. As can be seen, we have two algebraic equations for the determination of three unknown quantities v_2, β, and J. Newton noticed that the ratio of normal components of velocities does not depend on the velocity of motion of the colliding bodies or their geometry but on the materials they are made of. In our case the *coefficient of restitution* ν is described by the formula $(\mathbf{v}_2 \circ \mathbf{E}_2)\mathbf{E}_2 = -\nu(\mathbf{v}_1 \circ \mathbf{E}_2)\mathbf{E}_2$ or

$$\nu = -\frac{v_{2n}}{v_{1n}} = -\frac{v_{2x_2}}{v_{1x_1}} = -\frac{\mathbf{v}_2 \circ \mathbf{E}_2}{\mathbf{v}_1 \circ \mathbf{E}_2} = \frac{v_2 \cos\beta}{v_1 \cos\alpha}, \tag{5.29}$$

where $\mathbf{v}_2 \circ \mathbf{E}_2 = v_{2n}$ and $\mathbf{v}_1 \circ \mathbf{E}_2 = v_{1n}$, which follows from the hypothesis of Newton (the law of restitution of a normal velocity). However, it should be emphasized that the Newton hypothesis is too simple, and in many cases experimental investigations show that the restitution coefficient depends not only on the materials of impacting objects but also on their geometry and velocities [3].

In the preceding equation v_n denotes the normal velocity (in this case in the X_2 direction). The minus sign in (5.29) is a consequence of the fact that for the given sense of the normal, the velocities $\mathbf{E}_2 v_{1n}$ and $\mathbf{E}_2 v_{2n}$ always have opposite senses (signs). From the equation of conservation of momentum in the X_1 direction we have

$$v_{2x_1} = v_{1x_1} = v_1 \sin\alpha. \tag{5.30}$$

In turn, from the equation of restitution of a normal velocity (in the X_2 direction) we obtain

$$v_{2x_2} = -\nu v_{1n} = \nu v_1 \cos\alpha, \tag{5.31}$$

where $v_{2x_2} = v_{2n}$ and $v_{1x_2} = v_{1n}$, and vector $\mathbf{v}_{2n} = -\nu \mathbf{v}_{1n}$.

The desired velocity after the collision is equal to [see (5.30) and (5.31)]

$$v_2 = \sqrt{v_{2x_1}^2 + v_{2x_2}^2} = v_1 \sqrt{\sin^2\alpha + \nu^2 \cos^2\alpha}. \tag{5.32}$$

The angle formed between the velocity v_2 and the axis X_2 can be calculated from the following equation [dividing (5.30) by (5.31) by sides]:

$$\tan\beta = \frac{v_{2x_1}}{v_{2x_2}} = \frac{1}{\nu}\tan\alpha. \tag{5.33}$$

An impulse of the impact force calculated from the equation for the change in momentum in the X_2 direction [the second equation of system (5.27)] after taking into account (5.29) is equal to

$$\mathrm{J} = m v_1 (1 + \nu)\cos\alpha. \tag{5.34}$$

5.4 A Physical Interpretation of Impact

Let us consider a small ball (of negligible radius) hitting a stationary deformable obstacle (Fig. 5.3).

All equations of this section are formulated for collinear vectors, and therefore the scalar notation is used. The reaction $\mathbf{N}_1$ of the obstacle is a result of the deformation of its material, and during contact of two bodies the velocity of magnitude v_1 at the moment of impact decreases to zero for the time instant $t_0 + \tau_1$. The reaction force $\mathbf{N}_1$ generates the impulse of a force of the form

$$\mathrm{J}_1 = \int_{t_0}^{t_0+\tau_1} N_1 \mathrm{d}\tau = m v_1 \cos\alpha. \tag{5.35}$$

In turn, the deformed material produces the impulsive force of reaction $\mathbf{N}_2$, which generates the following impulse of a force:

$$\mathrm{J}_2 = \int_{t_0}^{t_0+\tau_2} N_2 \mathrm{d}\tau = m v_2 \cos\beta. \tag{5.36}$$

The ratio of impulses

$$\frac{\mathrm{J}_2}{\mathrm{J}_1} = \frac{m v_2 \cos\beta}{m v_1 \cos\alpha} = \nu \tag{5.37}$$

describes the coefficient of restitution. The limiting values of this coefficient can be determined easily on the basis of the following considerations. If the obstacle is a plastic body, then the velocity $v_2 = 0$ (the ball does not bounce off the obstacle) and then $\nu = 0$. If the angle of incidence is equal to the angle of rebound and $v_2 = v_1$, then the impact is perfectly elastic and $\nu = 1$. In general, $0 \leq \nu \leq 1$.

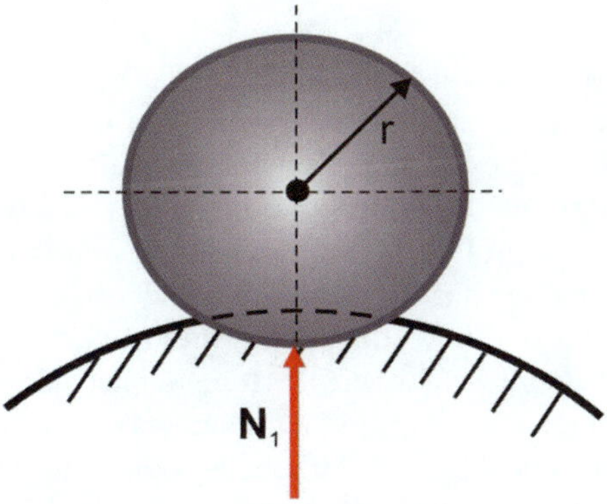

Fig. 5.3 Impact of a small ball of radius $r \to 0$ against a deformable obstacle

Table 5.1 Restitution coefficient values

Material	ν
Steel	0.56
Wood	0.26
Cast iron	0.66
Lead	0.20
Ivory	0.81
Glass	0.94

In Table 5.1 the values of the coefficients of restitution are given for an elastic particle, and on the assumption that both colliding bodies are made of the same material.

The model of a perfectly rigid body does not allow for an explanation of the phenomenon of rebound of a body during impact. In order to explain this phenomenon, it is necessary to introduce a model of a deformable body. If a particle hits an obstacle in a normal direction (Fig. 5.3), then from (5.37) for $\alpha = \beta = 0$ we obtain

$$\nu = \frac{v_2}{v_1}. \tag{5.38}$$

The difference in kinetic energy before and after impact is equal to

$$T_1 - T_2 = \frac{mv_1^2 - mv_2^2}{2} = \frac{mv_1^2 - mv_1^2\nu^2}{2} = \frac{mv_1^2}{2}\left(1 - \nu^2\right). \tag{5.39}$$

The smaller the coefficient of restitution, the more kinetic energy is lost because of conversion into heat. The kinetic energy is converted into heat completely during plastic impact, whereas during perfectly elastic impact the energy is completely preserved since $\nu = 0$.

5.5 Collision of Two Balls in Translational Motion

We will now consider the impact of two homogeneous balls of equal radii, but made of different materials, which are in translational motion, where the vectors of the velocities of their centers just before the instant of impact lie on a line passing through the centers of both balls (Fig. 5.4).

From Fig. 5.4 we can see that in order for the first ball to hit the second ball, the first one has to "chase" the other, that is, we have $v_1 > v_2$. It should be emphasized that in this section, subscripts 1 and 2 correspond to the numbering of bodies; the velocities of bodies before impact possess no superscripts, and those following impact are denoted by a superscript prime ($'$). After the collision, which lasts for a very short time, the balls move with unknown velocities v_1' and v_2'. In the time

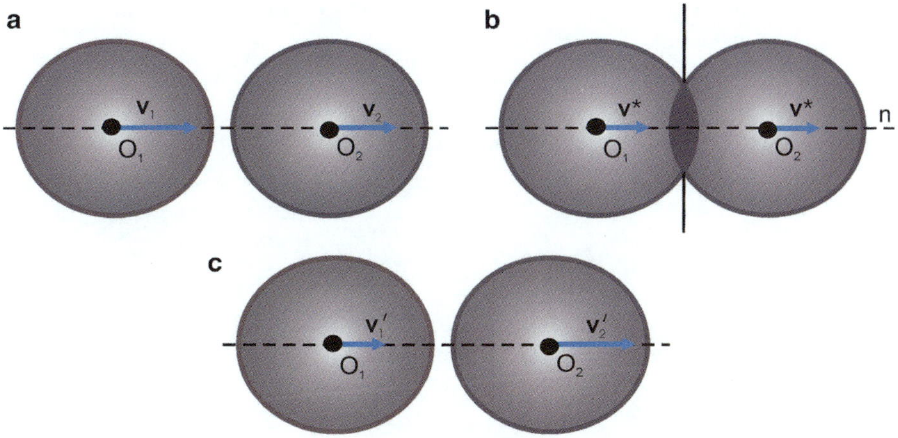

Fig. 5.4 Collision of two balls moving in translational motion: (**a**) the instant just before collision; (**b**) the collision; (**c**) the instant immediately following collision

interval, when the balls remain in contact with each other, their interactions are treated as internal forces of this mechanical system, and, according to Newton's third law, they cancel out. Therefore, these forces cannot produce any change in momentum of the considered mechanical system. It follows that the momentum of the system before and after impact remains unchanged, that is,

$$m_1 v_1 + m_2 v_2 = m_1 v_1' + m_2 v_2', \tag{5.40}$$

where in the preceding equation vector quantities are replaced by scalars since the vectors are collinear. Recall that impact is called a *central collision* if a normal to the surface of contact between the bodies (a line of impact) passes through the centers of the colliding bodies. The impact of two bodies is called a *direct collision* if the velocities of the bodies' points at which contact occurs are directed along the common normal to the surfaces of both bodies [4].

We adopt the following physical interpretation of the considered *direct* and *central collision* of two balls. The first stage, corresponding to the time interval τ_1, is associated with the build-up of a local deformation of both balls, where in the time interval $t_0 + \tau_1 - t_0$ the velocity of the first ball decreases, whereas the velocity of the second increases until the velocities of both balls become equal to $\mathbf{v}^*$. At this time instant, forces of mutual interaction reach their maximum magnitudes.

The deformations of the balls accumulate their kinetic energy in the form of a potential energy that is subsequently transferred to each of the balls during the so-called "second stage" of the collision in the time interval τ_2. At the end of this time interval, the balls recover from the elastic deformations and gain velocities $\mathbf{v}_1'$ and $\mathbf{v}_2'$, and starting from this moment they cease to act on each other.

The increments of the momentum of both balls in the time interval from t_0 to $t_0 + \tau_1$ are equal to

$$m_1 v^* - m_1 v_1 = -\int_{t_0}^{t_0+\tau_1} F \mathrm{d}\tau = -\mathrm{J}_1$$

$$m_2 v^* - m_2 v_2 = \int_{t_0}^{t_0+\tau_1} F \mathrm{d}\tau = \mathrm{J}_1, \tag{5.41}$$

where vector notation is not used for the reasons mentioned earlier, and **F** is the force of interaction of the balls on each other. Adding (5.41) to each other by sides we get

$$v^* = \frac{m_1 v_1 + m_2 v_2}{m_1 + m_2}. \tag{5.42}$$

We will now consider the collision of balls in the time interval from $t_0 + \tau_1$ to $t_0 + \tau_1 + \tau_2$. In this case we have

$$m_1 v_1' - m_1 v^* = -\int_{t_0+\tau_1}^{t_0+\tau_1+\tau_2} F \mathrm{d}\tau = -\mathrm{J}_2,$$

$$m_2 v_2' - m_2 v^* = \int_{t_0+\tau_1}^{t_0+\tau_1+\tau_2} F \mathrm{d}\tau = \mathrm{J}_2. \tag{5.43}$$

According to previous calculations, determination of the velocities of balls at the time instant just after collision is possible only by the introduction of the notion of a *coefficient of restitution*, that is,

$$\mathrm{J}_2 = \nu \mathrm{J}_1. \tag{5.44}$$

We must determine four unknowns v^*, J_1, v_1', v_2' from (5.41), (5.42), (5.43), and (5.44). Eventually, from those equations we obtain

$$v_1' = \frac{(m_1 - \nu m_2)v_1 + (1+\nu)m_2 v_2}{m_1 + m_2},$$

$$v_2' = \frac{(m_2 - \nu m_1)v_2 + (1+\nu)m_1 v_1}{m_1 + m_2}. \tag{5.45}$$

In the case of plastic collision ($\nu = 0$), from (5.45) we obtain

$$v_1' = v_2' = \frac{m_1 v_1 + m_2 v_2}{m_1 + m_2}, \tag{5.46}$$

and in the case of perfectly elastic collision ($\nu = 1$) from (5.45) we have

$$\begin{aligned} v_1' &= \frac{(m_1 - m_2)v_1 + 2m_2 v_2}{m_1 + m_2}, \\ v_2' &= \frac{(m_2 - m_1)v_2 + 2m_1 v_1}{m_1 + m_2}. \end{aligned} \tag{5.47}$$

If $m_1 = m_2$, then from (5.46) and (5.47) we respectively obtain

$$v_1' = v_2' = \frac{v_1 + v_2}{2}, \tag{5.48}$$

$$v_1' = v_2, \quad v_2' = v_1. \tag{5.49}$$

The obtained equations for the velocities of balls following the collision allow also for the analysis of the case of collision presented in Fig. 5.3. Into (5.45) one should substitute $v_1 = v$, $v_1' = v'$, $v_2 = 0$, $m_2 = \infty$. We divide the numerator and denominator of those equations by m_2, obtaining

$$\begin{aligned} v_1' &= \lim_{m_2 \to \infty} \frac{\left(\frac{m_1}{m_2} - \nu\right) v_1}{1 + \frac{m_1}{m_2}} = -\nu v_1, \\ v_2' &= \lim_{m_2 \to \infty} \frac{(1 + \nu)\frac{m_1}{m_2} v_1}{1 + \frac{m_1}{m_2}} = 0. \end{aligned} \tag{5.50}$$

From the first equation of (5.50) it follows that after the collision the ball has a velocity of smaller magnitude than before the collision and of opposite sense. If we release the ball from height h onto horizontal ground and measure the maximum height after the ball bounces off the ground, then

$$v' = \sqrt{2gh'}, \quad v = \sqrt{2gh}, \tag{5.51}$$

and making use of (5.50) we obtain

$$\nu = \left|\frac{v'}{v}\right| = \sqrt{\frac{h'}{h}}, \tag{5.52}$$

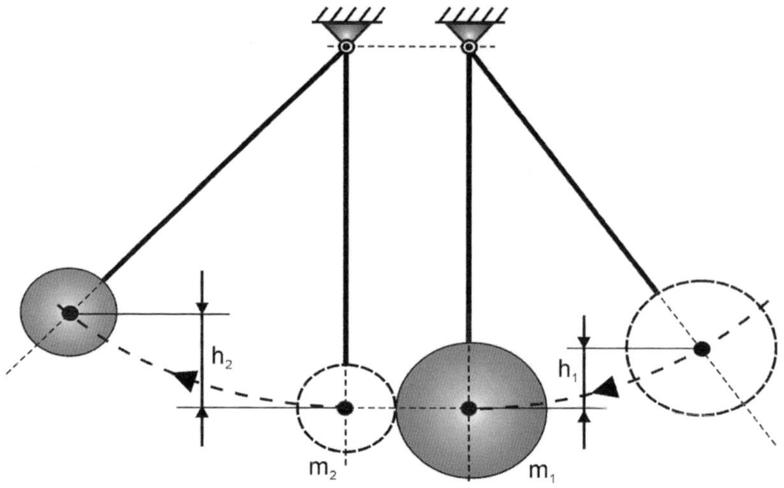

Fig. 5.5 Direct collision of two balls

which allows for the experimental determination of the coefficient of restitution ν. In the case of collision of two balls it is possible to determine the loss of the kinetic energy that occurs in the analyzed mechanical system

$$\begin{aligned} T - T' &= \frac{1}{2}\left(m_1 v_1^2 + m_2 v_2^2\right) - \frac{1}{2}\left(m_1 v_1'^2 + m_2 v_2'^2\right) \\ &= \frac{1}{2}\left(1 - \nu^2\right)\frac{m_1 m_2}{m_1 + m_2}(v_1^2 - v_2^2). \end{aligned} \tag{5.53}$$

From the preceding equation for $\nu = 0$ we get

$$T - T' = \frac{1}{2}\frac{m_1 m_2}{m_1 + m_2}(v_1^2 - v_2^2), \tag{5.54}$$

and for $\nu = 1$ we have $T = T'$.

Example 5.1. Two balls of masses m_1 and m_2 are suspended from two weightless rods. The ball of larger mass is deflected from the equilibrium position up to the height h_1 and then released with no initial velocity. Provided that the height h_2 reached by the second ball after impact is known, determine the coefficient of restitution (Fig. 5.5).

The potential energy of the first ball after its free release converts into kinetic energy, which makes it possible to determine the velocity of the first ball just before it hits the second ball, that is,

$$\frac{m_1 v_1^2}{2} = m_1 g h_1.$$

Since after the collision the balls are treated again as non-deformable, just after impact the velocity of the second ball is equal to v_2' and its kinetic energy turns into potential energy according to the equation

$$\frac{m_2 v_2'^2}{2} = m_2 g h_2.$$

From two preceding equations we obtain

$$v_1 = \sqrt{2gh_1}, \quad v_2' = \sqrt{2gh_2}.$$

Let us now make use of the second of equations (5.45) to obtain (for $v_2 = 0$)

$$\sqrt{2gh_2} = \frac{(1+\nu)m_1\sqrt{2gh_1}}{m_1 + m_2},$$

and following the transformation we have

$$\nu = \frac{m_1 + m_2}{m_1}\sqrt{\frac{h_2}{h_1}} - 1.$$

For $\nu = 1$ and $m_1 = m_2$ we obtain $h_1 = h_2$. □

5.6 Collision of Two Freely Moving Rigid Bodies

Let two rigid bodies of masses m_1 and m_2 start to make contact at point A at time instant t^i (Fig. 5.6). Let us assume that at time instant t^i the mass centers of the bodies have velocities $\mathbf{v}^i_{C_i}$, and their angular velocities are equal to $\boldsymbol{\omega}^i_{C_i}$, $i = 1, 2$. As can be seen from Fig. 5.6, the contact of bodies commences at point A, where we assume a smooth surface of contact in order to neglect friction forces. According to Newton's third law, impact forces $\mathbf{F}$ of opposite senses are created that are perpendicular to the surface of contact at point A. The bodies remain in contact during the short time interval τ, that is, $t^f = t^i + \tau$, where t^f denotes the instant when the contact between the bodies ceases.

Further motion of bodies after contact cessation can be analyzed on the basis of the initial conditions and equations of motion of bodies (which take into account the forces and the resistance of the medium) determined at time instant t^f. Our aim is to determine the velocities $\mathbf{v}^f_{C_i}$ and $\boldsymbol{\omega}^f_{C_i}$, $i = 1, 2$, of both rigid bodies at time instant t^f, which are essential for further analysis of motion. We will proceed after introducing the following assumptions [5]:

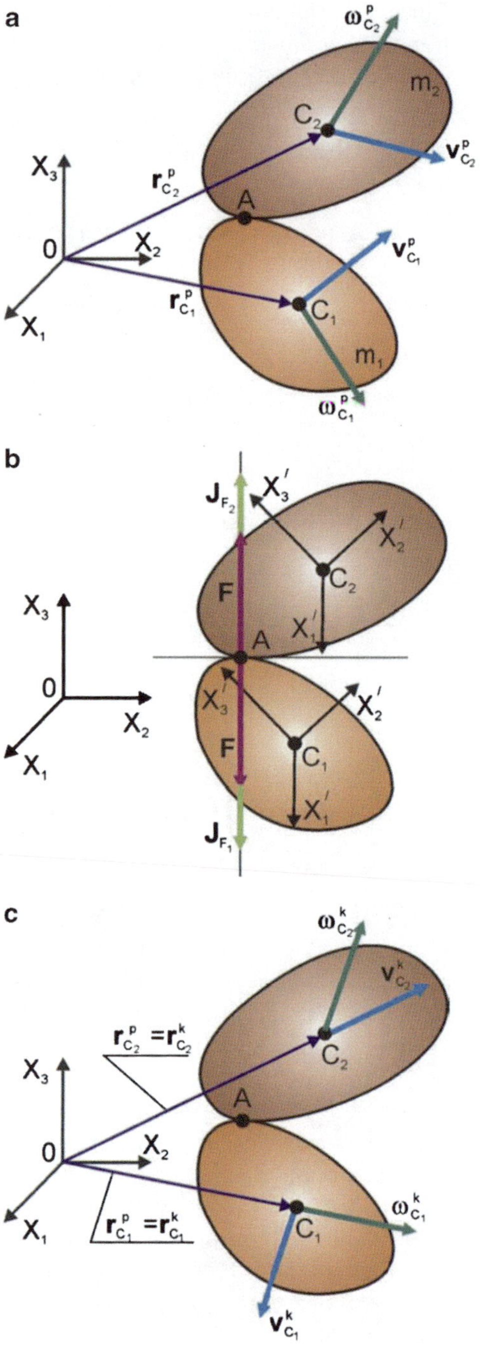

Fig. 5.6 Collision of two rigid bodies: (**a**) instant just before collision t^i; (**b**) collision $t^i < t < t^f$; (**c**) instant just after collision t^f

(1) Only impact forces **F** are taken into account (the remaining forces are assumed to be negligible).
(2) The positions of bodies in the system $OX_1X_2X_3$ are the same at instants t^i and t^f (they do not change during time interval τ).
(3) Impact forces emerge and develop during time interval τ along a common line perpendicular to the surface of contact.

Axes of parallel coordinate systems associated with the bodies are chosen such that one axis of each system (here number 1 axes) is parallel to the normal direction at the point of collision.

By Theorems 5.1 [see also (5.12)] and 5.2 [see also (5.22)], we can write

$$\Delta \mathbf{P}_i = \mathbf{J}_{F_i}, \quad \Delta \mathbf{K}_i = \overrightarrow{C_i A} \times \mathbf{J}_{F_i}, \tag{5.55}$$

where $i = 1, 2$.

According to the introduced parallel body systems $C_i\, X_1'X_2'X_3'$ we have

$$\mathbf{J}_{F_1} = \mathbf{E}_1' J_F, \quad \mathbf{J}_{F_2} = -\mathbf{J}_{F_1} = -\mathbf{E}_1' J_F. \tag{5.56}$$

From the first equation of (5.55) we have

$$m_1 \left(\mathbf{v}_{C_1}^f - \mathbf{v}_{C_1}^i \right) = \mathbf{J}_F, \tag{5.57}$$

and from the second

$$\Delta \mathbf{K}_1 = \begin{vmatrix} \mathbf{E}_1' & \mathbf{E}_2' & \mathbf{E}_3' \\ x_{1A_1}' & x_{2A_1}' & x_{3A_1}' \\ \mathrm{J}_F & 0 & 0 \end{vmatrix} = \mathbf{E}_2' J_F x_{3A_1}' - \mathbf{E}_3' J_F x_{2A_1}'. \tag{5.58}$$

From vector equations (5.57) and (5.58) we obtain the following six scalar equations:

$$\begin{aligned} m_1 \left(v_{1C_1}^f - v_{1C_1}^i \right) &= \mathrm{J}_F, \\ m_1 \left(v_{2C_1}^f - v_{2C_1}^i \right) &= 0, \\ m_1 \left(v_{3C_1}^f - v_{3C_1}^i \right) &= 0, \end{aligned} \tag{5.59}$$

$$\begin{bmatrix} I_{X_1'}^{C_1} & -I_{X_1'X_2'}^{C_1} & -I_{X_1'X_3'}^{C_1} \\ -I_{X_2'X_1'}^{C_1} & I_{X_2'}^{C_1} & -I_{X_2'X_3'}^{C_1} \\ -I_{X_3'X_1'}^{C_1} & -I_{X_3'X_2'}^{C_1} & I_{X_3'}^{C_1} \end{bmatrix} \begin{bmatrix} \omega_{11}^f - \omega_{11}^i \\ \omega_{12}^f - \omega_{12}^i \\ \omega_{13}^f - \omega_{13}^i \end{bmatrix} = \begin{bmatrix} 0 \\ \mathrm{J}_F x_{3A_1}' \\ -\mathrm{J}_F x_{2A_1}' \end{bmatrix}, \tag{5.60}$$

where, here and also in the next equation, ω_{ik} denotes the component k of vector $\boldsymbol{\omega}_i$.

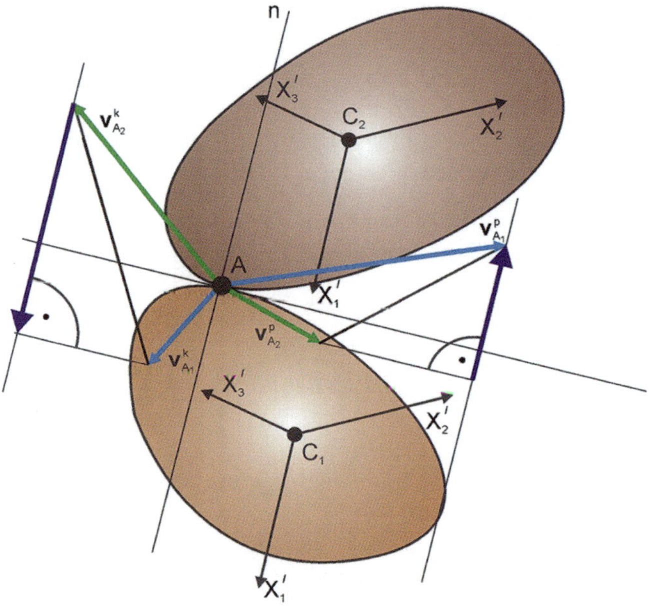

Fig. 5.7 Velocities of points A_1 and A_2 at time instants t^i and t^f

Applying a similar procedure for the second rigid body we get

$$m_2\left(v_{1C_2}^f - v_{1C_2}^i\right) = -\mathrm{J}_F,$$
$$m_2\left(v_{2C_2}^f - v_{2C_2}^i\right) = 0,$$
$$m_2\left(v_{3C_2}^f - v_{3C_2}^i\right) = 0, \tag{5.61}$$

$$\begin{bmatrix} I_{X_1'}^{C_2} & -I_{X_1'X_2'}^{C_2} & -I_{X_1'X_3'}^{C_2} \\ -I_{X_2'X_1'}^{C_2} & I_{X_2'}^{C_2} & -I_{X_2'X_3'}^{C_2} \\ -I_{X_3'X_1'}^{C_2} & -I_{X_3'X_2'}^{C_2} & I_{X_3'}^{C_2} \end{bmatrix} \begin{bmatrix} \omega_{21}^f - \omega_{21}^i \\ \omega_{22}^f - \omega_{22}^i \\ \omega_{23}^f - \omega_{23}^i \end{bmatrix} = \begin{bmatrix} 0 \\ -\mathrm{J}_F x_{3A_2}' \\ \mathrm{J}_F x_{3A_2}' \end{bmatrix}. \tag{5.62}$$

In total we have written 12 equations for the determination of 13 unknowns ω_{ik}^k, $v_{lC_m}^k$ $(i,k,l = 1,2,3, \quad m = 1,2)$, and J_F.

An additional equation can be obtained from the law of conservation of kinetic energy. Here, however, relying on the previous calculations, we make use of the vector calculus.

In Fig. 5.7 are shown the velocity vectors of point A at time instants t^i and t^f.

The velocities of points A_1 and A_2 are equal to

$$\mathbf{v}_{A_1} = v_{1A_1}\mathbf{E}'_1 + v_{2A_1}\mathbf{E}'_2 + v_{3A_1}\mathbf{E}'_3,$$
$$\mathbf{v}_{A_2} = v_{1A_2}\mathbf{E}'_1 + v_{2A_2}\mathbf{E}'_2 + v_{3A_2}\mathbf{E}'_3. \tag{5.63}$$

Because the force interaction takes place along the normal and there is no friction or slip, we have

$$v_{2A_1}\mathbf{E}'_2 + v_{3A_1}\mathbf{E}'_3 = v_{2A_2}\mathbf{E}'_2 + v_{3A_2}\mathbf{E}'_3. \tag{5.64}$$

Taking into account equality (5.64) in system (5.63) we obtain

$$\mathbf{v}_{A_1} - \mathbf{v}_{A_2} = \mathbf{E}'_1(v_{1A_1} - v_{1A_2}). \tag{5.65}$$

From (5.65) we obtain

$$-\frac{v^f_{1A_1} - v^f_{1A_2}}{v^i_{1A_1} - v^i_{1A_2}} = 1 \tag{5.66}$$

if we are dealing with a perfectly elastic rebound of bodies. On the other hand, if the rebound is perfectly plastic, then

$$\frac{v^f_{1A_1} - v^f_{1A_2}}{v^i_{1A_1} - v^i_{1A_2}} = 0. \tag{5.67}$$

In the end, according to the previous calculations, in the intermediate case we introduce the coefficient of restitution ν, and formulas (5.66) and (5.67) take the form

$$-\frac{v^f_{1A_1} - v^f_{1A_2}}{v^i_{1A_1} - v^i_{1A_2}} = \nu, \tag{5.68}$$

where $0 \le \nu \le 1$.

The obtained equations allow also for an analysis of the simple case where the second body m_2 is stationary and the first one is reduced to a particle of mass m (Fig. 5.2).

5.7 A Center of Percussion

Let a compound pendulum of mass m_1 be suspended at point O and let it be hit by a horizontally traveling bullet of mass m_2 (Fig. 5.8).

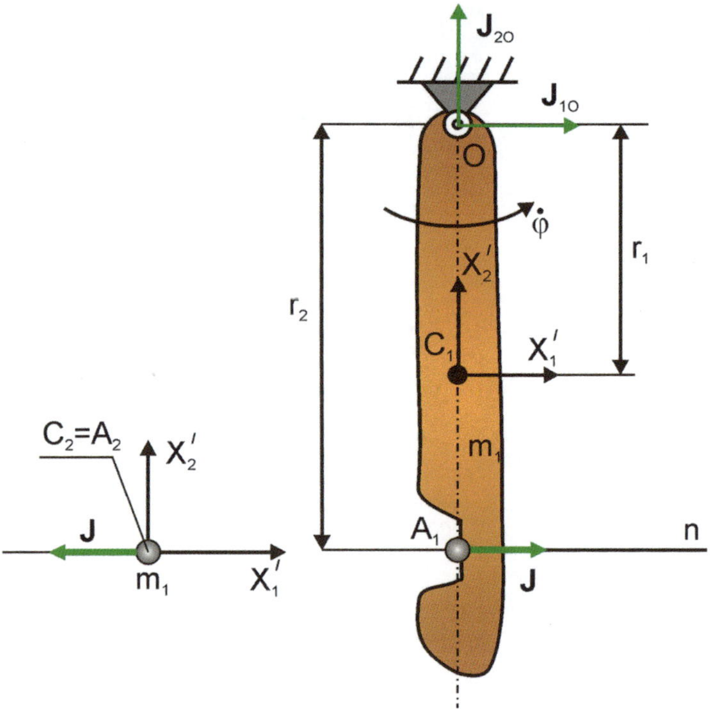

Fig. 5.8 A bullet's impact against a compound pendulum

By Theorems 5.1 and 5.2 we obtain the following equations for the pendulum:

$$m_1\left(v^{\prime f}_{1C_1} - v^{\prime i}_{1C_1}\right) = \mathrm{J} + \mathrm{J}_{1O},$$

$$m_1\left(v^{\prime f}_{2C_1} - v^{\prime i}_{2C_1}\right) = \mathrm{J}_{2O},$$

$$I_O(\dot{\varphi}^f - \dot{\varphi}^i) = J r_2, \tag{5.69}$$

and for the bullet of mass m_2

$$m_2\left(v^{\prime f}_{1C_2} - v^{\prime i}_{1C_2}\right) = -\mathrm{J}, \quad m_2\left(v^{\prime f}_{2C_2} - v^{\prime i}_{2C_2}\right) = 0, \tag{5.70}$$

and the law of restitution takes the form

$$-v^{\prime f}_{1A_1} + v^{\prime f}_{1C_2} = \nu\left(v^{\prime i}_{1A_1} - v^{\prime i}_{1C_2}\right). \tag{5.71}$$

At time instant t^i in the case of the pendulum we have $\varphi^i = 0$, $\dot{\varphi}^i = 0$, and in the case of the bullet $\dot{x}^{\prime i}_{1C_2} = v^{\prime i}_{1C_2} = v^i$ $\dot{x}^{\prime i}_{2C_2} = 0$. Because the pendulum cannot possibly move in the vertical direction, we also have $\dot{x}^f_{2C_1} = \dot{x}^i_{2C_1} = 0$, which enables us to obtain the result $\mathrm{J}_{2O} = 0$ from the second equation of system (5.69). In turn, from the second equation of (5.70) we get $v^f_{2C_2} = 0$ because $v^i_{2C_2} = 0$.

Eventually, the equations take the form

$$m_1\dot{\varphi}^f r_1 = \mathrm{J} + \mathrm{J}_{1O}, \qquad I_O\dot{\varphi}^f = Jr_2,$$

$$m_2\left(\dot{x}^f_{1C_2} - v_o\right) = -\mathrm{J}, \quad -\dot{\varphi}^k r_2 + \dot{x}^k_{1C_2} = -v_o v, \tag{5.72}$$

and their solutions

$$\dot{\varphi}^f = \frac{m_2 v_o r_2(1+v)}{I_O + m_2 r_2^2}, \qquad \dot{x}^f_{1C_2} = \frac{v_o\left(m_2 r_2^2 - vI_O\right)}{I_O + m_2 r_2^2},$$

$$\mathrm{J} = \frac{m_2 v_o I_O(1+v)}{I_O + m_2 r_2^2}, \quad \mathrm{J}_{1O} = \frac{m_2 v_o(1+v)}{I_O + m_2 r_2^2}(m_1 r_1 r_2 - I_O). \tag{5.73}$$

From the last equation of (5.73) it follows that $\mathrm{J}_{1O} = 0$, on condition that $r_2 = \frac{I_O}{m_1 r_1}$. This means that the impulse of a force is not generated at the pivot point of a pendulum if the mass m_2 hits the pendulum at a certain point located at distance $I_O m_1^{-1} r_1^{-1}$ from the pivot point; this point of a body is called a *center of percussion*.

References

1. J. Awrejcewicz, *Classical Mechanics: Statics and Kinematics* (Springer, New York, 2012)
2. V.F. Zhuravlev, *Fundamentals of Theoretical Mechanics* (Fizmatlit, Moscow, 2001)
3. P. Hagedorn, *Technische Mechanik: Dynamik* (Verlag Harri Deutsch, Frankfurt, 2008)
4. J. Leyko, *General Mechanics* (PWN, Warsaw, 1996)
5. J. Krodkiewski, *Dynamics of Mechanical Systems* (The University of Melbourne Press, Melbourne, 2006)

Chapter 6
Vibrations of Mechanical Systems

6.1 Introduction

Vibration theory belongs to well-developed branches of mechanics and physics. It cannot be understood without a good command of the fundamentals of mathematics. A large body of literature exists that is devoted to the theory of vibrations of discrete and continuous systems; it is not cited here in full; we mention only a few works [1–16], where an extensive bibliography covering the field can be found. This book will give certain basic information concerning the vibrations of discrete (or lumped) systems from the viewpoint of "mechanics." The vibrations of lumped mechanical systems are described by ordinary differential equations. We dealt with such equations in Chaps. 1–3.

Let us assume that particles of a lumped material system (Fig. 3.1) are connected to each other by means of massless spring-damper elements. Those connections generate forces and moments of forces (torques) that are dependent on the displacements and velocities of the particles. The imposition of initial conditions by means of, for example, the initial deflection and velocity of particles for certain known parameters of the system (masses, stiffnesses, damping coefficients, system geometry) causes vibrations of the mechanical system under consideration. In subsequent calculations we will confine ourselves only to small vibrations about a certain static configuration of the system (equilibrium position). Recall that in the case of non-linear systems, the system may have several distinct equilibrium positions. For small vibrations about the considered equilibrium position it is possible to conduct the linearization process, which consists in the expansion of certain functions into a Taylor[1] series (Maclaurin[2] series) and taking into account only linear terms (although linearization is not always possible). As a result, the problem boils down to the analysis of linear differential equations with constant

[1]Brook Taylor (1685–1731), English mathematician known for Taylor's theorem and Taylor series.

[2]Colin Maclaurin (1698–1746), Scottish mathematician working mainly in Edinburgh.

J. Awrejcewicz, *Classical Mechanics: Dynamics*, Advances in Mechanics and Mathematics 29, DOI 10.1007/978-1-4614-3740-6_6,

or variable coefficients. A natural step in subsequent calucations will be the use of Lagrange's equations of the second kind.

6.2 Motion Equation of Linear Systems with N Degrees of Freedom

In the general case, the vibrations of mechanical systems with many degrees of freedom can be derived immediately from Newton's second law (which is quite popular) or from Lagrange's equations of the second kind. Making use of the latter possibility in general case we have

$$\frac{\mathrm{d}}{\mathrm{d}t}\frac{\partial T}{\partial \dot{q}_n} - \frac{\partial T}{\partial q_n} = Q(t, \boldsymbol{q}, \dot{\boldsymbol{q}}), \quad n = 1, \ldots, N, \tag{6.1}$$

where $\boldsymbol{q} = (q_1, \ldots, q_N)$. Because we will consider small vibrations, it is possible to conduct linearization of the system in the neighborhood of a trivial equilibrium position $\boldsymbol{q} = \mathbf{0}$ (this is always possible after moving the coordinate system to the chosen equilibrium position). It turns out that, in the general case, for rheonomic constraints the kinetic energy $T = T(t, \boldsymbol{q}, \dot{\boldsymbol{q}})$ occurring in (6.1) possesses a certain characteristic structure described below. Because

$$\begin{aligned}
T &= \frac{1}{2}\int_m \mathbf{v}^2 \mathrm{d}m = \frac{1}{2}\int_m \left(\frac{\partial \mathbf{r}}{\partial t} + \sum_{n=1}^{N} \frac{\partial \mathbf{r}}{\partial q_n}\dot{q}_n\right)^2 \mathrm{d}m \\
&= \frac{1}{2}\int_m \frac{\partial \mathbf{r}}{\partial t}\frac{\partial \mathbf{r}}{\partial t}\mathrm{d}m + \sum_{n=1}^{N}\dot{q}_n \int_m \frac{\partial \mathbf{r}}{\partial t}\frac{\partial \mathbf{r}}{\partial q_n}\mathrm{d}m + \frac{1}{2}\int_m \left(\sum_{n=1}^{N}\frac{\partial \mathbf{r}}{\partial q_n}\dot{q}_n\right)^2 \mathrm{d}m \\
&= T_0 + T_1 + T_2,
\end{aligned} \tag{6.2}$$

where

$$\begin{aligned}
T_0 &= \frac{1}{2}\int_m \left(\frac{\partial \mathbf{r}}{\partial t}\right)^2 \mathrm{d}m \\
T_1 &= \sum_{n=1}^{N} b_n \dot{q}_n = \sum_{n=1}^{N}\left(\int_m \frac{\partial \mathbf{r}}{\partial t}\frac{\partial \mathbf{r}}{\partial q_n}\mathrm{d}m\right)\dot{q}_n, \\
T_2 &= \frac{1}{2}\sum_{n,j=1}^{N} m_{nj}\dot{q}_n\dot{q}_j = \frac{1}{2}\sum_{n,j=1}^{N}\left(\int_m \frac{\partial \mathbf{r}}{\partial q_n}\frac{\partial \mathbf{r}}{\partial q_j}\mathrm{d}m\right)\dot{q}_n\dot{q}_j,
\end{aligned} \tag{6.3}$$

and, as can be seen from (6.3), we have $T_0 = T_0(t, \boldsymbol{q})$, $b_n = b_n(t, \boldsymbol{q})$, $m_{nj} = m_{nj}(t, \boldsymbol{q})$. Next we will consider a mechanical system with time-independent (scleronomic) constraints, for which $T = T(\boldsymbol{q}, \dot{\boldsymbol{q}})$, and with a generalized force $Q = Q(\boldsymbol{q}, \dot{\boldsymbol{q}})$, which means that $T_0 = T_0(\boldsymbol{q})$, $b_n = b_n(\boldsymbol{q})$, $m_{nj} = m_{nj}(\boldsymbol{q})$. Let us expand the energy $T = T_2$ and generalized forces into a series about the equilibrium position $\boldsymbol{q} = \mathbf{0}$ making use of a Maclaurin series of the following form (for rheonomic constraints $\frac{\partial \mathbf{r}}{\partial t} = 0$, and hence $T_0 = T_1 = 0$):

$$T = T_2 = \frac{1}{2} m_{nj}(\boldsymbol{q}) \dot{q}_n \dot{q}_j = \frac{1}{2} m_{nj}(\mathbf{0}) \dot{q}_n \dot{q}_j + \ldots,$$
$$Q_n = Q_n(\boldsymbol{q}, \dot{\boldsymbol{q}}) = -c_{nj} \dot{q}_j - k_{nj} q_j, \tag{6.4}$$

where in (6.4) the summation convention introduced earlier applies.

Eventually, from Lagrange's equations we obtain a system of second-order linear ordinary differential equations, which in matrix notation has the form

$$\mathbf{M}\ddot{\mathbf{q}} + \mathbf{C}\dot{\mathbf{q}} + \mathbf{K}\mathbf{q} = \mathbf{0}, \tag{6.5}$$

where:

$\mathbf{M} = [m_{nj}]$ is a kinetic energy matrix or inertia matrix that is positive-definite and symmetric;
$\mathbf{C} = [c_{nj}]$ is an arbitrary square matrix of forces of resistance to motion, that is, forces dependent on velocity;
$\mathbf{K} = [k_{nj}]$ is an arbitrary square matrix of configuration (positional) forces, that is, forces dependent on displacement.

An important property of Lagrange's equations, regardless of whether the system is linear or non-linear, is their linearity with respect to accelerations. The obtained equations have the form

$$m_{nj} \ddot{q}_j + f_n(t, \boldsymbol{q}, \dot{\boldsymbol{q}}) = 0. \tag{6.6}$$

In the case of a linear scleronomic system we obtain

$$m_{nj} \ddot{q}_j + c_{nj} \dot{q}_j + k_{nj} q_j = 0, \tag{6.7}$$

which is the equivalent form of the matrix notation (6.5). It turns out that inertia matrixes are always non-singular, that is, $\det[m_{nj}] \neq 0$ (for a proof refer to [12]).

For $\mathbf{C} = \mathbf{0}$ from (6.5) we obtain

$$\mathbf{M}\ddot{\mathbf{q}} + \mathbf{K}\mathbf{q} = \mathbf{0}, \tag{6.8}$$

and let us consider two square forms

$$(\mathbf{Mq} \circ \mathbf{q}) = \sum_{n,j=1}^{N} m_{nj} q_n q_j,$$

$$(\mathbf{Kq} \circ \mathbf{q}) = \sum_{n,j=1}^{N} k_{nj} q_n q_j, \tag{6.9}$$

which are positive-definite. It is known from linear algebra that if at least one of the forms (6.9) is positive-definite, then there exists a real non-singular change of the variables

$$\mathbf{q} = \mathbf{U}\boldsymbol{\psi}, \tag{6.10}$$

where $\det \mathbf{U} \neq 0$, $\boldsymbol{\psi}^{\mathrm{T}} = (\psi_1, \psi_2, \ldots, \psi_N)^T$, which makes it possible to reduce (6.9) to their counterpart forms

$$(\mathbf{Mq} \circ \mathbf{q}) = \sum_{k=1}^{N} \psi_k^2, \qquad (\mathbf{Kq} \circ \mathbf{q}) = \sum_{k=1}^{N} \sigma_k \psi_k^2. \tag{6.11}$$

The change of variables

$$\mathbf{q} = \sum_{k=1}^{N} \psi_k \mathbf{u}_k, \tag{6.12}$$

allows us to introduce the normalization procedure

$$(\mathbf{Mu}_n \circ \mathbf{u}_j) = \delta_{nj}, \tag{6.13}$$

where δ_{nj} is the Kronecker symbol. Differentiation of (6.10) yields

$$\dot{\mathbf{q}} = \mathbf{U}\dot{\boldsymbol{\psi}}. \tag{6.14}$$

Using $\dot{\mathbf{q}}$ and $\dot{\boldsymbol{\psi}}$ instead of $\mathbf{q}$ and $\boldsymbol{\psi}$ in the first equation of (6.11), (6.11) can be read as the following positive definite forms:

$$2T = \sum_{k=1}^{N} \dot{\psi}_k^2, \qquad 2V = \sum_{k=1}^{N} \sigma_k \psi_k^2. \tag{6.15}$$

Owing to the introduced *normal coordinate* ψ_k, the obtained simple form of kinetic T and potential V energies allows to cast (6.8) in the form

$$\ddot{\psi}_k + \sigma_k \psi_k = 0, \qquad k = 1, \ldots, N, \tag{6.16}$$

where all σ_k are positive. This means that each equation of (6.16) governs the harmonic oscillations of a one-degree-of-freedom autonomous conservative oscillator, and its oscillations have the form

$$\psi_k = C_k \sin(\omega_k t + \beta_k), \qquad k = 1, \ldots, N, \tag{6.17}$$

where the frequency $\omega_k = \sqrt{\sigma_k}$, and C_k and β_k are arbitrary constants. Substituting (6.17) into (6.12) yields

$$\mathbf{q} = \sum_{k=1}^{N} C_k \mathbf{u}_k \sin(\omega_k + \beta_k). \tag{6.18}$$

Let us assume that only $C_i \neq 0$. Then (6.18) allows us to defined the so-called *ith normal form* or *ith mode* of oscillations

$$\mathbf{q}_i = C_i \mathbf{u}_i \sin(\omega_i t + \beta_i). \tag{6.19}$$

If we introduce the initial conditions in such a way that the ith mode of oscillations is realized, then all generalized coordinates oscillate harmonically with the same frequency ω_i. The geometrical properties of the modes are defined via the coefficients of normal modes as a ratio of vectors $\mathbf{u}_i$, which can be normalized. Assuming the solution

$$\mathbf{q} = \mathbf{u}e^{i\sigma t}, \tag{6.20}$$

and substituting it into (6.8) we get

$$(\mathbf{K} - \sigma\mathbf{M})\mathbf{u} = \mathbf{0}, \tag{6.21}$$

and hence the algebraic equation

$$\det(\mathbf{K} - \sigma\mathbf{M}) = 0 \tag{6.22}$$

yields $\sigma_j = j = 1, \ldots, N$, where $\sigma_j = \omega_j^2$. On the other hand, each of the obtained σ_j eigenvalues allows us to define the associated vector $\mathbf{u}_j$ from (6.21).

Here the problem of normal forms of linear vibrations has been only briefly addresed since it has been widely described in numerous books/monographs devoted to oscillations of lumped mechanical systems.

6.3 Classification and Properties of Linear Mechanical Forces

The matrix notation of linear equations of motion for lumped mechanical system (6.5) will be used to introduce the classification of forces.

The matrix of forces of resistance to motion $\mathbf{C}$ can always be represented in the form

$$\mathbf{C} = \mathbf{C}^S + \mathbf{C}^A, \tag{6.23}$$

where

$$\left(\mathbf{C}^S\right)^{\mathrm{T}} = \mathbf{C}^S, \qquad \left(\mathbf{C}^A\right)^{\mathrm{T}} = -\mathbf{C}^A, \tag{6.24}$$

which means that $\mathbf{C}^S$ ($\mathbf{C}^A$) is a symmetric (skew-symmetric) matrix.

Forces of the form $\mathbf{C}^S\dot{\mathbf{q}}$ generated by the symmetric matrix $\mathbf{C}^S$ are called *dissipative forces*, provided that $\dot{\mathbf{q}}^{\mathrm{T}}\mathbf{C}^S\dot{\mathbf{q}} \geq 0$ for an arbitrary vector $\dot{\mathbf{q}} \neq 0$ (a strict inequality describes the forces of *complete dissipation*).

Let us introduce the function

$$R = \frac{1}{2}\dot{\mathbf{q}}^{\mathrm{T}}\mathbf{C}^S\dot{\mathbf{q}}, \tag{6.25}$$

called a *Rayleigh*[3] *dissipation function* [10]. The presented quadratic form has a very important property, that is,

$$\frac{\partial R}{\partial \dot{\mathbf{q}}} = \mathbf{C}^S\dot{\mathbf{q}}, \tag{6.26}$$

which will be demonstrated for a two-element vector $\mathbf{q}$.

We have

$$\begin{aligned} R &= \frac{1}{2}[\dot{q}_1, \dot{q}_2]\begin{bmatrix} c_{11} & c_{12} \\ c_{21} & c_{22} \end{bmatrix}\begin{bmatrix} \dot{q}_1 \\ \dot{q}_2 \end{bmatrix} \\ &= \frac{1}{2}[\dot{q}_1 c_{11} + \dot{q}_2 c_{21}, \ \dot{q}_1 c_{12} + \dot{q}_2 c_{22}]\begin{bmatrix} \dot{q}_1 \\ \dot{q}_2 \end{bmatrix} \\ &= \frac{1}{2}[\dot{q}_1^2 c_{11} + \dot{q}_1\dot{q}_2 c_{21} + \dot{q}_1\dot{q}_2 c_{12} + \dot{q}_2^2 c_{22}], \end{aligned}$$

$$\frac{\partial R}{\partial \dot{\mathbf{q}}} = \begin{bmatrix} \dfrac{\partial R}{\partial \dot{q}_1} \\ \dfrac{\partial R}{\partial \dot{q}_2} \end{bmatrix} = \begin{bmatrix} c_{11} & c_{12} \\ c_{21} & c_{22} \end{bmatrix}\begin{bmatrix} \dot{q}_1 \\ \dot{q}_2 \end{bmatrix},$$

[3]John Rayleigh (1842–1919), English physicist awarded the Nobel prize in 1904.

hence

$$\frac{\partial R}{\partial \dot{q}_1} = \dot{q}_1 c_{11} + \frac{1}{2}\dot{q}_2(c_{21} + c_{12}) = \dot{q}_1 c_{11} + \dot{q}_2 c_{12},$$
$$\frac{\partial R}{\partial \dot{q}_2} = \dot{q}_2 c_{22} + \frac{1}{2}\dot{q}_1(c_{21} + c_{12}) = \dot{q}_1 c_{12} + \dot{q}_2 c_{22},$$

because for the considered matrix $\mathbf{C}^S = \mathbf{C}$, which means that $c_{12} = c_{21}$.

The forces $\mathbf{C}^A\dot{\mathbf{q}}$ are called *gyroscopic forces*. Below we present their properties:

1. The power of gyroscopic forces is equal to $N = \dot{\mathbf{q}}^{\mathrm{T}}\mathbf{C}^A\dot{\mathbf{q}} = 0$. Let us note that

$$N^T = \left(\dot{\mathbf{q}}^{\mathrm{T}}\mathbf{C}^A\dot{\mathbf{q}}\right)^{\mathrm{T}} = \left[\mathbf{q}^T\left(\mathbf{C}^A\dot{\mathbf{q}}\right)\right]^{\mathrm{T}} = \left(\mathbf{C}^A\dot{\mathbf{q}}\right)^{\mathrm{T}}\dot{\mathbf{q}}$$
$$= \dot{\mathbf{q}}^{\mathrm{T}}\left(\mathbf{C}^A\right)^{\mathrm{T}}\dot{\mathbf{q}} = -\dot{\mathbf{q}}^{\mathrm{T}}\mathbf{C}^A\dot{\mathbf{q}} = -N, \tag{6.27}$$

   where the definition of gyroscopic forces and the second equation of (6.24) were used. Equality (6.27) is satisfied only if $N = 0$.
2. Generalized potential of gyroscopic forces.

   Gyroscopic forces have the following generalized potential:

$$V(\mathbf{q}, \dot{\mathbf{q}}) = \frac{1}{2}\mathbf{q}^T\mathbf{C}^A\dot{\mathbf{q}}, \tag{6.28}$$

   since we obtain

$$\frac{\mathrm{d}}{\mathrm{d}t}\frac{\partial V}{\partial \dot{\mathbf{q}}} - \frac{\partial V}{\partial \mathbf{q}} = -\mathbf{C}^A\dot{\mathbf{q}}, \tag{6.29}$$

   which will be demonstrated for a two-element vector $\mathbf{q}$.

   We have

$$V = \frac{1}{2}[q_1, q_2]\begin{bmatrix} c_{11} & c_{12} \\ c_{21} & c_{22} \end{bmatrix}\begin{bmatrix} \dot{q}_1 \\ \dot{q}_2 \end{bmatrix}$$
$$= \frac{1}{2}[\dot{q}_1 q_1 c_{11} + \dot{q}_1 q_2 c_{21} + q_1 \dot{q}_2 c_{12} + q_2 \dot{q}_2 c_{22}],$$
$$\frac{\partial V}{\partial \mathbf{q}} = \begin{bmatrix} \frac{1}{2}(\dot{q}_1 c_{11} + \dot{q}_2 c_{12}) \\ \frac{1}{2}(\dot{q}_1 c_{21} + \dot{q}_2 c_{22}) \end{bmatrix},$$
$$\frac{\partial V}{\partial \dot{\mathbf{q}}} = \begin{bmatrix} \frac{1}{2}(q_1 c_{11} + q_2 c_{21}) \\ \frac{1}{2}(q_1 c_{12} + q_2 c_{22}) \end{bmatrix},$$
$$\frac{\mathrm{d}}{\mathrm{d}t}\frac{\partial V}{\partial \dot{\mathbf{q}}} = \begin{bmatrix} \frac{1}{2}(\dot{q}_1 c_{11} + \dot{q}_2 c_{21}) \\ \frac{1}{2}(\dot{q}_1 c_{12} + \dot{q}_2 c_{22}) \end{bmatrix},$$

$$\frac{\mathrm{d}}{\mathrm{d}t}\frac{\partial V}{\partial \dot{\mathbf{q}}} - \frac{\partial V}{\partial \mathbf{q}} = \begin{bmatrix} \frac{1}{2}\dot{q}_2(c_{21} - c_{12}) \\ \frac{1}{2}\dot{q}_1(c_{12} - c_{21}) \end{bmatrix},$$

$$\mathbf{C}^A\dot{\mathbf{q}} = \begin{bmatrix} c_{11} & c_{12} \\ c_{21} & c_{22} \end{bmatrix}\begin{bmatrix} \dot{q}_1 \\ \dot{q}_2 \end{bmatrix} = \begin{bmatrix} c_{11}\dot{q}_1 + c_{12}\dot{q}_2 \\ c_{21}\dot{q}_1 + c_{22}\dot{q}_2 \end{bmatrix},$$

which proves the validity of the formula, because in this case $c_{11} = -c_{11} = 0$, $c_{22} = -c_{22} = 0$, $c_{21} = -c_{12}$.

Also the matrix of positional forces $\mathbf{K}$ can be represented as the sum of a symmetric and skew-symmetric matrix of the form

$$\begin{aligned} \mathbf{K} &= \mathbf{K}^S + \mathbf{K}^A, \\ \left(\mathbf{K}^S\right)^{\mathrm{T}} &= \mathbf{K}^S, \\ \left(\mathbf{K}^A\right)^{\mathrm{T}} &= -\mathbf{K}^A. \end{aligned} \tag{6.30}$$

Forces $\mathbf{K}^S\mathbf{q}$ are called *conservative forces* or *potential forces*. Potential forces are associated with the potential in the following way:

$$\mathbf{K}^S\mathbf{q} = \frac{\partial V}{\partial \mathbf{q}}, \tag{6.31}$$

where

$$V = \frac{1}{2}\mathbf{q}^T\mathbf{K}^S\mathbf{q}. \tag{6.32}$$

They have the following property. The work of those forces in a configuration space along a certain closed curve is equal to zero, that is,

$$\oint K_{nj}q_j \mathrm{d}q_n = 0. \tag{6.33}$$

Forces $\mathbf{K}^A\mathbf{q}$ are called *circulatory forces*. They are orthogonal to the vector of generalized coordinates, that is,

$$\mathbf{q}^T\mathbf{K}^A\mathbf{q} = 0. \tag{6.34}$$

For the planar case (a two-element vector $\mathbf{q}$) a vector field of circulatory forces consists of circles, which is depicted in Fig. 6.1.

Example 6.1. Let us consider a particle with mass m connected to a ring by means of linear springs of stiffnesses k_1 and k_2 and viscous damping c, where the ring rotates with angular velocity ω (Fig. 6.2).

We assume small linear vibrations, and non-linear geometric relations are neglected.

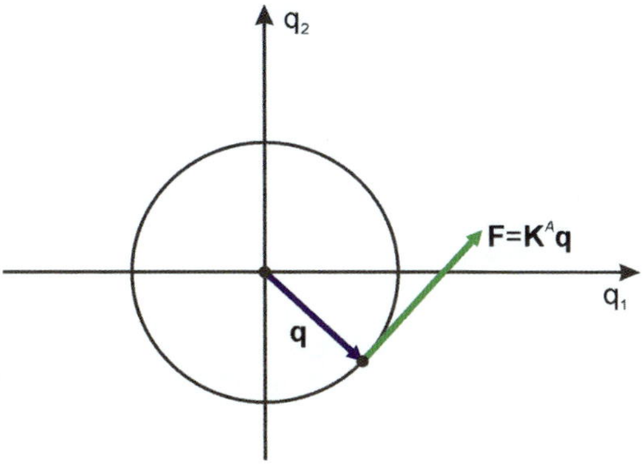

Fig. 6.1 Interpretation of property (6.34) of circulatory forces

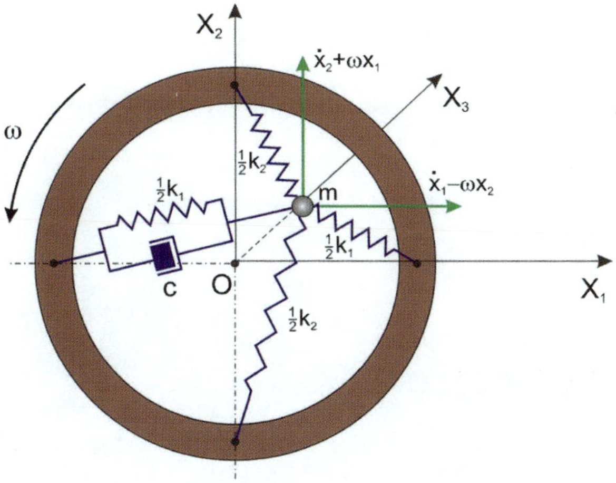

Fig. 6.2 Mass m fixed by means of spring-damper elements to ring rotating with angular velocity ω

Let us introduce the Cartesian coordinate system $OX_1X_2X_3$ rigidly connected to a ring, where the mass oscillates in the OX_1X_2 plane.

The kinetic energy of the mass in the system $OX_1X_2X_3$ is equal to

$$T = \frac{1}{2}m\left[(\dot{x}_1 - \omega x_2)^2 + (\dot{x}_2 + \omega x_1)^2\right] = T_2 + T_1 + T_0,$$

where

$$T_1 = m\omega(x_1\dot{x}_2 - \dot{x}_1 x_2),$$

$$T_2 = \frac{1}{2}m(\dot{x}_1^2 - \dot{x}_2^2)$$

$$T_0 = \frac{1}{2}m\omega^2(x_1^2 + x_2^2).$$

Potential energy accumulated in the springs is equal to

$$V = \frac{1}{2}k_1 x_1^2 + \frac{1}{2}k_2 x_2^2,$$

and the dissipation function has the form

$$R = \frac{1}{2}c\dot{x}_1^2.$$

Now, let us use Lagrange's equations of the form

$$\frac{\mathrm{d}}{\mathrm{d}t}\left(\frac{\partial L}{\partial \dot{q}_i}\right) + \frac{\partial R}{\partial \dot{q}_i} - \frac{\partial L}{\partial q_i} = Q_i, \quad i = 1, 2,$$

where

$$\begin{aligned} L = T - V =& \frac{1}{2}m(\dot{x}_1^2 + \dot{x}_2^2) + m\omega(x_1\dot{x}_2 - \dot{x}_1 x_2) \\ &+ \frac{1}{2}m\omega^2(x_1^2 + x_2^2) - \frac{1}{2}k_1 x_1^2 - \frac{1}{2}k_2 x_2^2 \end{aligned}$$

and

$$q_1 = x_1, \quad q_2 = x_2.$$

Let us successively calculate

$$\begin{aligned} &\frac{\partial L}{\partial \dot{x}_1} = m\dot{x}_1 - m\omega x_2, && \frac{\partial L}{\partial \dot{x}_2} = m\dot{x}_2 + m\omega x_1, \\ &\frac{\mathrm{d}}{\mathrm{d}t}\left(\frac{\partial L}{\partial \dot{x}_1}\right) = m\ddot{x}_1 - m\omega\dot{x}_2, && \frac{\mathrm{d}}{\mathrm{d}t}\left(\frac{\partial L}{\partial \dot{x}_2}\right) = m\ddot{x}_2 - m\omega\dot{x}_1, \\ &\frac{\partial R}{\partial \dot{x}_1} = c\dot{x}_1, && \frac{\partial R}{\partial \dot{x}_2} = 0, \\ &\frac{\partial L}{\partial x_1} = m\omega\dot{x}_2 + m\omega^2 x_1 - k_1 x_1, && \frac{\partial L}{\partial x_2} = -m\omega\dot{x}_1 + m\omega^2 x_2 - k_2 x_2, \end{aligned}$$

which after substitution into Lagrange equations gives

$$\begin{aligned} m\ddot{x}_1 - 2m\omega\dot{x}_2 + c\dot{x}_1 - m\omega^2 x_1 + k_1 x_1 &= 0, \\ m\ddot{x}_2 + 2m\omega\dot{x}_1 - m\omega^2 x_2 + k_2 x_2 &= 0. \end{aligned} \tag{$*$}$$

The obtained ODEs can be rewritten in the form (Sect. 6.3)

$$\mathbf{M}\ddot{\mathbf{q}} + \left(\mathbf{C}^A + \mathbf{C}^S\right)\dot{\mathbf{q}} + (\mathbf{K}_O + \mathbf{K}_S)\,\mathbf{q} = \mathbf{0},$$

where particular matrices have the forms

$$\mathbf{M} = \begin{bmatrix} m & 0 \\ 0 & m \end{bmatrix}, \quad \mathbf{C}^S = \begin{bmatrix} c & 0 \\ 0 & 0 \end{bmatrix}, \quad \mathbf{C}^A = \begin{bmatrix} 0 & -2m\omega \\ 2m\omega & 0 \end{bmatrix},$$

$$\mathbf{K}_S = \begin{bmatrix} k_1 & 0 \\ 0 & k_2 \end{bmatrix}, \quad \mathbf{K}_O = \begin{bmatrix} -m\omega^2 & 0 \\ 0 & -m\omega^2 \end{bmatrix} = -\omega^2\mathbf{M},$$

$$\mathbf{K}^S = \mathbf{K}_O + \mathbf{K}_S.$$

Matrix $\mathbf{M}$ is the mass matrix, matrix $\mathbf{C}^S$ is the damping matrix generated by the Rayleigh dissipation function, $\mathbf{C}^A$ is the skew-symmetric matrix generating gyroscopic forces that couples the motion of the mass in two directions, $\mathbf{K}_S$ is the symmetric stiffness matrix, and $\mathbf{K}_O = -\omega^2\mathbf{M}$ is a matrix generating circulatory forces (centrifugal forces).

Using the definitions of matrices introduced previously, by those matrices we can express particular terms of total energy of the system under investigation, that is,

$$T_2 = \tfrac{1}{2}\dot{\mathbf{q}}^{\mathrm{T}}\mathbf{M}\dot{\mathbf{q}}, \qquad T_1 = \frac{1}{2}\mathbf{q}^{\mathrm{T}}\mathbf{C}^A\dot{\mathbf{q}}, \qquad T_0 = \frac{\omega^2}{2}\mathbf{q}^T\mathbf{M}\mathbf{q},$$

$$V = \tfrac{1}{2}\mathbf{q}^T\mathbf{K}^S\mathbf{q}, \qquad R = \frac{1}{2}\dot{\mathbf{q}}^{\mathrm{T}}\mathbf{C}^S\dot{\mathbf{q}}.$$

In order to highlight the special properties of the analyzed system described by $(*)$ let us consider a special case for which $c = 0$, $k_1 = k_2 = k$.

From $(*)$ we obtain

$$\ddot{x}_1 - 2\omega\dot{x}_2 + (\alpha^2 - \omega^2)x_1 = 0,$$

$$\ddot{x}_2 + 2\omega\dot{x}_1 + (\alpha^2 - \omega^2)x_2 = 0, \qquad (**)$$

where $\alpha^2 = \frac{k}{m}$.

In this case gyroscopic forces play a stabilizing role in the investigated system, although at first glance it seems that the system would exhibit unstable properties since so-called negative stiffnesses appear in the system for $\omega^2 > \alpha^2$. To verify the system's behavior, we seek its solutions in the form

$$x_1 = X_1 e^{\sigma t}, \quad x_2 = X_2 e^{\sigma t}.$$

Substituting the preceding solutions into $(**)$ we obtain

$$\begin{bmatrix} \sigma^2 + (\alpha^2 - \omega^2) & -2\omega\sigma \\ 2\omega\sigma & \sigma^2 + (\alpha^2 - \omega^2) \end{bmatrix} \begin{bmatrix} X_1 \\ X_2 \end{bmatrix} = 0.$$

A characteristic equation takes the form

$$\sigma^4 + 2\sigma^2(\alpha^2 + \omega^2) + (\alpha^2 - \omega^2)^2 = 0.$$

This equation has the following roots:

$$\sigma_1^2 = -(\alpha - \omega)^2, \quad \sigma_2^2 = -(\alpha + \omega)^2.$$

From the preceding calculations it follows that four roots are always imaginary for any value of ω, and therefore there is no instability.

Let us assume now that the system has only one degree of freedom, that is, let the mass m move only along the axis OX_1. Setting $x_2 = \dot{x}_2 = \ddot{x}_2 = 0$ in $(**)$ from the first equation we have

$$\ddot{x}_1 + (\alpha^2 - \omega^2)x_1 = 0,$$

which describes the motion of the mass along the axis OX_1. However, in this case for $\omega > \alpha$ instabilities of vibrations appear in association with the so-called "negative stiffness."

It follows that enabling the mass to also move along the axis OX_2 allows for the elimination of vibrational instabilities.

6.4 Small Vibrations of Linear One-Degree-of-Freedom Systems

Let us consider the case of forced vibrations of a system with one degree of freedom (Fig. 6.3).

An equation of motion for the system takes the form

$$m\ddot{x} + c\dot{x} + kx = F_0 \cos \omega t \tag{6.35}$$

or

$$\ddot{x} + 2h\dot{x} + \alpha^2 x = q \cos \omega t, \tag{6.36}$$

where

$$\frac{c}{m} = 2h, \quad \alpha^2 = \frac{k}{m}, \quad q = \frac{F_0}{m}.$$

First, let us consider free vibrations of the system ($F_0 = 0$). We seek a solution in the form

$$x(t) = e^{rt}, \tag{6.37}$$

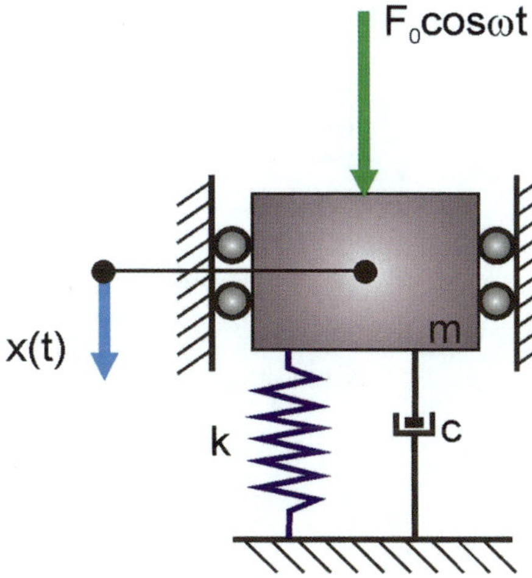

Fig. 6.3 Vibrations of a one-degree-of-freedom system with damping and harmonic excitation

which, substituted into (6.36), leads to the following characteristic equation:

$$r^2 + 2hr + \alpha^2 = 0. \tag{6.38}$$

The roots of this equation are given by

$$r_{1,2} = -h \pm \sqrt{h^2 - \alpha^2}. \tag{6.39}$$

Therefore, a general solution has the form

$$x(t) = A_1 e^{r_1 t} + A_1 e^{r_1 t}. \tag{6.40}$$

If $h > \alpha$, then the solution reads

$$x(t) = e^{-ht}\left(A_1 e^{\sqrt{h^2-\alpha^2 t}} + A_2 e^{-\sqrt{h^2-\alpha^2 t}}\right), \tag{6.41}$$

and as can be easily noticed, $\lim_{t\to\infty} x(t) = 0$. A time response $x(t)$ tends to zero without vibrations.

If $h < \alpha$, then (6.39) takes the form

$$r_{1,2} = -h \pm i\lambda, \qquad i^2 = -1, \tag{6.42}$$

where $\lambda = \sqrt{\alpha^2 - h^2}$, and the solution in this case has the form

$$x(t) = \frac{1}{2} e^{-ht} \mathrm{Re}\left[\bar{A}e^{i\lambda t} + Ae^{i\lambda t}\right], \tag{6.43}$$

where $\bar{A}$ and A are complex conjugates of each other.

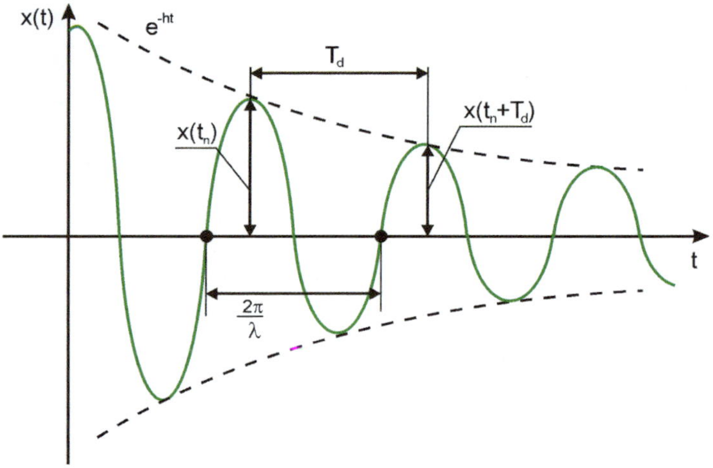

Fig. 6.4 Free damped vibrations of one-degree-of-freedom system

The transformation to a real form of the solution is enabled by the following Euler formula:

$$e^{i\lambda t} = \cos\lambda t + i\sin\lambda t. \tag{6.44}$$

We have

$$\begin{aligned} x(t) &= \frac{1}{2}e^{-ht}\mathrm{Re}[(A_R - A_I i)(\cos\lambda t + i\sin\lambda t) \\ &\quad + (A_R + A_I i)(\cos\lambda t - i\sin\lambda t)] \\ &= e^{-ht}(A_1\cos\lambda t + A_2\sin\lambda t), \end{aligned} \tag{6.45}$$

where $A = A_R + A_I i$, $A_R = A_1$, $A_I = A_2$.

The third case remains to be considered, that is, $h = \alpha$, which corresponds to a critical damping coefficient c_{cr} described by the equation

$$c_{cr} = 2m\sqrt{\frac{k}{m}} = 2\sqrt{km}. \tag{6.46}$$

In this case we are dealing with a double root of the characteristic equation (see [11]), and the solution has the form

$$x(t) = (A_1 + A_2 t)e^{-ht}. \tag{6.47}$$

In the case of critical damping we do not observe any vibrations. Solution (6.45) describes the process of damped harmonic vibrations whose time plot is shown in Fig. 6.4.

A characteristic of damped vibrations is that the maximum (minimum) deflections are reached periodically after the time $T_d = \frac{2\pi}{\lambda}$ called a *period of damped vibrations*. Therefore, we introduce the notion of a *logarithmic decrement* in the form

$$\begin{aligned}\delta &= \ln \frac{x(t)}{x(t+T_d)} \\ &= \ln \frac{(A_1 \cos \lambda t_n + A_2 \sin \lambda t_n)\mathrm{e}^{-ht_n}}{\left[A_1 \cos \lambda \left(t_n + \frac{2\pi}{\lambda}\right) + A_2 \sin \lambda \left(t_n + \frac{2\pi}{\lambda}\right)\right] \mathrm{e}^{-h\left(t_n + \frac{2\pi}{\lambda}\right)}} \\ &= \ln \frac{1}{\mathrm{e}^{-h\frac{2\pi}{\lambda}}} = \frac{2\pi}{\lambda} h. \end{aligned} \tag{6.48}$$

The logarithmic decrement can be used to determine a viscous damping coefficient of vibrations. Since, if we know δ, that is, the natural logarithm of the ratio of two consecutive maximum deflections and the time interval between occurrences of those deflections, from (6.48) we find the value of damping:

$$c = 2m \frac{\delta}{T_d}. \tag{6.49}$$

In the end, let us consider the case of forced vibrations, that is, $F_0 \neq 0$. To determine a solution we will exploit the notion of complex numbers. Equation (6.36) takes the form

$$\ddot{x} + 2h\dot{x} + \alpha^2 x = q(\cos \omega t + \mathrm{i} \sin \omega t) = q\mathrm{e}^{\mathrm{i}\omega t}. \tag{6.50}$$

The preceding second-order differential equation is non-homogeneous. Its solution is the sum (superposition) of a general solution of the homogeneous differential equation [i.e., (6.50) for $q = 0$] and a particular solution of non-homogeneous (6.50). This latter solution is sought in the form

$$x = \bar{A}\mathrm{e}^{\mathrm{i}\omega t}. \tag{6.51}$$

Substituting (6.51) into (6.50) we obtain

$$\left(-\omega^2 + 2h\omega i + \alpha^2\right) \bar{A} = q, \tag{6.52}$$

where $\bar{A}$ is the complex number conjugate to $A = A_R + A_I i$.

From the preceding equation we get

$$\bar{A} = A_R - A_I i = \frac{q}{(\alpha^2 - \omega^2) + 2h\omega i}. \tag{6.53}$$

In order to determine A_R and A_I, let us multiply the numerator and the denominator of the right-hand side of (6.53) by $[(\alpha^2-\omega^2)-2h\omega i]$, thereby obtaining

$$A_R - A_I i = \frac{q(\alpha^2-\omega^2)}{(\alpha^2-\omega^2)^2+4h^2\omega^2} - \frac{2h\omega q}{(\alpha^2-\omega^2)^2+4h^2\omega^2} i, \tag{6.54}$$

and hence we have

$$A_R = \frac{q(\alpha^2-\omega^2)}{(\alpha^2-\omega^2)^2+4h^2\omega^2}, \qquad A_I = \frac{2h\omega q}{(\alpha^2-\omega^2)^2+4h^2\omega^2}. \tag{6.55}$$

Solution (6.51) has a real interpretation in the form

$$\begin{aligned} \operatorname{Re} x(t) &= \operatorname{Re}[(A_R - A_I i)(\cos\omega t + \mathrm{i}\sin\omega t)] \\ &= A_R\cos\omega t + A_I\sin\omega t = a\cos(\omega t - \beta). \end{aligned} \tag{6.56}$$

Because

$$A_R\cos\omega t + A_I\sin\omega t = a\cos\beta\cos\omega t + a\sin\beta\sin\omega t, \tag{6.57}$$

we have

$$A_R = a\cos\beta, \qquad A_I = a\sin\beta, \tag{6.58}$$

and hence

$$a = \sqrt{A_R^2 + A_I^2}, \qquad \tan\beta = \frac{A_I}{A_R}. \tag{6.59}$$

Because the general solution of the homogeneous equation of the form (6.45) vanishes for $t \to \infty$, only the particular solution of the non-homogeneous equation of the form (6.56) remains. As can be seen, the response of the system is harmonic and shifted in phase by the angle β with respect to a driving force. From (6.59) we obtain

$$a = \frac{q}{\sqrt{(\alpha^2-\omega^2)^2+4h^2\omega^2}} = \frac{q}{\alpha^2}\frac{1}{\sqrt{\left(1-\left(\frac{\omega}{\alpha}\right)^2\right)^2 + 4\left(\frac{h}{\alpha}\right)^2\left(\frac{\omega}{\alpha}\right)^2}},$$

$$\beta = \arctan\frac{2h\omega}{\alpha^2-\omega^2} = \arctan\frac{2\frac{h}{\alpha}\frac{\omega}{\alpha}}{1-\left(\frac{\omega}{\alpha}\right)^2}. \tag{6.60}$$

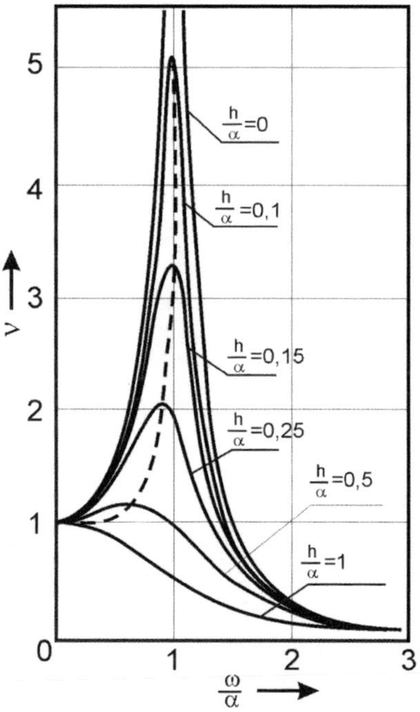

Fig. 6.5 Amplitude response of an oscillator with harmonic excitation and damping for different values of ratio h/α

The coefficient $q/\alpha^2 \equiv F_0/k = x_{st}$ describes a static deflection, and the first of (6.60) is transformed into

$$\nu \equiv \nu\left(\frac{\omega}{\alpha}\right) = \frac{a}{x_{st}} = \frac{1}{\sqrt{\left(1-\left(\frac{\omega}{\alpha}\right)^2\right)^2 + 4\left(\frac{h}{\alpha}\right)^2\left(\frac{\omega}{\alpha}\right)^2}}. \tag{6.61}$$

The preceding function $\nu = \nu(\frac{\omega}{\alpha})$ describes the *amplitude response* (Fig. 6.5).

The second equation of system (6.60) describes a *phase response* (Fig. 6.6).

We obtain the case of vibrations of an undamped oscillator with a harmonic excitation after setting $h = 0$. According to (6.56) we have

$$x(t) = a\cos\omega t \tag{6.62}$$

for $\beta = 0$ and $\omega < \alpha$ and

$$x(t) = -a\cos\omega t \tag{6.63}$$

for $\omega > \alpha$. This means that before resonance the forced vibrations are in phase with the external driving force, whereas following resonance the vibrations of the mass are out of phase with the driving force.

The *resonance* takes place when $\omega = \alpha$, and then the amplitude goes to infinity. Damping decreases amplitudes of resonant vibrations. Also the case $h = 0$ is

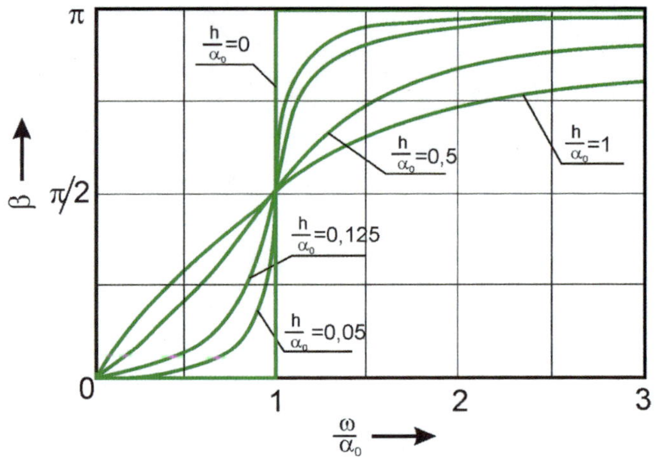

Fig. 6.6 Phase response of an oscillator with harmonic excitation and damping for different values of ratio h/α

depicted in Figs. 6.5 and 6.6. Now we will show the method of application of a function of a complex variable to solving the problem of vibrations of the system depicted in Fig. 6.3 when it is subjected to two driving forces $F\cos\omega t$ and $F\sin\omega t$.

The equations of vibrations of the oscillator have the form

$$\ddot{x}_1 + 2h\dot{x}_1 + \alpha^2 x_1 = q\cos\omega t,$$
$$\ddot{x}_2 + 2h\dot{x}_2 + \alpha^2 x_2 = q\sin\omega t. \tag{6.64}$$

The solutions of the preceding equations are, respectively,

$$x_1 = a\cos(\omega t - \beta),$$
$$x_2 = a\sin(\omega t - \beta) = a\cos\left(\omega t - \frac{\pi}{2} - \beta\right), \tag{6.65}$$

where for both cases the amplitude and phase are described by (6.60).

Let us now introduce a *complex excitation* of the form

$$F = F_R + \mathrm{i}F_I = F(\cos\omega t + \mathrm{i}\sin\omega t) = F\mathrm{e}^{\mathrm{i}\omega t}, \tag{6.66}$$

and then the equation of vibrations of the analyzed oscillator takes the form

$$\ddot{x} + 2h\dot{x} + \alpha^2 x = q\mathrm{e}^{\mathrm{i}\omega t}, \tag{6.67}$$

where now x is the complex variable and $q = F/m$.

A particular solution of (6.67) has the form

$$x = a\mathrm{e}^{\mathrm{i}(\omega t-\beta)} = a\mathrm{e}^{\mathrm{i}\omega t}\mathrm{e}^{-\mathrm{i}\beta}, \tag{6.68}$$

and substituting expression (6.68) into (6.67) we obtain

$$a(-\omega^2 + 2h\omega i + \alpha^2 q)\mathrm{e}^{-\mathrm{i}\beta} = q \tag{6.69}$$

or, in expanded form,

$$a(-\omega^2 + 2h\omega i + \alpha^2 q)(\cos\beta - i\sin\beta) = q. \tag{6.70}$$

Splitting the preceding equation into real and imaginary parts, we obtain

$$\begin{aligned} a(-\omega^2\cos\beta + 2h\omega\sin\beta + \alpha^2 q\cos\beta) &= q,\\ \omega^2\sin\beta + 2h\omega\cos\beta - \alpha^2 q\sin\beta &= 0. \end{aligned} \tag{6.71}$$

The solution of this system gives the values of the amplitude and phase angle β described by (6.60).

From (6.68) it follows that

$$x = x_R + \mathrm{i}x_I, \tag{6.72}$$

where

$$x_R = a\cos(\omega t - \beta), \quad x_I = a\sin(\omega t - \beta). \tag{6.73}$$

From those calculations it follows that the solution of the first (second) of (6.64) is x_R (x_I). In other words the real part of solution (6.72) corresponds to the real part of the force (excitation) and the imaginary part of solution (6.72) corresponds to the imaginary part of the force.

We will now present the geometric interpretation of solution (6.72). Let vector $\boldsymbol{a}$ rotate around point O with a constant angular velocity ω.

According to Fig. 6.7 we have

$$\begin{aligned} \boldsymbol{a} = \boldsymbol{a}_R + \mathrm{i}\boldsymbol{a}_I &= a\cos(\omega t - \beta)\mathbf{E}_R + \mathrm{i}a\sin(\omega t - \beta)\mathbf{E}_I\\ &= x_R\mathbf{E}_R + ix_I\mathbf{E}_I. \end{aligned} \tag{6.74}$$

From (6.74) it follows that the projection of vector $\boldsymbol{a}$ onto the horizontal real axis represents the real part of the solution, whereas the projection of the rotating vector $\boldsymbol{a}$ onto the vertical axis represents the imaginary part of the solution.

Two successive differentiations of the rotating vector $\boldsymbol{a}$ lead to the determination of the velocity and acceleration of the analyzed oscillator, and their projections onto the real and imaginary axes give the velocities and accelerations of the oscillator with driving forces $F\cos\omega t$ and $F\sin\omega t$.

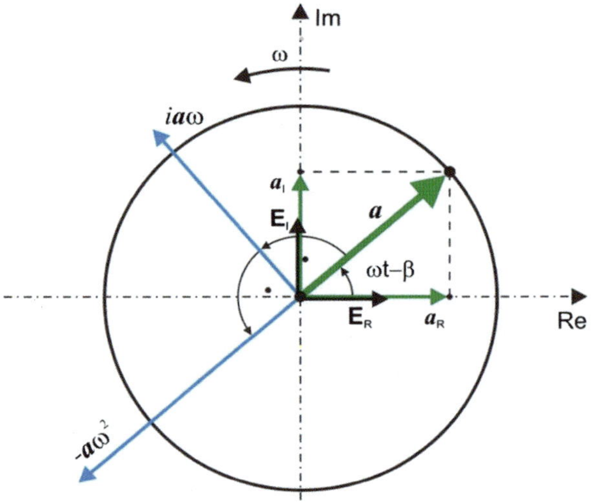

Fig. 6.7 Vector $\boldsymbol{a}$ rotating with constant angular velocity ω

According to relation (6.74) we have

$$\dot{\boldsymbol{a}} = \mathrm{i}\boldsymbol{a}\omega \mathrm{e}^{\mathrm{i}(\omega t-\beta)} = -a\omega\sin(\omega t-\beta)\mathbf{E}_R + \mathrm{i}a\omega\cos(\omega t-\beta)\mathbf{E}_I$$
$$= -\dot{x}_R\mathbf{E}_R + \mathrm{i}\dot{x}_I\mathbf{E}_I. \tag{6.75}$$

We will demonstrate that vectors $\dot{\boldsymbol{a}}$ and $\boldsymbol{a}$ are perpendicular to each other. We have

$$\boldsymbol{a}\circ\dot{\boldsymbol{a}} = (x_R\mathbf{E}_R + \mathrm{i}x_I\mathbf{E}_I)\circ(-\dot{x}_R\mathbf{E}_R + \mathrm{i}\dot{x}_I\mathbf{E}_I)$$
$$= -\,x_R\dot{x}_R - x_I\dot{x}_I = -a\cos(\omega t-\beta)(-a\omega\sin(\omega t-\beta))$$
$$-\,a\sin(\omega t-\beta)(a\omega\cos(\omega t-\beta)) = 0. \tag{6.76}$$

Moreover, vector $\dot{\boldsymbol{a}}$ leads vector $\boldsymbol{a}$ through an angle $\frac{\pi}{2}$. A similar situation occurs for the acceleration $\ddot{\boldsymbol{a}}$, which leads vector $\dot{\boldsymbol{a}}$ through an angle $\frac{\pi}{2}$, and projections of this vector onto the axes give the accelerations of the oscillator for the case of $F\cos\omega t$ and $F\sin\omega t$. The relative orientation of vectors $\boldsymbol{a}$, $\dot{\boldsymbol{a}}$, and $\ddot{\boldsymbol{a}}$ is shown in Fig. 6.7.

In order to apply the described approach based on the introduction of a complex variable, we will consider transverse vibrations of a disk mounted on a flexible steel shaft of length l and circular cross section [14] (Fig. 6.8a) and disk imbalance $O'C = e$, where C is the disk mass center. The mass of the shaft is negligible as compared to the mass m of the disk. According to Newton's second law we have

$$m\ddot{x}_{1C} = -kx_1, \quad m\ddot{x}_{2C} = -kx_2, \tag{6.77}$$

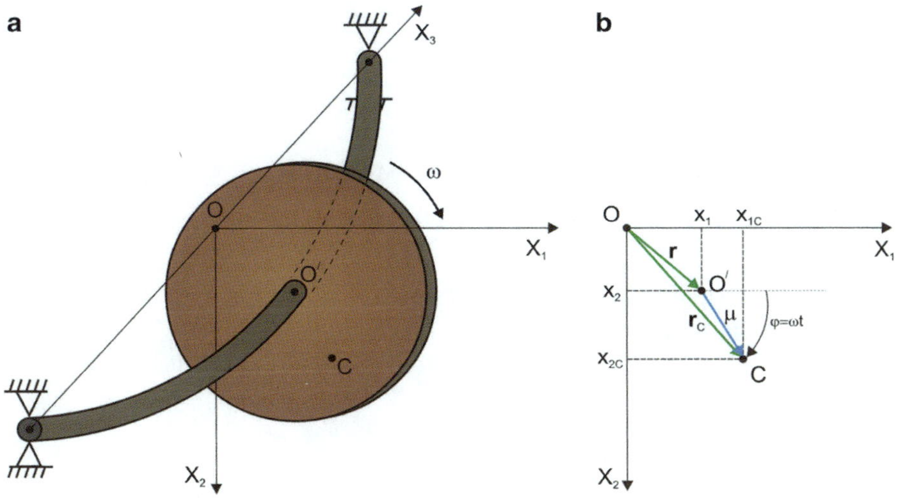

Fig. 6.8 Transverse vibrations of the disk (**a**) and the geometry of motion (**b**)

where

$$x_{1C} = x_1 + \mu \cos\varphi, \quad x_{2C} = x_2 + \mu \cos\varphi. \tag{6.78}$$

The coefficient $k = \frac{l^3}{48EI}$ is the bending stiffness of the shaft, and the third degree of freedom of the disk associated with the possible occurrence of torsional vibrations in its plane is neglected.

Let us introduce the complex representation in the following form:

$$x = x_1 + \mathrm{i}x_2, \quad x_C = x_{1C} + \mathrm{i}x_{2C}. \tag{6.79}$$

Multiplying the second equation of (6.77) by i and adding these equations to each other we obtain

$$\ddot{x}_C + \alpha^2 x = 0, \tag{6.80}$$

where now x and x_C are complex variables.

In turn, according to relation (6.78), we have

$$\begin{aligned} x &= (x_{1C} - \mu\cos\varphi) + \mathrm{i}(x_{2C} - \mu\sin\varphi) \\ &= x_{1C} + \mathrm{i}x_{2C} - \mu(\cos\varphi + \mathrm{i}\sin\varphi) = x_C - \mu \mathrm{e}^{\mathrm{i}\varphi}, \end{aligned} \tag{6.81}$$

which, substituted into (6.80), gives

$$\ddot{x}_C + \alpha^2 x_C = \mu\alpha^2 \mathrm{e}^{\mathrm{i}\varphi}. \tag{6.82}$$

We can obtain the equation of vibrations of point O' directly as

$$\frac{d^2}{dt^2}\left(x + \mu e^{i\varphi}\right) + \alpha^2 x = 0, \tag{6.83}$$

which following transformations leads to the equation

$$\ddot{x} + \alpha^2 x = \mu\omega^2 e^{i\varphi}, \tag{6.84}$$

where $\varphi = \omega t$. A single equation in the complex plane describes transverse vibrations of the disk.

The solution of (6.82) represents the displacement of mass center C of the disk and has the form

$$x_C = \frac{\mu}{1-(\omega/\alpha)^2} e^{i\omega t}, \tag{6.85}$$

and vector $\mathbf{r}_C$ of magnitude

$$r_C = \mu \left| \frac{1}{1-(\omega/\alpha)^2} \right| \tag{6.86}$$

rotates with angular velocity ω in the direction of shaft rotation. In turn, the deflection of the shaft determined by point O' can be obtained from the solution of (6.84) in the form

$$x = \frac{\mu(\omega/\alpha)^2}{1-(\omega/\alpha)^2} e^{i\omega t}, \tag{6.87}$$

which allows for the determination of the vector of deflection of the shaft $\mathbf{r}$ in the form

$$r = \mu \frac{(\omega/\alpha)^2}{|1-(\omega/\alpha)^2|}, \tag{6.88}$$

which rotates in the direction of shaft rotation with angular velocity ω. The mass center C remains in the plane of deflection of the shaft during vibrations.

In both cases [formulas (6.85) and (6.87)] it can be seen that if $\omega \to \alpha$, then we are dealing with resonance and the velocity $\alpha \equiv \omega \equiv \omega_{cr}$ is called a *critical speed.*

For large angular velocities of the shaft $\omega \gg \alpha$ its deflection following application of the l'Hospital's[4] rule in (6.88) is equal to μ, whereas the position of the mass center of the disk $\mathbf{r}_C \to 0$ (i.e., it tends to the position of the axis of rotation) which follows from (6.86).

So far we have considered the strongly idealized case where damping is not present in the system. Following the introduction of damping replacing external

[4]Guillame de l'Hospital (1661–1704), French mathematician, taught by Leibnitz and Johann Bernoulli, who published a l'Hospital rule that was, in fact, discovered by J. Bernoulli.

resistances, the equation of vibrations (6.84) takes the form

$$\ddot{x} + 2h\dot{x} + \alpha^2 x = \mu\omega^2 e^{i\omega t}. \tag{6.89}$$

For this equation we seek a solution in the form

$$x = Re^{i(\omega t - \beta)}. \tag{6.90}$$

Substituting relation (6.90) into (6.89) we obtain

$$-\omega^2 R + 2h\omega i R + \alpha^2 R = \mu\omega^2 e^{i\beta}. \tag{6.91}$$

From formula (6.91) it follows that

$$\begin{aligned}(\alpha^2 - \omega^2)R &= \mu\omega^2 \cos\beta,\\ 2h\omega R &= \mu\omega^2 \sin\beta.\end{aligned} \tag{6.92}$$

From (6.92) we obtain the desired quantities

$$\begin{aligned}\frac{R}{\mu} &= \frac{(\omega/\alpha)^2}{\sqrt{[1-(\omega/\alpha)^2]^2 + 4(h/\alpha)^2(\omega/\alpha)^2}},\\ \tan\beta &= \mu\frac{2(h/\alpha)(\omega/\alpha)}{1-(\omega/\alpha)^2}.\end{aligned} \tag{6.93}$$

The plot of $\frac{R}{\mu}$ is a function of $\left(\frac{\omega}{\alpha}\right)$ and $\left(\frac{h}{\alpha}\right)$ and is depicted in Fig. 6.9.

The relative position of the centroid O' and the mass center C of the disk is shown in Fig. 6.10 [11]. For small values of the viscous damping coefficient points O' and C lie approximately on a line perpendicular to the originally straight axis of the shaft (away from resonance).

The increase of ω (with values of α and h kept fixed) causes the increase of the angle β, which for $\omega \gg \alpha$ leads to $\beta = 180°$, and we are dealing with the phenomenon of the *self-centering of the shaft.*

If in Fig. 6.3 instead of the harmonic excitation we assume an arbitrary time-dependent driving force of the form $F(t)$, then the equation of motion of this system takes the form

$$m\ddot{x} + c\dot{x} + kx = F(t) \tag{6.94}$$

or

$$\ddot{x} + 2h\dot{x} + \alpha^2 x = q(t), \tag{6.95}$$

where now $q(t) = \frac{1}{m}F(t)$.

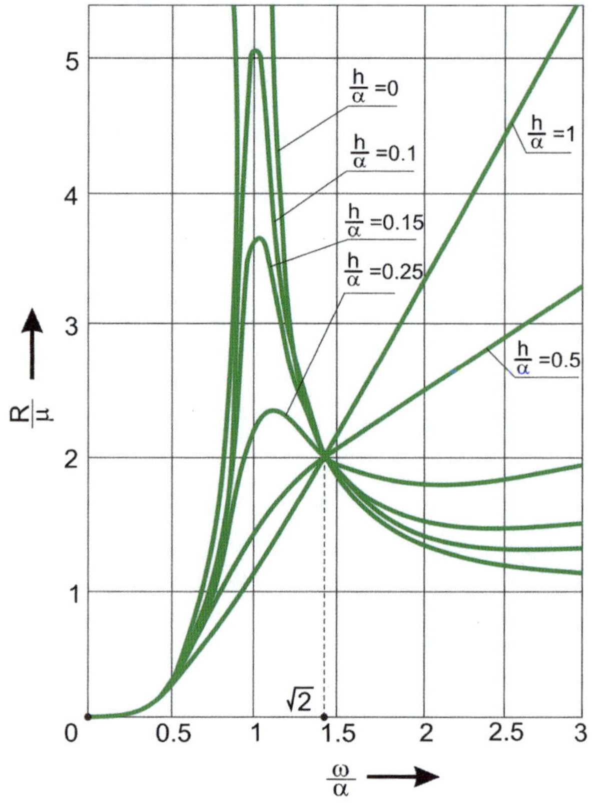

Fig. 6.9 Resonance plot $\frac{R}{\mu}\left(\frac{\omega}{\alpha}\right)$ for different values of $\left(\frac{h}{\alpha}\right)$

In order to solve (6.95), we apply a Laplace[5] transform, that is, we map the function $X(s)$ of a complex variable to the function $x(t)$ in the time domain, by means of the following integral transformation:

$$X(s) = \int_{-\infty}^{+\infty} x(t)\mathrm{e}^{-st}\mathrm{d}t, \tag{6.96}$$

where $s = c + \mathrm{i}\omega$, and further we determine the initial conditions for (6.95) for $t_0 = 0$. By means of algebraic transformations associated with (6.95) and variable s in the complex domain, we can conduct the inverse transformation

$$x(t) = L^{-1}\left[X(s)\right] = \frac{1}{2\pi i}\int_{c-\mathrm{i}\omega}^{c+\mathrm{i}\omega} X(s)\mathrm{e}^{-st}\mathrm{d}t, \tag{6.97}$$

[5]Pierre-Simon Laplace (1749–1827), French mathematician and astronomer.

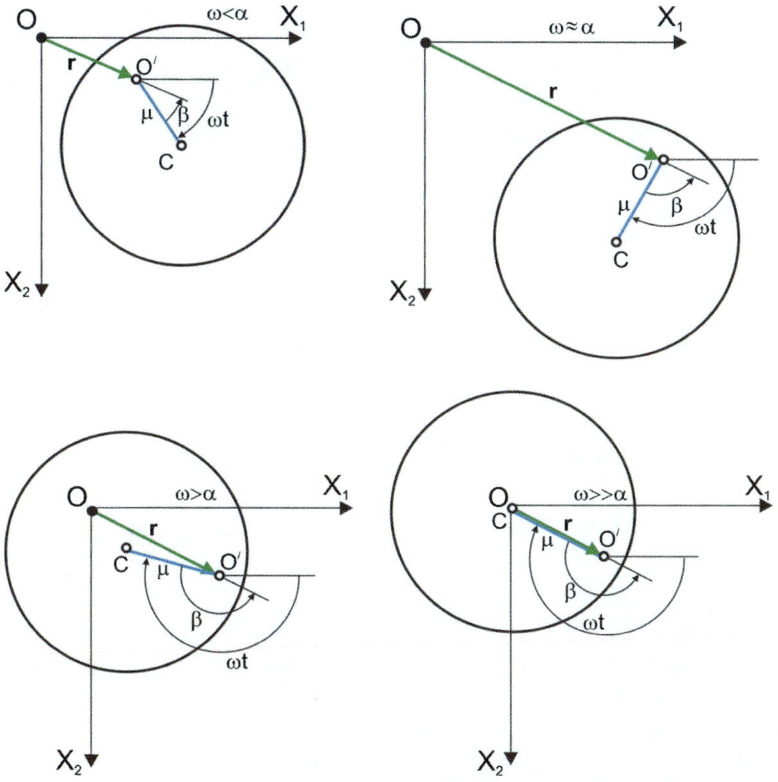

Fig. 6.10 Relative position of geometric center O' and mass center C of disk depending on the relation between ω and α

that is, from the complex domain to the time domain. In practice we do not calculate such integrals, but we use a table of original functions and their transforms. Because we are interested in a particular solution, during the analysis of (6.95) we assume the following initial conditions: $x(0) = \dot{x}(0) = 0$. Applying the Laplace transform to (6.95) we obtain

$$\left(s^2 + 2hs + \alpha^2\right) X(s) = Q(s), \tag{6.98}$$

and following transformation

$$X(s) = \frac{Q(s)}{\alpha^2} \frac{1}{\frac{1}{\alpha^2}s^2 + 2\frac{h}{\alpha}\frac{1}{\alpha}s + 1}. \tag{6.99}$$

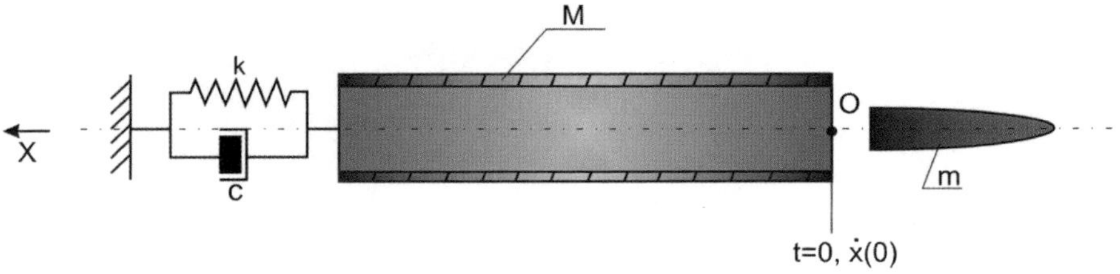

Fig. 6.11 Firing of the gun barrel situated in a horizontal position

The right-hand side of (6.99) is the ratio of two transforms. By the *convolution theorem* we have

$$f_1(t) * f_2(t) = \int_0^t f_1(\tau) f_2(t-\tau)\mathrm{d}\tau = \int_0^t f_1(t-\tau) f_2(\tau)\mathrm{d}\tau,$$

$$L\left[f_1(t) * f_2(t)\right] = F_1(s)F_2(s),$$

$$L^{-1}\left[F_1(s)F_2(s)\right] = f_1(t) * f_2(t), \tag{6.100}$$

and hence an original function of such a product is a convolution of time functions, which are original functions of the aforementioned transforms.

On the other hand,

$$L^{-1}\left[\frac{1}{\alpha^2}Q(s)\right] = \frac{1}{\alpha^2}L^{-1}\left[Q(s)\right] = \frac{1}{\alpha^2}q(t),$$

$$L^{-1}\left[\frac{1}{T^2s^2 + 2\xi T s + 1}\right] = \frac{1}{T\sqrt{1-\xi^2}}\mathrm{e}^{\frac{\xi t}{T}} \sin\left(\sqrt{1-\xi^2}\frac{t}{T}\right), \tag{6.101}$$

where $\xi = \frac{h}{\alpha}$, $T = \frac{1}{\alpha}$.

According to (6.100) we obtain

$$x(t) = \int_0^t \frac{1}{\alpha^2}q(\tau)\frac{\alpha^2}{\sqrt{\alpha^2 - h^2}}\mathrm{e}^{-h(t-\tau)} \sin\left(\sqrt{\alpha^2 - h^2}(t-\tau)\right)\mathrm{d}\tau$$

$$= \int_0^t \frac{q(\tau)}{\lambda}\mathrm{e}^{-h(t-\tau)} \sin \lambda(t-\tau)\mathrm{d}\tau, \tag{6.102}$$

where $\lambda = \sqrt{\alpha^2 - h^2}$ and (6.102) is valid for $c < c_{\mathrm{cr}}$.

Example 6.2. In a gun barrel situated horizontally (Fig. 6.11) there is a projectile of mass m, and the barrel of mass M is supported by means of a recoil mechanism (a spring and damper). Determine the recoil, that is, the displacement of the barrel after the projectile leaves the barrel. Assume that the time in which the projectile moves in the barrel is negligibly small, and the velocity of the projectile at the moment when it leaves the barrel is equal to $\mathbf{v}_{\mathrm{p}}$.

At the moment when the projectile leaves the gun barrel, let us assume $x(0) = 0$, $\dot{x}(0) = \dot{x}_0$.

The equation of motion of the barrel has the form

$$M\ddot{x} + c\dot{x} + kx = 0,$$

and its vibrations are excited by the initial velocity $\dot{x}_0$. The motion of the barrel is described by (6.45), from which, following the introduction of the initial conditions mentioned previously, we obtain $A_1 = 0$, $A_2 = \frac{\dot{x}_0}{\lambda}$. Eventually, the equation of motion of the barrel has the form

$$x(t) = \frac{\dot{x}_0}{\lambda}\mathrm{e}^{-ht}\sin\lambda t.$$

The initial velocity of the projectile $\dot{x}_0$ remains to be determined. To this end we exploit the theorem of the conservation of momentum for a system composed of a barrel and projectile (Sect. 1.1.4).

The momentum of the system at the instant when the projectile leaves the barrel is equal to

$$p = M\dot{x}_0 + m(\dot{x}_0 - v_p).$$

Under the conditions of the problem, before firing, the barrel and projectile have velocities equal to zero. The momentum of the system at the moment just before firing is equal to zero.

Because the momentum of the system has to be conserved, the desired magnitude of the velocity is equal to

$$\dot{x}_0 = \frac{m}{m+M}v_p.$$

In [15] it is shown that a solution obtained in this way differs only slightly from a solution that takes into account the time of motion of the projectile in the barrel.

6.5 Non-Linear Conservative 1DOF System and Dimensionless Equations

Let us consider the system from Fig. 6.3 with no damping and no excitation, but with non-linear stiffness $k = k(x)$ (see [16]). The equation of motion of mass m has the form

$$m\ddot{x} + k(x) = 0, \tag{6.103}$$

and dividing by sides by the mass m we get

$$\ddot{x} + s(x) = 0, \tag{6.104}$$

and we assume the initial conditions

$$x(0) = x_0, \qquad \dot{x}(0) = v_0. \tag{6.105}$$

Let us note that

$$\ddot{x} = \frac{\mathrm{d}\dot{x}}{\mathrm{d}x}\frac{\mathrm{d}x}{\mathrm{d}t} = v\frac{\mathrm{d}v}{\mathrm{d}x}, \tag{6.106}$$

and substituting this relationship into (6.104) we obtain

$$v\,\mathrm{d}v + s(x)\mathrm{d}x = 0. \tag{6.107}$$

Following integration we have

$$\int\limits_{v_0}^{\dot{x}} v\mathrm{d}v = -\int\limits_{x_0}^{x} s(\eta)\mathrm{d}\eta, \tag{6.108}$$

and further

$$\frac{\dot{x}^2 - v_0^2}{2} = -\int\limits_{x_0}^{x} s(\eta)\mathrm{d}\eta, \tag{6.109}$$

which allows for the calculation of the velocity

$$v \equiv \frac{\mathrm{d}x}{\mathrm{d}t} = \pm\sqrt{v_0^2 - 2\int\limits_{x_0}^{x} s(\eta)\mathrm{d}\eta}. \tag{6.110}$$

Separating the variables in (6.110) and integrating again, we get

$$t = \int\limits_{x_0}^{x} \frac{\mathrm{d}\xi}{\pm\sqrt{v_0^2 - 2\int\limits_{\xi_0}^{\xi} s(\eta)\mathrm{d}\eta}}. \tag{6.111}$$

We determine equilibrium positions from the algebraic equation $s(x) = 0$, and then choosing one of the positions (x_0) we place there the origin of the axis OX, that is, we will consider small vibrations about the equilibrium position $x_0 = 0$. Our aim is to determine the period of oscillations T of the considered conservative system. To this end we impose the following initial conditions on the system: $x(0) = A_0$, $\dot{x}(0) = 0$. Following the passage of time $\frac{T}{2}$ we obtain $x\left(\frac{T}{2}\right) = A_1$, $\dot{x}\left(\frac{T}{2}\right) = 0$. According to (6.111) we have

$$\frac{T}{2} = \int_{A_0}^{A} \frac{d\xi}{-\sqrt{-2\int_{A_0}^{\xi} s(\eta)d\eta}}. \tag{6.112}$$

The upper limit of integration A_1 is determined from (6.110), that is, we have

$$\sqrt{-2\int_{A_0}^{A_1} s(\eta)d\eta} = 0. \tag{6.113}$$

In this case, for an odd function $s(\eta)$, from (6.112) we get

$$T = 4\int_{A_0}^{0} \frac{d\xi}{-\sqrt{-2\int_{A_0}^{\xi} s(\eta)d\eta}}. \tag{6.114}$$

Let us consider an odd function of the form

$$s(x) = ax^3, \tag{6.115}$$

that is, our aim is to determine the vibrations of the oscillator described by the equation

$$\ddot{x} + ax^3 = 0. \tag{6.116}$$

From formula (6.114) we obtain

$$T = 4\int_{A_0}^{0} \frac{d\xi}{-\sqrt{-2a\int_{A_0}^{\xi} \eta^3 d\eta}} = -4\sqrt{\frac{2}{a}}\int_{A_0}^{0} \frac{d\xi}{\sqrt{A_0^4 - \xi^4}}. \tag{6.117}$$

Introducing a new variable of the form

$$A_0 u = \xi, \tag{6.118}$$

from (6.117) we obtain

$$T=-4\sqrt{\frac{2}{a}}\int_{1}^{0}\frac{\mathrm{d}\xi}{\sqrt{A_0^4-u^4A_0^4}}=4\sqrt{\frac{2}{a}}\frac{1}{A_0}\int_{0}^{1}\frac{\mathrm{d}u}{\sqrt{1-u^4}}. \tag{6.119}$$

Unfortunately, the integral in formula (6.119) cannot be expressed in terms of elementary functions. However, it can be expressed through the function Γ, because we have

$$\int_{0}^{1}\frac{\mathrm{d}u}{\sqrt{1-u^4}}=\frac{1}{4\sqrt{2\pi}}\Gamma\left(\frac{1}{4}\right)^2. \tag{6.120}$$

The function $\Gamma(x)$ is tabulated for $1<x<2$, and additionally $\Gamma(x)=\frac{\Gamma(x+1)}{x}$. Hence $\Gamma(\frac{1}{4})=4\Gamma(1.25)$, and from (6.119) we obtain the period

$$T=7.418\left(A_0\sqrt{a}\right)^{-1}. \tag{6.121}$$

Now, let us turn to problems of reduction of equations to a dimensionless form. Usually, during the idealization process it is necessary to introduce a certain order of various elements by comparing them with each other and with respect to characteristic quantities chosen in advance. For example, if one element has a length of 1 cm, then a natural question arises: is this quantity small or large? The answer to this question can be provided only by the initial formulation of the problem. It is clear that if we investigate the motion of a satellite in near-Earth orbit, then we can assume the length of 1 cm to be negligibly small. On the other hand, if we consider a distance between molecules, then the length of 1 cm is an extremely large value.

Let us give another example. It is well known that air is compressible. But do we always have to take into account the compressibility of the air? It depends on the initial formulation of the problem. If an investigated object moves through the air with small velocity V, then compressibility can be neglected while constructing the mathematical model. However, if the velocity of the object is large, and even close to or greater than the speed of sound, then, inevitably, compressibility has to be taken into account. In this case, it is very convenient to introduce a dimensionless quantity $M=V/a$ called the *Mach number*, which plays an important role in aerodynamics. For instance, at $M\ll 1$ it is possible to make use of the idealized mathematical model of incompressible gas, whereas for greater values of the Mach number one should take into account air compressibility. A similar situation occurs during the construction of mathematical models in other fields of science and technology, where an important role is played by some other characteristic dimensionless numbers, which, in general, are created by a combination of three dimensionless quantities: length L, time T, and mass M. For the sake of convenience it is assumed that the dimension of the combination FT^2/ML is equal to 1 (where F is a force). In other words, one of the quantities F, T, L, or M can be chosen as the independent one.

To recapitulate, all the variable quantities of a process under consideration should be reduced to dimensionless quantities. This can be achieved by dividing the process input variables into certain corresponding (in the sense of dimension) characteristic quantities or their combinations: length L, velocity V, a viscous damping coefficient c, spring stiffness k, dynamic viscosity ν, etc. Therefore, in various branches of science and technology the application of dimensional analysis to every problem yields a set of characteristic dimensionless numbers – the similarity parameters – whose values qualitatively describe the nature of the investigated processes (similarity laws for processes). As examples we may cite the numbers of Mach,[6] Nusselt,[7] Reynolds,[8] Strouhal,[9] Froude,[10] Biot,[11] and many others. They can be very small or, conversely, very large.

The reduction of equations to a dimensionless form has other important features as well. Usually after rescaling of the parameters, their number decreases and the analyzed problem gains a metascientific character, that is, the description is valid in distinct fields of science, such as mechanics, electrical systems, atomic physics, etc.

We will consider a particle mounted to a base by means of a certain massless spring with non-linear characteristics (the aerodynamic drag and friction are neglected). Let, at time instant $t = 0$, the point mass be deflected from an equilibrium position by a certain value x_0^* and then released. The deflection of the body from the original position at time instant t^* is denoted by $x^*(t^*)$. The equation of motion reads

$$\frac{d^2x^*}{dt^{*2}} + f(x^*) = 0, \tag{6.122}$$

and we assume that $f(x^*) = 0$ for $x^* = 0$.

---

[6]Ernst Mach (1838–1916), Czech/Austrian physicist and philosopher; the Mach number $M = \frac{v_0}{v_*}$, where v_0 is the velocity of an object and v_* the velocity of sound in the considered medium.

[7]Wilhelm Nusselt (1882–1957), German engineer; the Nusselt number $N = \frac{hL}{k_f}$, where L is the characteristic length, h the convective heat transfer coefficient, and k_f the thermal conductivity of a liquid.

[8]Osborne Reynolds (1842–1912), Irish professor who studied fluid dynamics; the Reynolds number $\mathrm{Re} = \frac{v_* L}{\nu}$, where v_* is the mean fluid velocity, L the characteristic length, and ν the kinematic viscosity of a fluid.

[9]Vincent Strouhal (1850–1922), Czech physicist; the Strouhal number $Sr = \frac{fc^3}{U}$, where f is the frequency, c the coefficient of expansion, and U the flow rate.

[10]William Froude (1810–1879), English engineer; the Froude number (dimensionless) $Fr = \frac{v}{c}$, where v is the characteristic velocity and c the characteristic velocity of water wave propagation.

[11]Jean-Baptiste Biot (1774–1862), French physicist, astronomer, and mathematician; the Biot number $Bi = \frac{hl}{k}$, where h is the heat transfer coefficient, l the characteristic length of a body, and k the coefficient of thermal conductivity of the body.

Let us expand the non-linear function $f(x^*)$ into the Maclaurin series

$$f(x^*) = x^* \frac{\mathrm{d}f}{\mathrm{d}x^*} + \frac{x^{*2}}{2!} \frac{\mathrm{d}^2 f}{\mathrm{d}x^{*2}} + \frac{x^{*3}}{3!} \frac{\mathrm{d}^3 f}{\mathrm{d}x^{*3}} + \cdots , \tag{6.123}$$

where the derivatives $\frac{\mathrm{d}f_i}{\mathrm{d}x^{*i}}$ are calculated at the point $x^* = 0$.

Substituting relationships (6.123) into (6.122) we obtain

$$\frac{\mathrm{d}^2 x^*}{\mathrm{d}t^{*2}} + x^* \frac{\mathrm{d}f}{\mathrm{d}x^*} + \frac{x^{*3}}{6} \frac{\mathrm{d}^3 f}{\mathrm{d}x^{*3}} = 0, \tag{6.124}$$

where it is assumed that $f(0) = \frac{\mathrm{d}^2 f}{\mathrm{d}x^{*2}}(0) = 0$, $\frac{\mathrm{d}f}{\mathrm{d}x^*}(0) > 0$.

Let us introduce the dimensionless quantities

$$x = \frac{x^*}{l}, \quad t = \frac{t^*}{T}, \tag{6.125}$$

where l and T are characteristic constants of length and time.

Let the time constant be described by the equation

$$\left[T^2 \frac{\mathrm{d}f}{\mathrm{d}x^*}(0) \right] = [1]. \tag{6.126}$$

Let us note that the preceding equation corresponds to

$$\left[\frac{FT^2}{ML} \right] = [1], \tag{6.127}$$

where F is the force, T denotes time, M is the mass, and L denotes length. Introducing dimensionless quantities (6.125) into (6.124) we obtain

$$\frac{l}{T^2} \frac{\mathrm{d}^2 x}{\mathrm{d}t^2} + xl \frac{\mathrm{d}f}{\mathrm{d}x^*}(0) + \frac{1}{6} x^3 l^3 \frac{\mathrm{d}^3 f}{\mathrm{d}x^{*3}}(0) = 0, \tag{6.128}$$

and multiplying this equation through by T^2/l we have

$$\frac{\mathrm{d}^2 x}{\mathrm{d}t^2} + xT^2 \frac{\mathrm{d}f}{\mathrm{d}x^*}(0) + \frac{1}{6} x^3 T^2 l^2 \frac{\mathrm{d}^3 f}{\mathrm{d}x^{*3}}(0) = 0. \tag{6.129}$$

Following introduction of the quantities

$$\beta_0 = T^2 \frac{\mathrm{d}f}{\mathrm{d}x^*}(0), \quad \beta_1 = \frac{1}{6} T^2 l^2 \frac{\mathrm{d}^3 f}{\mathrm{d}x^{*3}}(0), \tag{6.130}$$

(6.129) takes the form

$$\frac{\mathrm{d}^2 x}{\mathrm{d}t^2} + \beta_0 x + \beta_1 x^3 = 0. \tag{6.131}$$

This equation plays an important role in non-linear mechanics and is called a *Duffing equation.*

The initial condition $x^*(0) = x_0^*$ assumes the dimensionless form of $x(0) = \frac{x_0^*}{l} = x_0$.

6.6 One-Degree-of-Freedom Mechanical Systems with a Piecewise Linear and Impulse Loading

In this section we will consider the vibrations of a one-degree-of-freedom system, depicted in Fig. 6.12, loaded with a piecewise linear loading $F(t)$.

The considered loads $F(t)$ are shown in Fig. 6.13.

For the case presented in Fig. 6.13a, the equation of motion of the system shown in Fig. 6.12 has the form

$$m\ddot{x} + kx = \frac{F_0}{t_0} t \tag{6.132}$$

for $0 \le t \le t_0$.

Dividing by mass m we have

$$\ddot{x} + \alpha^2 x = \frac{F_0}{m} \frac{t}{t_0} = q_0 \frac{t}{t_0}. \tag{6.133}$$

The solution of the preceding equation is sought in the form

$$x = C_1 \cos \alpha t + C_2 \sin \alpha t + \frac{x_0}{t_0} t. \tag{6.134}$$

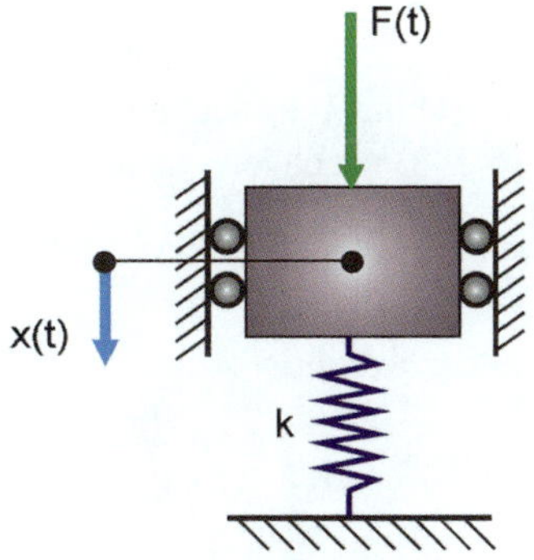

Fig. 6.12 A model of the investigated system

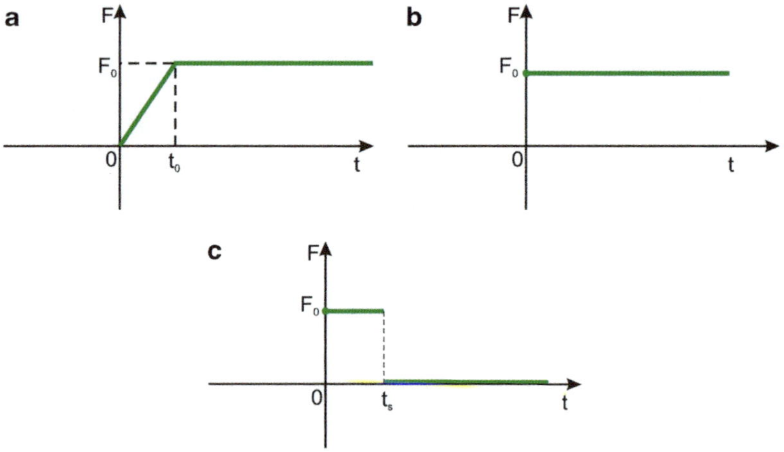

Fig. 6.13 Models of "piecewise linear" loading

Substituting (6.134) into (6.133) we get

$$\alpha^2 x_0 = q_0. \tag{6.135}$$

In order to determine the constants C_1 and C_2, one has to specify initial conditions, which we assume to be in the form

$$x(0) = x^-, \quad \dot{x}(0) = \dot{x}^-. \tag{6.136}$$

Taking into account the first equation of system (6.136) in relation (6.134) we obtain

$$C_1 = x^-, \tag{6.137}$$

and differentiating (6.134) we have

$$\dot{x} = -\alpha C_1 \sin \alpha t + \alpha C_2 \cos \alpha t + \frac{x_0}{t_0}. \tag{6.138}$$

Taking into account the second initial condition of system (6.136) in relation (6.138) we have

$$\dot{x}^- = C_2 \alpha + \frac{x_0}{t_0}, \tag{6.139}$$

hence we find

$$C_2 = \frac{1}{\alpha}\left(\dot{x}^- - \frac{x_0}{t_0}\right). \tag{6.140}$$

Substituting the determined constants into (6.134) and (6.138) we get

$$x = x^{-} \cos \alpha t + \frac{1}{\alpha}\left(\dot{x}^{-} - \frac{x_0}{t_0}\right) \sin \alpha t + \frac{x_0}{t_0} t, \tag{6.141}$$

$$x = -\alpha x^{-} \sin \alpha t + \left(\dot{x}^{-} - \frac{x_0}{t_0}\right) \cos \alpha t + \frac{x_0}{t_0}. \tag{6.142}$$

At the instant t_0 the deflection and velocity, according to formulas (6.141) and (6.142), are equal to

$$\bar{x}(t_0) = x^{-} \cos \alpha t_0 + \frac{1}{\alpha}\left(\dot{x}^{-} - \frac{x_0}{t_0}\right) \sin \alpha t_0 + x_0, \tag{6.143}$$

$$\dot{\bar{x}}(t_0) = -\alpha x^{-} \sin \alpha t_0 + \left(\dot{x}^{-} - \frac{x_0}{t_0}\right) \cos \alpha t_0 + \frac{x_0}{t_0}. \tag{6.144}$$

For time $t > t_0$ (6.133) takes the form

$$\ddot{x} + \alpha^2 x = q_0 \equiv \frac{F_0}{m}, \tag{6.145}$$

and we seek its solution in the form

$$x = x_0 + \bar{C}_1 \cos \alpha t + \bar{C}_2 \sin \alpha t, \tag{6.146}$$

which means that at the instant t_0 the time is measured from zero, and the initial conditions are described by (6.143) and (6.144).

Substituting $t = 0$ into (6.146) and its derivative, and taking into account relations (6.143) and (6.144), we obtain

$$x_0 + \bar{C}_1 = x^{-} \cos \alpha t_0 + \frac{1}{\alpha}\left(\dot{x}^{-} - \frac{x_0}{t_0}\right) \sin \alpha t_0 + x_0, \tag{6.147}$$

$$\alpha \bar{C}_2 = -\alpha x^{-} \sin \alpha t_0 + \left(\dot{x}^{-} - \frac{x_0}{t_0}\right) \cos \alpha t_0 + \frac{x_0}{t_0}. \tag{6.148}$$

Substituting the constants $\bar{C}_1$ and $\bar{C}_2$ into (6.146) we have

$$\begin{aligned} x = {} & x_0 + \left[x^{-} \cos \alpha t_0 + \frac{1}{\alpha}\left(\dot{x}^{-} - \frac{x_0}{t_0}\right) \sin \alpha t_0\right] \cos \alpha t \\ & + \left[-x^{-} \sin \alpha t_0 + \frac{1}{\alpha}\left(\dot{x}^{-} - \frac{x_0}{t_0}\right) \cos \alpha t_0 + \frac{x_0}{\alpha t_0}\right] \sin \alpha t, \end{aligned} \tag{6.149}$$

and following further transformations we get

$$x = x_0 + \left[x^- \cos\alpha t_0 + \frac{\dot{x}^-}{\alpha}\sin\alpha t_0 - x_0\frac{\sin\alpha t_0}{\alpha t_0}\right]\cos\alpha t$$
$$+ \left[-x^- \sin\alpha t_0 + \frac{\dot{x}^-}{\alpha}\cos\alpha t_0 + x_0\left(\frac{1-\cos\alpha t_0}{\alpha t_0}\right)\right]\sin\alpha t. \qquad (6.150)$$

Let us now consider the case where $\alpha t_0 \to 0$ and we assume the following approximations:

$$\sin\alpha t_0 \approx 0, \qquad \cos\alpha t_0 \approx 1, \qquad \frac{1-\cos\alpha t_0}{\alpha t_0} \approx \frac{\alpha t_0}{2}. \qquad (6.151)$$

We will show how the last approximation of (6.151) is obtained. We have

$$\cos\alpha t_0 = \cos 0 - (\cos 0)\frac{(\alpha t_0)^2}{2} = 1 - \frac{(\alpha t_0)^2}{2},$$

which substituted into the left-hand side of the third equation of (6.151) leads to the result presented there.

Taking into account relation (6.151) in (6.150) we have

$$x = x_0 + (x^- - x_0)\cos\alpha t + \left(\frac{\dot{x}^-}{\alpha} + \frac{\alpha x_0 t_0}{2}\right)\sin\alpha t, \qquad (6.152)$$

and differentiating this equation we obtain

$$\dot{x} = -\alpha(x^- - x_0)\sin\alpha t + \left(\dot{x}^- + \frac{\alpha^2 x_0 t_0}{2}\right)\cos\alpha t. \qquad (6.153)$$

If $\alpha t_0 \to 0$, then from (6.152) we obtain

$$x = x_0 + (x^- - x_0)\cos\alpha t + \frac{\dot{x}^-}{\alpha}\sin\alpha t. \qquad (6.154)$$

Let us now consider the case presented in Fig. 6.13b for initial conditions equal to zero.

The equation of motion of the system has the form

$$\ddot{x} + \alpha^2 x = q_0, \qquad (6.155)$$

where $q_0 = \frac{F_0}{m}$. The solution of (6.155) has the form

$$x = x_0 + A\sin\alpha t + B\cos\alpha t, \qquad (6.156)$$

hence for $x(0) = 0$, $\dot{x}(0) = 0$ we obtain

$$\begin{aligned} x(0) &\equiv B + x_0 = 0, \\ \dot{x}(0) &\equiv A\alpha = 0, \end{aligned} \tag{6.157}$$

and eventually we have

$$x = x_0(1 - \cos\alpha t) = \frac{q_0}{\alpha^2}(1 - \cos\alpha t). \tag{6.158}$$

This result can be obtained immediately from (6.154) after substituting $x^- = \dot{x}^- = 0$.

Moreover, for $\cos\alpha t_n = -1$ we have $x = \frac{2q_0}{\alpha^2} = \frac{2F_0}{k} = 2x_0$, which means that the dynamic action of the force F_0 causes a deflection that is two times greater.

Let us now turn to the solution of the case shown in Fig. 6.13c. For $t \geq t_s$ there is no loading and the system oscillates according to the equation

$$x(t) = C_1 \cos\alpha t + C_2 \sin\alpha t. \tag{6.159}$$

At the time instant $0 \leq t \leq t_s$ the system was loaded with force F_0 of constant magnitude (the case from Fig. 6.13b), and its vibrations were described by (6.158). For the instant $t = t_s$ from (6.158) and (6.159) it follows that

$$x(t_s) \equiv x_0(1 - \cos\alpha t_s) = C_1 \cos\alpha t_s + C_2 \sin\alpha t_s, \tag{6.160}$$

hence

$$\dot{x}(t_s) \equiv x_0\alpha \sin\alpha t_s = -C_1\alpha \sin\alpha t_s + C_2\alpha \cos\alpha t_s. \tag{6.161}$$

Let us multiply (6.160) by $\alpha \sin\alpha t_s$ and (6.161) by $\cos\alpha t_s$ and then respectively by $-\alpha \cos\alpha t_s$ and $\sin\alpha t_s$. Adding the obtained results by sides we obtain the values of the desired constants

$$\begin{aligned} C_1 &= x_0(\cos\alpha t_s - 1), \\ C_2 &= x_0 \sin\alpha t_s. \end{aligned} \tag{6.162}$$

Substituting these constants into (6.159) we obtain

$$\begin{aligned} x(t) &= -x_0 \cos\alpha t + x_0 \cos\alpha t_s \cos\alpha t + x_0 \sin\alpha t_s \sin\alpha t \\ &= -x_0 \cos\alpha t + x_0 \cos(\alpha t_s - \alpha t) \\ &= x_0 \left[\cos(\alpha t_s - \alpha t) - \cos\alpha t\right] \\ &= x_0 \left(-2 \sin\frac{\alpha t_s - \alpha t + \alpha t}{2} \sin\frac{\alpha t_s - \alpha t - \alpha t}{2}\right) \end{aligned} \tag{6.163}$$

because

$$\cos\alpha - \cos\beta = -2\sin\frac{\alpha+\beta}{2}\sin\frac{\alpha-\beta}{2}.$$

Eventually relation (6.163) takes the form

$$\begin{aligned} x(t) &= -2x_0 \sin\frac{\alpha t_s}{2}\sin\left(\frac{\alpha t_s - 2\alpha t}{2}\right) \\ &= 2x_0\sin\frac{\alpha t_s}{2}\sin\alpha\left(t - \frac{t_s}{2}\right). \end{aligned} \tag{6.164}$$

To recapitulate, for the time instant $t \le t_s$ we have solution (6.160), and for $t > t_s$ we have (6.164).

We introduce the following ratio into the calculations:

$$\frac{t_s}{T} = \frac{t_s}{2\pi}\alpha, \tag{6.165}$$

which taken into account in (6.164) yields

$$x(t) = 2x_0 \sin\pi\left(\frac{t_s}{T}\right)\sin\pi\left(\frac{2t}{T} - \frac{t_s}{T}\right). \tag{6.166}$$

Let us note that

$$\sin\pi\left(\frac{t_s}{T}\right) = 0, \tag{6.167}$$

which means that $x(t) = 0$ for $\pi(\frac{t_s}{T}) = n\pi$, $n = 1, 2, \ldots$, that is, for $t_s = nT$.

Now, let us estimate the value of $x(t)$ for $t_s = T/2$. From (6.166) we have

$$x(t) = 2x_0\sin\frac{\pi}{2}\left(\frac{2t}{T} - \frac{1}{2}\right) = -2x_0\cos 2\pi\frac{t}{T}, \tag{6.168}$$

which means that in this case (for $t_s = T/2$) we have $x_{\max}$ for $2\pi\frac{t_n}{T} = 2\pi n$, that is, for $t_n = Tn$.

In general, the time response $x(t)$ depends on the duration of action of the force F_0 relative to a period of free vibrations of the system $T = 2\pi/\alpha$, namely:

1. For $t_s \ge 0.5T$ the maximum displacement $x_{\max}(t)$ appears during application of the loading and is equal to $2x_0$.
2. For $t_s < 0.5T$ the maximum displacement appears after the force F_0 ceases to act and is always smaller than $2x_0$.
3. For $t_s = T$ at the same time we have $x(t_s) = \dot{x}(t_s) = 0$, which means that the system does not move.

Let us consider now a certain special case of the loading depicted in Fig. 6.13c, that is, for $t_s \to 0$. According to the calculation of Sect. 5.1, the impulse of a force is described by (5.4), and if $F = F_0 = \text{const}$ acts during time interval $(0, t_s)$, then from (5.4) we obtain

$$J = F_0 t_s. \tag{6.169}$$

The impulse of a force in this case is equal to the area of a rectangle of height F_0 and base t_s. Thus we can take various values of F_0 and t_s keeping the same value of impulse J. A special case of the impulse of a force is an *instantaneous impulse*, where we set $J = \text{const}$ and the time of its duration $t_s \to 0$, which results in $F_0 \to \infty$.

Using the notion of the impulse of a force, (6.164) can be written in the form

$$x(t) = \frac{J}{m\alpha} \frac{\sin \frac{\alpha t_s}{2}}{\frac{\alpha t_s}{2}} \sin \alpha \left(t - \frac{t_s}{2} \right), \tag{6.170}$$

and proceeding to the limit as $t_s \to 0$ we obtain

$$\lim_{t_s \to 0} x(t) = \frac{J}{m\alpha} \sin \alpha t. \tag{6.171}$$

In the preceding calculations it is assumed that the instantaneous impulse acts at the time instant $t_s = 0$, but if the impulse acts at an arbitrary instant t_s, then the motion of the considered one-degree-of-freedom system is described by the following equation:

$$x(t) = \begin{cases} 0 & \text{for} \quad t < t_s, \\ \dfrac{J}{m\alpha} \sin \alpha (t - t_s) & \text{for} \quad t \geq t_s. \end{cases} \tag{6.172}$$

Figure 6.14, in turn, shows a series of instantaneous impulses.

Because we are dealing with a linear system, for which the superposition principle is applicable, the motion of the analyzed system is described by

$$x(t) = \sum_{n=0}^{N} \frac{J_n}{m\alpha} \sin \alpha (t - t_n). \tag{6.173}$$

The notion of an instantaneous impulse of a force makes it possible to explain the physical meaning of the response of a one-degree-of-freedom system subjected to an arbitrary driving force $F(t)$ and described by (6.102). Because we consider the vibrations of the system without damping, substituting $h = 0$ into (6.102) yields

$$x(t) = \frac{1}{m\alpha} \int_0^t F(t_s) \sin \alpha (t - t_s) \mathrm{d}t_s. \tag{6.174}$$

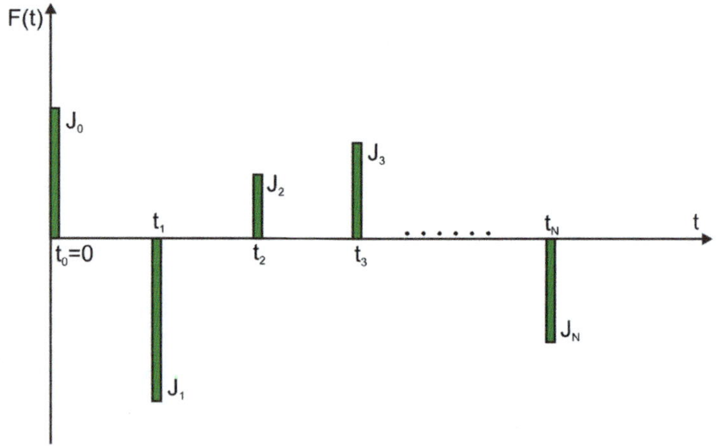

Fig. 6.14 A series of instantaneous impulses acting on a one-degree-of-freedom system

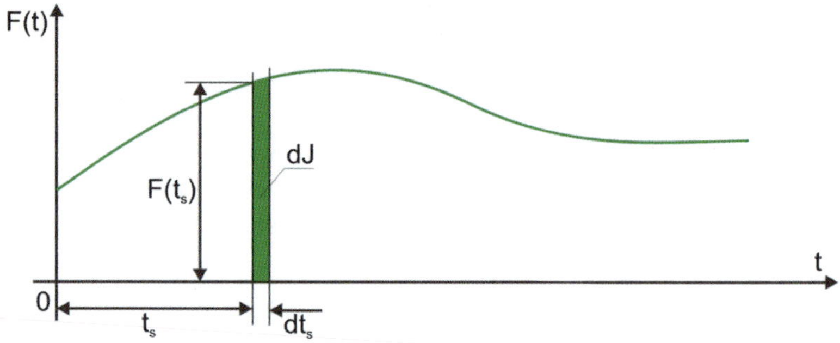

Fig. 6.15 Time plot of an arbitrary force $F(t)$

Let an arbitrary force $F(t)$ have the form presented in Fig. 6.15, where an elementary impulse $\mathrm{d}J = F(t_s)\mathrm{d}t_s$ is shown.

If the system is acted upon only by an elementary impulse, then for initial conditions equal to zero and $t \geq t_s$, according to (6.172), we have

$$\mathrm{d}x(t) = \frac{\mathrm{d}J}{m\alpha}\sin\alpha(t - t_s) = \frac{F(t_s)}{m\alpha}\sin\alpha(t - t_s)\mathrm{d}t_s. \qquad (6.175)$$

An extension of the action of an elementary impulse to the interval $0 \leq t_s \leq t$ makes it possible to determine the total displacement $x(t)$ through the integration of elementary displacements (6.175), which results in (6.174).

At the end of this section we will consider the case of excitation of the system shown in Fig. 6.12 by a series of impulses of short duration $J(t)$ and repeating periodically, which is depicted in Fig. 6.16.

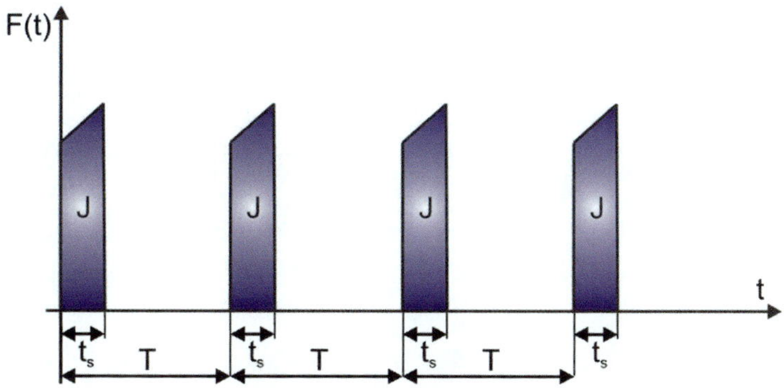

Fig. 6.16 Excitation of a one-degree-of freedom conservative system by a series of periodic impulses of force

A general solution of the differential equation

$$\ddot{x} + \alpha^2 x = q(t) \tag{6.176}$$

during the action of only one impulse has the form

$$x(t) = x_0 \cos \alpha t + \frac{\dot{x}_0}{\alpha} \sin \alpha t + \frac{1}{m\alpha} \int_0^{t_s} F(t_s) \sin \alpha (t - t_s) \mathrm{d}t_s, \tag{6.177}$$

which is valid for $t > t_s$.

The instantaneous impulse of a force preserves the value of the integral, that is,

$$\lim_{t_s \to 0} \int_0^{t_s} F(t) \mathrm{d}t = J = \text{const.} \tag{6.178}$$

In order to preserve the constant value of J and for $t_s \to 0$ the force $F(t^*) \to \infty$, where $t = t^*$ denotes the time instant when the instantaneous impulse is applied. This phenomenon can be described using the Dirac delta function $\delta(t_s)$ by the equation

$$F(t_s) = J\delta(t_s). \tag{6.179}$$

Substituting relations (6.179) into (6.177) we get

$$x(t) = x_0 \cos \alpha t + \frac{\dot{x}_0}{\alpha} \sin \alpha t + \frac{J}{m\alpha} \int_0^{t_s} \delta(t_s) \sin \alpha (t - t_s) \mathrm{d}t_s. \tag{6.180}$$

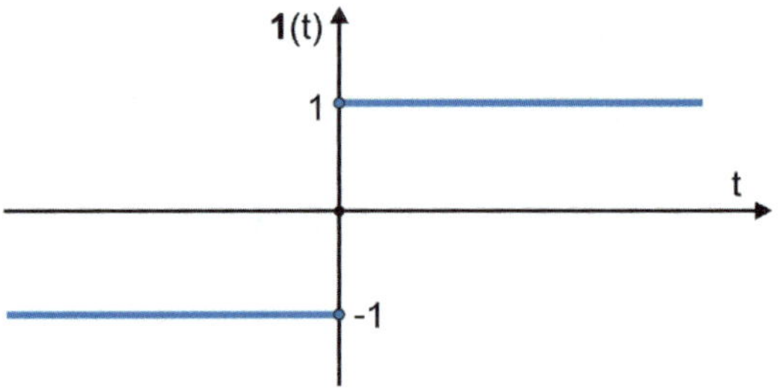

Fig. 6.17 A unit step function *sign (t)*

For linear systems the superposition principle applies, and thus the response of the analyzed system to the sum of excitations by distinct forces is equal to the sum of system responses to each of those excitations. Thus, the problem boils down to the determination of the system response to an arbitrary excitation $F(t)$, provided that we know the so-called time functions $g(t)$ corresponding to a *transfer function* $G(s)$ in the domain of complex variable s, which can be represented by the equation

$$x(t) = \int_0^t g(t-\tau)F(\tau)\mathrm{d}\tau = \int_0^t g(t)F(t-\tau)\mathrm{d}\tau. \tag{6.181}$$

Calculation of the preceding integral is substantially simplified for the case $F(t) = \delta(t)$ by virtue of certain properties of the function $\delta(t)$, which will be briefly described below [10].

The unit step function is described by the equation

$$\mathbf{1}(t) = \frac{1}{2}(1 + \mathrm{sgn}(t)), \tag{6.182}$$

where

$$\mathrm{sgn}(t) = \begin{cases} 1 & \text{for} \quad t > 0, \\ 0 & \text{for} \quad t = 0, \\ -1 & \text{for} \quad t < 0, \end{cases} \tag{6.183}$$

and the function (6.183) is presented in Fig. 6.17.

If we introduce a parameter α, then we can describe a family of functions

$$\mathbf{1}(\alpha, t) = \frac{1}{2} + \frac{1}{\pi}\arctan \alpha t, \tag{6.184}$$

where

$$\lim_{\alpha \to \infty} \mathbf{1}(\alpha, t) = \mathbf{1}(t). \tag{6.185}$$

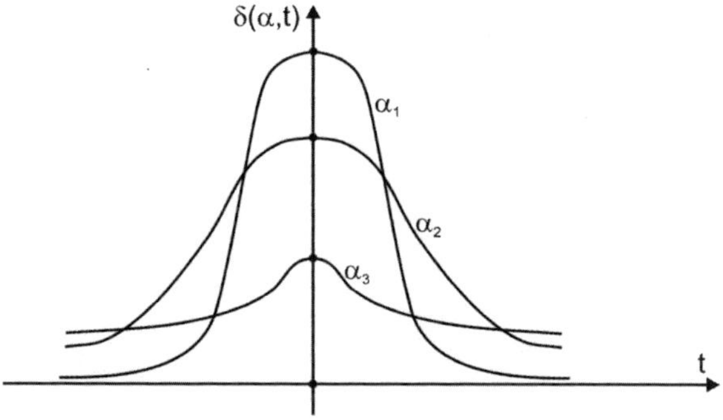

Fig. 6.18 Areas bounded by curves $\delta(\alpha_i, t)$ for different $\alpha_i, i = 1, 2, 3$

Let us introduce the following function with parameter

$$\delta(\alpha, t) = \frac{\mathrm{d}\mathbf{1}(\alpha, t)}{\mathrm{d}t} = \frac{1}{\pi}\frac{1}{1+\alpha^2 t^2}. \tag{6.186}$$

Let us note that the area bounded by the curve $\delta(\alpha, t)$ does not depend on the parameter α since

$$\int_{-\infty}^{\infty} \delta(\alpha, t)\mathrm{d}t = [\mathbf{1}(\alpha, t)]_{-\infty}^{\infty} = \frac{1}{\pi}\left[\frac{\pi}{2} + \frac{\pi}{2}\right] = 1, \tag{6.187}$$

which is also illustrated in Fig. 6.18.

For $\alpha_1 > \alpha_2 > \alpha_3$ a maximum of the function $\delta(\alpha, t)$ moves upward along the vertical axis. An impulse function $\delta(t)$ is defined as

$$\delta(t) = \lim_{\alpha \to \infty} \delta(\alpha, t). \tag{6.188}$$

The impulse function is equal to zero for $t \neq 0$ and its value for $t = 0$ is equal to infinity, and the area bounded by this function is equal to 1 ($\int_{-\infty}^{\infty} \delta(t-\tau)\mathrm{d}t = 1$). According to (6.186) we have

$$\delta(t) = \frac{\mathrm{d}\mathbf{1}(t)}{\mathrm{d}t}. \tag{6.189}$$

Below we present two important properties of the *Dirac delta function* $\delta(t)$:

1. The following equation holds true:

$$x(t)\delta(t-\tau) = x(\tau)\delta(t-\tau), \tag{6.190}$$

since the function $\delta(t-\tau)$ is different from zero only for $t = \tau$.

2. The following equation holds true:

$$\int_{-\infty}^{\infty} x(t)\delta(t-\tau)\mathrm{d}t = x(\tau). \tag{6.191}$$

Exploiting property 1 we obtain

$$\int_{-\infty}^{\infty} x(t)\delta(t-\tau)\mathrm{d}t = \int_{-\infty}^{\infty} x(\tau)\delta(t-\tau)$$
$$= x(\tau)\int_{-\infty}^{\infty} \delta(t-\tau)\mathrm{d}t = x(\tau). \tag{6.192}$$

Infinite limits of integration in (6.191) can be replaced with finite limits, but in such a way that the argument of the function δ should be equal to zero within these limits, that is,

$$\int_{-\infty}^{\infty} x(t)\delta(t-\tau)\mathrm{d}t = \int_{\tau-\varepsilon}^{\tau+\varepsilon} x(t)\delta(t-\tau)\mathrm{d}t = x(\tau). \tag{6.193}$$

If we now set $F(t) = \delta(t)$ in (6.181), then taking into account property 2 we have

$$x(t) = \int_{0}^{t} g(t-\tau)\delta(\tau)\mathrm{d}\tau = \int_{0}^{t} g(t)\delta(t-\tau)\mathrm{d}\tau = g(t), \tag{6.194}$$

and in control theory the function $g(t)$ is traditionally called an *impulse response* of a system. Using property (6.177), (6.194) takes the form

$$x(t) = x_0\cos\alpha t + \frac{\dot{x}_0}{\alpha}\sin\alpha t + \frac{J}{m\alpha}\sin\alpha t, \tag{6.195}$$

where we further assume $x_0 = x(0) = 0$, $\dot{x}_0 = \dot{x}(0) = 0$ to simplify the calculations.

The motion of a particle of mass m is described by the equation

$$x(t) = \frac{J}{m\alpha}\sin\alpha t, \quad \dot{x}(t) = \frac{J}{m}\cos\alpha t. \tag{6.196}$$

For the assumed initial conditions $x(0) = \dot{x}(0) = 0$ from (6.196) it follows that they satisfy the following initial conditions: $x(0) = 0$, $\dot{x}(0) = J/m$. This leads to the important conclusion that the action of an instantaneous impulse J is equivalent to subjecting an autonomous system to a kinematic excitation $\dot{x}(0) = J/m$ while retaining $x(0) = 0$.

A motion of mass m described by the initial conditions $x(0) = 0$, $\dot{x}(0) = J/m$ has the form

$$x(t) = x(0)\cos\alpha t + \frac{\dot{x}(0)}{\alpha}\sin\alpha t = \frac{J}{m\alpha}\sin\alpha t. \tag{6.197}$$

Let us now return to the solution of our problem of excitation of the mass by a series of impulses presented in Fig. 6.16, where this time we will exploit the *superposition principle*.

The solution in the time interval $T < t < 2T$ has the form

$$x(t) = \frac{J}{m\alpha}\left[\sin\alpha t + \sin\alpha(t-T)\right], \tag{6.198}$$

and in the time interval $2T < t < 3T$ the form

$$x(t) = \frac{J}{m\alpha}\left[\sin\alpha t + \sin\alpha(t-T) + \sin\alpha(t-2T)\right], \tag{6.199}$$

and, finally, in the time interval $T_N < t < T_{N+1}$ it takes the form

$$\begin{aligned} x(t) = &\frac{J}{m\alpha}\left[\sin\alpha t + \sin\alpha(t-T) + \sin\alpha(t-2T) + \cdots\right. \\ &\left.\cdots + \sin\alpha(t-nT)\right] = \frac{J}{m\alpha}\sum_{n=0}^{N}\sin\alpha(t-nT). \end{aligned} \tag{6.200}$$

If we consider the one-degree-of-freedom mechanical system shown in Fig. 6.2, that is, with viscous damping of the coefficient c, then, depending on the relation between c and its critical value c_{cr}, the impulse response of the system has the following form:

1. For $c < c_{cr}$:

$$g(t) = \begin{cases} 0 & \text{for} \quad t < 0, \\ \dfrac{1}{m\sqrt{\alpha^2 - h^2}}\mathrm{e}^{-ht}\sin\left(\sqrt{\alpha^2 - h^2}t\right) & \text{for} \quad t \geq 0; \end{cases} \tag{6.201}$$

2. For $c > c_{cr}$:

$$g(t) = \begin{cases} 0 & \text{for} \quad t < 0, \\ \dfrac{1}{m\sqrt{h^2 - \alpha^2}} e^{-ht} \sinh\left(\sqrt{h^2 - \alpha^2} t\right) & \text{for} \quad t \geq 0; \end{cases}$$

3. For $c = c_{cr}$:

$$g(t) = \begin{cases} 0 & \text{for} \quad t < 0, \\ \dfrac{1}{m} e^{-ht} t & \text{for} \quad t \geq 0. \end{cases}$$

The example of vibrations of the previously mentioned non-autonomous oscillator with harmonic excitation described by (6.35) together with (6.102) make it possible to explain the physical consequences of introducing the driving force $F_0 \cos \omega t$.

Let the driving force be introduced in the following way:

$$F(t) = \begin{cases} 0 & \text{for} \quad t < 0, \\ F_0 \cos \omega t & \text{for} \quad t \geq 0. \end{cases} \tag{6.202}$$

Apart from the components of the solution discussed previously, that is, the one associated with excitation by the initial conditions $x(0) = x_0$ and $\dot{x}(0) = \dot{x}_0$ and the one associated with the action of the driving force at the time $t > 0$ described by (6.56), a component describing the contribution from vibrations caused by a sudden application of the force $F_0 \cos \omega t$ at the time instant $t = 0$ appears.

This component also describes the free vibrations of the system, which are, however, completely independent of the initial conditions.

Both components describe the so-called *transient process* and for $t \to \infty$ and when the damping $h \neq 0$ is present they decay.

The solution of (6.35), which is the sum of the general solution of a homogeneous equation without a driving force and two particular solutions with a driving force at the time instant $t = 0$ in the form F_0 and for $t > 0$ in the form $F_0 \cos \omega t$, reads

$$x(t) = e^{-ht} \left[x_0 \cos \lambda t + \frac{\dot{x}_0 + h x_0}{\lambda} \sin \lambda t \right] + \frac{q}{\lambda} \int_0^t e^{-h(t-\tau)} \cos \omega \tau \sin \lambda (t - \tau) d\tau, \tag{6.203}$$

and following integration we have

$$\begin{aligned} x(t) &= \mathrm{e}^{-ht}\left[x_0\cos\lambda t + \frac{\dot{x}_0 + hx_0}{\lambda}\sin\lambda t\right] \\ &\quad + \frac{\mathrm{e}^{-ht}q}{\sqrt{(\alpha^2-\omega^2)^2+4h^2\omega^2}}\left(\cos\beta\cos\lambda t + \frac{h\cos\beta+\omega\sin\beta}{\lambda}\sin\lambda t\right) \\ &\quad + \frac{q}{\sqrt{(\alpha^2-\omega^2)^2+4h^2\omega^2}}\cos(\omega t-\beta), \end{aligned} \tag{6.204}$$

where the angle β is described by (6.58) and (6.59).

If we consider a transient state (i.e., we are interested in the initial process of vibrations), then all three terms of solution (6.204) should be taken into account. However, if we are interested in a steady state of vibrations, that is, when the contribution from free vibrations is removed by viscous damping, then the first two terms can be dropped to leave only the third term of solution (6.204), which was analyzed earlier in detail in Sect. 6.4.

In the end, it is worth noting that the steady state of vibrations with no damping ($c = 0$) can be periodic or quasiperiodic because it depends on the ratio of frequencies ω/α. If $\omega/\alpha = k/l$, where k and l are commensurable natural numbers, then the response is periodic. However, if the ratio ω/α is an irrational number, for example, $\sqrt{2}$, then the so-called quasiperiodic solution appears, where two incommensurable frequencies ω and α appear. In a linear system with one or many degrees of freedom, chaotic vibrations cannot appear.

Example 6.3. Determine the equation of motion of a one-degree-of-freedom linear oscillator with viscous damping and a piecewise linear excitation of the form

$$F(t) = F_0\left(1 - \frac{t}{t_s}\right) \qquad \text{for} \quad t < t_s$$

and

$$F(t) = 0 \qquad \text{for} \quad t \geq t_s$$

for initial conditions equal to zero.

The equation of motion of the oscillator has the form

$$m\ddot{x} + c\dot{x} + kx = \begin{cases} F_0\left(1 - \dfrac{t}{t_s}\right) & \text{for} \quad t < t_s, \\ 0 \quad \text{for} \quad t \geq t_s, \end{cases}$$

where $x(0) = 0$, $\dot{x}(0) = 0$.

According to (6.181) we have

$$x(t) = \int_0^t F(\tau)g(t-\tau)\mathrm{d}\tau,$$

where: $g(t) = \frac{1}{m\lambda}\mathrm{e}^{-ht}\sin\lambda t$, $\lambda = \sqrt{\alpha^2 - h^2}$, $2hm = c$, $\alpha^2 = \sqrt{\frac{k}{m}}$.

The desired solution is determined by the following equations:

1. In the case $t < t_s$:

$$x(t) = \frac{F_0}{m\lambda}\int_0^{t_s}\left(1 - \frac{\tau}{t_s}\right)\mathrm{e}^{-h(t-\tau)}\sin\lambda(t-\tau)\mathrm{d}\tau$$

2. In the case $t \geq t_s$:

$$x(t) = \frac{F_0}{m\lambda}\int_0^{t}\left(1 - \frac{\tau}{t_s}\right)\mathrm{e}^{-h(t-\tau)}\sin\lambda(t-\tau)\mathrm{d}\tau.$$

Let us consider case 1, in which the problem boils down to the calculation of the following integral:

$$x(t) = \frac{F_0}{m\lambda}\left(I_1^{(i)} + I_2^{(i)}\right),$$

$$I_1^{(i)} = \int_0^t \mathrm{e}^{-h(t-\tau)}\sin\lambda(t-\tau)\mathrm{d}\tau$$

$$= \mathrm{e}^{-ht}\left[\sin\lambda t\int_0^t \mathrm{e}^{h\tau}\cos\lambda\tau\mathrm{d}\tau - \cos\lambda t\int_0^t \mathrm{e}^{h\tau}\sin\lambda\tau\mathrm{d}\tau\right],$$

and

$$I_2^{(i)} = -\frac{1}{t_s}\int_0^t \tau\mathrm{e}^{-h(t-\tau)}\sin\lambda(t-\tau)\mathrm{d}\tau$$

$$= \frac{\mathrm{e}^{-ht}}{t_s}\left[\sin\lambda t\int_0^t \tau\mathrm{e}^{h\tau}\cos\lambda\tau\mathrm{d}\tau - \cos\lambda t\int_0^t \tau\mathrm{e}^{h\tau}\sin\lambda\tau\mathrm{d}\tau\right].$$

First, let us calculate the terms of the integral $I_1^{(i)}$, where during the calculations we will make use of the method of integration by parts based on the formula

$$\int f(x)g'(x)\mathrm{d}x = f(x)g(x) - \int f'(x)g(x)\mathrm{d}x,, \tag{$*$}$$

on the assumption that the derivatives of functions f and g are continuous.

Using the preceding equation we obtain

$$A \equiv \int_0^t \mathrm{e}^{h\tau}\cos\lambda\tau \mathrm{d}\tau = \left[\frac{1}{h}\mathrm{e}^{h\tau}\cos\lambda t\right]_0^t + \frac{\lambda}{h}B,$$

$$B \equiv \int_0^t \mathrm{e}^{h\tau}\sin\lambda\tau \mathrm{d}\tau = \left[\frac{1}{h}\mathrm{e}^{h\tau}\sin\lambda t\right]_0^t - \frac{\lambda}{h}A.$$

Solving the preceding system of equations with respect to unknowns A and B we obtain

$$A = \frac{\mathrm{e}^{ht}}{h^2+\lambda^2}(h\cos\lambda t + \lambda\sin\lambda t) - \frac{h}{h^2+\lambda^2},$$

$$B = \frac{\mathrm{e}^{ht}}{h^2+\lambda^2}(h\sin\lambda t - \lambda\cos\lambda t) + \frac{h}{h^2+\lambda^2}.$$

Taking into account the preceding calculations we obtain

$$\begin{aligned} I_1^{(i)} &= \mathrm{e}^{-ht}\left[A\sin\lambda t - B\cos\lambda t\right] \\ &= \frac{\lambda}{h^2+\lambda^2}\left(1-\mathrm{e}^{-ht}\cos\lambda t\right) - \frac{h\mathrm{e}^{-ht}}{h^2+\lambda^2}\sin\lambda t. \end{aligned}$$

Let us now consider the case $I_2^{(i)}$, in which the problem boils down to the calculation of two integrals

$$\int_0^t \tau\mathrm{e}^{h\tau}\cos\lambda\tau\mathrm{d}\tau \quad \text{and} \quad \int_0^t \tau\mathrm{e}^{h\tau}\sin\lambda\tau\mathrm{d}\tau.$$

Also in this case we apply the method of integration by parts. Let us set $f = \tau$ and $g' = \mathrm{e}^{h\tau}\cos\lambda\tau$, and from formula ($*$) we obtain

$$\int_0^t \tau e^{h\tau}\cos\lambda\tau d\tau = tA - \int_0^t A d\tau = \frac{te^{ht}}{h^2+\lambda^2}(h\cos\lambda t + \lambda\sin\lambda t) - \frac{ht}{h^2+\lambda^2}$$

$$-\int_0^t \frac{e^{h\tau}}{h^2+\lambda^2}(h\cos\lambda\tau + \lambda\sin\lambda\tau)\,d\tau + \frac{ht}{h^2+\lambda^2}$$

$$= \frac{te^{ht}}{h^2+\lambda^2}(h\cos\lambda t + \lambda\sin\lambda t) - \frac{h}{h^2+\lambda^2}\int_0^t e^{h\tau}\cos\lambda\tau d\tau$$

$$-\frac{\lambda}{h^2+\lambda^2}\int_0^t e^{h\tau}\sin\lambda\tau d\tau = \frac{te^{ht}}{h^2+\lambda^2}(h\cos\lambda t + \lambda\sin\lambda t)$$

$$-\frac{h}{h^2+\lambda^2}\left[\frac{e^{ht}}{h^2+\lambda^2}(h\cos\lambda t + \lambda\sin\lambda t) - \frac{h}{h^2+\lambda^2}\right]$$

$$-\frac{\lambda}{h^2+\lambda^2}\left[\frac{e^{ht}}{h^2+\lambda^2}(h\sin\lambda t - \lambda\cos\lambda t) + \frac{\lambda}{h^2+\lambda^2}\right]$$

$$= \frac{te^{ht}}{h^2+\lambda^2}(h\cos\lambda t + \lambda\sin\lambda t)$$

$$+\frac{e^{ht}}{(h^2+\lambda^2)^2}\left[(\lambda^2-h^2)\cos\lambda t - 2h\lambda\sin\lambda t\right] + \frac{h^2-\lambda^2}{(h^2+\lambda^2)^2}.$$

Similarly we calculate

$$\int_0^t \tau e^{h\tau}\sin\lambda\tau d\tau = \frac{te^{ht}}{h^2+\lambda^2}(h\sin\lambda t - \lambda\cos\lambda t)$$

$$+\frac{e^{ht}}{(h^2+\lambda^2)}\left[(\lambda^2-h^2)\sin\lambda t - 2h\lambda\cos\lambda t\right] - \frac{2h\lambda}{(h^2+\lambda^2)^2}.$$

Substituting the results obtained earlier, for case 1 we obtain

$$I_1^{(i)} = e^{-ht}\Bigg\{\sin\lambda t\left[\frac{e^{ht}}{h^2+\lambda^2}(h\cos\lambda t + \lambda\sin\lambda t) - \frac{h}{h^2+\lambda^2}\right]$$

$$-\cos\lambda t\left[\frac{e^{ht}}{h^2+\lambda^2}(h\sin\lambda t + \lambda\cos\lambda t) + \frac{\lambda}{h^2+\lambda^2}\right]\Bigg\},$$

$$
\begin{aligned}
I_2^{(i)} = -\frac{e^{-ht}}{t_s}\Bigg\{ & \sin\lambda t \Bigg\{ \frac{t e^{ht}}{h^2+\lambda^2}(h\cos\lambda t + \lambda\sin\lambda t) \\
& + \frac{e^{ht}}{(h^2+\lambda^2)^2}\left[(\lambda^2-h^2)\cos\lambda t - 2h\lambda\sin\lambda t\right] + \frac{h^2-\lambda^2}{(h^2+\lambda^2)^2}\Bigg\} \\
& - \cos\lambda t \Bigg\{ \frac{t e^{ht}}{h^2+\lambda^2}(h\sin\lambda t - \lambda\cos\lambda t) \\
& + \frac{e^{ht}}{(h^2+\lambda^2)^2}\left[(\lambda^2-h^2)\sin\lambda t - 2h\lambda\cos\lambda t\right] - \frac{2h\lambda}{(h^2+\lambda^2)^2}\Bigg\}\Bigg\}.
\end{aligned}
$$

In a similar way we carry out the calculations for case 2, that is, for $t \geq t_s$. In this case following the transformations we eventually obtain

$$
\begin{aligned}
x(t) = & \frac{F_0}{m\lambda}\left\{ \frac{1}{(h^2+\lambda^2)^2 t_s} e^{-h(t-t_s)} \left[(h^2-\lambda^2)\sin\lambda(t-t_s) + 2h\lambda\cos\lambda(t-t_s) \right] \right\} \\
& - \frac{F_0}{m}\Bigg\{ \frac{1}{(h^2+\lambda^2)^2} e^{-ht} \Bigg[\left(1 + \frac{2h}{h^2+\lambda^2}\right)\cos\lambda t \\
& + \frac{1}{\lambda}\left(h + \frac{h^2-\lambda^2}{(h^2+\lambda^2)t_s}\right)\sin\lambda t \Bigg]\Bigg\}.
\end{aligned}
$$

References

1. V.I. Arnold, *Mathematical Methods of Classical Mechanics* (Springer, Berlin, 1989)
2. N.V. Butenin, N.A. Fufayev, *Introduction to Analytical Mechanics* (Nauka, Moscow, 1991) (in Russian)
3. S.T. Thornton, J.B. Marion, *Classical Dynamics of Particles and Systems* (Saunders College Publishers, New York, 1995)
4. C.D. Murray, S.F. Dermott, *Solar System Dynamics.* (Cambridge University Press, Cambridge, 1999)
5. M.D. Ardema, *Newton-Euler Dynamics* (Springer, Berlin, 2005)
6. R. Abraham, J.E. Marsden, *Foundations of Mechanics* (AMS Chelsea Publishing, Chelsea, 2008)
7. A.J. Brizard, *An Introduction to Lagrangian Mechanics* (World Scientific, Singapore, 2008)
8. C. Gignoux, B. Silvestre-Brac, *Solved Problems in Lagrangian and Hamiltonian Mechanics* (Springer, Dordrecht, 2009)
9. R.D. Gregory, *Classical Mechanics* (Cambridge University Press, Cambridge, 2011)
10. J.Z. Tsypkin, *Introduction to the Theory of Automatic Systems* (Nauka, Moscow, 1977) (in Russian)
11. Z. Parszewski, *Vibration and Dynamics of Machines* (WNT, Warsaw, 1982) (in Polish)
12. J. Awrejcewicz, *Oscillations of Lumped Deterministic Systems* (Technical University of Lodz Press, Lodz, 1996) (in Polish)
13. J. Leyko, *General Mechanics* (PWN, Warsaw, 1996) (in Polish)

14. V.E. Zhuravlev, *Fundamentals of Theoretical Mechanics* (Fizmatlit, Moscow, 2001) (in Russian)
15. Z. Koruba, J.W. Osiecki, *Elements of Advanced Mechanics* (Kielce University of Technology Press, Kielce, 2007) (in Polish)
16. J. Awrejcewicz, *Mathematical Modelling of Systems* (WNT, Warsaw, 2007) (in Polish)

Chapter 7
Elements of Dynamics of Planets

7.1 Introduction

For a long time the structure of the universe has aroused admiration and dread, and for a long time people have tried to explain the phenomena of the cosmos. However, not much has changed for hundreds of years in this field of human activity.

If a body moves in space as a real element of the Universe (a planet) or an object artificially introduced into space by humans (e.g., an artificial satellite, spacecraft), then it is subjected to, for example, gravitational interactions with other bodies (planets), forces of resistance associated with atmospheres surrounding the planets (bodies) that can be non-homogeneous, and forces caused by the pressure of the solar wind.

A simple problem of planetary dynamics is the so-called two-body problem, which boils down to the analysis of the dynamics of two particles. Despite the simplicity of the statement, this problem is capable of modeling the motion of planets of the Solar System or artificial satellites because the forces of resistance mentioned earlier are small compared to the forces of gravitational interaction of a planet and the Sun, or an artificial satellite and the Earth.

From such a point of view a "particle" is a certain "asymptotic" approximation of the object whose dimensions are negligibly small compared to its distances from other "particles" of the considered system. Let us note that, apart from the mass, such a particle can have an electric charge, be situated in a gravitational force field, etc. In a Newtonian description of motion of planets, the notion of *free motion* means that an arbitrary particle in the $\mathbf{R}^3$ space is acted upon by each of the remaining particles. The movements of these particles are the results of the particles' mutual interactions, for instance, in the form of gravitational forces.

In order to observe the motion of particles in such a physical space, we have to introduce certain *reference systems*. These are real or virtual physical objects that either move or are stationary and from where a subject makes observations or carries out measurements of motion.

J. Awrejcewicz, *Classical Mechanics: Dynamics*, Advances in Mechanics and Mathematics 29, DOI 10.1007/978-1-4614-3740-6_7,

Much like during the analysis of DMSs, also here it is convenient to introduce a certain right-handed Cartesian coordinate system and determine the positions of particles by their position vectors $\mathbf{r}_n$, $n = 1, \ldots, N$. These vectors are time dependent $\mathbf{r}_n = \mathbf{r}_n(t)$, which follows from the force interactions affecting those particles.

There might be infinitely many reference systems, but they are related by certain relationships described by *Galileo's principle of relativity*, which includes the following elements.

1. All mechanical phenomena take place in $\mathbf{R}^3$ space, which is common (the same) for all objects (bodies) moving in this space.
2. The phenomena taking place in this space are ordered in time, and this order is independent of the motion of bodies.
3. Descriptions of the phenomena (or the motion of bodies) are identical in each of the adopted coordinate systems, on the condition that they remain at rest or move in uniform (rectilinear) motion.

If we take two Cartesian coordinate systems $OX_1X_2X_3$ and $OX_1'X_2'X_3'$, then the relationships between the position vectors of a particle and time in both systems read

$$\begin{aligned} \mathbf{r}' &= \mathbf{r} - \mathbf{v}t, \\ t' &= t, \end{aligned} \tag{7.1}$$

where $\mathbf{v} = \text{const}$ is the velocity of the system $OX_1'X_2'X_3'$ with respect to $OX_1X_2X_3$. Both systems $OX_1X_2X_3$ and $OX_1'X_2'X_3'$ are inertial (their axes are constantly mutually parallel and move relative to each other in uniform rectilinear motion with velocity $\mathbf{v}$). At the initial time instant $t' = t = 0$ we have $O' = O$.

Differentiating (7.1) we obtain

$$\begin{aligned} \dot{\mathbf{r}}' &= \dot{\mathbf{r}} - \mathbf{v}, \\ \ddot{\mathbf{r}}' &= \ddot{\mathbf{r}}. \end{aligned} \tag{7.2}$$

In other words, the laws of dynamics should be according to Galileo's principle of relativity *invariant* under transformations (7.1) and (7.2). Let us note that the notion introduced by Newton of force being a vector does not represent certain attributes of particles, such as gravitational mass or electric charge.

On an arbitrary particle A_n of a material system, in general, acts a force that is dependent on all the remaining particles at instant t of the form $\mathbf{F}_n(\mathbf{r}_1, \ldots, \mathbf{r}_N, \dot{\mathbf{r}}_1, \ldots, \dot{\mathbf{r}}_N, t)$, which means that it also depends on the position and velocity of the analyzed particle A_n. According to Newton's second law, the motion of every particle in the space $\mathbf{R}^3$ is described by a system of N second-order vector differential equations of the form

$$m_n\ddot{\mathbf{r}}_n(t) = \mathbf{F}_n(\mathbf{r}_1, \ldots, \mathbf{r}_N, \dot{\mathbf{r}}_1, \ldots, \dot{\mathbf{r}}_N, t), \quad n = 1, \ldots, N. \tag{7.3}$$

Systems whose motion is described by (7.3) are called *inertial systems*. An arbitrary method leading to the determination of the relations $\mathbf{r}_n(t)$ is called the *integration of equations of motion* (7.3).

In the general case the obtained system of non-linear differential equations (7.3) is not integrable. If exclusively by means of analytical operations (i.e., not using numerical procedures) we are able to determine the solution of system (7.3), then we say that system (7.3) is integrable by *quadratures*.

It turns out that systems like (7.3) written in an arbitrary way, in general, do not model the motion of a given mechanical systems. That is because they have to satisfy Galileo's principle of relativity as well as certain properties of time and space. Taking into account relations (7.2), the invariance of (7.3) under a Galilean transformation (7.1) leads to the determination of the same magnitudes of forces in the form

$$\begin{aligned}\frac{1}{m_n'}\mathbf{F}_n'(\mathbf{r}_1',\ldots,\dot{\mathbf{r}}_N' t') &= \frac{1}{m_n'}\mathbf{F}_n'(\mathbf{r}_1 - \mathbf{v}t,\ldots,\dot{\mathbf{r}}_N - \mathbf{v},t)\\ &= \frac{1}{m_n}\mathbf{F}_n(\mathbf{r}_1,\ldots,\dot{\mathbf{r}}_N,t).\end{aligned} \tag{7.4}$$

If in the considered space we are faced with small speeds compared to the speed of light, then masses of bodies do not change, and the forces $\mathbf{F}_n(\mathbf{r}_1,\ldots,\mathbf{r}_n,\ldots,\mathbf{r}_N,\dot{\mathbf{r}}_1,\ldots,\dot{\mathbf{r}}_n,\ldots,\dot{\mathbf{r}}_N,t)$ depend on the radius vector $\mathbf{r}_n$ of particle A_n, and the velocity of this vector $\dot{\mathbf{r}}_n$ through the difference $\mathbf{r}_n - \mathbf{r}_j$, where $j = 1,\ldots,N$, $j \neq n$, because $|\mathbf{r}_n - \mathbf{r}_j|$ denotes the distance between particle n and each of the remaining particles j.

The considered $\mathbf{R}^3$ space should be physically *isotropic*. In other words, motion along a straight line of a body subjected to a certain resultant force is the same regardless of the direction in which the force acts. *Homogeneity of* $\mathbf{R}^3$ *space* means that phenomena proceed identically in different locations in space at the same time instant. *Homogeneity of time* means that the phenomena proceed identically regardless of the time instant at which they occur. Homogeneity and isotropy imply an invariance of (7.3) under arbitrary *translations* and *rotations* of the coordinate system. The homogeneity of time implies an invariance of (7.3) under arbitrary translations of a particle along the time axis.

Space isotropy means that the field of forces $\mathbf{F}_n(\mathbf{r}_1,\ldots,\dot{\mathbf{r}}_N,t)$ is a vector field, that is, the positions of particle A_n and the forces acting on it in the coordinate systems $OX_1X_2X_3$ and $OX_1'X_2'X_3'$ are given by the following formulas:

$$\begin{aligned}\mathbf{r}_n' &= \mathbf{A}\mathbf{r}_n,\\ \mathbf{F}_n'(\mathbf{r}_1',\ldots,\dot{\mathbf{r}}_N',t') &= \mathbf{A}\mathbf{F}_n(\mathbf{r}_1,\ldots,\dot{\mathbf{r}}_N,t),\end{aligned} \tag{7.5}$$

where, as was shown earlier, $\mathbf{A}$ is the constant rotation matrix. Relations (7.5) indicate that the quantities occurring there are vectors.

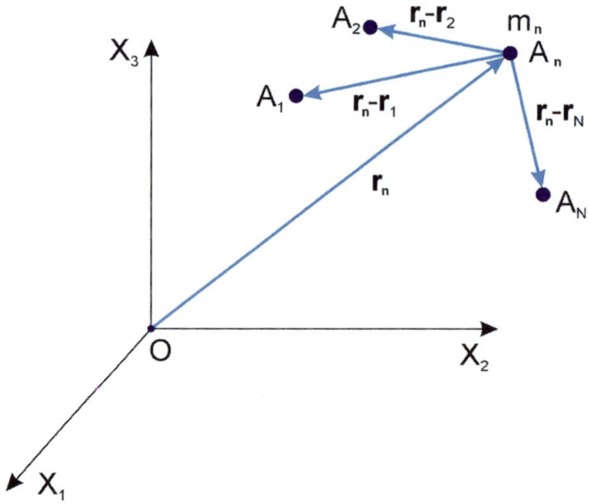

Fig. 7.1 A schematic of relation (7.8) with vectors $\mathbf{r}_1$, $\mathbf{r}_2$, $\mathbf{r}_N$ not drawn for clarity

The homogeneity of space implies that the following relation is satisfied:

$$\mathbf{F}_n(\mathbf{r}_1 + \boldsymbol{c}, \ldots, \mathbf{r}_N + \boldsymbol{c}, \dot{\mathbf{r}}_1, \ldots, \dot{\mathbf{r}}_N, t) = \mathbf{F}_n(\mathbf{r}_1, \ldots, \mathbf{r}_N, \dot{\mathbf{r}}_1, \ldots, \dot{\mathbf{r}}_N, t),$$
$$\mathbf{r}_n - \mathbf{r}_j = \boldsymbol{c}, \quad n, j = 1, \ldots, N. \tag{7.6}$$

Likewise, the homogeneity of time implies that the following relation is satisfied:

$$\mathbf{F}_n(\mathbf{r}_1, \ldots, \dot{\mathbf{r}}_N, t + t_0) = \mathbf{F}_n(\mathbf{r}_1, \ldots, \dot{\mathbf{r}}_N, t). \tag{7.7}$$

Because the forces acting on particles are generated by these particles, they should be dependent on the relative positions and velocities of those particles, that is,

$$\mathbf{F}_n = \mathbf{F}_n(\mathbf{r}_n - r_1, \ldots, \mathbf{r}_n - \mathbf{r}_j, \ldots, \dot{\mathbf{r}}_n - \dot{\mathbf{r}}_2, \ldots, \dot{\mathbf{r}}_n - \dot{\mathbf{r}}_N), \quad n = 1, \ldots, N, \tag{7.8}$$

which is presented in Fig. 7.1.

From relations (7.8) it follows that the force acting on a particle of mass m_n depends on the interactions with each of the remaining particles of the system through vectors $\overrightarrow{A_n A_1}$, $\overrightarrow{A_n A_2}, \ldots$, but it does not depend on the mutual interactions of other particles, that is, neither on vectors $\overrightarrow{A_n A_j}$, $n, j = 1, \ldots, N$, and $n \neq j$, nor on their time derivatives. As a result we obtain

$$\mathbf{F}_n(\ldots, \mathbf{r}_n - \mathbf{r}_j, \ldots) = \sum_{\substack{n,j=1\ldots N \\ (n \neq j)}} \mathbf{F}_{nj}(\mathbf{r}_n - \mathbf{r}_j, \dot{\mathbf{r}}_n - \dot{\mathbf{r}}_j). \tag{7.9}$$

If in (7.9) we subsequently introduce the symmetry principle, that is, $\mathbf{F}_{nj} = \mathbf{F}_{jn}$, which means that

$$\mathbf{F}_{nj}(\mathbf{r}_n - \mathbf{r}_j, \dot{\mathbf{r}}_n - \dot{\mathbf{r}}_j) = -\mathbf{F}_{jn}(\mathbf{r}_j - \mathbf{r}_n, \dot{\mathbf{r}}_j - \dot{\mathbf{r}}_n), \tag{7.10}$$

then we obtain Newton's third law.

From (7.10) it follows that $\mathbf{F}_{nj} + \mathbf{F}_{jn} = \mathbf{0}$, that is, the forces act along the same direction, which, however, is generally different from the direction of $\mathbf{r}_n - \mathbf{r}_j$. The forces will act along a straight line connecting particles A_n and A_j if they are independent of the velocity $\dot{\mathbf{r}}_n - \dot{\mathbf{r}}_j$.

7.2 Potential Force Fields

The previously mentioned gravitational or electrostatic forces are generated by so-called *potential force fields*. Those forces do not depend on the velocities of particles, and the work done by them on a closed path vanishes (i.e., there is no dissipation or delivery of energy to the system). Such forces have certain corresponding functions called *potentials of forces*. For example, the potential (the potential energy) is the energy corresponding to compression or elongation of a spring.

Let the force from an arbitrary point of the $\mathbf{R}^3$ space, whose position is described by the radius vector $\mathbf{r}$ (Fig. 7.1), act on particle A_n, which is determined by vector $\mathbf{r}_n$:

$$\mathbf{F}_n(\mathbf{r} - \mathbf{r}_n) = -\nabla_r V_n(|\mathbf{r} - \mathbf{r}_n|), \tag{7.11}$$

where $V_n(|\mathbf{r} - \mathbf{r}_n|)$ is the potential of a force field and differentiation is carried out with respect to vector $\mathbf{r}$.

Let us now consider any two points n and j of the field. According to Newton's third law we have

$$\nabla_{rj} V_n(|\mathbf{r}_j - \mathbf{r}_n|) + \nabla_{rn} V_j(|\mathbf{r}_n - \mathbf{r}_j|) = 0. \tag{7.12}$$

Because $|\mathbf{r}_j - \mathbf{r}_n| = |\mathbf{r}_n - \mathbf{r}_j|$, so $V_n = V_j \equiv V_{nj}(|\mathbf{r}|) = V_{jn}(|\mathbf{r}|)$, where $\mathbf{r}$ describes the position of an arbitrary point of the potential force field.

Because every particle $n = 1, \ldots, N$ of a force field moves with the velocity $\dot{\mathbf{r}}_n$, so the kinetic energy of the system of particles

$$T = \frac{1}{2}\sum_{n=1}^{N} m_n \dot{\mathbf{r}}_n^2, \tag{7.13}$$

and the potential energy

$$V = \frac{1}{2}\sum_{\substack{n,j=1 \\ n \neq j}}^{N} V_{nj}(|\mathbf{r}_n - \mathbf{r}_j|). \tag{7.14}$$

The total energy of the system remains unchanged during its motion, that is, the integral of motion of the system has the form

$$E = T + V = \text{const.} \tag{7.15}$$

7.3 Dynamics of Two Particles

This chapter was written on the basis of [1–5] and the author's own studies. Let us limit ourselves in our calculations in physical space to $N = 2$. The problem boils down to the following. Knowing the positions and velocities of particles of masses m_n $(n = 1, 2)$ in the adopted coordinate system, determine the motion of these particles on the assumption that between them there exist gravitational forces that are consistent with Newton's law of gravitation (see Markeev [1]).

Let us introduce an absolute Cartesian coordinate system whose origin O lies at the mass center of the Solar System and whose axes are directed toward certain fixed stars. The introduction of such a coordinate system requires some commentary. Galileo's principle of relativity (the homogeneity of time, space, and the $\mathbf{R}^3$ space isotropy) reduced our calculations to the *internal* interactions of the particles. In the Solar System $\sum_{n=1}^{N} m_n \sim 10^{-3} M_S$, where M_S is the mass of the Sun.

The action of all the planets of the Solar System on the Sun is negligibly small because of the Sun's large mass, and we can assume that the Sun remains at rest and take as its mass center the origin of the absolute coordinate system $OX_1X_2X_3$. With point $A_1 = O'$ we associate the coordinate system $O'X_1'X_2'X_3'$ of axes parallel to the absolute system $OX_1X_2X_3$ (Fig. 7.2) because the dimensions of the body tend to zero. Another point A_2 is described in the absolute coordinate system by radius vector $\mathbf{r}_2$.

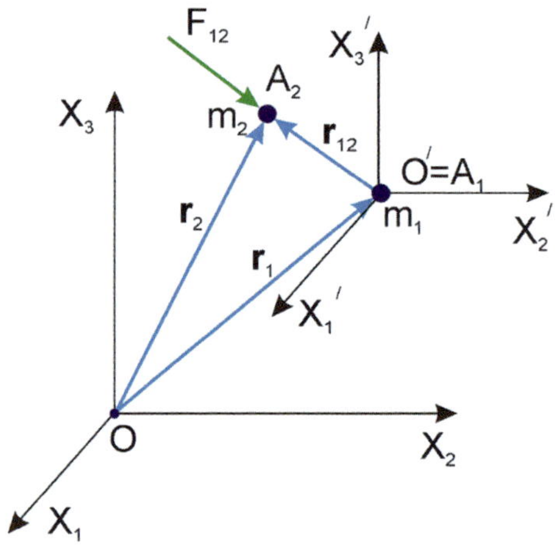

Fig. 7.2 Motion of two particles in $\mathbf{R}^3$ space

The force acting on mass m_2 from the side of mass m_1 has the following form [see (7.11)]:

$$\mathbf{F}_2(\mathbf{r}-\mathbf{r}_1) \equiv \mathbf{F}_{12} = -\nabla_{r_1} V_2(|\mathbf{r}-\mathbf{r}_2|) = -G\frac{m_1 m_2}{r_{12}^3}\mathbf{r}_{12},$$

and the force acting on mass m_1 from the side of mass m_2 has the form

$$\mathbf{F}_{21} = -G\frac{m_1 m_2}{r_{21}^3}\mathbf{r}_{21} = G\frac{m_1 m_2}{r_{12}^3}\mathbf{r}_{12},$$

where G is the universal gravitational constant.

Particles m_1 and m_2 remain in equilibrium under the action of the following forces:

$$m_2\frac{\mathrm{d}^2\mathbf{r}_2}{\mathrm{d}t^2} = \mathbf{F}_{12}, \quad m_1\frac{\mathrm{d}^2\mathbf{r}_1}{\mathrm{d}t^2} = \mathbf{F}_{21},$$

that is,

$$m_1\ddot{\mathbf{r}}_1 = G\frac{m_2 m_1}{r_{12}^2}\frac{\mathbf{r}_{12}}{|\mathbf{r}_{12}|},$$
$$m_2\ddot{\mathbf{r}}_2 = G\frac{m_1 m_2}{r_{12}^2}\frac{\mathbf{r}_{12}}{|\mathbf{r}_{12}|}. \tag{7.16}$$

Then, from Fig. 7.2 it follows that

$$\mathbf{r}_1 + \mathbf{r}_{12} = \mathbf{r}_2, \tag{7.17}$$

that is, differentiating and taking into account relation (7.16) we have

$$\ddot{\mathbf{r}}_{12} = \ddot{\mathbf{r}}_2 - \ddot{\mathbf{r}}_1 = -k\frac{\mathbf{r}_{12}}{r_{12}^3}, \tag{7.18}$$

where $k = G(m_1 + m_2)$.

The preceding equation describes the motion of a system of one degree of freedom, that is, the motion of a body of mass m_2 with respect to point O of mass m_1. Integrating (7.18) we determine $\mathbf{r}_{12}(t)$, and consequently we determine $\mathbf{r}_1(t)$ and $\mathbf{r}_2(t)$ after integrating the now linear (7.16) whose right-hand sides are known.

It is also possible to determine the motion of the mass center of particles A_1 and A_2. Let point C be the mass center of the moving particles A_1 and A_2, and let its position be described by radius vector $\mathbf{r}_C$. By definition we have

$$\mathbf{r}_C = \frac{m_1\mathbf{r}_1 + m_2\mathbf{r}_2}{m_1 + m_2},$$

and solving a system consisting of the preceding equation and (7.17) we obtain

$$\mathbf{r}_2 = \mathbf{r}_C + \frac{m_1}{m_1 + m_2}\mathbf{r}_{12},$$
$$\mathbf{r}_1 = \mathbf{r}_C - \frac{m_2}{m_1 + m_2}\mathbf{r}_{12}. \tag{7.19}$$

The mass center C moves in uniform rectilinear motion. Knowing vectors $\mathbf{r}_{12}(t)$, $\mathbf{r}_1(t)$, and $\mathbf{r}_2(t)$ we can determine $\mathbf{r}_C(t)$. The obtained differential (7.18) can be interpreted as the motion of a reduced mass with respect to a fixed attracting center O'.

Subtracting (7.16) by sides we have

$$m_2\ddot{\mathbf{r}}_2 - m_1\ddot{\mathbf{r}}_1 = -2G\frac{m_1 m_2}{r_{12}^2}\frac{\mathbf{r}_{12}}{|\mathbf{r}_{12}|},$$

and from (7.18) we obtain

$$\ddot{\mathbf{r}}_2 - \ddot{\mathbf{r}}_1 = -G\frac{m_1 + m_2}{r_{12}^2}\frac{\mathbf{r}_{12}}{|\mathbf{r}_{12}|}.$$

Dividing the two preceding equations by sides we have

$$\frac{m_2\ddot{\mathbf{r}}_2 - m_1\ddot{\mathbf{r}}_1}{\ddot{\mathbf{r}}_2 - \ddot{\mathbf{r}}_1} = 2\frac{m_1 m_2}{m_1 + m_2},$$

that is,

$$2\frac{m_1 m_2}{m_1 + m_2}(\ddot{\mathbf{r}}_2 - \ddot{\mathbf{r}}_1) = -2G\frac{m_1 m_2}{r_{12}^2}\frac{\mathbf{r}_{12}}{|\mathbf{r}_{12}|}.$$

Eventually, the equation of vibrations of such a system with one degree of freedom has the form

$$\frac{m_1 m_2}{m_1 + m_2}\ddot{\mathbf{r}}_{12} = -G\frac{m_1 m_2}{r_{12}^2}\frac{\mathbf{r}_{12}}{|\mathbf{r}_{12}|} = -\nabla V_{12}(\mathbf{r}). \tag{7.20}$$

In the considered system we are dealing with the following three integrals of motion:

1. Total energy:

$$E = \frac{1}{2}m_1\dot{\mathbf{r}}_1^2 + \frac{1}{2}m_2\dot{\mathbf{r}}_2^2 + V\left(|\mathbf{r}_1 - \mathbf{r}_2|\right); \tag{7.21}$$

2. Total system momentum:

$$\mathbf{P} = m_1\dot{\mathbf{r}}_1 + m_2\dot{\mathbf{r}}_2; \tag{7.22}$$

3. Total system angular momentum:

$$\mathbf{K} = \mathbf{r}_1 \times m_1 \dot{\mathbf{r}}_1 + \mathbf{r}_2 \times m_2 \dot{\mathbf{r}}_2. \tag{7.23}$$

Equation (7.22) proves the linear dependence of vectors $\mathbf{r}_1$ and $\mathbf{r}_2$, which was used during the derivation of (7.18). Integrating (7.18) we obtain

$$\dot{\mathbf{r}}_{12} = \dot{\mathbf{r}}_2 - \dot{\mathbf{r}}_1 = -k \int \frac{\mathbf{r}_{12}}{r_{12}^3} \mathrm{d}t,$$

and multiplying (cross product) by $\mathbf{r}_{12}$ we have

$$\mathbf{r}_{12} \times \dot{\mathbf{r}}_{12} = -k \mathbf{r}_{12} \times \int \frac{\mathbf{r}_{12}}{r_{12}^3} \mathrm{d}t = \mathbf{b}(t).$$

Differentiation with respect to time the right-hand side of the last equation yields

$$\frac{\mathrm{d}\mathbf{b}}{\mathrm{d}t} = \mathbf{0}.$$

Thus we have demonstrated the validity of the following relation:

$$\mathbf{r} \times \mathbf{v} = \mathbf{b} = \text{const}, \tag{7.24}$$

which is also called a *surface integral* (setting $\mathbf{r}_{12} = \mathbf{r}$, $\dot{\mathbf{r}}_{12} = \mathbf{v}$). This name is justified by the following physical interpretation. According to (7.24) we have

$$\mathbf{b} \equiv b_1 \mathbf{E}_1' + b_2 \mathbf{E}_2' + b_3 \mathbf{E}_3' = \begin{vmatrix} \mathbf{E}_1' & \mathbf{E}_2' & \mathbf{E}_3' \\ r_1' & r_2' & r_3' \\ v_1' & v_3' & v_3' \end{vmatrix}$$

$$= \mathbf{E}_1'(r_2' v_3' - r_3' v_2') + \mathbf{E}_2'(r_3' v_1' - r_1' v_3') + \mathbf{E}_3'(r_1' v_2' - r_2' v_1'). \tag{7.25}$$

Because $\mathbf{b} = \text{const}$, the magnitude of $\mathbf{b}$ can be calculated for an arbitrary time instant, for example, $t = t_0$. If $\mathbf{b} = \mathbf{0}$, from (7.25) we obtain

$$r_2' v_3' = r_3' v_2', \quad r_3' v_1' = r_1' v_3', \quad r_1' v_2' = r_2' v_1'. \tag{7.26}$$

Recall that an analogous problem was already considered earlier (Chap. 5). In a similar way one can show that

$$\frac{v_1'}{r_1'} = \frac{v_2'}{r_2'} = \frac{v_3'}{r_3'} = \sigma'. \tag{7.27}$$

Equation (7.27) describes a straight line in the coordinate system $O'X_1'X_2'X_3'$, where $r'_n = x'_n - x_{nO'}$, $n = 1, 2, 3$. In this case during the motion of two particles A_1 and A_2 the tip of vector $\mathbf{r}$ moves along a straight line.

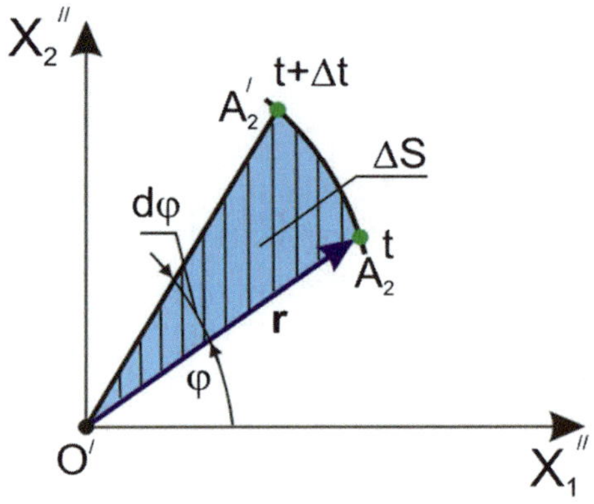

Fig. 7.3 Geometric interpretation of the *surface integral*

In the general case, the motion of particles takes place on a plane determined by vectors $\mathbf{r}$ and $\mathbf{v}$. According to (7.24), the plane is perpendicular to vector $\mathbf{b}$. In other words, vector $\mathbf{b}$ is normal to the aforementioned plane. An equation of the plane has the form

$$\mathbf{b} \circ \mathbf{r} = 0 \tag{7.28}$$

or, in expanded form,

$$b_1 x_1' + b_2 x_2' + b_3 x_3' = 0, \tag{7.29}$$

where $A_2 = A_2(x_1', x_2', x_3')$ (Fig. 7.2).

A hodograph of vector $\mathbf{r}$ (of particle A_2) is a curve lying on the plane given by (7.29). Because we know that the path of particle A_2 is a plane curve, let us take a new coordinate system $O''X_1''X_2''X_3''$ such that the path of motion of a particle lies, for instance, on the plane $O'X_1''X_2''$. Vector $\mathbf{b}$, as a vector normal to the plane, has the form $\mathbf{b} = [0, 0, b_3'']^T$. On the other hand, vector $\mathbf{b}$ can be expressed in terms of the coordinates of $\mathbf{r}$ and $\dot{\mathbf{r}}$, and according to formula (7.25) we have

$$b_3'' = x_1''\dot{x}_2'' - x_2''\dot{x}_1''. \tag{7.30}$$

Let us now introduce polar coordinates (r, φ) on the plane $O'X_1''X_2''$ (Fig. 7.3). According to Fig. 7.3 we have

$$\begin{aligned} x_1'' &= r\cos\varphi, & \dot{x}_1'' &= \dot{r}\cos\varphi - r\dot{\varphi}\sin\varphi, \\ x_2'' &= r\sin\varphi, & \dot{x}_2'' &= \dot{r}\sin\varphi - r\dot{\varphi}\cos\varphi. \end{aligned} \tag{7.31}$$

A surface integral, described by (7.24), in this case reduces to the equation

$$r\cos\varphi(\dot{r}\sin\varphi + r\dot{\varphi}\cos\varphi) - r\sin\varphi(\dot{r}\cos\varphi - r\dot{\varphi}\sin\varphi) = b_3'',$$

or, following transformation, to

$$r^2\dot{\varphi} = b_3''. \tag{7.32}$$

If the particle moves from point A_2 at time instant t to point A_2' at time instant $t + \Delta t$, then the position vector of the particle sweeps the area of a curvilinear triangle $O'A_2A_2'$. The area of this triangle is approximately equal to

$$\Delta S = \frac{1}{2}r^2\Delta\varphi,$$

and dividing by Δt and proceeding to the limit as $\Delta t \to 0$ we obtain

$$\frac{\mathrm{d}S}{\mathrm{d}t} = \frac{1}{2}r^2\frac{\Delta\varphi}{\Delta t} = \frac{1}{2}b_3''. \tag{7.33}$$

The quantity $\frac{\mathrm{d}S}{\mathrm{d}t}$ is called a sector velocity, and, as was shown, it is constant. The obtained equation justifies Kepler's second law, discussed in Chap. 1 of [3].

Premultiplying (7.18) by vector $\mathbf{b}$ and using relation (7.24) we obtain

$$\mathbf{b} \times \ddot{\mathbf{r}} = -\frac{k}{r^3}(\mathbf{r} \times \dot{\mathbf{r}}) \times \mathbf{r}. \tag{7.34}$$

Because

$$\begin{aligned}(\mathbf{r} \times \dot{\mathbf{r}}) \times \mathbf{r} &= \dot{\mathbf{r}}(\mathbf{r} \circ \mathbf{r}) - \mathbf{r}(\mathbf{r} \circ \dot{\mathbf{r}}) \\ &= \dot{\mathbf{r}}r^2 - \mathbf{r}r\dot{r} = r^3\frac{\dot{\mathbf{r}} \circ \mathbf{r} - \mathbf{r} \circ \dot{\mathbf{r}}}{r^2} = r^3\frac{\mathrm{d}}{\mathrm{d}t}\left(\frac{\mathbf{r}}{r}\right)\end{aligned}$$

and

$$\mathbf{b} \times \ddot{\mathbf{r}} = \frac{\mathrm{d}}{\mathrm{d}t}(\mathbf{b} \times \mathbf{v}),$$

so substituting the preceding relations into (7.34) we obtain

$$\frac{\mathrm{d}}{\mathrm{d}t}(\mathbf{b} \times \mathbf{v}) = -k\frac{\mathrm{d}}{\mathrm{d}t}\left(\frac{\mathbf{r}}{r}\right), \tag{7.35}$$

hence, integrating we get

$$\mathbf{b} \times \mathbf{v} + k\frac{\mathbf{r}}{r} = -\mathbf{b}_{\mathrm{L}}, \tag{7.36}$$

where $\mathbf{b}_{\mathrm{L}} = \mathrm{const}$ is called the *Laplace vector*, and relation (7.36) is called a *Laplace integral* (the minus sign is formally introduced as required by further transformations).

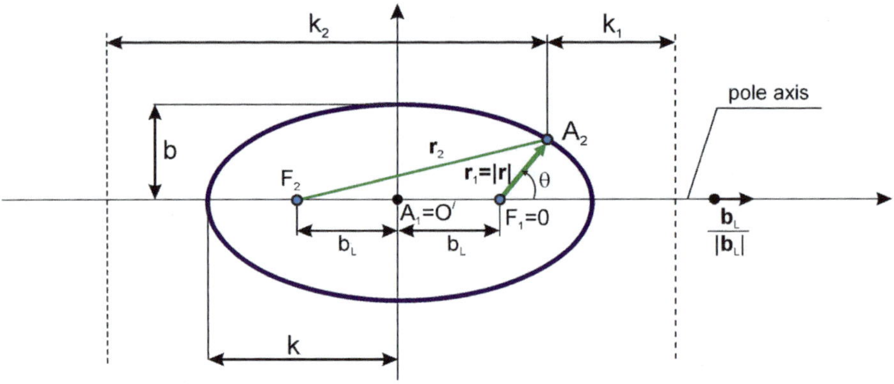

Fig. 7.4 Properties of the path of point A_2 and the geometric interpretation of vectors **r** and $\mathbf{b}_{\mathrm{L}}$

It turns out that the Laplace vector lies in the plane of the orbit. Multiplying (7.36) through by **b** we have

$$\mathbf{b} \circ (\mathbf{b} \times \mathbf{v}) + \frac{k}{r}(\mathbf{b} \circ \mathbf{r}) = \mathbf{b} \circ \mathbf{b}_{\mathrm{L}}. \tag{7.37}$$

Taking into account (7.28) from relation (7.37) we obtain

$$\mathbf{b} \circ \mathbf{b}_{\mathrm{L}} = 0. \tag{7.38}$$

This means that the Laplace vector and the vector normal to the surface sector are perpendicular to each other. Following scalar multiplication of (7.36) by **v** we have

$$\mathbf{r} \circ (\mathbf{b} \times \mathbf{v}) + \frac{k}{r}(\mathbf{r} \circ \mathbf{r}) = -\mathbf{r} \circ \mathbf{b}_{\mathrm{L}}. \tag{7.39}$$

Using the rule of cyclic permutation of factors

$$\mathbf{r} \circ (\mathbf{b} \times \mathbf{v}) = \mathbf{b} \circ \mathbf{v} \times \mathbf{r} = -\mathbf{b} \circ \mathbf{r} \times \mathbf{v} = -\mathbf{b} \circ \mathbf{b} = -b^2, \tag{7.40}$$

from relation (7.39) we obtain

$$-b^2 + kr = -b_{\mathrm{L}} r \cos\theta. \tag{7.41}$$

The right-hand side follows from the fact that vectors $\mathbf{b}_{\mathrm{L}}$ and **r** lie in one plane (Fig. 7.4). The angle θ is called an angle of *true anomaly*.

Because the Laplace vector $\mathbf{b}_{\mathrm{L}} = \text{const}$, the position of particle A_2 is described by the polar coordinates $(r,\ \theta)$.

From relation (7.41) we get

$$r = \frac{b^2}{k + b_{\mathrm{L}} \cos\theta} = \frac{(b^2/k)}{(k + b_{\mathrm{L}} \cos\theta)/k} = \frac{p}{1 + e\cos\theta}, \tag{7.42}$$

where

$$p = b^2 k^{-1}, \quad e = b_{\mathrm{L}} k^{-1}. \tag{7.43}$$

Recall that p is called a *semilatus rectum* (it is a positive value of an ordinate corresponding to the focus of the ellipse), and the quantity e an *eccentricity of an ellipse*.

The polar equation of an ellipse (7.42) is obtained on the assumption that point $A_1 = O'$ is located at the center of an ellipse, and a polar axis lies on a major axis and is directed toward the nearest vertex of the ellipse.

Let us recall some properties of an ellipse (Fig. 7.4).

The path (an orbit) of point A_2 with respect to point O' is an ellipse.

The position vectors of an ellipse are equal to $r_1 = k + ex'$ and $r_2 = k - ex'$, where x' is an abscissa of an arbitrary point A_2, and hence we get $r_1 + r_2 = 2k$.

The directrices of an ellipse $k_{1,2} = \pm k^2/b_{\mathrm{L}}$. An *ellipse* is a locus of all points for which the ratio of their distances from a focus to a directrix is constant and equal to eccentricity of an ellipse $e < 1$. If $e = 1$, then the path of point A_2 is a parabola. If $e > 1$, then the path of point A_2 is a hyperbola, and for $e = 0$ the orbit is a circle.

7.4 Kepler's First Law

Planets move along ellipses with the Sun at one focus.

In this case, referring back to our previous calculations, a body of mass $m_1 \gg m_2$, and let the Sun, that is, the body of mass m_1, be situated at point O'.

From (7.19) we approximately get

$$\mathbf{r}_2 = \mathbf{r}_{\mathrm{C}} + \mathbf{r}_{12}, \quad \mathbf{r}_1 = \mathbf{r}_{\mathrm{C}},$$

and hence it follows that the mass center of the system of these two bodies is located at the center of the Sun.

As in the case of the pendulum, the orbit of point A_2 depends on the initial velocity, which will be shown below.

The kinetic and potential energy of a particle with respect to O' are expressed by the formulas

$$T = \frac{1}{2}mv^2, \quad V = -\frac{mk}{r},$$

where we assumed $m = m_1$ and $r = r_1$.

Because in a potential force field there is no energy dissipation, the total energy is conserved. On this basis, we obtain an integral of motion, which is equal to

$$C = v_0^2 - \frac{2k}{r_0} = \text{const}, \tag{7.44}$$

where v_0 and r_0 are the magnitudes respectively of the velocity and position vector at the initial time instant.

From the formula

$$v^2 - \frac{2k}{r} = C \tag{7.45}$$

it follows that if the distance between points O' and A_2 decreases, then the speed v of point A_2 decreases accordingly, and vice versa. In other words, for $C \geq 0$ particle A_2 can move an arbitrary distance away from point O'.

According to (7.36) we have

$$\left(\mathbf{b} \times \mathbf{v} + k\frac{\mathbf{r}}{r}\right)^2 = b_{\mathrm{L}}^2,$$

that is,

$$b^2 v^2 + \frac{2k}{r}(\mathbf{b} \times \mathbf{v}) \circ \mathbf{r} + k^2 = b_{\mathrm{L}}^2,$$

because vector $\mathbf{b}$ is perpendicular to $\mathbf{v}$.

Taking into account relation (7.40) we obtain

$$b^2\left(v^2 - \frac{2k}{r}\right) + k^2 = b_{\mathrm{L}}^2,$$

and taking into account (7.44) we eventually arrive at

$$b^2 C + k^2 = b_{\mathrm{L}}. \tag{7.46}$$

From (7.43) and (7.46) we have

$$e = \frac{b_{\mathrm{L}}}{k} = \sqrt{1 + \left(\frac{b}{k}\right)^2 \left(v_0^2 - \frac{2k}{r_0}\right)}. \tag{7.47}$$

From the preceding relationship it follows that an elliptical orbit ($e < 1$) appears in the case where $v_0 < \sqrt{\frac{2k}{r_0}}$. Such allowable initial velocities are called *elliptic velocities* and are denoted by the symbol v_I.

In the case $v_0 = v_{II} = \sqrt{\frac{2k}{r_0}}$ we are dealing with a parabolic orbit ($e = 1$). Finally, for the velocity $v_0 > \sqrt{\frac{2k}{r_0}}$ particle A_2 moves along a hyperbola, and such velocities are called *hyperbolic velocities* (v_{III}).

The *first cosmic velocity* is associated with the motion of a satellite near the surface of Earth. It can be determined from the equation

$$m\frac{v_1^2}{r_0} = mg = G\frac{mM}{r_0^2}, \tag{7.48}$$

where m is the mass of the satellite and M denotes the mass of Earth.

From the preceding equation we obtain

$$v_1 = \sqrt{gr_0} = \sqrt{\frac{GM}{r_0}} \cong \sqrt{\frac{k}{r_0}} \tag{7.49}$$

because for $m \ll M$ we have $k = G(m+M) \approx GM$.

The *second cosmic velocity* v_{II} (a parabolic velocity) is equal to

$$v_{II} = \sqrt{\frac{2k}{r_0}} = \sqrt{2}v_1 \cong 11.2\,\text{km/s}, \tag{7.50}$$

because for the radius of Earth $r_0 = 6371$ km and $g = 9.81\,\text{m/s}^2$ we have $v_I \equiv 7.91$ km/s.

It is known from analytical geometry that for an ellipse of axes a, b $(a > b)$ and the focus c the following relations hold true:

$$p = \frac{b^2}{a}, \quad c = \sqrt{a^2 - b^2}, \quad e = \frac{c}{a}.$$

From these relations one may easily obtain

$$a^2 = b^2 + c^2 = c^2 + ap = e^2a^2 + pa,$$

hence

$$a = \frac{p}{1 - e^2}. \tag{7.51}$$

In turn, we have

$$b = \sqrt{pa} = \frac{p}{\sqrt{1 - e^2}}. \tag{7.52}$$

Formulas (7.51) and (7.52) describe semiaxes of an ellipse in terms of its semilatus rectum p and eccentricity e.

One may also introduce the notion of an *apocenter* (or *apoapsis*) (α), that is, the point on an ellipse that is the most distant from its focus, and a *pericenter* (or *periapsis*) (π), that is, the point on an ellipse that is the closest to the focus (Fig. 7.5).

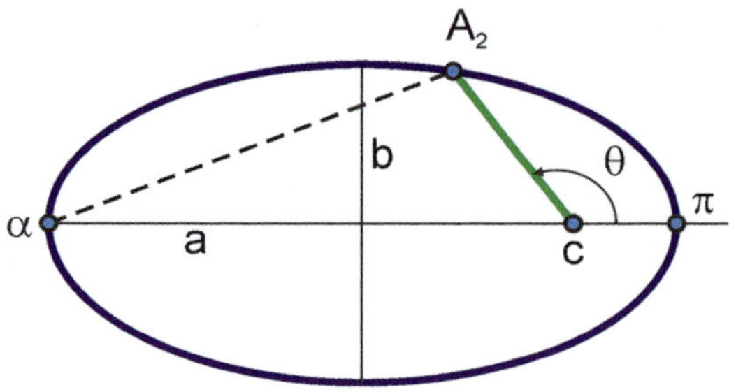

Fig. 7.5 Basic parameters of an ellipse

According to (7.33) the area of an ellipse S is equal to

$$S = \frac{1}{2}\int_{0}^{T} b_3'' \mathrm{d}t = \frac{1}{2} b_3'' T = \pi ab,$$

where T is a period of revolution of particle A_2 on the ellipse.

According to relation (7.43) we have that $b_3'' = \sqrt{pk}$, that is,

$$T = \frac{2\pi}{\omega} = \frac{2\pi ab}{\sqrt{pk}} = \frac{2\pi ab}{\sqrt{k\frac{b^2}{a}}} = \frac{2\pi a^{\frac{3}{2}}}{\sqrt{k}}. \tag{7.53}$$

The obtained quantity,

$$\omega = \frac{\sqrt{k}}{a^{\frac{3}{2}}}, \tag{7.54}$$

is called an *average angular velocity (frequency)* of motion of particle A_2 on the ellipse.

Let us now consider two particles A_1 and A_2 respectively of masses m_1 and m_2 that move on elliptical orbits.

If a gravitational interaction between particles A_1 and A_2 is neglected, then periods of their motion along elliptical orbits are equal to

$$T_1 = \frac{2\pi a_1^{\frac{3}{2}}}{\sqrt{G(m_1 + M)}}, \quad T_2 = \frac{2\pi a_2^{\frac{3}{2}}}{\sqrt{G(m_2 + M)}}, \tag{7.55}$$

where M is the mass of the Sun.

Dividing (7.55) by sides we get

$$\left(\frac{T_1}{T_2}\right)^2 = \frac{m_2 + M}{m_1 + M}\left(\frac{a_1}{a_2}\right)^3, \tag{7.56}$$

and assuming $m_1 + M \approx M, m_2 + M \approx M$ we obtain *Kepler's third law*:

$$\left(\frac{T_1}{T_2}\right)^2 = \left(\frac{a_1}{a_2}\right)^3. \tag{7.57}$$

According to this law *the ratio of squares of a planets periods of revolution around the Sun is equal to the ratio of cubes of the semimajor axes of their elliptical paths.*

References

1. A.D. Markeev, *Theoretical Mechanics* (Nauka, Moscow, 1990) (in Russian)
2. H.R. Harrison, T. Nettleton, *Advanced Engineering Dynamics* (Arnold, London, 1997)
3. C.D. Murray, S.F. Dermott, *Solar System Dynamics* (Cambridge University Press, Cambridge, 1999)
4. V.I. Arnold, V.V. Kozlov, A.I. Neishtadt, *Mathematical Aspects of Classical and Celestial Mechanics* (Springer, Berlin, 2006)
5. J. Awrejcewicz, *Classical Mechanics: Statics and Kinematics* (Springer, New York, 2012)

Chapter 8
Dynamics of Systems of Variable Mass

8.1 Introduction

So far we have considered DMSs and CMSs in which masses of particles m_n and their number have not changed. In nature and technology, however, phenomena are commonly known where the number of particles of a system or their mass change over time.

If floating icebergs are heated by the Sun's rays, then the ice melts and their mass decreases. If the falling snow becomes frozen to the floating icebergs, then their mass increases. Earth's mass increases when meteorites fall on its surface. In turn, the mass of the meteorites before they reach Earth's surface decreases as a result of burning in Earth's atmosphere. The mass of rockets decreases as the fuel they contain burns. The mass of elements transported on a conveyor belt changes as a result of their loading and unloading.

8.2 Change in Quantity of Motion and Angular Momentum

Let the mass of a mechanical system $m(t)$ be changing in time according to the equation

$$m(t) = m_0 - m_1(t) + m_2(t), \tag{8.1}$$

where $m(t) = m(t_0)$, $m_1(t) \geq 0$, ($m_2(t) \geq 0$) denotes the mass of particles leaving (entering) the system (Fig. 8.1).

Let us choose a time instant t during motion of the system, and let for this instant the momentum $\mathbf{p}$ of the considered system of particles increase by $\Delta\mathbf{p}$ during time Δt. Then, by $\mathbf{p}^*$ let us denote the momentum of analogous system, but of a constant mass. At the instant $t + \Delta t$ the quantity of motion of a system of variable mass is equal to

$$\mathbf{p} + \Delta\mathbf{p} = \mathbf{p}^* + \Delta\mathbf{p}^* - \Delta\mathbf{p}_1 + \Delta\mathbf{p}_2. \tag{8.2}$$

J. Awrejcewicz, *Classical Mechanics: Dynamics*, Advances in Mechanics and Mathematics 29, DOI 10.1007/978-1-4614-3740-6_8,

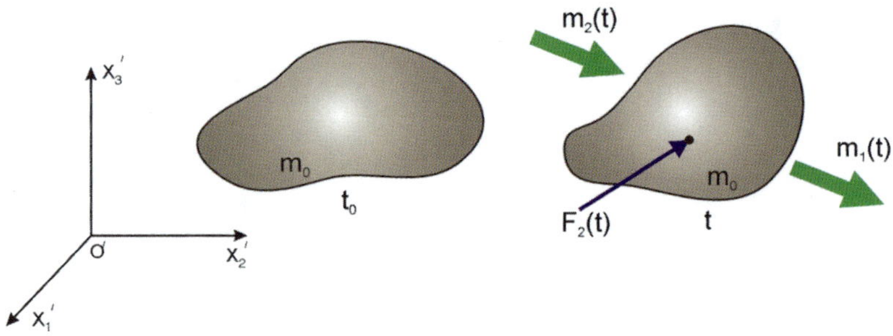

Fig. 8.1 Motion of a body of variable mass with respect to the inertial coordinate system $O'X_1'X_2'X_3'$

This means that the increment of momentum of the investigated system follows from the increment of momentum of a system of constant mass and the additional quantity of motion delivered ($\Delta\mathbf{p}_2$) and removed ($\Delta\mathbf{p}_1$) to/from the system during time Δt.

From the preceding equation we obtain

$$\Delta\mathbf{p} = \Delta\mathbf{p}^* - \Delta\mathbf{p}_1 + \Delta\mathbf{p}_2 \tag{8.3}$$

because at the instant t we have

$$\mathbf{p} = \mathbf{p}^*. \tag{8.4}$$

Dividing by Δt and on the assumption that $\Delta t \to 0$ we get

$$\lim_{\Delta t\to 0}\frac{\Delta\mathbf{p}}{\Delta t} = \lim_{\Delta t\to 0}\frac{\Delta\mathbf{p}^*}{\Delta t} - \lim_{\Delta t\to 0}\frac{\Delta\mathbf{p}_1}{\Delta t} + \lim_{\Delta t\to 0}\frac{\Delta\mathbf{p}_2}{\Delta t}, \tag{8.5}$$

hence

$$\frac{\mathrm{d}\mathbf{p}}{\mathrm{d}t} = \mathbf{F} + \mathbf{F}_1^R + \mathbf{F}_2^R, \tag{8.6}$$

where

$$\mathbf{F} = \lim_{\Delta t\to 0}\frac{\Delta\mathbf{p}^*}{\Delta t} = \frac{\mathrm{d}\mathbf{p}^*}{\mathrm{d}t},$$

$$\mathbf{F}_1^R = -\lim_{\Delta t\to 0}\frac{\Delta\mathbf{p}_1}{\Delta t}, \qquad \mathbf{F}_2^R = \lim_{\Delta t\to 0}\frac{\Delta\mathbf{p}_2}{\Delta t}, \tag{8.7}$$

and $\mathbf{F}$ is a main vector of a system of external forces acting at the time instant t.

Equation (8.6) extends the well-known theorem concerning the change in the quantity of motion (momentum) of a system. On its right-hand side additionally appear the so-called *thrust forces*, $\mathbf{F}_1^R$ and $\mathbf{F}_2^R$.

In a similar way one can generalize the theorem regarding the change in angular momentum (moment of momentum) of a system. Applying an argument analogous to the previous one, we obtain

$$\mathbf{K} + \Delta\mathbf{K} = \mathbf{K}^* + \Delta\mathbf{K}^* - \Delta\mathbf{K}_1 + \Delta\mathbf{K}_2, \tag{8.8}$$

where $\mathbf{K}$ is the moment of momentum of the system with respect to a certain arbitrary chosen fixed pole in the coordinate system $O'X_1'X_2'X_3'$, and $\Delta\mathbf{K}_{1(2)}$ denotes the sum of moments of a quantity of motion for those particles that left (entered) the considered system of variable mass during the time interval Δt. Dividing the preceding equation by Δt and proceeding to the limit as $\Delta t \to 0$ we have

$$\frac{\mathrm{d}\mathbf{K}}{\mathrm{d}t} = \mathbf{M} + \mathbf{M}_1^R + \mathbf{M}_2^R, \tag{8.9}$$

where

$$\mathbf{M} = \lim_{\Delta t \to 0} \frac{\Delta\mathbf{M}^*}{\Delta t} = \frac{\Delta\mathbf{K}^*}{\mathrm{d}t},$$
$$\mathbf{M}_1^R = -\lim_{\Delta t \to 0} \frac{\Delta\mathbf{K}_1}{\Delta t}, \qquad \mathbf{M}_2^R = \lim_{\Delta t \to 0} \frac{\Delta\mathbf{K}_2}{\Delta t}. \tag{8.10}$$

Equation (8.9) is a generalization of a theorem concerning changes in the angular momentum of a mechanical system. On its right-hand side additionally appear *moments of a thrust force*, $\mathbf{M}_1^R$ and $\mathbf{M}_2^R$.

8.3 Motion of a Particle of a Variable Mass System

Let us consider a particle A belonging to the investigated system of variable mass, and let the mass of this particle be described by (8.11) in the form

$$m_A(t) = m_A(t_0) - m_{A_1}(t) + m_{A_2}(t). \tag{8.11}$$

The kinematics of a particle of variable mass is presented in Fig. 8.2. In Fig. 8.2 the absolute velocity of a piece of mass Δm_2 is denoted by u_2, whereas the absolute velocity of a piece of mass Δm_1 is denoted by u_1. We will assume that $\Delta m_{A_1} \ll m_A(t_0)$ and $\Delta m_{A_2} \ll m_A(t_0)$.

In order to derive the differential equation of the motion of a particle of variable mass $m(t)$, we will make use of (8.6). A quantity of motion (momentum) of particle A at an arbitrary time instant t reads

$$\mathbf{p}(t) = m(t)\,\mathbf{v}(t), \tag{8.12}$$

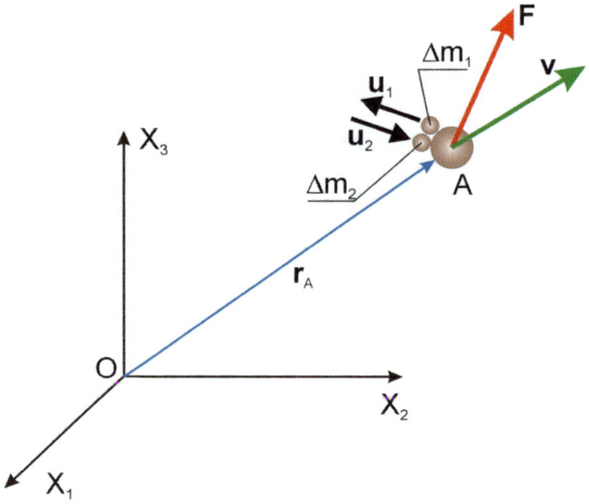

Fig. 8.2 Particle of mass m_0 in absolute system $OX_1X_2X_3$ at time instant t and piece of mass Δm_1 expelled (absorbed Δm_2) from (by) particle A

and the changes in momentum that follow from absorbing mass Δm_2 and expelling mass Δm_1 by particle A during the time interval Δt are respectively equal to

$$\Delta \mathbf{p}_i = \Delta m_i \mathbf{u}_i, \quad i = 1, 2. \tag{8.13}$$

According to (8.7) we have

$$\mathbf{F}_1^R = -\lim_{\Delta t \to 0} \frac{\Delta \mathbf{p}_1}{\Delta t} = -\lim_{\Delta t \to 0} \frac{\Delta m_1 \mathbf{u}_1}{\Delta t} = -\mathbf{u}_1 \frac{dm_1}{dt},$$
$$\mathbf{F}_2^R = \lim_{\Delta t \to 0} \frac{\Delta \mathbf{p}_2}{\Delta t} = \lim_{\Delta t \to 0} \frac{\Delta m_2 \mathbf{u}_2}{\Delta t} = \mathbf{u}_2 \frac{dm_2}{dt}. \tag{8.14}$$

Substituting (8.12) and (8.14) into (8.6) we obtain

$$\frac{d}{dt}\left[m(t)\,\mathbf{v}(t)\right] = \mathbf{F} - \frac{dm_1}{dt}\mathbf{u}_1 + \frac{dm_2}{dt}\mathbf{u}_2, \tag{8.15}$$

and following the transformations we have

$$m\ddot{\mathbf{r}}_A = \mathbf{F} - \dot{m}_1(\mathbf{u}_1 - \mathbf{v}) + \dot{m}_2(\mathbf{u}_2 - \mathbf{v}). \tag{8.16}$$

The obtained (8.16) is called a *generalized Meshcherskiy*[1] *equation*, and it describes the motion of a particle of variable mass. If the mass of particle A does not change, then $\dot{m}_1 = \dot{m}_2 = 0$, and from (8.16) we obtain Newton's second law on the motion of particle A of constant mass m.

[1]Ivan Meshcherskiy (1859–1935), professor working mainly in Saint Petersburg.

In previous calculations the dynamics of a particle of variable mass was presented descriptively during the derivation of (8.6) and (8.9). Presently we will proceed in a different way (see [1, 2]) by taking into account only the change in momentum of particle A. Let an elementary mass $\mathrm{d}m_2$ of velocity $\mathbf{u}_2(t)$ be added, and an elementary mass $\mathrm{d}m_1$ of velocity $\mathbf{u}_1(t)$ be removed, to/from particle A of mass $m(t)$ and velocity $\mathbf{v}(t)$. The momenta at the time instants t and $t + \mathrm{d}t$ are equal to

$$\mathbf{p}\,(t) = (m + \mathrm{d}m_1)\,\mathbf{v} + \mathrm{d}m_2\mathbf{u}_2,$$

$$\mathbf{p}\,(t + \mathrm{d}t) = (m + \mathrm{d}m_2)\,(\mathbf{v} + \mathrm{d}\mathbf{v}) + \mathrm{d}m_1\mathbf{u}_1.$$

The increment of momentum is equal to

$$\mathbf{p}\,(t + \mathrm{d}t) - \mathbf{p}\,(t) = \mathbf{v}\mathrm{d}m_2 + \mathrm{d}\mathbf{v}\mathrm{d}m_2 + m\mathrm{d}\mathbf{v} + \mathrm{d}m_1\mathbf{u}_1 - \mathbf{v}\mathrm{d}m_1 - \mathrm{d}m_2\mathbf{u}_2,$$

and neglecting differentials of the second order and dividing by dt we obtain the generalized Meshcherskiy equation (8.16), where $\mathbf{F} = \mathrm{d}\mathbf{p}/\mathrm{d}t$. Following the introduction of relative velocities,

$$\mathbf{w}_i = \mathbf{u}_i - \mathbf{v}, \quad i = 1, 2. \tag{8.17}$$

Respectively expelling and absorbing the mass by particle A (8.16) takes the form

$$m\ddot{\mathbf{r}}_A = \mathbf{F} - \dot{m}_1\mathbf{w}_1 + \dot{m}_2\mathbf{w}_2. \tag{8.18}$$

Taking into account relation (8.17), (8.18) is identical to (8.16). If the case of separation of mass from particle A is considered alone, then from (8.11) for $m_{A_2} \equiv 0$ we obtain

$$m(t) = m(t_0) - m_1(t), \tag{8.19}$$

hence

$$\dot{m}(t) = -\dot{m}_1(t). \tag{8.20}$$

Substituting (8.20) into (8.18) we get

$$m\ddot{\mathbf{r}}_A = \mathbf{F} + \mathbf{F}_1^R. \tag{8.21}$$

The preceding equation is called a *Meshcherskiy equation*. From (8.21) it follows that the effect of separation of mass is equivalent to the action of an additional force $\mathbf{F}_1^R = \dot{m}\mathbf{w}_1$ on particle A, called a *thrust force*. The thrust force $\mathbf{F}_1^R$ (removal of mass) has a sense opposite to the sense of velocity $\mathbf{w}_1$, whereas the thrust force $\mathbf{F}_2^R$ (addition of mass) has the same sense as the sense of the relative velocity $\mathbf{w}_2$. The quantity $\dot{m}_1$ ($\dot{m}_2$) is called the *mass removal (addition) per second.*

In a special case, where the absolute velocity of the mass that separates is $\mathbf{u}_1 = 0$, (8.21) takes the form

$$m\frac{\mathrm{d}\mathbf{v}}{\mathrm{d}t} = \mathbf{F} - \frac{\mathrm{d}m}{\mathrm{d}t}\mathbf{v} \tag{8.22}$$

or

$$\frac{\mathrm{d}(m\mathbf{v})}{\mathrm{d}t} = \mathbf{F}. \tag{8.23}$$

We have shown that if the absolute velocity of the mass that separates is equal to zero, then the derivative of momentum of particle A balances the external forces acting on this particle. If, in turn, the relative velocity of the mass that separates is $\mathbf{w}_1 = \mathbf{u}_1 - \mathbf{v} = \mathbf{0}$, then from (8.21) we obtain

$$m(t)\frac{\mathrm{d}\mathbf{v}}{\mathrm{d}t} = \mathbf{F}. \tag{8.24}$$

In this case we obtained an equation that is formally consistent with Newton's second law on the motion of a particle of constant mass.

8.4 Motion of a Rocket (Two Problems of Tsiolkovsky)

Let us now consider two problems of Tsiolkovsky.[2]

8.4.1 First Tsiolkovsky Problem

Let a rocket, treated further as a particle, be moving in space, and let the action of external forces on it be negligibly small. The initial conditions of motion are as follows: $\mathbf{v}(0) = \mathbf{v}_0$, $m(t) = m_0 + m_1(t)$, where m_0 is the mass of the rocket and $m_1(t)$ is the mass of fuel ($m_1(0) = m_{10}$).

In the considered case, the Meshcherskiy equation, (8.21), takes the form

$$m\frac{\mathrm{d}\mathbf{v}}{\mathrm{d}t} = \frac{\mathrm{d}m}{\mathrm{d}t}\mathbf{w}_1. \tag{8.25}$$

Let us assume that the relative velocity of combustion gases $\mathbf{w}_1 = \mathbf{u}_1 - \mathbf{v} = \text{const}$ and its sense are opposite to those of velocity vector $\mathbf{v}$. It follows that a rocket moves along a straight line according to the sense of vector $\mathbf{v}$ (Fig. 8.3).

Following the projection (multiplication by $\mathbf{E}_1$) of (8.25) onto the axis OX we obtain

$$m\frac{\mathrm{d}v}{\mathrm{d}t} = -\frac{\mathrm{d}m}{\mathrm{d}t}w_1$$

[2] Konstantin Tsiolkowsky (1857–1935), Russian teacher of mathematics and physics of Polish origin; precursor to the theory of rocket flight.

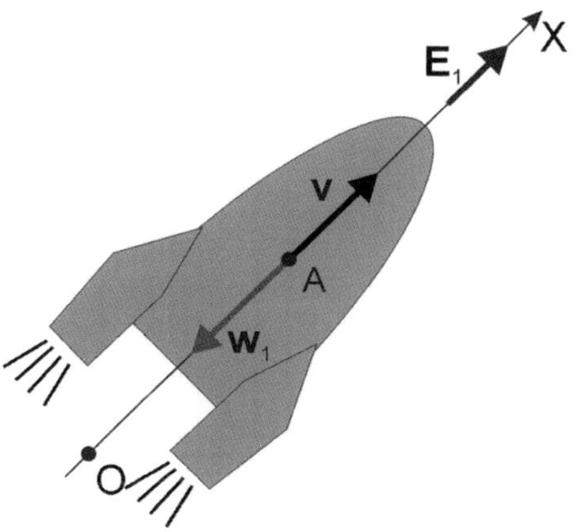

Fig. 8.3 Motion of rocket in a force-free field

or

$$dv = -w_1 \frac{dm}{m}. \tag{8.26}$$

Integrating (8.26) we have

$$v(t) = -w_1 \ln m + C, \tag{8.27}$$

where C is the constant of integration equal to $C = v_0 + w_1 \ln(m_0 + m_{10})$.

Finally, the time change in the velocity of a rocket is described by the scalar equation

$$\frac{dx}{dt} \equiv v(t) = v_0 + w_1 \ln\left(\frac{m_0 + m_{10}}{m(t)}\right). \tag{8.28}$$

The maximum velocity is reached by the rocket after the fuel is completely spent, that is, when $m(t_*) = m_0$, and it is equal to

$$v(t_*) = v_0 + w_1 \ln\left(1 + \frac{m_{10}}{m_0}\right). \tag{8.29}$$

The obtained equation is called a *rocket equation*. The maximum velocity of a rocket does not depend on the process of fuel combustion, that is, whether combustion proceeds slowly or quickly. The constant quantity m_{10}/m_0 is also known as a *Tsiolkovsky constant*.

In contrast, the trajectory of motion of a rocket does depend on the process of fuel combustion. Integrating (8.28), for the initial condition $x(0) = 0$, we have

$$x(t) = v_0 t + w_1 \int_0^t \ln \frac{m_0'}{m(\tau)} d\tau, \tag{8.30}$$

where $m_0' = m_0 + m_{10}$.

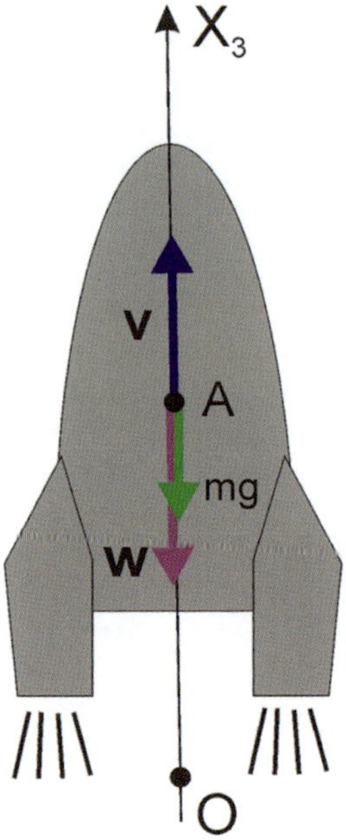

Fig. 8.4 Vertical motion of rocket in Earth's gravitational field

8.4.2 *Second Tsiolkovsky Problem*

Let a rocket, treated further as a particle, move vertically upward in a uniform gravitational field of Earth, its resistance to motion being neglected. The relative velocity of ejection of fuel combustion products is constant and directed vertically downward (Fig. 8.4).

In this case after projection of the Meshcherskiy equation (8.21) onto the axis OX_3 we get

$$m\frac{\mathrm{d}v}{\mathrm{d}t} = -mg - \frac{\mathrm{d}m}{\mathrm{d}t}w_1, \tag{8.31}$$

or, separating the variables,

$$d(v + gt) = -w\frac{\mathrm{d}m}{m}. \tag{8.32}$$

Integrating the preceding equation we have

$$v + gt = -w\ln m + C. \tag{8.33}$$

The constant C is equal to

$$C = v_0 + w\ln m_0'. \tag{8.34}$$

Substituting the obtained value of C into (8.33) we have

$$\frac{dx_3}{dt} \equiv v(t) = v_0 - gt + w \ln\left(\frac{m_0'}{m(t)}\right). \tag{8.35}$$

If we assume the initial conditions to be $x_3(0) = 0$, $v_0 = 0$, then following the integration of (8.35) we get

$$x_3(t) = w \int_0^t \ln\left(\frac{m_0'}{m(t)}\right) d\tau - \frac{1}{2} g t^2. \tag{8.36}$$

Let the fuel combustion take place according to the following process:

$$m(t) = m_0' e^{-\alpha t}, \tag{8.37}$$

where α is a constant coefficient characterizing the speed of fuel combustion.

The mass of combustion products $m_1(t)$ can be calculated from

$$m(t) + m_1(t) = m_0 + m_{10} \equiv m_0'$$

and is equal to

$$m_1(t) = m_0 + m_{10} - (m_0 + m_{10}) e^{-\alpha t} = m_0' \left(1 - e^{-\alpha t}\right). \tag{8.38}$$

The thrust force is equal to

$$F_1^R = \dot{m}_1 w_1 = m_0' w_1 \alpha e^{-\alpha t} = m(t) w_1 \alpha, \tag{8.39}$$

where αw_1 is the acceleration imposed on the rocket due to fuel combustion. Because we assumed certain combustion process described by (8.37), from (8.35) we have

$$v(t) = v_0 - gt + w \ln\left(\frac{m_0'}{e^{-\alpha t}}\right),$$

and for $v_0 = 0$ we obtain

$$v(t) = (\alpha w - g)\, t. \tag{8.40}$$

In turn, from (8.36) (or by integrating (8.40)) we have

$$x_3(t) = (\alpha w - g) \frac{t^2}{2}. \tag{8.41}$$

From the last equation it follows that the launch of the rocket is possible if $\alpha w > g$, that is, the acceleration coming from a thrust force F_1^R should exceed the acceleration of gravity.

If the fuel is burned completely at the time instant $t = t_f$, then according to (8.37) we have

$$m(t_f) = m_0 + m_1(t_f) = m_0' e^{-\alpha t_f},$$

that is,

$$m_0 = m_0' e^{-\alpha t_f}, \tag{8.42}$$

because at the instant t_k we have no more fuel, that is, $m_1(t_f) = 0$.

From (8.42) we can determine the time required for complete combustion of fuel by a rocket, which is equal to

$$t_f = \frac{\gamma}{\alpha}, \tag{8.43}$$

where

$$\gamma = \ln\left(1 + \frac{m_{10}}{m_0}\right).$$

From (8.40) and (8.41) one can determine the velocity and ceiling height of a rocket corresponding to the time instant when the fuel is spent:

$$v_f = \frac{\gamma}{\alpha}(\alpha w - g), \tag{8.44}$$

$$x_{3f} = \frac{\gamma^2(\alpha w - g)}{2\alpha^2}. \tag{8.45}$$

Because at the instant when the fuel has run out $t = t_f$ and $v_f = v(t_f)$, for such initial conditions a rocket of mass $m(t_f) = m_0$ additionally climbs in Earth's gravitational field at the height

$$h_d = \frac{v_f^2}{2g} = \frac{\gamma^2}{2\alpha^2 g}(\alpha w - g)^2. \tag{8.46}$$

We obtain the maximum height h of the rocket using (8.45) and (8.46):

$$h = h_d + x_{3f} = \frac{\gamma^2 w}{2}\left(\frac{w}{g} - \frac{1}{\alpha}\right). \tag{8.47}$$

The height reached by a rocket depends on the coefficient of the fuel combustion rate α. For example, at a rapid (explosive) rate of fuel combustion the height attained is equal to

$$h_{\max} = \frac{\gamma^2 w^2}{2g}. \tag{8.48}$$

8.5 Equations of Motion of a Body with Variable Mass

A group of particles $n = 1, \ldots, N$, between which the mutual distances do not change and at least one of which is a particle with variable mass, is called a rigid body of variable mass [3].

According to the previous calculations, let the particles of the body (the material system) change their mass according to (8.1), that is,

$$m_n(t) = m_{0n} - m_{1n}(t) + m_{2n}(t), \quad n = 1, \ldots, N, \tag{8.49}$$

where $m_{1n}(t)$ is the total mass lost by particle n at time t, and $m_{2n}(t)$ is the total mass gained by the particle at time t.

Let us further consider the case of motion of a rigid body with variable mass about a certain fixed point O (motion about a point of a system with variable mass). The angular momentum $\mathbf{K}_O$ of the system about point O is equal to (in the system rigidly connected to the body $OX_1''X_2''X_3''$)

$$\frac{\mathrm{d}\mathbf{K}_O}{\mathrm{d}t} + \boldsymbol{\omega} \times \mathbf{K}_O = \mathbf{M}_O^Z + \mathbf{M}_O^R, \tag{8.50}$$

where $\mathbf{M}_O^Z$ is the main moment of external forces acting on the system with respect to point O, and $\mathbf{M}_O^R$ is the additional moment of a thrust force that needs to be determined.

According to relation (8.8) we have

$$\begin{aligned} \mathrm{d}\mathbf{K}_{1O} &= \sum_{n=1}^{N} \mathrm{d}m_{1n}\mathbf{r}_n \times \mathbf{u}_{1n}, \\ \mathrm{d}\mathbf{K}_{2O} &= \sum_{n=1}^{N} \mathrm{d}m_{2n}\mathbf{r}_n \times \mathbf{u}_{2n}, \end{aligned} \tag{8.51}$$

where $\mathbf{r}_n$ is a radius vector of particle n, and on that basis a moment of thrust forces is equal to

$$\mathbf{M}_O^R = \mathbf{M}_{1O}^R + \mathbf{M}_{2O}^R, \tag{8.52}$$

where

$$\begin{aligned} \mathbf{M}_{1O}^R &= -\sum_{n=1}^{N} \frac{\mathrm{d}m_{1n}}{\mathrm{d}t}\mathbf{r}_n \times \mathbf{u}_{1n}, \\ \mathbf{M}_{2O}^R &= -\sum_{n=1}^{N} \frac{\mathrm{d}m_{2n}}{\mathrm{d}t}\mathbf{r}_n \times \mathbf{u}_{2n}. \end{aligned}$$

Introducing the notion of relative velocity $\mathbf{w}_n$ according to equations

$$\mathbf{u}_{1n} = \mathbf{v}_n + \mathbf{w}_{1n},$$
$$\mathbf{u}_{2n} = \mathbf{v}_n + \mathbf{w}_{2n}, \tag{8.53}$$

we have

$$\begin{aligned}
\mathbf{M}_O^R &= -\sum_{n=1}^{N} \frac{dm_{1n}}{dt}\mathbf{r}_n \times (\mathbf{v}_n + \mathbf{w}_{1n}) + \sum_{n=1}^{N} \frac{dm_{2n}}{dt}\mathbf{r}_n \times (\mathbf{v}_n + \mathbf{w}_{2n}) \\
&= \sum_{n=1}^{N} \mathbf{r}_n \times \left(\frac{dm_{2n}}{dt} - \frac{dm_{1n}}{dt}\right) \times \mathbf{v}_n \\
&\quad + \sum_{n=1}^{N} \mathbf{r}_n \times \left(-\frac{dm_{1n}}{dt}\mathbf{w}_{1n} + \frac{dm_{2n}}{dt}\mathbf{w}_{2n}\right) \\
&= \sum_{n=1}^{N} \mathbf{r}_n \times \left(\frac{dm_{2n}}{dt}\mathbf{w}_{2n} - \frac{dm_{1n}}{dt}\mathbf{w}_{1n}\right) + \sum_{n=1}^{N} \mathbf{r}_n \times \frac{dm_n}{dt}\mathbf{v}_n,
\end{aligned} \tag{8.54}$$

where (8.49) was used.

Eventually we obtain

$$\mathbf{M}_O^R = \mathbf{M}_O^W + \frac{d\mathbf{I}}{dt}\boldsymbol{\omega}, \tag{8.55}$$

where

$$\mathbf{M}_O^W = \sum_{n=1}^{N} \mathbf{r}_n \times \left(\frac{dm_{2n}}{dt}\mathbf{w}_{2n} - \frac{dm_{1n}}{dt}\mathbf{w}_{1n}\right),$$
$$\frac{d\mathbf{I}}{dt}\boldsymbol{\omega} = \sum_{n=1}^{N} \mathbf{r}_n \times \frac{dm_n}{dt}(\boldsymbol{\omega} \times \mathbf{r}_n), \tag{8.56}$$

and $\mathbf{I}$ is the matrix of the inertia tensor of a body for point O, and in this case the matrix depends on time. Because $\mathbf{K}_O = \mathbf{I}\boldsymbol{\omega}$, from (8.50) and taking into account (8.55) we obtain

$$\frac{d\mathbf{I}}{dt}\boldsymbol{\omega} + \mathbf{I}\frac{d\boldsymbol{\omega}}{dt} + \boldsymbol{\omega} \times \mathbf{I}\boldsymbol{\omega} = \mathbf{M}_O^Z + \mathbf{M}_O^W + \frac{d\mathbf{I}}{dt}\boldsymbol{\omega},$$

hence

$$\mathbf{I}\frac{d\boldsymbol{\omega}}{dt} + \boldsymbol{\omega} \times \mathbf{I}\boldsymbol{\omega} = \mathbf{M}_O^Z + \mathbf{M}_O^W. \tag{8.57}$$

If the axes of a coordinate system during the process of gaining and losing mass remain the principal axes of inertia, then (8.57) has the following scalar representation:

$$I_1(t)\frac{\mathrm{d}\omega_1}{\mathrm{d}t} + (I_3(t) - I_2(t))\,\omega_2\omega_3 = M_1 + M_1^W,$$

$$I_2(t)\frac{\mathrm{d}\omega_2}{\mathrm{d}t} + (I_1(t) - I_3(t))\,\omega_1\omega_3 = M_2 + M_2^W,$$

$$I_3(t)\frac{\mathrm{d}\omega_3}{\mathrm{d}t} + (I_2(t) - I_1(t))\,\omega_1\omega_2 = M_3 + M_3^W, \tag{8.58}$$

where $I_i(t)$ are the moments of inertia of the body with respect to the axes OX_i, M_i are the projections of a main vector of external forces onto these axes, and $\boldsymbol{\omega} = \omega_1\mathbf{E}_1 + \omega_2\mathbf{E}_2 + \omega_3\mathbf{E}_3$.

In the case of rotation of the body about a fixed axis (let it be the axis OX_3), we have $\boldsymbol{\omega} = \omega_3\mathbf{E}_3$, and from the last equation of (8.58) we obtain

$$I_3(t)\frac{\mathrm{d}\omega_3}{\mathrm{d}t} = M_3 + M_3^W. \tag{8.59}$$

As distinct from the previously considered case of the rotation of a rigid body about a fixed axis, on the right-hand side additionally appeared the moment of a thrust force, and on the left-hand side the mass moment of inertia of a body changing in time.

Example 8.1. Figure 8.5 shows a drum having moment of inertia I_0 with respect to the axis OX_3 perpendicular to the plane of the drawing and passing through point O, onto which a rope of length S and mass m is wound. Determine the angular velocity of the drum on the assumption that the rope started to reel out from the drum at an initial velocity of zero and the drum axis was horizontal.

For the solution of the problem we make use of (8.59). In this case

$$M_3^W = (\omega_1 - \omega)\,\frac{\mathrm{d}I}{\mathrm{d}t},$$

where ω_1 is the angular velocity of an elementary moment of inertia $\mathrm{d}I$ of a rope separating from a drum that is rotating with angular velocity ω. The element of the rope leaving the drum has a velocity equal to the peripheral speed of the drum, that is, $r\omega_1 = r\omega$, i.e., $M_3^W = 0$.

The equation of motion of the investigated system is analogous to (8.24) for the rotational motion. The problem reduces to the analysis of equation

$$I(\varphi(t))\frac{\mathrm{d}\omega}{\mathrm{d}t} = M^Z,$$

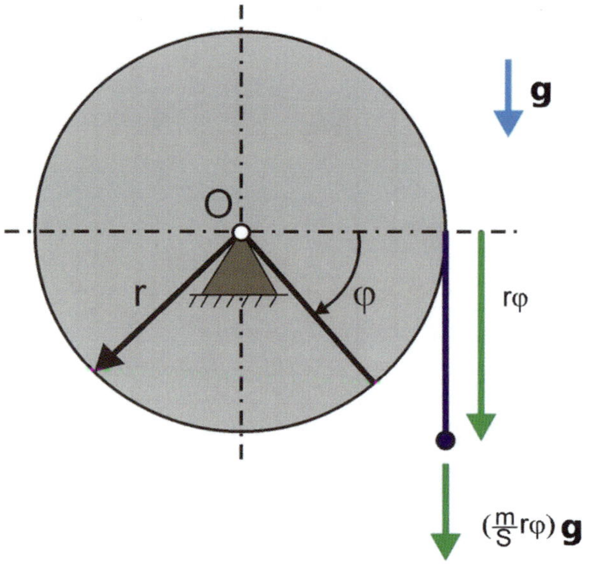

Fig. 8.5 Rope reeling out of a drum

where

$$I(\varphi(t)) = I_0 + mr^2 - \frac{m}{S}(r\varphi)r^2.$$

In turn, the moment M^Z follows from the action of the force coming from the rope reeling out from the drum and is equal to

$$M^Z(\varphi(t)) = \frac{m}{S}(r\varphi)gr.$$

Because

$$\frac{\mathrm{d}\omega}{\mathrm{d}t} = \frac{\mathrm{d}\omega}{\mathrm{d}\varphi}\frac{\mathrm{d}\varphi}{\mathrm{d}t} = \omega\frac{\mathrm{d}\omega}{\mathrm{d}\varphi},$$

from the equation of motion we have

$$\left(I_0 + mr^2 - \frac{m}{S}r^3\varphi\right)\omega\frac{\mathrm{d}\omega}{\mathrm{d}\varphi} = \frac{m}{l}r^2 g\varphi,$$

and separating the variables we get

$$\omega\mathrm{d}\omega = \frac{mr^2g}{S}\frac{\varphi}{I_0 + mr^2 - \frac{m}{S}r^3\varphi}\mathrm{d}\varphi.$$

Setting $I_0 = 3mr^2$ we have

$$\frac{\omega^2}{2} = \frac{g}{S}\left(\int \frac{\varphi}{4 - \frac{r}{S}\varphi}\mathrm{d}\varphi + C\right).$$

The obtained indefinite integral is calculated by substitution:

$$t = 4 - \frac{r}{S}\varphi,$$

hence

$$\mathrm{d}\varphi = -\frac{S}{r}\mathrm{d}t, \qquad \varphi = \frac{S}{r}(4 - t).$$

We have then

$$\int \frac{\varphi}{4 - \frac{r}{S}\varphi} = -\int \frac{4-t}{t}\mathrm{d}t = \frac{S^2}{r^2}(4\ln|t| + t)$$
$$= \frac{4S^2}{r^2}\ln\left|4 - \frac{r}{S}\varphi\right| - \frac{S}{r}\varphi + C\frac{S}{r},$$

that is,

$$\frac{\omega^2}{2} = \frac{g}{r}\left(-\frac{4S}{r}\ln\left|4 - \frac{r}{S}\varphi\right| - \varphi + C\right).$$

The integration constant is determined from the initial condition $\omega(0) = 0$ and is equal to

$$C = \frac{4S}{r}\ln 4.$$

The desired function

$$\omega \equiv \omega[\varphi(t)] = \left[\frac{2g}{r}\left(-\frac{4S}{r}\ln\left[4 - \frac{r}{S}\varphi\right] - \varphi + \frac{4S}{r}\ln 4\right)\right]^{\frac{1}{2}}.$$

The maximum angular velocity $\omega = \omega_{\mathrm{max}}$ is obtained after the rope has been completely unwound, that is, substituting $\varphi \equiv \varphi_{\mathrm{max}} = \frac{S}{r}$ into the preceding formula. □

Example 8.2. A body of mass m is thrown upward with initial speed v_0, and there is a chain of unit mass ρ stacked on a horizontal plane and attached to the body. Determine the maximum height attained by the chain (Fig. 8.6).

During the motion of the chain its links are successively lifted from the stationary stack, that is, their absolute velocity is equal to zero. The problem is therefore described by (8.23), which in our case takes the form

$$\frac{\mathrm{d}}{\mathrm{d}t}\left[(m + \rho x)\,\dot{x}\right] = -mg.$$

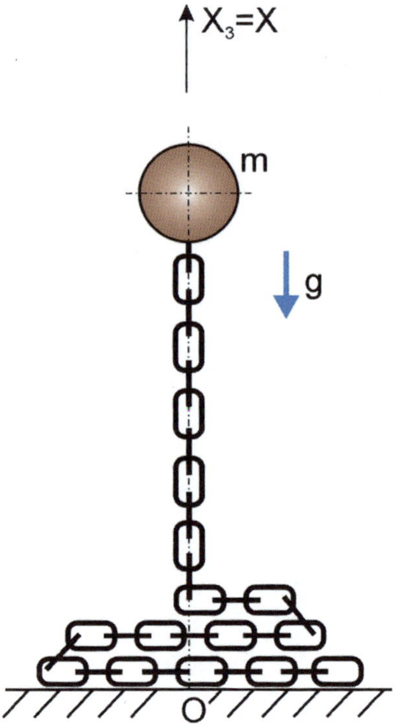

Fig. 8.6 Projection of ball of mass m with attached chain

The first integral of the preceding equation reads

$$m\dot{x} + \rho x\dot{x} = -mgt + C_1$$

or

$$\frac{\mathrm{d}}{\mathrm{d}t}\left[mx + \frac{\rho x^2}{2}\right] = -mgt + C_1.$$

The second integral is equal to

$$mx + \frac{\rho x^2}{2} = -mg\frac{t^2}{2} + C_1 t + C_2.$$

Let $x(0) = x_0$, $\dot{x}(0) = v_0$. The constant C_1 is found from the equation

$$C_1 = (m + \rho x_0)\, v_0,$$

and the constant C_2 reads

$$C_2 = \left(m + \frac{\rho x_0}{2}\right) x_0.$$

The maximum height is attained for a velocity of mass equal to $\dot{x} \equiv v(t_*) = 0$, that is, for the time instant

$$t_* = \frac{C_1}{mg} = \frac{(m + \rho x_0)v_0}{mg}.$$

The desired quantity $x(t_*) = x_*$ is determined from the equation

$$mx_* + \frac{\rho x_*^2}{2} = -\frac{(m + \rho x_0)^2 v_0^2}{2mg} + \frac{(m + \rho x_0)^2 v_0^2}{mg} + \left(m + \frac{\rho x_0}{2}\right) x_0$$

or, following transformation,

$$x_*^2 + \frac{2m}{\rho} x_* - \left[\frac{(m + \rho x_0)^2 v_0^2}{mg\rho} + \left(\frac{2m}{\rho} + x_0\right) x_0\right] = 0.$$

For $x_0 = 0$ we have

$$x_*^2 + \frac{2m}{\rho} x_* - \frac{m v_0^2}{g\rho} = 0.$$

Solving the preceding quadratic equation and rejecting the negative root we obtain

$$x_* = -\frac{m}{\rho} + \frac{1}{2}\sqrt{\frac{4m}{\rho}\left(\frac{m}{\rho} + \frac{v_0^2}{g}\right)}. \qquad \square$$

Finally, this chapter can be supplemented by the classic works [4–9].

References

1. J. Nizioł, *Methods of Solving Mechanical Tasks* (WNT, Warsaw, 2002) (in Polish)
2. W. Kurnik, *Lectures on General Mechanics* (Warsaw Technological University Press, Warsaw, 2005) (in Polish)
3. A.D. Markeev, *Theoretical Mechanics* (Nauka, Moscow, 1990) (in Russian)
4. S.T. Thornton, J.B. Marion, *Classical Dynamics of Particles and Systems* (Saunders College Publishers, New York, 1995)
5. A.P. Arya, *Introduction to Classical Mechanics* (Addison-Wesley, San Francisco, 1998)
6. J.V. José, E.J. Saletan, *Classical Dynamics: A Contemporary Approach* (Cambridge University Press, Cambridge, 1998)
7. H. Goldstein, C. Poole, J. Safko, *Classical Mechanics*, 3rd edn. (Addison-Wesley, San Francisco, 2002)
8. T.W.B. Kibble, F.H. Berkshire, *Classical Mechanics*, 5th edn. (Imperial College Press, London, 2004)
9. R.D. Gregory, *Classical Mechanics* (Cambridge University Press, Cambridge, 2011)

Chapter 9
Body and Multibody Dynamics

9.1 Rotational Motion of a Rigid Body About a Fixed Axis

The kinematics and statics of particles of a body supported by a thrust bearing and radial bearing have already been considered in Chaps. 2 and 5 of [1].

Let a rigid body have two fixed points O and $\tilde{O}$, and let the reactions of constraints at these points be equal to $\mathbf{R}$ and $\tilde{\mathbf{R}}$. The system of external forces acting on the body is replaced by a main force $\mathbf{F}$ and main moment $\mathbf{M}_O$ applied at point O (the pole), which is the origin of both the stationary $OX_1X_2X_3$ and non-stationary coordinate systems $O'X_1'X_2'X_3'$, where $O = O'$ (Fig. 9.1).

The body has only one degree of freedom described by an angle $\varphi(t)$. In order to obtain equations of motion of the body we will make use of the laws of conservation of momentum and angular momentum, which were stated in Chap. 1 of [1]. Here they are related to the mass center C of the body assuming the following form:

$$M\frac{\mathrm{d}\mathbf{v}_C}{\mathrm{d}t} = \mathbf{F} + \mathbf{R} + \tilde{\mathbf{R}}, \tag{9.1}$$

$$\frac{\mathbf{K}_C}{\mathrm{d}t} = \mathbf{M}_O + \overrightarrow{O\tilde{O}} \times \tilde{\mathbf{R}}, \tag{9.2}$$

valid in the coordinate system $OX_1X_2X_3$.

We are interested in the reactions acting on the body, and that is why we will express (9.1) and (9.2) in the body system $O'X_1'X_2'X_3'$. Some vectors used in subsequent calculations will have the form

$$\begin{aligned}
\mathbf{F} &= F_1\mathbf{E}_1' + F_2\mathbf{E}_2' + F_3\mathbf{E}_3',\\
\mathbf{R} &= R_1\mathbf{E}_1' + R_2\mathbf{E}_2' + R_3\mathbf{E}_3',\\
\tilde{\mathbf{R}} &= \tilde{R}_1\mathbf{E}_1' + \tilde{R}_2\mathbf{E}_2' + \tilde{R}_3\mathbf{E}_3',\\
\boldsymbol{\omega} &= \dot{\varphi}\mathbf{E}_3',
\end{aligned}$$

J. Awrejcewicz, *Classical Mechanics: Dynamics*, Advances in Mechanics and Mathematics 29, DOI 10.1007/978-1-4614-3740-6_9,

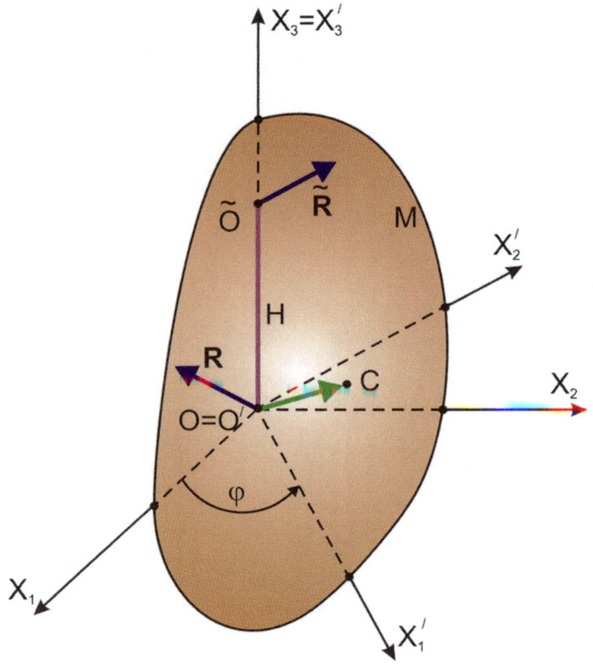

Fig. 9.1 Rigid body rotation measured with angle $\varphi = \varphi(t)$ about fixed axis OX_3 $(\overrightarrow{O\tilde{O}} = H)$

$$\overrightarrow{OC} = x'_{1C}\mathbf{E}'_1 + x'_{2C}\mathbf{E}'_2 + x'_{3C}\mathbf{E}'_3, \tag{9.3}$$

and all vectors are expressed in the body system $O'X'_1X'_2X'_3$ $(\tilde{R}_3 = 0)$.

According to previous calculations we have $\mathbf{K}_O = \mathbf{I}\boldsymbol{\omega}$, that is,

$$\begin{aligned} K_{O_1} &= I_1\omega'_1 - I_{12}\omega'_2 - I_{13}\omega'_3, \\ K_{O_2} &= -I_{12}\omega'_1 + I_2\omega'_2 - I_{23}\omega'_3, \\ K_{O_3} &= -I_{13}\omega'_1 - I_{23}\omega'_2 + I_3\omega'_3, \end{aligned} \tag{9.4}$$

and because in our case $\omega'_1 = \omega'_2 = 0$, we have

$$K_{O_1} = -I_{13}\dot{\varphi}, \qquad K_{O_2} = -I_{23}\dot{\varphi}, \qquad K_{O_3} = I_3\dot{\varphi}, \tag{9.5}$$

where

$$\mathbf{K}_O = K_{O_1}\mathbf{E}'_1 + K_{O_2}\mathbf{E}'_2 + K_{O_3}\mathbf{E}'_3. \tag{9.6}$$

In (9.4) $I_{ik} = I_{ik}(t)$, $\omega'_i = \omega'_i(t)$, because in the space (global) coordinate system the body changes its position and, consequently, changes both its mass moments of inertia and components of angular velocity vector. Therefore, below we present the equations of motion of a rigid body in a local coordinate system. Equations (9.1) and (9.2) in the system $O'X_1'X_2'X_3'$ have the form

$$M\frac{\tilde{\mathrm{d}}\mathbf{v}_C}{\mathrm{d}t} + M\boldsymbol{\omega} \times \mathbf{v}_C = \mathbf{F} + \mathbf{R} + \tilde{\mathbf{R}}, \tag{9.7}$$

$$\frac{\tilde{\mathrm{d}}\mathbf{K}_\mathbf{O}}{\mathrm{d}t} + \boldsymbol{\omega} \times \mathbf{K}_O = \mathbf{M}_O + \overrightarrow{O\tilde{O}} \times \tilde{\mathbf{R}}. \tag{9.8}$$

Recall that the operator $\tilde{\mathrm{d}}/\mathrm{d}t$ is a relative differential operator (as opposed to an absolute differential operator $\mathrm{d}/\mathrm{d}t$), that is, it describes the differentiation of a vector in the coordinate system $O'X_1'X_2'X_3'$.

An arbitrary vector $\mathbf{a}$ obeys the following differentiation: $\mathrm{d}\mathbf{a}/\mathrm{d}t = \tilde{\mathrm{d}}\mathbf{a} + \boldsymbol{\omega} \times \mathbf{a}$, where $\boldsymbol{\omega}$ denotes the rotational velocity of the system $O'X_1'X_2'X_3'$, and $\mathrm{d}\mathbf{a}/\mathrm{d}t$ ($\tilde{\mathrm{d}}\mathbf{a}/\mathrm{d}t$) denotes an absolute (relative) derivative. In other words, the absolute velocity of the tip of angular momentum vector $\mathbf{K}_O$ is equal to the geometric sum of its relative velocity $\tilde{\mathrm{d}}\mathbf{K}_O/\mathrm{d}t$ and the velocity of transportation $\boldsymbol{\omega} \times \mathbf{K}_O$. Because

$$\mathbf{v}_C = \boldsymbol{\omega} \times \overrightarrow{OC}, \tag{9.9}$$

taking into account relations (9.3)–(9.6) we successively calculate

$$\mathbf{v}_C = \begin{vmatrix} \mathbf{E}_1' & \mathbf{E}_2' & \mathbf{E}_3' \\ 0 & 0 & \dot{\varphi} \\ x_{1C}' & x_{2C}' & x_{3C}' \end{vmatrix} = -\dot{\varphi}x_{2C}'\mathbf{E}_1' + x_{1C}'\dot{\varphi}\mathbf{E}_2', \tag{9.10}$$

$$\boldsymbol{\omega} \times \mathbf{K}_O = \begin{vmatrix} \mathbf{E}_1' & \mathbf{E}_2' & \mathbf{E}_3' \\ 0 & 0 & \dot{\varphi} \\ K_{O1} & K_{O2} & K_{O3} \end{vmatrix} = -\dot{\varphi}K_{O2}\mathbf{E}_1' + K_{O1}\dot{\varphi}\mathbf{E}_2', = I_{23}\dot{\varphi}^2\mathbf{E}_1' - I_{13}\dot{\varphi}^2\mathbf{E}_{21}' \tag{9.11}$$

$$\boldsymbol{\omega} \times \mathbf{v}_C = \begin{vmatrix} \mathbf{E}_1' & \mathbf{E}_2' & \mathbf{E}_3' \\ 0 & 0 & \dot{\varphi} \\ -\dot{\varphi}x_{2C}' & \dot{\varphi}x_{1C}' & 0 \end{vmatrix} = -x_{1C}'\dot{\varphi}^2\mathbf{E}_1' - \dot{\varphi}^2x_{2C}'\mathbf{E}_2'. \tag{9.12}$$

According to formulas (9.6) and (9.10) we have

$$\frac{\tilde{\mathrm{d}}\mathbf{v}_C}{\mathrm{d}t} = -\ddot{\varphi}x_{2C}'\mathbf{E}_1' + x_{1C}'\ddot{\varphi}\mathbf{E}_2', \tag{9.13}$$

$$\frac{\tilde{\mathrm{d}}\mathbf{K}_O}{\mathrm{d}t} = -I_{13}\ddot{\varphi}\mathbf{E}_1' - I_{23}\ddot{\varphi}\mathbf{E}_2' + I_3\ddot{\varphi}\mathbf{E}_3'. \tag{9.14}$$

From (9.7) and (9.8), taking into account relations (9.10)–(9.14), we obtain

$$\begin{aligned} -Mx'_{2C}\ddot{\varphi} - Mx'_{1C}\dot{\varphi}^2 &= F'_1 + R'_1 + \tilde{R}'_1, \\ Mx'_{1C}\ddot{\varphi} - Mx'_{2C}\dot{\varphi}^2 &= F'_2 + R'_2 + \tilde{R}'_2, \\ 0 &= F'_3 + R'_3 + \tilde{R}'_3, \end{aligned} \tag{9.15}$$

$$\begin{aligned} -I_{13}\ddot{\varphi} + I_{23}\dot{\varphi}^2 &= M_{O1} - H\tilde{R}_2, \\ -I_{23}\ddot{\varphi} - I_{13}\dot{\varphi}^2 &= M_{O2} + H\tilde{R}_1, \\ I_3\ddot{\varphi} &= M_{O3}. \end{aligned} \tag{9.16}$$

The obtained differential equations have the following properties. The third equation of (9.15) is the algebraic sum $R'_3 + \tilde{R}'_3 = -F'_3$, and because it does not occur in the remaining equations, it is not possible to determine the reactions R'_3 and $\tilde{R}'_3$ separately. Moreover, their sum is independent of body motion. The remaining transverse reactions can be determined from the remaining equations of systems (9.15) and (9.16). The third equation of system (9.16) describes the rotational motion of a rigid body about the axis OX'_3 driven by the moment M_{O3}.

In Example 2.4 of [1] we determined the reactions in the bearings in a static case. Now we will try to determine the conditions for which the dynamic reactions of the system (that is, the reactions during body rotation with angular velocity $\dot{\varphi}\mathbf{E}'_3$) are equal to the static reactions.

Transverse static reactions are determined from (9.15) and (9.16), and setting $\dot{\varphi} = \ddot{\varphi} = 0$ we obtain

$$\begin{aligned} R_1 + \tilde{R}_1 &= -F_1, \\ R_2 + \tilde{R}_2 &= -F_2, \\ \tilde{R}_2 &= \frac{M_{O1}}{H}, \\ \tilde{R}_1 &= -\frac{M_{O2}}{H}, \end{aligned} \tag{9.17}$$

that is, $R_1 = -F_1 + M_{O2}/H$, $R_2 = -F_2 - M_{O1}/H$.

The same magnitudes of reactions are obtained for $\dot{\varphi} \neq 0$ and $\ddot{\varphi} \neq 0$, on the condition that the following equations are satisfied:

$$\begin{cases} x'_{2C}\ddot{\varphi} + x'_{1C}\dot{\varphi}^2 = 0, \\ x'_{1C}\ddot{\varphi} - x'_{2C}\dot{\varphi}^2 = 0, \end{cases} \tag{9.18}$$

$$\begin{cases} I_{13}\ddot{\varphi} - I_{23}\dot{\varphi}^2 = 0, \\ I_{23}\ddot{\varphi} + I_{13}\dot{\varphi}^2 = 0. \end{cases} \tag{9.19}$$

System (9.18) can be treated as a system of homogeneous equations of the form

$$\begin{bmatrix} \dot{\varphi}^2 & \ddot{\varphi} \\ \ddot{\varphi} & -\dot{\varphi}^2 \end{bmatrix} \begin{bmatrix} x'_{1C} \\ x'_{2C} \end{bmatrix} = \begin{bmatrix} 0 \\ 0 \end{bmatrix}, \tag{9.20}$$

and, in turn, system (9.19) can be represented in the form

$$\begin{bmatrix} \dot{\varphi}^2 & \ddot{\varphi} \\ \ddot{\varphi} & -\dot{\varphi}^2 \end{bmatrix} \begin{bmatrix} I_{13} \\ I_{23} \end{bmatrix} = \begin{bmatrix} 0 \\ 0 \end{bmatrix}. \tag{9.21}$$

In both cases the determinant of system $W = -(\dot{\varphi}^4 + \ddot{\varphi}^2) \neq 0$, which leads to two conditions: $x'_{1C} = x'_{2C} = 0$ and $I_{13} = I_{23} = 0$. This means that the transverse dynamic reactions in the bearings during rigid-body rotation about a fixed axis are equal to the static reactions if and only if the axis of rotation is a principal centroidal axis of inertia of the body.

9.2 Motion of a Rigid Body About a Fixed Point

In order to obtain equations of motion for the present case it suffices to make use of the equation

$$\frac{\mathrm{d}\mathbf{K}_O}{\mathrm{d}t} = \mathbf{M}_O, \tag{9.22}$$

where $\mathbf{K}_O$ denotes the angular momentum (moment of momentum) calculated with respect to point O, and $\mathbf{M}_O$ is the main moment of forces acting on the body with respect to the same point. Point O is fixed, and there we locate the origins of the non-stationary $O'X'_1X'_2X'_3$ and stationary $OX_1X_2X_3$ coordinate systems.

Equation (9.22) in the body coordinate system $O'X'_1X'_2X'_3$ takes the form

$$\frac{\tilde{\mathrm{d}}\mathbf{K}_O}{\mathrm{d}t} + \boldsymbol{\omega} \times \mathbf{K}_O = \mathbf{M}_O, \tag{9.23}$$

where $\mathbf{K}_O$ is described by (9.6) and (9.4), and $\tilde{\mathrm{d}}/\mathrm{d}t$ is a local derivative.

Because

$$\boldsymbol{\omega} \times \mathbf{K}_O = \begin{vmatrix} \mathbf{E}'_1 & \mathbf{E}'_2 & \mathbf{E}'_3 \\ \omega'_1 & \omega'_2 & \omega'_3 \\ K_{O1} & K_{O2} & K_{O3} \end{vmatrix} = \mathbf{E}'_1\left(\omega'_2 K_{O3} - K_{O2}\omega'_3\right)$$
$$+ \mathbf{E}'_2\left(K_{O1}\omega'_3 - K_{O3}\omega'_1\right) + \mathbf{E}'_3\left(\omega'_1 K_{O2} - \omega'_2 K_{O1}\right), \tag{9.24}$$

from (9.23) we obtain

$$
\begin{aligned}
& I_1\dot{\omega}_1' - I_{12}\dot{\omega}_2' - I_{13}\dot{\omega}_3' + \omega_2'\left(-I_{13}\omega_1' - I_{23}\omega_2' + I_3\omega_3'\right) \\
& - \omega_3'\left(-I_{12}\omega_1' + I_2\omega_2' - I_{23}\omega_3'\right) = M_{O1}, \\
& - I_{12}\dot{\omega}_1' + I_2\dot{\omega}_2' - I_{13}\dot{\omega}_3' + \omega_3'\left(I_1\omega_1' - I_{12}\omega_2' - I_{13}\omega_3'\right) \\
& - \omega_1'\left(-I_{13}\omega_1' - I_{23}\omega_2' - I_3\omega_3'\right) = M_{O2}, \\
& - I_{13}\dot{\omega}_1' - I_{23}\dot{\omega}_2' + I_3\dot{\omega}_3' + \omega_1'\left(-I_{12}\omega_1' - I_2\omega_2' - I_{23}\omega_3'\right) \\
& - \omega_2'\left(I_1\omega_1' - I_{12}\omega_2' - I_{13}\omega_3'\right) = M_{O3},
\end{aligned}
\tag{9.25}
$$

and following transformations we have

$$
\begin{aligned}
& I_1\dot{\omega}_1' - I_{12}\dot{\omega}_2' - I_{13}\dot{\omega}_3' + (I_3 - I_2)\,\omega_2'\omega_3' \\
& + I_{23}\left(\omega_3'^2 - \omega_2'^2\right) + \omega_1'\left(I_{12}\omega_3' - I_{13}\omega_2'\right) = M_{O1}, \\
& - I_{12}\dot{\omega}_1' + I_2\dot{\omega}_2' - I_{13}\dot{\omega}_3' + (I_1 - I_3)\,\omega_1'\omega_3' \\
& + I_{13}\left(\omega_1'^2 - \omega_2'^2\right) + \omega_2'\left(I_{23}\omega_1' - I_{12}\omega_3'\right) = M_{O2}, \\
& - I_{13}\dot{\omega}_1' - I_{23}\dot{\omega}_2' + I_3\dot{\omega}_3' + (I_2 - I_1)\,\omega_1'\omega_2' \\
& + I_{12}\left(\omega_2'^2 - \omega_1'^2\right) + \omega_3'\left(I_{13}\omega_2' - I_{23}\omega_1'\right) = M_{O3}.
\end{aligned}
\tag{9.26}
$$

Let us choose the axes $O'X_1'$, $O'X_2'$, and $O'X_3'$ so that they coincide with the principal axes of inertia of the body associated with point $O' = O$. In this case $I_{12} = I_{13} = I_{23} = 0$ and (9.26) take the much simpler form

$$
\begin{aligned}
I_1\dot{\omega}_1' + (I_3 - I_2)\,\omega_2'\omega_3' &= M_{O1}, \\
I_2\dot{\omega}_2' + (I_1 - I_3)\,\omega_1'\omega_3' &= M_{O2}, \\
I_3\dot{\omega}_3' + (I_2 - I_1)\,\omega_2'\omega_1' &= M_{O3}.
\end{aligned}
\tag{9.27}
$$

Equations (9.27) are called *Euler's dynamic equations*. If the known components of a main moment $M_{Oi} = M_{Oi}(\omega_1', \omega_2', \omega_3', t)$, $i = 1, 2, 3$, then in order to obtain $\omega_1'(t)$, $\omega_2'(t)$, and $\omega_3'(t)$, one should integrate (e.g., numerically) (9.27). Recall that the position of a body in an absolute coordinate system $OX_1X_2X_3$ is associated with the introduction of three Euler angles, derived in [1] we have

$$
\begin{aligned}
\omega_1''' \equiv \omega_1' &= \dot{\psi}\sin\phi\sin\theta + \dot{\theta}\cos\phi, \\
\omega_2''' \equiv \omega_2' &= \dot{\psi}\cos\phi\sin\theta - \dot{\theta}\sin\phi, \\
\omega_3''' \equiv \omega_3' &= \dot{\psi}\sin\theta + \dot{\phi},
\end{aligned}
\tag{9.28}
$$

and in order to simplify the notation, the symbols ($'''$) are dropped in this section. In other words, the transition from the coordinate system $OX_1X_2X_3$ to the system $O'X_1'X_2'X_3'$ takes place through Euler's angles.

Substituting the determined functions of time $\omega_i' = \omega_i'(t)$, $i = 1, 2, 3$, into *Euler's kinematic equations* (9.28) we are able to solve them with respect to Euler's angles, that is, to determine $\psi = \psi(t)$, $\theta = \theta(t)$, and $\phi = \phi(t)$.

If, however, $M_{Oi} = M_{Oi}(\omega_1', \omega_2', \omega_3', \phi, \psi, \theta, \dot{\phi}, \dot{\psi}, \dot{\theta}, t)$, then one should solve simultaneously six differential equations given by (9.27) and (9.28). If a rigid body is not acted upon by a main moment, that is, $M_O = 0$, then we are dealing with the so-called *Euler case* of rigid-body motion about a fixed point O, and for this case (9.27) take the form

$$\begin{aligned} I_1\dot{\omega}_1' + (I_3 - I_2)\,\omega_2'\omega_3' &= 0, \\ I_2\dot{\omega}_2' + (I_1 - I_3)\,\omega_1'\omega_3' &= 0, \\ I_3\dot{\omega}_3' + (I_2 - I_1)\,\omega_2'\omega_1' &= 0. \end{aligned} \tag{9.29}$$

The preceding equations constitute a system of three first-order non-linear ordinary differential equations. If the aforementioned body is in the gravitational field, then the condition $M_O = 0$ is satisfied, provided that the fixed point about which the body rotates is coincident with the mass center of the body, that is, $O = C$.

From (9.22) for the *Euler case* we have

$$\mathbf{K}_O = \text{const.} \tag{9.30}$$

Vector $\mathbf{K}_O$'s property of being *constant* means that during motion of the rigid body both its direction and magnitude are constant. From (9.4) and (9.6) for the Euler case we have

$$\mathbf{K}_O = I_1\omega_1'\mathbf{E}_1' + I_2\omega_2'\mathbf{E}_2' + I_3\omega_3'\mathbf{E}_3', \tag{9.31}$$

where now unit vectors $\mathbf{E}_1'$, $\mathbf{E}_2'$, $\mathbf{E}_3'$ are the unit vectors of the principal axes of inertia of the body at point $O = O'$.

From conditions (9.30) and (9.31) it follows that

$$K_O^2 = I_1^2\omega_1'^2 + I_2^2\omega_2'^2 + I_3^2\omega_2'^2 = \text{const.} \tag{9.32}$$

The increment of kinetic energy of the rigid body is equal to

$$\mathrm{d}T = \mathbf{M}_O \circ \boldsymbol{\omega}\mathrm{d}t + \mathbf{F} \circ \mathbf{v}_O\mathrm{d}t = 0 \tag{9.33}$$

because in the considered *Euler case* $\mathbf{M}_O = \mathbf{0}$ and $\mathbf{v}_O = \mathbf{0}$. From that we have

$$T = \frac{1}{2}\left(I_1\omega_1'^2 + I_2\omega_2'^2 + I_3\omega_3'^2\right) = \text{const.} \tag{9.34}$$

From the preceding calculations it follows that Euler's equations (9.29) possess two first integrals in the forms (9.32) and (9.34). The mentioned integrals can also be obtained immediately from (9.29).

The Euler case can be further simplified if the vector of angular velocity of the body $\boldsymbol{\omega}$ is constant with respect to the body. This means that $\dot{\omega}_1' = \dot{\omega}_2' = \dot{\omega}_3' = 0$, and from (9.29) we obtain

$$
\begin{aligned}
(I_3 - I_2)\,\omega_2'\omega_3' &= 0,\\
(I_1 - I_3)\,\omega_1'\omega_3' &= 0,\\
(I_2 - I_1)\,\omega_1'\omega_2' &= 0.
\end{aligned}
\tag{9.35}
$$

Because by assumption $\omega_1' \neq 0$, $\omega_2' \neq 0$, and $\omega_3' \neq 0$, all algebraic equations of (9.35) are satisfied for an arbitrary vector $\boldsymbol{\omega}$ when $I_1 = I_2 = I_3$. In this case the ellipsoid of inertia at point O' changes into the surface of a ball. This special Euler case is called the *stationary spinning* of a rigid body ($\boldsymbol{\omega} = \text{const}$).

Let any two principal moments of inertia with respect to point O' be equal, for instance, $I_1 = I_2 \neq I_3$. Also in this case (9.35) are satisfied, but this time not for an arbitrary vector $\boldsymbol{\omega}$. They are satisfied for (1) $\omega_1' = \omega_2' = 0$ and $\omega_3' \neq 0$ (spinning about a principal axis of inertia OX_3') or for (2) $\omega_3' = 0$, and arbitrary ω_1', ω_2' (in this case spinning takes place about an arbitrary axis passing through point O and lying in an equatorial plane of the ellipsoid of inertia).

Finally, if I_1, I_2 and I_3 are all distinct, then (9.35) can also be satisfied. Then two out of the three quantities ω_1', ω_2', and ω_3' are equal to zero, and the third one is arbitrary. The body rotates about the principal axis of inertia associated with the non-zero component of vector $\boldsymbol{\omega}$.

Let us consider the second dynamical simplification of the Euler case described by (9.29). If $I_1 = I_2 \neq I_3$, then the rigid body is called a *dynamically symmetric body* and the axis OX_3' an *axis of dynamical symmetry of a body*. If we introduce the system $OX_1X_2X_3$ in such way that the axis OX_3 is directed along the angular momentum vector $\mathbf{K}_O = \text{const}$, then, using Fig. 9.2 and projecting vector $\mathbf{K}_O$ onto the axes of the system $O'X_1'X_2'X_3'$, we obtain

$$
\begin{aligned}
I_1\omega_1' &= K_O \sin\theta \sin\phi,\\
I_2\omega_2' &= K_O \sin\theta \cos\phi,\\
I_3\omega_3' &= K_O \cos\theta.
\end{aligned}
\tag{9.36}
$$

If $I_1 = I_2$, then from the third equation of (9.29) we have

$$
\omega_3' \equiv \omega_{3O}' = \text{const}. \tag{9.37}
$$

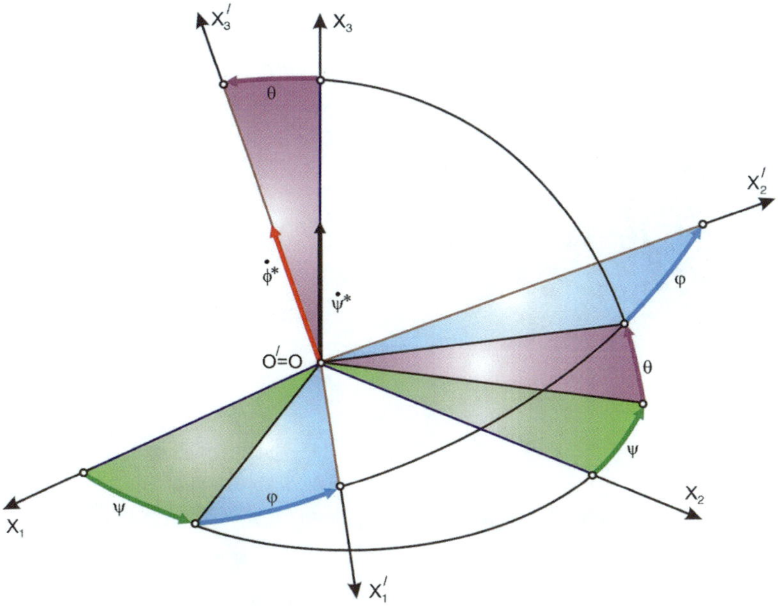

Fig. 9.2 Projecting angular momentum vector $\mathbf{K}_O$ onto axes of a body system, and vectors of angular velocity of spin $\dot{\phi}^*$ and angular velocity of precession $\dot{\psi}^*$

This means that the projection of vector $\boldsymbol{\omega}$ onto the *axis of dynamical symmetry of the body* is constant. From (9.37) and the third equation of system (9.36) we obtain

$$\cos\theta_O = \frac{I_3\omega'_{3O}}{K_O} = \text{const}, \tag{9.38}$$

that is, the angle of nutation θ is constant. In this case *Euler's kinematic equations* (9.28) take the form

$$\begin{aligned} \omega'_1 &= \dot{\psi}\sin\phi\sin\theta_O, \\ \omega'_2 &= \dot{\psi}\cos\phi\sin\theta_O, \\ \omega'_3 &= \dot{\psi}\cos\theta_O + \dot{\phi}. \end{aligned} \tag{9.39}$$

Substituting the first equation of (9.39) into the first equation of (9.36) we obtain

$$I_1\dot{\psi} = K_O, \tag{9.40}$$

hence

$$\dot{\psi} = \frac{K_O}{I_1} = \dot{\psi}^* = \text{const}. \tag{9.41}$$

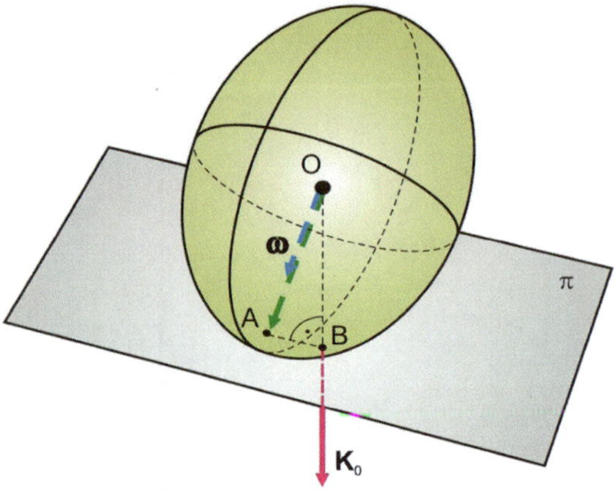

Fig. 9.3 Poinsot's geometric interpretation of motion of a rigid body with one point O fixed

The component $\dot{\psi}^*$ is called the *rate of precession*. One quantity remains to be described – $\dot{\phi}$, which we determine from the third equation of (9.38). Using relations (9.37), (9.38), and (9.41) we calculate successively

$$\dot{\phi} = \omega'_{3O} - \dot{\psi}^* \cos\theta_O = \omega'_{3O} - \frac{K_O}{I_1}\cos\theta_O$$

$$= \omega'_{3O} - \frac{I_3}{I_1}\omega'_{3O} = \frac{I_1 - I_3}{I_1}\omega'_{3O} = \dot{\phi}^* = \text{const.} \tag{9.42}$$

The quantity $\dot{\phi}^*$ is called the *rate of spin* of a body.

In the general case, the motion of a body about a fixed point O that is composed of the rotational motion about the axis associated with the body (in the present case $O'X'_3$) and the rotational motion of this axis ($O'X'_3$) around a fixed axis (OX_3) is called the *precession of a rigid body*. A special case of precession is the *steady precession*, where the two previously mentioned angular velocities (i.e., $\dot{\phi}^*$ and $\dot{\psi}^*$) are constant.

The Euler case for a dynamically symmetric body describes its *steady precession*. In this case, during motion the symmetry axis of the body OX'_3 describes the cone of circular cross section and opening angle $2\theta_O$ at the apex O, where $\cos\theta_O = \cos(\mathbf{E}'_3, \mathbf{K}_O)$. Likewise, the revolving motion of body symmetry axis of unit vector $\mathbf{E}'_3$ takes place with a constant velocity $\dot{\psi}^*$ and is accompanied by (simultaneous) spinning of the body with a constant angular velocity $\dot{\phi}^*$ about its symmetry axis of unit vector $\mathbf{E}'_3$.

The geometric interpretation associated with the considered Euler case proposed by Poinsot is worth noting.

Let a rigid body in a gravitational field have its center of mass at point O and let it be supported at this point. The motion about a point of the body in this case is called the *inertial motion of a body* with its mass center fixed. Figure 9.3 shows the ellipsoid of inertia of a body like that in arbitrary motion about a fixed point O. The center of the ellipsoid is located at point O.

Let the body rotate with angular velocity $\boldsymbol{\omega}\,[\omega_1', \omega_2', \omega_3']$ given in the body system $O'X_1'X_2'X_3'$. The ellipsoid of inertia related to point O has the form

$$I_1 x_1'^2 + I_2 x_2'^2 + I_3 x_3'^2 = 1. \tag{9.43}$$

Let the axis determined by vector $\boldsymbol{\omega}$ have one point A in common with the surface of the ellipsoid. The plane passing through point A and tangent to the ellipsoid is called an *invariable plane* and denoted by π. Below we will cite proofs of three characteristics of the Euler case presented by Poinsot (see also [2]).

1. We will prove that $\boldsymbol{\omega} \parallel \overrightarrow{OA}$.

   From Fig. 9.3 it follows that vectors $\boldsymbol{\omega}$ and $\overrightarrow{OA}$ are collinear, that is,

$$\overrightarrow{OA}\left[x_{1A}', x_{2A}', x_{3A}'\right] = \lambda\boldsymbol{\omega}\left[\omega_1', \omega_2'\omega_3'\right]. \tag{9.44}$$

   It should be demonstrated that $\lambda = \text{const}$. From (9.44) we obtain

$$x_{iA}' = \lambda\omega_i', \qquad i = 1, 2, 3, \tag{9.45}$$

and since point A belongs to the ellipsoid of inertia, from (9.43) and (9.44) we obtain

$$\lambda^2\left(I_1\omega_1'^2 + I_2\omega_2'^2 + I_3\omega_3'^2\right) = 2T\lambda^2 = 1, \tag{9.46}$$

where (9.34) was used. From relation (9.46) we obtain

$$\lambda = \frac{1}{\sqrt{2T}} = \text{const}. \tag{9.47}$$

2. We will demonstrate that $\pi \perp \mathbf{K}_O$.

   The equation of the surface of the ellipsoid follows.

$$f\left(x_1', x_2', x_3'\right) = I_1 x_1'^2 + I_2 x_2'^2 + I_3 x_3'^2 - 1 = 0, \tag{9.48}$$

and the vector normal to this surface at point A is given by

$$\begin{aligned}\operatorname{grad}_A f &= \left.\frac{\partial f}{\partial x_1'}\right|_A \mathbf{E}_1' + \left.\frac{\partial f}{\partial x_2'}\right|_A \mathbf{E}_2' + \left.\frac{\partial f}{\partial x_3'}\right|_A \mathbf{E}_3' \\ &= 2I_1 x_{1A}'\mathbf{E}_1' + 2I_2 x_{2A}'\mathbf{E}_2' + 2I_3 x_{3A}'\mathbf{E}_3' \\ &= 2\lambda\left[I_1\omega_1'\mathbf{E}_1' + I_2\omega_2'\mathbf{E}_2' + I_3\omega_3'\mathbf{E}_3'\right] = 2\lambda\mathbf{K}_O. \end{aligned} \tag{9.49}$$

   Because, by assumption, the normal vector $\operatorname{grad}_A f$ is perpendicular to π, from relation (9.49) it follows that $\mathbf{K}_O \perp \pi$.

3. We will demonstrate that the projection of the radius vector $\mathbf{r}'_A = \overrightarrow{OA}$ of point A onto the direction of vector $\mathbf{K}_O$ is constant. From Fig. 9.3, (9.45) and (9.47), and from the relation $2T = \mathbf{K}_O \circ \boldsymbol{\omega}$ it follows successively that

$$OB = \frac{\mathbf{K}_O \circ \overrightarrow{OA}}{K_O} = \frac{\lambda\,(\boldsymbol{\omega} \circ \mathbf{K}_O)}{K_O} = \frac{2\lambda T}{K_O}\frac{\sqrt{2T}}{K_O} = \text{const.} \tag{9.50}$$

Vector $\mathbf{K}_O = \text{const}$, which means that it has a constant direction in the system $OX_1X_2X_3$ during the time that a rigid body is in motion. In the system $O'X_1'X_2'X_3'$ its direction changes, but the magnitude is preserved, which follows from (9.32). Let point $A(t_A)$ of the body become $A_1(t_{A_1})$ after a certain time. Projections of vectors $\overrightarrow{OA}$ and $\overrightarrow{OA_1}$ onto the direction of $\mathbf{K}_O$ are equal. This means that the plane $\pi \perp \mathbf{K}_O$ during the motion of point A described by the radius vector is always at the same distance OB from point O, that is, the plane π is fixed in space. The velocity of point A is equal to zero because it lies on an instantaneous axis of rotation $\boldsymbol{\omega}$.

From the foregoing calculations it follows that in the Euler case the motion of the rigid body has the following geometric interpretation proposed by Poinsot.

The ellipsoid of inertia related to the fixed point of a body rolls without sliding on the plane π, which is fixed in space. The plane π is continuously perpendicular to the vector of angular momentum $\mathbf{K}_O$ and the vectors of the angular velocity of the body $\boldsymbol{\omega}$ and the radius vector of the point of contact between the ellipsoid and the plane $\mathbf{r}'_A$ are collinear and proportional to each other. The plane π is called the *invariant plane*, and the axis along which is directed vector $\mathbf{K}_O$ is called the *invariant axis*. Point A, which is the tip of radius vector $\overrightarrow{OA}$, during motion belongs simultaneously to the plane π and the ellipsoid of inertia that rolls on this plane. The motion of point A takes place along a curve on the plane π called the *herpolhode*. In turn, on the surface of the ellipsoid, point A moves along a curve called the *polhode*.

According to previous calculation, during motion, the instantaneous axis of rotation $\boldsymbol{\omega}$ describes a conical surface, that is, a *moving axode* in the system $O'X_1'X_2'X_3'$ and a *fixed axode* in the system $OX_1X_2X_3$. The moving axode rolls around the fixed axode along a generatrix that belongs at the given instant to both conical surfaces, and vector $\boldsymbol{\omega}$ lies on the generatrix. The herpolhode belongs to the fixed axode and the polhode to the moving axode.

At the end of the calculations, we will determine the orientation in space of a body for the Euler case, that is, in the stationary system $OX_1X_2X_3$. In the coordinate system presented in Fig. 9.3, that is, where $\mathbf{K}_O$ lies on the axis OX_3, and for the general case, that is, the dynamically asymmetric case ($I_1 \neq I_2 \neq I_3$), relations between Euler's angles and the coordinates of vector $\boldsymbol{\omega}$ are given by (9.36). If the coordinates of vectors $\boldsymbol{\omega}\,[\omega_1', \omega_2', \omega_3']$ are known, then from (9.36) we calculate

$$\tan\phi = \frac{I_1\omega_1'}{I_2\omega_2'}, \qquad \cos\theta = \frac{I_3\omega_3'}{K_O}. \tag{9.51}$$

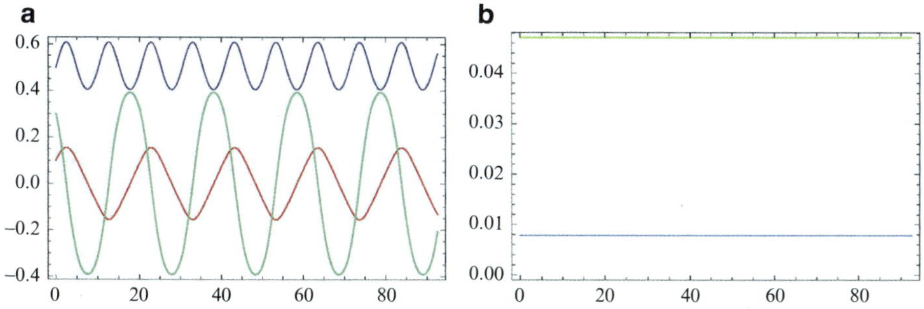

Fig. 9.4 Values of $\omega_i(t)$ (**a**) and K_o^2 and 2T (**b**) for $M_i(t) = 0,\ i = 1, 2, 3$ obtained as a result of the solution to (9.27)

Let us multiply the first equation of system (9.28) by $\sin\phi$ and the second by $\cos\phi$, then, adding them to each other, we obtain

$$\dot{\psi} = \frac{\omega_1' \sin\phi + \omega_2' \cos\phi}{\sin\theta}, \tag{9.52}$$

and from (9.36) we have

$$\sin\phi = \frac{I_1\omega_1'}{K_O \sin\theta}, \qquad \cos\phi = \frac{I_2\omega_2'}{K_O \sin\theta}. \tag{9.53}$$

Substituting (9.53) into (9.52) we obtain

$$\begin{aligned} \dot{\psi} &= \frac{\omega_1'^2 I_1 + \omega_2'^2 I_2}{K_O \sin^2\theta} = \frac{K_O\left(\omega_1'^2 I_1 + \omega_2'^2 I_2\right)}{K_O^2 \sin^2\theta} = \frac{K_O\left(\omega_1'^2 I_1 + \omega_2'^2 I_2\right)}{K_O^2\left(1 - \cos^2\theta\right)} \\ &= \frac{K_O\left(\omega_1'^2 I_1 + \omega_2'^2 I_2\right)}{\left(K_O^2 - I_3^2\omega_3'^2\right)} = \frac{K_O\left(\omega_1'^2 I_1 + \omega_2'^2 I_2\right)}{I_1^2\omega_1'^2 + I_2^2\omega_2'^2}, \end{aligned} \tag{9.54}$$

where during transformations the second equation of (9.51) and (9.32) were used.

To illustrate the conducted theoretical calculations, we perform some numerical simulations of the dynamics of a rigid body about a fixed point situated at the mass center of the body.

Euler's dynamic equations (9.27) are solved numerically.

In addition, as can be seen from the preceding algorithm, values of invariants (9.32) and (9.34) were estimated numerically.

Figure 9.4 presents time plots of $\omega_i(t)$ (panel **a**) and runs of values of K_O^2 and $2T$ (panel **b**).

In the next step we use Euler's kinematic equations (9.28) in order to determine Euler's angles $\phi = \phi(t)$, $\psi = \psi(t)$, $\theta = \theta(t)$. The left-hand sides of those equations are described by the solutions to Euler's dynamic equations.

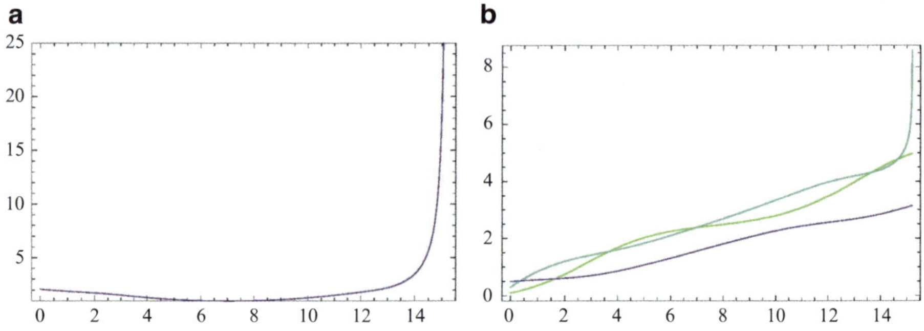

Fig. 9.5 Euler's angles as functions of time numerically estimated using Euler's dynamic and kinematic equations (**a**) and a graph of the function $[\sin\theta(t)]^{-1}$ (**b**)

The results of our computations (angles $\phi(t)$, $\psi(t)$, and $\theta(t)$) are given in Fig. 9.5a, and additionally Fig. 9.5b shows a graph of function $[\sin\theta(t)]^{-1}$, where it is seen that for $t = 15.2$ we are dealing with a singularity ($\sin\theta(t) \to 0$).

Until now we have considered systems of autonomous kinematic and dynamic Euler differential equations. Now we present two examples of solutions to non-autonomous Euler dynamic equations, that is, when $M_i \equiv M_i(t) = M_{Oi}\cos(\sqrt{B_i}t)$. In all calculations we use the same initial conditions and parameters given in the description of the algorithm for a solution to Euler's equations.

From Fig. 9.6 it follows that $K_O^2(t)$ and $2T(t)$ change in time, and additionally with an increase in the amplitude of the forcing moments we observe a transition from regular (periodic) to non-regular (chaotic) dynamics.

9.3 Dynamics of Rigid-Body Motion About a Fixed Point in a Gravitational Field

Let us introduce the coordinate system as shown in Fig. 9.2, where the fixed point $O = O'$ is the origin of both the stationary coordinate system $OX_1X_2X_3$ (the axis OX_3 is vertical) and the system $O'X_1'X_2'X_3'$ rigidly connected to a body (the axis $O'X_3'$ is the axis about which the body spins); it does not coincide with the mass center of the body $C(x_{1C}', x_{2C}', x_{3C}')$. We determine the orientation of the body with respect to space, that is, to the coordinate system $OX_1X_2X_3$, by means of Euler's angles ψ, ϕ, and θ (Fig. 9.7).

The principal moments of inertia of the body with respect to axes $O'X_1'$, $O'X_2'$, and $O'X_3'$ are denoted respectively by I_1, I_2 and I_3, and at the mass center of the body we apply its weight $\mathbf{G} = m\mathbf{g}$. The unit vector $\mathbf{E}_3$ of axis OX_3 is denoted by $\mathbf{N}$ ($\mathbf{E}_3 = \mathbf{N}$), which has the coordinates $\mathbf{N} = \mathbf{N}[n_1', n_2', n_3']$ in the body coordinate system. From Euler's kinematic equations (9.28) and Fig. 9.7 it follows that the

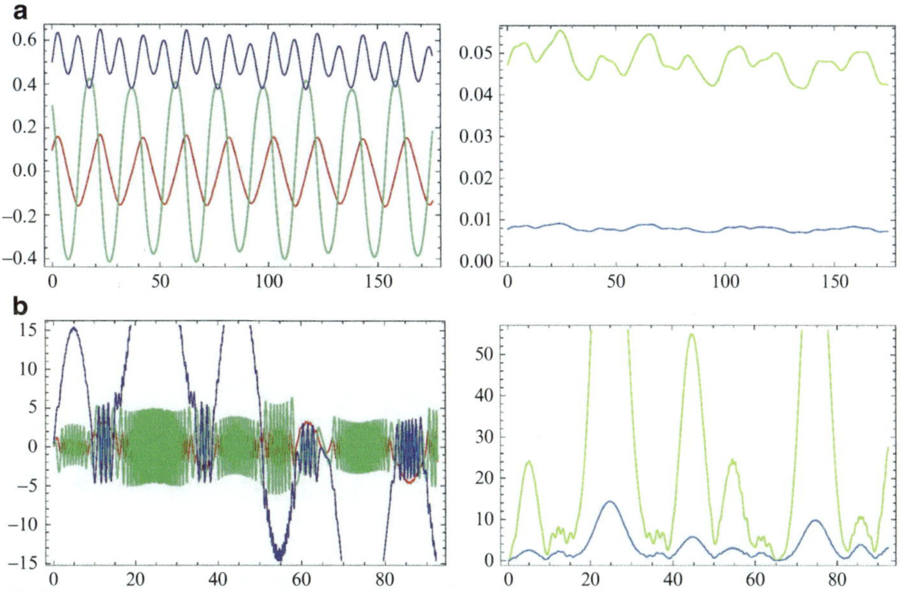

Fig. 9.6 Time plots $\phi(t)$, $\psi(t)$, and $\theta(t)$ as a result of the solution to non-autonomous Euler dynamic equations (**a**) and plots of $K_O^2(t)$ and $2T(t)$ (**b**) for $M_{Oi} = 0.001$ (**a**) and $M_{Oi} = 0.5$ (**b**)

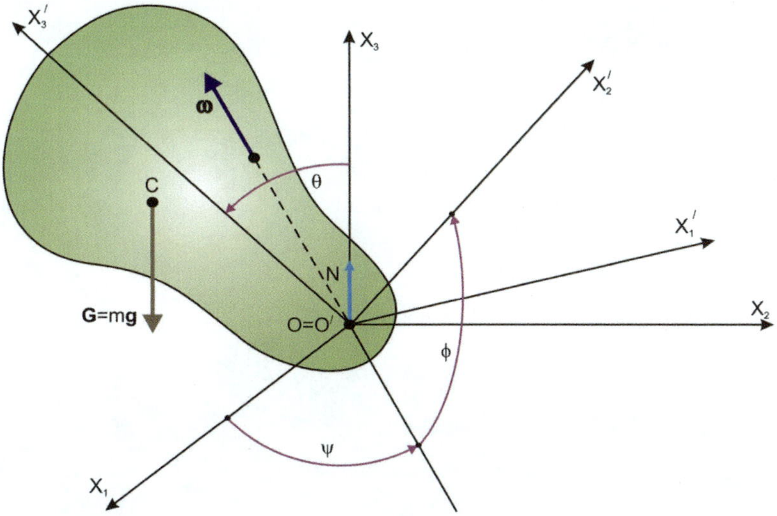

Fig. 9.7 A rigid body in a uniform gravitational field and Euler's angles

components of unit vector **N** in the coordinate system $O'X_1'X_2'X_3'$ are equal to multipliers at $\dot{\psi}$, and hence we have

$$n_1' = \sin\phi \sin\theta, \qquad n_2' = \cos\phi \sin\theta, \qquad n_3' = \cos\theta. \tag{9.55}$$

Because the normal unit vector **N** is in a stationary system, we have

$$\frac{\mathrm{d}\mathbf{N}}{\mathrm{d}t} = 0, \tag{9.56}$$

and in a body system (local system) (9.56) takes the form called a *Poisson equation*:

$$\frac{\tilde{\mathrm{d}}}{\mathrm{d}t}\mathbf{N} + \boldsymbol{\omega} \times \mathbf{N} = \mathbf{0}, \tag{9.57}$$

where $\boldsymbol{\omega}$ is the vector of the angular velocity of the body and $\tilde{\mathrm{d}}/\mathrm{d}t$ denotes the local derivative.

Because, according to (9.55), we have

$$\mathbf{N} = \mathbf{E}_1' n_1' + \mathbf{E}_2' n_2' + \mathbf{E}_3' n_3' \tag{9.58}$$

and

$$\begin{aligned}\boldsymbol{\omega} \times \mathbf{N} = \begin{vmatrix} \mathbf{E}_1' & \mathbf{E}_2' & \mathbf{E}_3' \\ \omega_1' & \omega_2' & \omega_3' \\ n_1' & n_2' & n_3' \end{vmatrix} &= \mathbf{E}_1'\left(n_3'\omega_2' - n_2'\omega_3'\right) \\ &+ \mathbf{E}_2'\left(n_1'\omega_3' - \omega_1' n_3'\right) + \mathbf{E}_3'\left(\omega_1' n_2' - n_1'\omega_2'\right),\end{aligned} \tag{9.59}$$

from (9.57), and taking into account relations (9.58) and (9.59), we obtain three differential equations of first order:

$$\begin{aligned}\frac{\mathrm{d}n_1'}{\mathrm{d}t} &= -\omega_2' n_3' + \omega_3' n_2', \\ \frac{\mathrm{d}n_2'}{\mathrm{d}t} &= -\omega_3' n_1' + \omega_1' n_3', \\ \frac{\mathrm{d}n_3'}{\mathrm{d}t} &= -\omega_1' n_2' + n_1'\omega_2'.\end{aligned} \tag{9.60}$$

In the considered case, the right-hand sides of Euler's dynamic equations (9.27) need to be specified, that is, the moment of forces $\mathbf{M}_O$ should be determined. This moment is determined only by the gravity force of the body **G** and is equal to

$$\mathbf{M}_O = \overrightarrow{OC} \times \mathbf{G} = \overrightarrow{OC} \times (-\mathbf{N}G) = G\mathbf{N} \times \overrightarrow{OC}, \tag{9.61}$$

and, according to the assumptions adopted earlier, we have

$$\mathbf{M}_O = \begin{vmatrix} \mathbf{E}'_1 & \mathbf{E}'_2 & \mathbf{E}'_3 \\ n'_1 G & n'_2 G & n'_3 G \\ x'_{1C} & x'_{2C} & x'_{3C} \end{vmatrix} = G\mathbf{E}'_1 \left(n'_2 x'_{3C} - n'_3 x'_{2C}\right)$$
$$+ G\mathbf{E}'_2 \left(n'_3 x'_{1C} - n'_1 x'_{3C}\right) + G\mathbf{E}'_3 \left(n'_1 x'_{2C} - n'_2 x'_{1C}\right). \tag{9.62}$$

Euler's dynamic equations for the considered system take the form

$$I_1 \frac{\mathrm{d}\omega'_1}{\mathrm{d}t} + (I_3 - I_2)\,\omega'_2\omega'_3 = G\left(n'_2 x'_{3C} - n'_3 x'_{2C}\right),$$
$$I_2 \frac{\mathrm{d}\omega'_2}{\mathrm{d}t} + (I_1 - I_3)\,\omega'_1\omega'_3 = G\left(n'_3 x'_{1C} - n'_1 x'_{3C}\right),$$
$$I_3 \frac{\mathrm{d}\omega'_3}{\mathrm{d}t} + (I_2 - I_1)\,\omega'_1\omega'_2 = G\left(n'_1 x'_{2C} - n'_2 x'_{1C}\right). \tag{9.63}$$

Complete knowledge concerning dynamics of a rigid body with one point fixed and located in the uniform gravitational field boils down to the integration of six first-order non-linear differential equations given by (9.60) and (9.63). If the integration can be successfully carried out, then we obtain the desired solutions $n'_i = n'_i(t)$, $\omega'_i = \omega'_i(t)$, $i = 1, 2, 3$. This means that we determine the vectors $\boldsymbol{\omega} = \boldsymbol{\omega}(t)$ and $\mathbf{N} = \mathbf{N}(t)$ in a body system. In turn, knowledge of $n'_i(t)$ makes it possible to determine the position of the body in the space system because from (9.55) one is able to determine $\phi = \phi(t)$ and $\theta = \theta(t)$. The angle $\psi = \psi(t)$ is yet to be determined, and it can be found using one of Euler's kinematic equations (9.28). The analysis of the problem boils down to the determination of first integrals of differential equations (9.60) and (9.63).

We obtain the first integrals of (9.60) and (9.63) from the observation that $|\mathbf{N}| = 1$, which means that

$$\left(n'_1(t)\right)^2 + \left(n'_2(t)\right)^2 + \left(n'_3(t)\right)^2 = 1. \tag{9.64}$$

In the considered case neither vector $\mathbf{G}$ nor the reactions at point O produce a moment with respect to the axis OX_3. This means that the projection of an angular momentum $\mathbf{K}_O$ onto the axis OX_3 is constant and equal to

$$\mathbf{K}_O \circ \mathbf{N} = \text{const}, \tag{9.65}$$

and because in the non-stationary system $\mathbf{K}_O = \mathbf{K}_O[I_1\omega'_1, I_2\omega'_2, I_3\omega'_3]$ and $\mathbf{N} = \mathbf{N}[n'_1, n'_2, n'_3]$, (9.65) in the system $O'X'_1X'_2X'_3$ takes the form

$$I_1\omega'_1 n'_1 + I_2\omega'_2 n'_2 + I_3\omega'_3 n'_3 = \text{const}. \tag{9.66}$$

Equation (9.66) describes the second first integral of the considered problem.

During the motion of a rigid body with one point fixed in a uniform gravitational field the total energy is conserved, that is,

$$T + V = \text{const}, \tag{9.67}$$

where

$$V = G\overrightarrow{OC} \circ \mathbf{N} = G\left(x_{1C}n'_1 + x_{2C}n'_2 + x_{3C}n'_3\right),$$
$$T = \frac{1}{2}\left(I_1\omega_1'^2 + I_2\omega_2'^2 + I_3\omega_3'^2\right). \tag{9.68}$$

Therefore the third first integral of the analyzed problem has the form

$$\frac{1}{2}\left(I_1\omega_1'^2 + I_2\omega_2'^2 + I_3\omega_3'1^2\right)$$
$$+ G\left(x_{1C}n'_1 + x_{2C}n'_2 + x_{3C}n'_3\right) = \text{const}. \tag{9.69}$$

In the considered case for the illustration of the theoretical calculations some numerical simulations of systems of Euler's dynamic equations (9.63) and differential equations (9.60) are performed.

Figure 9.8 presents graphs of $\omega_i(t)$ and first integrals (invariants) for the case of rigid-body motion about a fixed point in a uniform gravitational field for three values of body weight G. It is evident that the increase of G causes an increase of $\omega_i(t)$; note that the runs have an irregular character, and the invariants preserve constant values in time.

In the monograph [2] one finds the proof that in order for a problem to be integrable by quadratures, it is sufficient to know four independent first integrals of (9.60) and (9.63). For this reason, yet another first integral is left to be determined. However, until now it has not been done except for certain special cases, that is, for certain select regions of initial conditions.

The fourth *first integral* has been determined only for the first three special cases of motion of a given body. Those cases will be briefly described below.

1. *The Euler case*. This case was considered in detail in Sect. 9.2. It is characterized by the coincidence of the mass center with the pivot point, that is, $C = O$ (right-hand sides of (9.63) are then equal to zero since $x'_{1C} = x'_{2C} = x'_{3C} = 0$).
2. *The Lagrange case*. The mass center of a body C lies on the axis of rotation, and the ellipsoid of inertia of the body with respect to the fixed point $O \neq C$ coincides with the ellipsoid of spin of the body. We are dealing with the Lagrange case, for instance, when $I_1 = I_2 \neq I_3$, $x'_{1C} = x'_{2C} = 0$, $x'_{3C} \neq 0$. From the third equation of system (9.63) we determine the fourth first integral of the form

$$\omega'_3 = \text{const}, \tag{9.70}$$

   that is, the projection of the vector of angular velocity $\boldsymbol{\omega}$ onto the axis of rotation of the body is conserved

$$\mathbf{n}'_3 \circ \boldsymbol{\omega} = \text{const}. \tag{9.71}$$

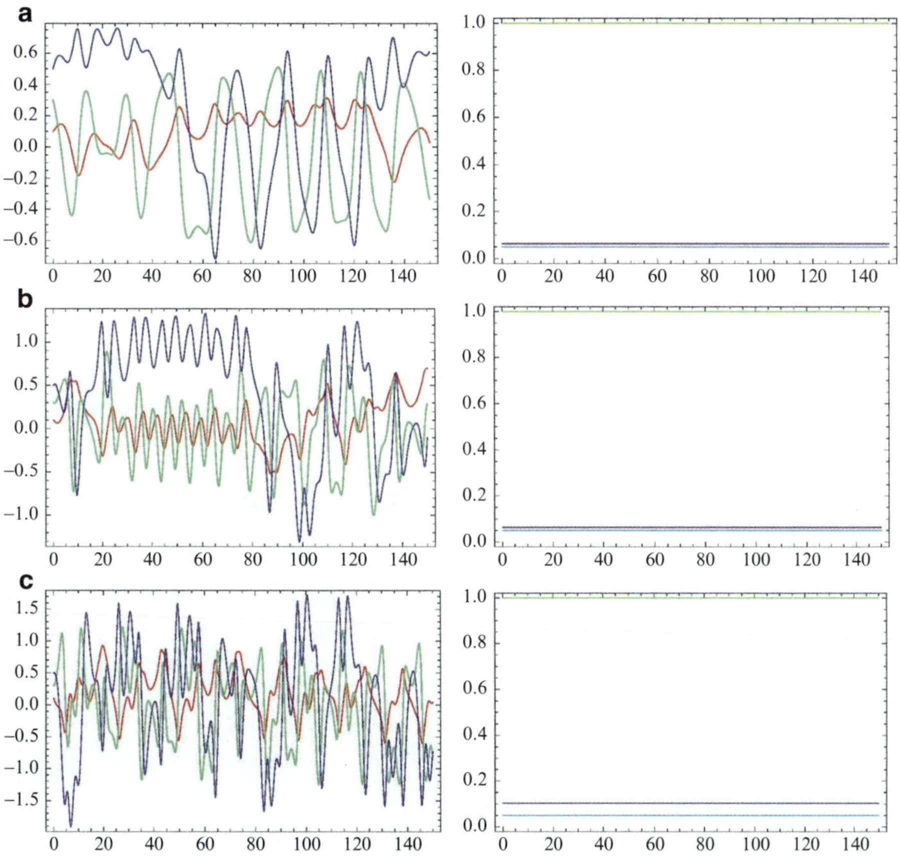

Fig. 9.8 Functions $\omega_i(t)$, $i = 1, 2, 3$ and three invariants determined by (9.64), (9.66), and (9.69) for $G = 0.2N$ (**a**), $G = 1.0N$ (**b**), and $G = 2.0N$ (**c**)

3. *The Kovalevskaya*[1] *case*. In this case an ellipsoid of inertia with respect to the fixed point $O' = O$ is the ellipsoid of spin. Let this spinning occur with respect to the axis $O'X'_3$, let $I_1 = I_2 = 2I_3$, and let the mass center C be located on an equatorial plane of the ellipsoid of inertia ($x'_{3C} = 0$). It can be demonstrated that when the ellipsoid of spin is coincident with the ellipsoid of inertia, an arbitrary axis lying in the equatorial plane and passing through point O is a principal axis of inertia. If we take OX'_1 as the second axis of inertia, then $x'_{2C} = 0$. Euler's dynamic equations (9.63) for the Kovalevskaya case take the form

$$2\frac{\mathrm{d}\omega'_1}{\mathrm{d}t} - \omega'_2\omega'_3 = 0, \qquad 2\frac{\mathrm{d}\omega'_2}{\mathrm{d}t} + \omega'_1\omega'_3 = an'_3, \qquad \frac{\mathrm{d}\omega'_3}{\mathrm{d}t} = an'_2 = 0, \quad (9.72)$$

[1]Zophia Kovalevskaya (1850–1891), Russian mathematician of Polish origin who worked on differential equations.

where $a = (Gx'_{1C})/I_3$. In the Kovalevskaya case the fourth first integral has the following algebraic form:

$$\left(\omega_1'^2 - \omega_2'^2 - an_1'\right)^2 + \left(2\omega_1'\omega_2' - an_2'\right)^2 = \text{const.} \tag{9.73}$$

One may be convinced of that after differentiating (9.73) and using (9.60) and (9.63).

9.4 General Free Motion of a Rigid Body

From the calculations of kinematics it follows that the general motion of a rigid body can be described through the motion of its arbitrary point (the pole) and the motion of the body with respect to this point treated as a fixed point. However, such an arbitrary choice leads to complex equations of motion for the body. If we take the mass center of the rigid body as the pole, then the problem is substantially simplified. The motion of the rigid body can be treated as being composed of the motion of its mass center C and the motion of the body about point C. The motion of the mass center is treated as the motion of a particle acted upon by forces and moments of force applied to the considered rigid body. Equations of free motion of a rigid body have the form

$$M\frac{\mathrm{d}\mathbf{v}_C}{\mathrm{d}t} = \mathbf{F}, \qquad \frac{\mathrm{d}\mathbf{K}_C}{\mathrm{d}t} = \mathbf{M}_C, \tag{9.74}$$

where the first one describes the motion of the mass center subjected to the action of the main force vector applied at point C and the second one describes the change in the angular momentum of the body with respect to point C caused by the main moment of force about that point.

The first vector differential equation of (9.74) has the following form in an absolute system:

$$M\frac{\mathrm{d}^2 x_{1C}}{\mathrm{d}t^2} = F_1, \qquad M\frac{\mathrm{d}^2 x_{2C}}{\mathrm{d}t^2} = F_2, \qquad M\frac{\mathrm{d}^2 x_{3C}}{\mathrm{d}t^2} = F_3, \tag{9.75}$$

where

$$\mathbf{F} = \mathbf{E}_1 F_1 + \mathbf{E}_2 F_2 + \mathbf{E}_3 F_3.$$

The second vector differential equation of (9.74) is usually represented in a body system and has the form

$$\frac{\tilde{\mathrm{d}}\mathbf{K}_C}{\mathrm{d}t} + \boldsymbol{\omega} \times \mathbf{K}_C = \mathbf{M}_C, \tag{9.76}$$

where $\tilde{\mathrm{d}}/\mathrm{d}t$ denotes the local derivative of the vector. In scalar form (9.76) becomes system (9.23), where on the right-hand side the subscript O should be replaced with C.

However, this is not a good choice of non-stationary coordinate system either. If the axes of the system $CX_1'X_2'X_3'$ become coincident with the principal axes of inertia passing through point C, that is, they become principal centroidal axes of inertia, then (9.76) are Euler's dynamic equations in the form

$$I_1\frac{\mathrm{d}\omega_1'}{\mathrm{d}t} + (I_3 - I_2)\,\omega_2'\omega_3' = M_{C_1},$$

$$I_2\frac{\mathrm{d}\omega_2'}{\mathrm{d}t} + (I_1 - I_3)\,\omega_1'\omega_3' = M_{C_2},$$

$$I_3\frac{\mathrm{d}\omega_3'}{\mathrm{d}t} + (I_2 - I_1)\,\omega_1'\omega_2' = M_{C_3}, \tag{9.77}$$

where the moments of inertia and the moments of forces are described in a local coordinate system.

As was already mentioned, in order to trace motion in an absolute system $CX_1X_2X_3$ one should determine the dependencies of Euler's angles on time, that is, simultaneously (or additionally) solve the system of Euler's kinematic equations of the form

$$\omega_1' = \dot{\psi}\sin\theta\sin\phi + \dot{\theta}\cos\phi,$$

$$\omega_2' = \dot{\psi}\sin\theta\cos\phi - \dot{\theta}\sin\phi,$$

$$\omega_3' = \dot{\psi}\sin\theta + \dot{\phi}. \tag{9.78}$$

In problems involving the simulation of the motion of a rigid body, kinematic differential equations are used in a form inverse to (9.78), that is, a form that expresses the time derivatives of Euler's angles in terms of the components of angular velocity in the body system. However, for any choice of Euler's angles this inverse relationship has a singular position, which was mentioned previously.

In the general case, the right-hand sides of systems of non-linear differential equations (9.75) and (9.77) have the forms $F_i = F_i(x_{1C}, x_{2C}, x_{3C}, \psi, \theta, \phi, t)$ and $M_{Ci} = M_{Ci}(x_{1C}, x_{2C}, x_{3C}, \psi, \theta, \phi, t)$, and because of that the three systems of differential equations (9.75), (9.77) and (9.78) should be solved simultaneously.

In one of the special cases, where $M_{Ci} = 0$ and $F_1 = F_2 = 0$, $F_3 = Mg$, that is, when the only force acting on the body is its weight, equations (9.75) take the form

$$\frac{\mathrm{d}^2x_{1C}}{\mathrm{d}t^2} = 0, \qquad \frac{\mathrm{d}^2x_{2C}}{\mathrm{d}t^2} = 0, \qquad \frac{\mathrm{d}^2x_{3C}}{\mathrm{d}t^2} = -g, \tag{9.79}$$

and the remaining two systems of (9.77) and (9.78) are independent of them; they were the subject of our discussion during the analysis of motion of a rigid body in the Euler case.

9.5 Motion of a Homogeneous Ball on a Horizontal Plane in Gravitational Field with Coulomb Friction

Let us consider the motion of a homogeneous ball of mass M and radius r on a horizontal rough plane, shown in Fig. 9.9 (the motion takes place in the absence of rolling resistance, and the only active force is the gravity force).

Let us introduce the absolute coordinate system $OX_1X_2X_3$ and the system $CX_1X_2X_3$ of axes parallel to the absolute system, but with its origin at the center of the ball C. Let $\boldsymbol{\omega}$ denote the angular velocity of the ball, and if $\mathbf{v}_C$ is the translational velocity of the mass center C of the ball, then the velocity of the point of contact of the ball with the surface reads

$$\mathbf{v}_A = \mathbf{v}_C + \boldsymbol{\omega} \times \overrightarrow{CA}. \tag{9.80}$$

The reaction at point A has the form

$$\mathbf{R} = \mathbf{N} + \mathbf{T}, \tag{9.81}$$

where the friction force

$$\mathbf{T} = -\mu N \mathbf{v}_0, \tag{9.82}$$

where $\mathbf{v}_0$ is the unit vector of velocity at point A (of the absolute velocity of point A at which the ball makes contact with the ground at the given instant), that is,

$$\mathbf{v}_A = v_A \mathbf{v}_0, \tag{9.83}$$

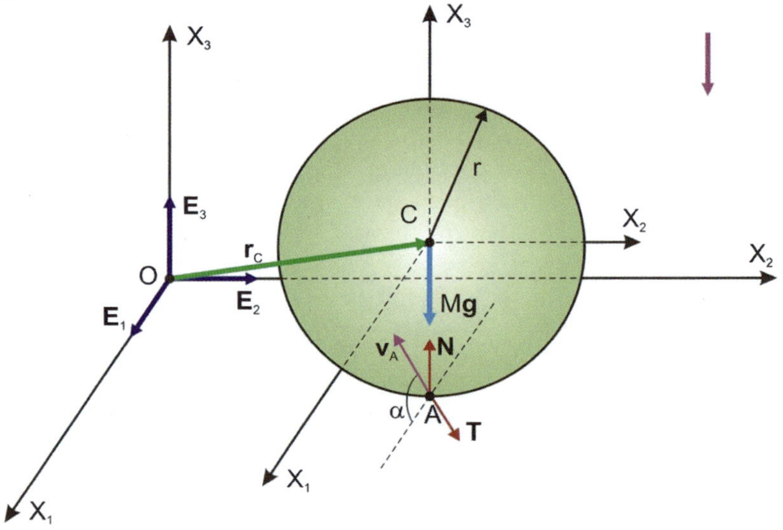

Fig. 9.9 Motion of a ball on a horizontal plane in the gravitational field

which also means that we are dealing with a case of a ball rolling with sliding.

Moreover, the angular momentum of the ball with respect to its center is equal to

$$\mathbf{K}_C = I_C\boldsymbol{\omega} = \frac{2}{5}Mr^2\boldsymbol{\omega}. \tag{9.84}$$

Equations (9.74) for the considered case take the form

$$M\frac{\mathrm{d}\mathbf{v}_C}{\mathrm{d}t} = M\mathbf{g} + \mathbf{R}, \tag{9.85}$$

$$I_C\frac{\mathrm{d}\boldsymbol{\omega}}{\mathrm{d}t} = \overrightarrow{CA} \times \mathbf{R}. \tag{9.86}$$

If we assume

$$\begin{aligned}
\mathbf{r}_C &= \mathbf{E}_1x_{1C} + \mathbf{E}_2x_{2C} + \mathbf{E}_3x_{3C},\\
\mathbf{T} &= \mathbf{E}_1T_1 + \mathbf{E}_2T_2,\\
\mathbf{N} &= \mathbf{E}_3N,
\end{aligned} \tag{9.87}$$

then (9.85) take the scalar form

$$M\frac{\mathrm{d}^2x_{1C}}{\mathrm{d}t^2} = T_1, \qquad M\frac{\mathrm{d}^2x_{2C}}{\mathrm{d}t^2} = T_2, \qquad M\frac{\mathrm{d}^2x_{3C}}{\mathrm{d}t^2} = -Mg + N, \tag{9.88}$$

and since during motion $x_{3C} = \text{const} = r$, from the third equation of system (9.88) we obtain $N = Mg$.

Then,

$$\overrightarrow{CA} \times \mathbf{R} = \begin{vmatrix} \mathbf{E}_1 & \mathbf{E}_2 & \mathbf{E}_3\\ 0 & 0 & -r\\ T_1 & T_2 & N \end{vmatrix} = \mathbf{E}_1rT_2 - \mathbf{E}_2T_1r, \tag{9.89}$$

and from (9.86) we obtain

$$\frac{\mathrm{d}\omega_1}{\mathrm{d}t} = \frac{5T_2}{2Mr}, \qquad \frac{\mathrm{d}\omega_2}{\mathrm{d}t} = \frac{5T_1}{2Mr}, \qquad \frac{\mathrm{d}\omega_3}{\mathrm{d}t} = 0. \tag{9.90}$$

From the last equation it follows that

$$\boldsymbol{\omega} \circ \mathbf{E}_3 = \text{const}, \tag{9.91}$$

regardless of whether the ball rolls without sliding ($\mathbf{v}_A = 0$) or with sliding ($\mathbf{v}_A \neq 0$). The magnitude of the friction force for the case with sliding (i.e., for $\mathbf{v}_A = v_A\mathbf{v}_0 \neq 0$), according to relation (9.82), is equal to $T = \mu N = \mu Mg = \text{const}$. We will demonstrate that the direction of vector $\mathbf{T}$ is conserved.

Differentiating (9.80) with respect to time we obtain

$$\frac{\mathrm{d}\mathbf{v}_A}{\mathrm{d}t} = \frac{\mathrm{d}\mathbf{v}_C}{\mathrm{d}t} + \dot{\boldsymbol{\omega}} \times \overrightarrow{CA} + \dot{\overrightarrow{CA}} \times \boldsymbol{\omega}$$
$$= \mathbf{g} + \frac{(-M\mathbf{g} + \mathbf{T})}{M} + \frac{5\mathbf{T}r}{2Mr} = \frac{7\mathbf{T}}{2M}, \tag{9.92}$$

where during the transformations relations (9.81), (9.82), (9.85), and (9.90) were taken into account.

From (9.92) we obtain

$$\frac{\mathrm{d}v_A}{\mathrm{d}t}\mathbf{v}_0 + v_A\frac{\mathrm{d}\mathbf{v}_0}{\mathrm{d}t} = -\frac{7}{2}\mu g \mathbf{v}_0, \tag{9.93}$$

where (9.83) was used. Because $\mathbf{v}_0 \perp \frac{d\mathbf{v}_0}{\mathrm{d}t}$, from relation (9.93) we obtain

$$\frac{\mathrm{d}\mathbf{v}_0}{\mathrm{d}t} = 0, \qquad \frac{\mathrm{d}v_A}{\mathrm{d}t} = -\frac{7}{2}\mu g. \tag{9.94}$$

From the first equation of (9.94) it follows that $\mathbf{v}_0 = \text{const}$. According to (9.82) this means that also the friction force $\mathbf{T}$ has constant direction. The integration of the second equation of (9.94) yields

$$v_A(t) = v_A(0) - \frac{7}{2}\mu g t, \tag{9.95}$$

which means that the plot of $v_A(t)$ on the $(v,\ t)$ plane is a straight line.

The motion of point A takes place in the horizontal plane OX_1X_2, and the motion of the mass center of the body C takes place in the plane CX_1X_2.

Integrating the first two equations of motion (9.88) we obtain the trajectory of motion of point C. We have successively

$$\begin{aligned} v_{1C} &= \frac{\mathrm{d}x_{1C}}{\mathrm{d}t} = \frac{T_1}{M}t + v_{1C}(0), \\ v_{2C} &= \frac{\mathrm{d}x_{2C}}{\mathrm{d}t} = \frac{T_2}{M}t + v_{2C}(0), \\ x_{1C} &= \frac{T_1}{2M}t^2 + v_{1C}(0)t + x_{1C}, \\ x_{2C} &= \frac{T_2}{2M}t^2 + v_{2C}(0)t + x_{2C}. \end{aligned} \tag{9.96}$$

Let the velocity vector, which according to formula (9.95) is equal to

$$\mathbf{v}_A(t) = \left(v_A(0) - \frac{7}{2}\mu g t\right)\mathbf{v}_0 = v_A(t)\mathbf{v}_0, \tag{9.97}$$

form an angle α with the axis OX_1. Then, multiplying relation (9.97) successively by $\mathbf{E}_1$ and $\mathbf{E}_2$ we obtain

$$\mathbf{v}_A(t) = v_A(t)\cos\alpha\mathbf{E}_1 + v_A(t)\sin\alpha\mathbf{E}_2. \tag{9.98}$$

Integrating (9.90) we obtain

$$\begin{aligned}\omega_1'(t) &= \frac{5T_2}{2Mr}t + \omega_1'(0),\\ \omega_2'(t) &= -\frac{5T_1}{2Mr}t + \omega_2'(0).\end{aligned} \tag{9.99}$$

According to relations (9.82) and (9.87) we have

$$-\mu N\mathbf{v}_0 = \mathbf{E}_1T_1 + \mathbf{E}_2T_2, \tag{9.100}$$

and multiplying this equation by $\mathbf{E}_1$ and $\mathbf{E}_2$ we obtain

$$\begin{aligned}T_1 &= -\mu N\mathbf{v}_0 \circ \mathbf{E}_1 = -\mu Mg\cos\alpha,\\ T_2 &= -\mu N\mathbf{v}_0 \circ \mathbf{E}_2 = -\mu Mg\sin\alpha.\end{aligned} \tag{9.101}$$

Taking into account formulas (9.101) in (9.99) we have

$$\begin{aligned}\omega_1'(t) &= -\frac{5\mu gt}{2r}\cos\alpha + \omega_1'(0),\\ \omega_2'(t) &= -\frac{5\mu gt}{2r}\sin\alpha + \omega_2'(0),\end{aligned} \tag{9.102}$$

and inserting quantities from (9.101) into relations (9.96) we obtain

$$\begin{aligned}v_{1C} &= -\mu gt\cos\alpha + v_{1C}(0),\\ v_{2C} &= -\mu gt\sin\alpha + v_{2C}(0),\\ x_{1C} &= -\frac{\mu gt^2}{2}\cos\alpha + v_{1C}(0)t + x_{1C}(0),\\ x_{2C} &= -\frac{\nu gt^2}{2}\sin\alpha + v_{2C}(0)t + x_{2C}(0).\end{aligned} \tag{9.103}$$

From the first two equations of (9.103) it follows that the parametric equations of the velocity of point C describe a line, and parametric equations of motion of point C describe a parabola, on the assumption that vectors $\mathbf{v}_A$ and $\mathbf{v}_C$ are not collinear, and the motion of point A on the plane OX_1X_2 takes place with sliding. Such motion lasts until the time instant $t_* = \frac{2v_A(0)}{7\mu g}$ [see (9.95)]. At the instant $t = t_*$ we have $\mathbf{v}_A(t_*) = 0$, so the ball motion with sliding comes to an end, and its rolling motion with simultaneous spinning starts. Then, because $\mathbf{v}_A = \mathbf{0}$, from (9.83) and (9.82) it follows that $\mathbf{T} = \mathbf{0}$, that is, $T_1 = T_2 = 0$. Substituting these values into differential equations (9.88), which describe the motion of the mass center of the ball, we obtain

$$\frac{\mathrm{d}^2 x_{1C}}{\mathrm{d}t^2} = 0, \qquad \frac{\mathrm{d}^2 x_{2C}}{\mathrm{d}t^2} = 0, \tag{9.104}$$

and from (9.90) we have

$$\frac{\mathrm{d}\omega_1'}{\mathrm{d}t} = 0, \qquad \frac{\mathrm{d}\omega_2'}{\mathrm{d}t} = 0. \tag{9.105}$$

From (9.105) it follows that during rolling with spinning $\boldsymbol{\omega} = \mathbf{E}_1\omega_1(0) + \mathbf{E}_2\omega_2(0) = \text{const}$. Integrating (9.104) we obtain

$$\begin{aligned} v_{1C} &= \frac{\mathrm{d}x_{1C}}{\mathrm{d}t} = v_{1C}(0), \qquad x_{1C} = v_{1C}(0)t + x_{1C}(0), \\ v_{2C} &= \frac{\mathrm{d}x_{2C}}{\mathrm{d}t} = v_{2C}(0), \qquad x_{2C} = v_{2C}(0)t + x_{2C}(0). \end{aligned} \tag{9.106}$$

The motion of point A can be determined using (9.80), where

$$\boldsymbol{\omega} \times \overrightarrow{CA} = \begin{vmatrix} \mathbf{E}_1 & \mathbf{E}_2 & \mathbf{E}_3 \\ \omega_1 & \omega_2 & \omega_3 \\ 0 & 0 & 0 \end{vmatrix} = \mathbf{E}_1\omega_2 r - \mathbf{E}_2\omega_1 r, \tag{9.107}$$

hence we obtain

$$\begin{aligned} x_{1A}(t) &= v_{1C}(0)t + x_{1C}(0) + r\omega_2(0), \\ x_{2A}(t) &= v_{2C}(0)t + x_{2C}(0) - r\omega_1(0). \end{aligned} \tag{9.108}$$

The solution of the problem of a ball's motion on a plane with Coulomb friction allows for a wider interpretation of the friction phenomenon within the framework of the classic Amontons–Coulomb model. Let an arbitrary rigid body, bounded by a convex surface π' in a region of contact with the plane π, move on the plane π with the point of contact at A (Fig. 9.10). The plane π' is tangent to the surface π during the motion of a rigid body. The velocity $\mathbf{v}_A$ of motion of point A lies in the plane π. If $\mathbf{v}_A = 0$, then the rigid body moves without sliding on the plane π. If $\mathbf{v}_A \neq 0$, then the rigid body moves with sliding on the plane π, and $\mathbf{v}_A$ is called the *velocity of sliding* [2].

If point A is taken as a pole, then the velocity of motion of an arbitrary point of the surface π' (described by radius vector $\mathbf{r}$) with respect to pole A is the geometric

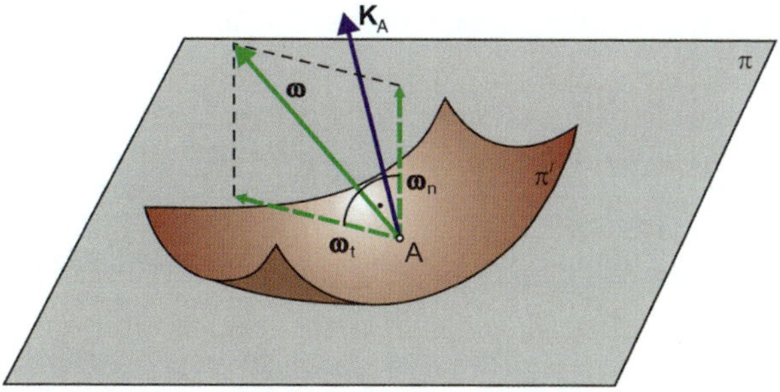

Fig. 9.10 Motion of a rigid body bounded by a convex surface π' on the horizontal plane π (with marked vectors of angular velocity $\boldsymbol{\omega}$ and angular momentum $\mathbf{K}_A$)

sum of the velocity of pole $\mathbf{v}_A$ and the velocity resulting from the rotation caused by $\boldsymbol{\omega}$ of the form (Fig. 9.10)

$$\mathbf{v} = \mathbf{v}_A + \boldsymbol{\omega} \times \mathbf{r}, \tag{9.109}$$

where

$$\boldsymbol{\omega} = \boldsymbol{\omega}_n + \boldsymbol{\omega}_t. \tag{9.110}$$

From (9.110) it follows that vector $\boldsymbol{\omega}$ is resolved into two components at the point of contact. Vector $\boldsymbol{\omega}_n$ (normal) is perpendicular to the plane π, and vector $\boldsymbol{\omega}_t$ (tangent) lies on the plane π, hence we have $\boldsymbol{\omega}_n \perp \boldsymbol{\omega}_t$. Vector $\boldsymbol{\omega}_n$ is called the *angular velocity of turning* and vector $\boldsymbol{\omega}_t$ the *angular velocity of rolling*.

Below we will present various types of motion of a rigid body on the plane π depending on the velocities $\mathbf{v}_A$, $\boldsymbol{\omega}_n$, and $\boldsymbol{\omega}_t$.

1. If $\mathbf{v}_A = \mathbf{0}$, $\boldsymbol{\omega}_n = \mathbf{0}$, and $\boldsymbol{\omega}_t \neq \mathbf{0}$, then the surface π' rolls on the plane π.
2. If $\mathbf{v}_A = \mathbf{0}$, $\boldsymbol{\omega}_n \neq \mathbf{0}$, and $\boldsymbol{\omega}_t = \mathbf{0}$, then we are dealing with the phenomenon of the turning of the surface π' on the plane π.
3. If $\mathbf{v}_A \neq \mathbf{0}$, $\boldsymbol{\omega}_n = \mathbf{0}$, and $\boldsymbol{\omega}_t = \mathbf{0}$, then the surface π' slides on the plane π.
4. If $\mathbf{v}_A \neq \mathbf{0}$, $\boldsymbol{\omega}_n \neq \mathbf{0}$, and $\boldsymbol{\omega}_t \neq \mathbf{0}$, then the surface π' slides, rolls, and turns on the stationary plane π.

Force interaction between surfaces π' and π includes a force called the *normal reaction* $N \geq 0$ perpendicular to both surfaces, acting from (the side of) the plane π on the surface π' and directed from the plane π toward the surface π'. Then, additionally, the friction force $\mathbf{T}$ lying in the plane π at point A acts on the rigid body (if at least one of the surfaces π or π' is a rough surface). If $\mathbf{v}_A = \mathbf{0}$, then the friction force is a force of rolling friction, which is usually smaller than the friction force $T = \mu N$, where μ is the *coefficient of sliding friction* (see Sect. 2.8 in [1]). If the contact surfaces are smooth at point A, then there is no friction force, and the reaction acting on the rigid body reduces to the normal force $\mathbf{N}$.

The following equation is valid in the body system [see (9.76)]

$$\frac{\tilde{\mathrm{d}}\mathbf{K}_A}{\mathrm{d}t} + \boldsymbol{\omega} \times \mathbf{K}_A = \mathbf{M}_A, \tag{9.111}$$

where $\mathbf{M}_A$ is the moment of force with respect to pole A, and then

$$\begin{aligned}\boldsymbol{\omega} \times \mathbf{K}_A &= \boldsymbol{\omega}_n \times \mathbf{K}_A + \boldsymbol{\omega}_t \times \mathbf{K}_A \\ &= \omega_n K'_{nA} \boldsymbol{\omega}_n^0 + \omega_t K'_{tA} \boldsymbol{\omega}_t^0 = \mathbf{M}_n + \mathbf{M}_t. \end{aligned} \tag{9.112}$$

Taking into account relations (9.112) in (9.111) we obtain

$$\begin{aligned}\frac{\tilde{\mathrm{d}}\mathbf{K}_A}{\mathrm{d}t} &= \left(M_{nA} - \omega_n K'_{nA}\right) \boldsymbol{\omega}_n^0 + \left(M_{tA} - \omega_t K'_{tnA}\right) \boldsymbol{\omega}_t^0 \\ &= \tilde{M}_{nA} \boldsymbol{\omega}_n^0 + \tilde{M}_{tA} \boldsymbol{\omega}_t^0. \end{aligned} \tag{9.113}$$

From (9.113) it follows that the force $\mathbf{N} + \mathbf{T}$ and the moment of force $\mathbf{M}_n$ along the unit vector of the normal velocity $\boldsymbol{\omega}_n^0$ and the moment of force $\mathbf{M}_t$ along the unit vector of the tangential velocity $\boldsymbol{\omega}_t^0$ act at point A on the rigid body act. A similar situation occurs for a main force $\mathbf{F}$ acting on a body and the force $\mathbf{R} = \mathbf{N} + \mathbf{T}$, which can be resolved in a similar manner, that is,

$$\mathbf{F} + \mathbf{N} + \mathbf{T} = F_n \boldsymbol{\omega}_n^0 + F_t \boldsymbol{\omega}_t^0. \tag{9.114}$$

The moment of force $\tilde{\mathbf{M}}_{nA}$ is called the *turning moment of force* and the moment $\tilde{\mathbf{M}}_{tA}$ the *rolling moment of force*. The main turning moment of force is carried by a turning moment of friction forces (a couple of forces) only if the rigid body makes contact with the stationary plane π over a certain (small) surface instead of a point.

9.6 Motion of a Rigid Body with an Arbitrary Convex Surface on a Horizontal Plane

We will consider a generalization of the motion of a ball on a rough plane in the case of motion of an arbitrary rigid body, on the assumption that its contact with a point of the horizontal plane is always in the convex region of the body [2]. The modeling of such contact was conducted in Sect. 9.5, and we will make use of the calculations presented there.

Let M denote the mass of a rigid body, $\boldsymbol{\omega}$ and $\mathbf{v}_C$ respectively the angular velocity of the body and translational velocity of the mass center C, $\mathbf{K}_C$ the moment of momentum with respect to point C, and $\mathbf{R} = \mathbf{N} + \mathbf{T}$ the reaction of the plane π on the rigid body. Equations of motion have the form described by (9.7) and (9.8), which in the present case are

$$\dot{\mathbf{v}}_C + \boldsymbol{\omega} \times \mathbf{v}_C = -g\mathbf{n} + \frac{1}{M}\mathbf{R}, \tag{9.115}$$

$$\dot{\mathbf{K}}_C + \boldsymbol{\omega} \times \mathbf{K}_C = \mathbf{r} \times \mathbf{R}, \tag{9.116}$$

where $\mathbf{r}$ is the vector connecting the mass center C with the point of contact of the bodies A, and the dot denotes $\tilde{\mathrm{d}}/\mathrm{d}t$, that is, a local differential operator.

Because, vector $\mathbf{n}$ is constant in a space system, according to the calculations of Sect. 9.3 we have

$$\dot{\mathbf{n}} + \boldsymbol{\omega} \times \mathbf{n} = 0. \tag{9.117}$$

Equation (9.117), called a *Poinsot equation*, holds true for all cases (1)÷(4) considered toward the end of Sect. 9.5. In order to consider specific cases it is necessary to include additional equations relevant to a given.

Let the motion of a body on the plane π take place without sliding (then $\mathbf{v}_A = \mathbf{0}$, where A is the point of contact of the body with the plane π). According to (9.80) in this case we obtain

$$\mathbf{v}_C + \boldsymbol{\omega} \times \mathbf{r} = \mathbf{0}. \tag{9.118}$$

According to Example 9.10 we have

$$f\left(x_1', x_2', x_3'\right) = 0, \qquad \mathbf{n} = -\frac{\operatorname{grad} f}{|\operatorname{grad} f|}, \tag{9.119}$$

which defines vector $\mathbf{n}$. The 12 desired quantities

$$\begin{aligned}
\mathbf{v}_C &= v_{1C}\mathbf{E}_1 + v_{2C}\mathbf{E}_2 + v_{3C}\mathbf{E}_3,\\
\boldsymbol{\omega} &= \omega_{1C}'\mathbf{E}_1' + \omega_{2C}'\mathbf{E}_2' + \omega_{3C}'\mathbf{E}_3',\\
\mathbf{r}_C &= x_{1C}\mathbf{E}_1 + x_{2C}\mathbf{E}_2 + x_{3C}\mathbf{E}_3,\\
\mathbf{R} &= R_1'\mathbf{E}_1' + R_2'\mathbf{E}_2' + R_3'\mathbf{E}_3' = R_1\mathbf{E}_1 + R_2\mathbf{E}_2 + R_3\mathbf{E}_3
\end{aligned} \tag{9.120}$$

can be determined from (9.115)–(9.117) and (9.119) which total 12 as well.

If there is no sliding, then there is no dissipation of energy, that is, the mechanical energy of the rigid body is conserved. This means that

$$E = \frac{1}{2}mv_C^2 + \frac{1}{2}(\mathbf{K} \circ \boldsymbol{\omega}) - Mg\,(\mathbf{r}_C \circ \mathbf{n}) = \text{const.} \tag{9.121}$$

If there is no friction in the system (contact surfaces are perfectly smooth), then

$$\mathbf{R} = \mathbf{N} = N\mathbf{n}. \tag{9.122}$$

Following scalar multiplication of (9.118) by $\mathbf{n}$ we obtain the equation of constraints of the form

$$\mathbf{n} \circ (\mathbf{v}_C + \boldsymbol{\omega} \times \mathbf{r}) = \mathbf{0}. \tag{9.123}$$

On the other hand, in an absolute system an equation of geometric constraints has the form (we do not consider the case of a body losing contact with the surface π)

$$x_{3C} = -\left(\mathbf{r} \circ \mathbf{n}\right), \tag{9.124}$$

from which after differentiation we obtain

$$\dot{x}_{3C} = -(\dot{\mathbf{r}} \circ \mathbf{n} + \mathbf{r} \circ \dot{\mathbf{n}}) = -\dot{\mathbf{r}} \circ \mathbf{n} = -(\mathbf{r} \times \boldsymbol{\omega}) \circ \mathbf{n}. \tag{9.125}$$

Equation of geometric constraints (9.123), after taking into account relation (9.125), acquires the form

$$\mathbf{n} \circ \mathbf{v}_C \equiv \mathbf{E}_3 \circ \mathbf{v}_C \equiv \dot{x}_{3C} = -\mathbf{n} \circ (\mathbf{r} \times \boldsymbol{\omega}). \tag{9.126}$$

From (9.115), (9.118), and (9.126) it follows that in the absolute system $OX_1X_2X_3$ equation (9.115) takes the form

$$\dot{\mathbf{v}} = 0 \cdot \mathbf{E}_1 + 0 \cdot \mathbf{E}_2 + \ddot{x}_{3C} \cdot \mathbf{E}_3 = \left(-g + \frac{N}{M}\right)\mathbf{E}_3. \tag{9.127}$$

According to the preceding equation we have

$$\dot{x}_{iC} = C_i = \text{const}, \quad i = 1, 2, \tag{9.128}$$

and the projections of the velocity of mass center C onto the plane π are constant. It follows that

$$x_{1C} = C_1 t, \qquad x_{2C} = C_2 t, \tag{9.129}$$

hence

$$x = \sqrt{x_{1C}^2 + x_{2C}^2} = \sqrt{C_1^2 + C_2^2}\,t. \tag{9.130}$$

This means that the mass center moves in uniform rectilinear motion on the plane OX_1X_2. Let us determine reaction $\mathbf{N}$ during such motion of a rigid body. Equating the coefficients at $\mathbf{E}_3$ in (9.127) we obtain

$$\begin{aligned} N &= M(g + \ddot{x}_{3C}) = M\left[g - \mathbf{n} \circ \frac{\mathrm{d}}{\mathrm{d}t}(\mathbf{r} \times \boldsymbol{\omega})\right] \\ &= M[g - \mathbf{n} \circ (\dot{\mathbf{r}} \times \boldsymbol{\omega} + \mathbf{r} \times \dot{\boldsymbol{\omega}})] \\ &= M\{g - \mathbf{n} \circ [(\boldsymbol{\omega} \times \mathbf{r}) \times \boldsymbol{\omega} + \mathbf{r} \times \dot{\boldsymbol{\omega}}]\}. \end{aligned} \tag{9.131}$$

Finally, we should also determine quantities ω_1', ω_2', ω_3', x_{1C}', x_{2C}', and x_{3C}', which are obtained after solving (9.116) and (9.117) by taking into account additional equations (9.119), (9.122), and (9.131). Let us now consider the case with sliding, that is, where $\mathbf{v}_A \neq \mathbf{0}$. Then the reaction at point A acting on the body is equal to

$$\mathbf{R} = \mathbf{N} + \mathbf{T} = N\left(\mathbf{n} - \mu\frac{\mathbf{v}_C}{v_C}\right), \tag{9.132}$$

where (9.126) and (9.131) hold true, and the whole course of our calculations was valid for the case $N > 0$.

9.7 Equations of Vibrations of a System of N Rigid Bodies Connected with Cardan Universal Joints

The subject of this section will be the vibrations of N compound pendulums, which are axisymmetrical rigid bodies connected to each other by means of Cardan universal joints (see Table 4.1 of [1]). These joints allow two degrees of freedom of motion between bodies. The first of the bodies in the chain is excited in a kinematic way also through a Cardan universal joint (the vector of angular velocity $\boldsymbol{\omega}$ comes from a vertically mounted motor). Each of the axisymmetrical bodies of numbers $j = 1, \ldots, N$ has three mass moments of inertia $I_{j1} = I_{j2} \neq I_{j3}$, where, according to the convention adopted in this work, the system of Cartesian coordinates rigidly connected to body j is denoted by $O_j X_1''' X_2''' X_3'''$, and moments I_3 are calculated with respect to the axis $O_j X_3'''$. The moments of inertia I_{jk}, where j denotes a rigid body number, and k denotes an axis number ($k = 1, 2, 3$), are calculated with respect to the coordinates $C_j X_1''' X_2''' X_3'''$ (axes $O_j X_k'''$ are parallel to principal axes $C_j X_k'''$), where C_j denote positions of the body mass centers.

In order to derive equations of motion of the connected rigid bodies (compound pendulums), we will make use of the previously introduced Euler angles to describe the positions of rigid bodies in space and Lagrange equations of the second kind. Figure 9.11a shows how rotational motion is transmitted from the motor S to the system of pendulums. Figure 9.11b presents the first rigid body connected by means of a Cardan universal joint to the motor, and the center of the joint is denoted by O_1. The convention adopted for labeling bodies, the positions of their mass centers C_j, and the distances between the successive centers of joints is depicted in Fig. 9.11c

Body 1 has two possibilities for moving with respect to a system rotating with velocity ω ($\psi_1 = \omega t$) described by the angles θ_1 and φ_1.

The appropriate notions and method for using Euler's angles are described in Sect. 5.5.4 of [1] on the basis of schematic diagrams presented in Fig. 5.56 [1].

The schematic diagram shown in Fig. 9.11b indicates that these angles are introduced in a slightly different way. However, it is worth emphasizing that all three intermediate Cartesian coordinate systems should be of one type, for instance, right-handed.

We will now show, using the vector calculus, how to carry out a transformation of the coordinates of a point in the Cartesian coordinate systems $O_j X_1 X_2 X_3$, $O_j X_1' X_2' X_3'$, $O_j X_1'' X_2'' X_3''$, and $O_j X_1''' X_2''' X_3'''$ making use of the introduced Euler's angles.

By rotating the coordinate system $O_j X_1''' X_2''' X_3'''$ successively through angles φ_1, θ_1, and ψ_1, the radius vector of an arbitrary point of the body having coordinates (x_1''', x_2''', x_3'''), that is, $\mathbf{E}_1''' x_1''' + \mathbf{E}_2''' x_2''' + \mathbf{E}_3''' x_3'''$, will be expressed in the $O_j X_1 X_2 X_3$ coordinate system (Fig. 9.12).

Rotation of the system $O_1 X_1''' X_2''' X_3'''$ through the angle φ_1 (Fig. 9.12a) is equivalent to the following transformation:

$$
\begin{aligned}
\mathbf{E}_1'' x_1'' + \mathbf{E}_3'' x_3'' &= \mathbf{E}_1''' x_1''' + \mathbf{E}_3''' x_3''', \\
\mathbf{E}_2'' &= \mathbf{E}_2''',
\end{aligned} \tag{9.133}
$$

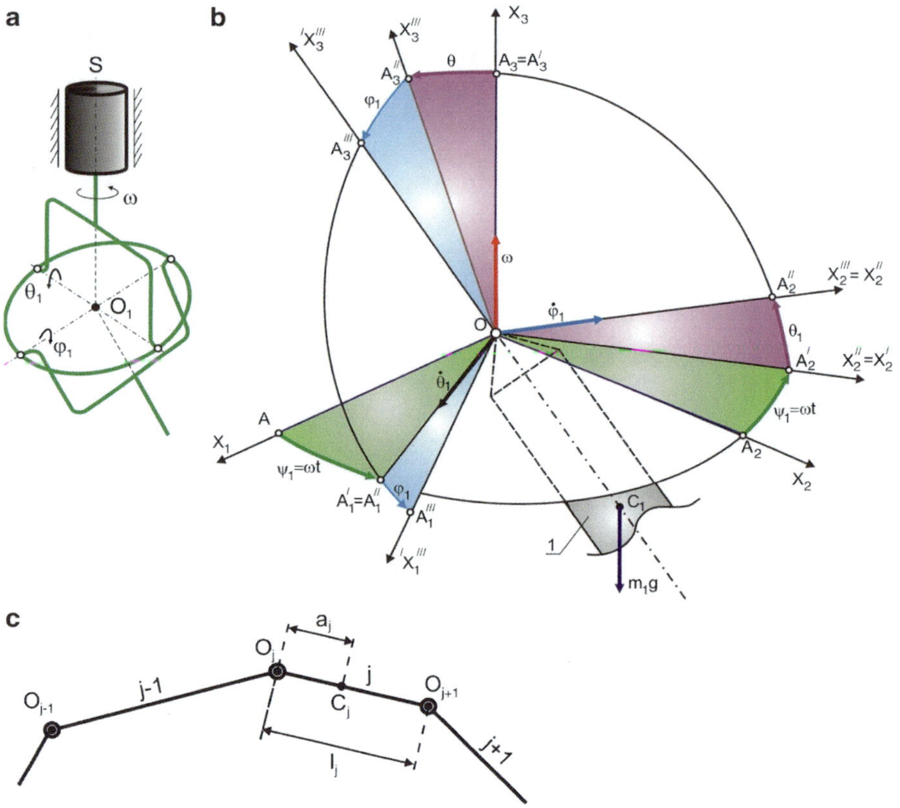

Fig. 9.11 (**a**) Motor S transmitting motion to a system of rigid bodies by means of a Cardan universal joint. (**b**) Space system $OX_1X_2X_3$ $(O = O_1)$, body system of body number 1 $O_1X_1'''X_2'''X_3'''$, body number 1, and Euler's angles. (**c**) Schematic diagram showing labeling convention for the geometry of the bodies

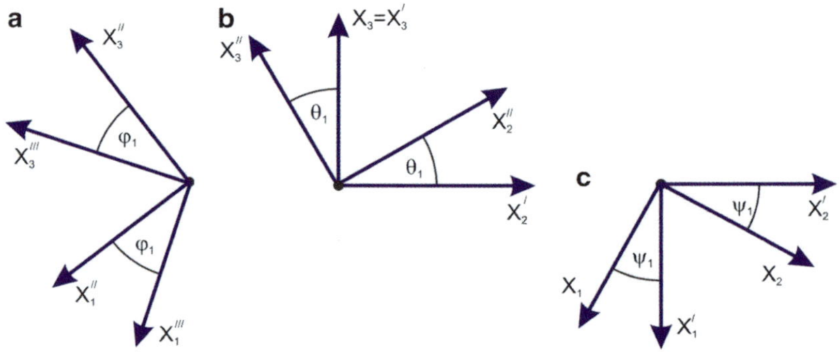

Fig. 9.12 Successive rotations of coordinate systems through Euler's angles φ_1 (**a**), θ_1 (**b**), and ψ_1 (**c**)

and hence, multiplying respectively by $\mathbf{E}_1''$ and $\mathbf{E}_3''$, we have

$$\begin{aligned} x_1'' &= \mathbf{E}_1''' \circ \mathbf{E}_1'' x_1''' + \mathbf{E}_3''' \circ \mathbf{E}_1'' x_3''', \\ x_3'' &= \mathbf{E}_1''' \circ \mathbf{E}_3'' x_1''' + \mathbf{E}_3''' \circ \mathbf{E}_3'' x_3''', \\ x_2'' &= x_2''', \end{aligned} \tag{9.134}$$

and eventually we obtain

$$\begin{aligned} x_1'' &= x_1''' \cos\varphi_1 + x_3''' \sin\varphi_1, \\ x_3'' &= -x_1''' \sin\varphi_1 + x_3''' \cos\varphi_1, \\ x_2'' &= x_2'''. \end{aligned} \tag{9.135}$$

The next rotation through angle θ_1 (Fig. 9.12b) leads to the relation

$$\begin{aligned} \mathbf{E}_2' x_2' + \mathbf{E}_3' x_3' &= \mathbf{E}_2'' x_2'' + \mathbf{E}_3'' x_3'', \\ \mathbf{E}_1'' &= \mathbf{E}_1', \end{aligned} \tag{9.136}$$

hence, multiplying by $\mathbf{E}_2'$ and $\mathbf{E}_3'$, we obtain

$$\begin{aligned} x_2' &= \mathbf{E}_2'' \circ \mathbf{E}_2' x_2'' + \mathbf{E}_3'' \circ \mathbf{E}_2' x_3'', \\ x_3' &= \mathbf{E}_2'' \circ \mathbf{E}_3' x_2'' + \mathbf{E}_3'' \circ \mathbf{E}_3' x_3'', \\ x_1'' &= x_1', \end{aligned} \tag{9.137}$$

or, in equivalent form,

$$\begin{aligned} x_2' &= x_2'' \cos\theta_1 - x_3'' \sin\theta_1, \\ x_3' &= x_2'' \sin\theta_1 + x_3'' \cos\theta_1, \\ x_1'' &= x_1'. \end{aligned} \tag{9.138}$$

Then the rotation through angle ψ_1 (Fig. 9.12c) is described by the relations

$$\begin{aligned} \mathbf{E}_1 x_1 + \mathbf{E}_2 x_2 &= \mathbf{E}_1' x_1' + \mathbf{E}_2' x_2', \\ \mathbf{E}_3' &= \mathbf{E}_3, \end{aligned} \tag{9.139}$$

from which we obtain

$$\begin{aligned} x_1 &= \mathbf{E}_1' \circ \mathbf{E}_1 x_1' + \mathbf{E}_2' \circ \mathbf{E}_1 x_2', \\ x_2 &= \mathbf{E}_1' \circ \mathbf{E}_2 x_1' + \mathbf{E}_2' \circ \mathbf{E}_2 x_2', \\ x_3' &= x_3, \end{aligned} \tag{9.140}$$

or, in equivalent form,

$$\begin{aligned}
x_1 &= x_1' \cos\psi_1 - x_2' \sin\psi_1,\\
x_2 &= x_1' \sin\psi_1 + x_2' \cos\psi_1,\\
x_3 &= x_3'.
\end{aligned} \tag{9.141}$$

Relations (9.135), (9.138), and (9.141) lead to the following equations:

$$\begin{aligned}
x_1 &= x_1'' \cos\psi_1 - \left(x_2'' \cos\theta_1 - x_3'' \sin\theta_1\right)\sin\psi_1,\\
x_2 &= x_1'' \sin\psi_1 + \left(x_2'' \cos\theta_1 - x_3'' \sin\theta_1\right)\cos\psi_1,\\
x_3 &= x_2'' \sin\theta_1 + x_3'' \cos\theta_1,
\end{aligned} \tag{9.142}$$

$$\begin{aligned}
x_1 &= \left(x_1''' \cos\varphi_1 + x_3''' \sin\varphi_1\right)\cos\psi_1\\
&\quad - \left[x_2''' \cos\theta_1 - \left(-x_1''' \sin\varphi + x_3''' \cos\varphi_1\right)\sin\theta_1\right]\sin\psi_1,\\
x_2 &= \left(x_1''' \cos\varphi_1 + x_3''' \sin\varphi_1\right)\sin\psi_1\\
&\quad - \left[x_2''' \cos\theta_1 - \left(-x_1''' \sin\varphi + x_3''' \cos\varphi_1\right)\sin\theta_1\right]\cos\psi_1,\\
x_3 &= x_2''' \sin\theta_1 + \left(-x_1''' \sin\varphi + x_3''' \cos\varphi_1\right)\cos\theta_1,
\end{aligned} \tag{9.143}$$

$$\begin{aligned}
x_1 &= x_1''' \left[\cos\varphi_1 \cos\psi_1 - \sin\varphi_1 \sin\theta_1 \sin\psi_1\right] - x_2''' \cos\theta_1 \sin\psi_1\\
&\quad + x_3''' \left[\sin\varphi \cos\psi_1 + \cos\varphi_1 \sin\theta_1 \sin\psi_1\right],\\
x_2 &= x_1''' \left[\cos\varphi_1 \sin\psi_1 + \sin\varphi_1 \sin\theta_1 \cos\psi_1\right] + x_2''' \cos\theta_1 \cos\psi_1\\
&\quad + x_3''' \left[\sin\varphi \sin\psi_1 - \cos\varphi_1 \sin\theta_1 \cos\psi_1\right],\\
x_3 &= -x_1''' \sin\varphi_1 \cos\theta_1 + x_2''' \sin\theta_1 + x_3''' \cos\varphi_1 \cos\theta_1,
\end{aligned} \tag{9.144}$$

and eventually we obtain the relationship between the coordinates of the point in the stationary system $\mathbf{x}$ and the body system $\mathbf{x}'''$ of the form

$$\mathbf{x} = \mathbf{A}_1 \mathbf{x}''', \tag{9.145}$$

where $\mathbf{A}_1$ is the matrix of rotation of body 1 of the form

$$\mathbf{A}_1 = \begin{bmatrix}
\cos\varphi_1 \cos\psi_1 - \sin\varphi_1 \sin\theta_1 \sin\psi_1 & -\cos\theta_1 \sin\psi_1 & \sin\varphi_1 \cos\psi_1 + \cos\varphi_1 \sin\theta_1 \sin\psi_1\\
\cos\varphi_1 \sin\psi_1 + \sin\varphi_1 \sin\theta_1 \cos\psi_1 & \cos\theta_1 \cos\psi_1 & \sin\varphi_1 \sin\psi_1 - \cos\varphi_1 \sin\theta_1 \cos\varphi_1\\
-\sin\varphi_1 \cos\theta_1 & \sin\theta_1 & \cos\varphi_1 \cos\theta_1
\end{bmatrix} \tag{9.146}$$

Body 1 during its motion is acted upon by the angular velocity vectors $\boldsymbol{\omega}$, $\dot{\boldsymbol{\varphi}}_1$, and $\dot{\boldsymbol{\theta}}_1$. We would like to determine the resultant vector of angular velocity $\boldsymbol{\omega}_1$ of the body system $O_1X_1'''X_2'''X_3'''$, that is, we have to project the vectors of angular velocity, mentioned previously, onto the axes of this coordinate system. To this end we determine the relationships between the coordinates of the aforementioned vectors in the systems $O_1X_1X_2X_3$, $O_1X_1'X_2'X_3'$, $O_1X_1''X_2''X_3''$, and $O_1X_1'''X_2'''X_3'''$.

According to the previous schematic diagrams we have

$$\begin{aligned} x_1' &= x_1\mathbf{E}_1 \circ \mathbf{E}_1' + x_2\mathbf{E}_2 \circ \mathbf{E}_1', \\ x_2' &= x_1\mathbf{E}_1 \circ \mathbf{E}_2' + x_2\mathbf{E}_2 \circ \mathbf{E}_2', \\ x_3' &= x_3, \end{aligned} \tag{9.147}$$

that is,

$$\begin{aligned} x_1' &= x_1 \cos\psi_1 + x_2 \sin\psi_1, \\ x_2' &= -x_1 \sin\psi_1 + x_2 \cos\psi_1, \\ x_3' &= x_3, \end{aligned} \tag{9.148}$$

and then

$$\begin{aligned} x_2'' &= x_2'\mathbf{E}_2' \circ \mathbf{E}_2'' + x_3'\mathbf{E}_3' \circ \mathbf{E}_2'', \\ x_3'' &= x_2'\mathbf{E}_2' \circ \mathbf{E}_2'' + x_3'\mathbf{E}_3' \circ \mathbf{E}_3'', \\ x_1'' &= x_1', \end{aligned} \tag{9.149}$$

that is,

$$\begin{aligned} x_2'' &= x_2' \cos\theta_1 + x_3' \sin\theta_1, \\ x_3'' &= -x_2' \sin\theta_1 + x_3' \cos\theta_1, \\ x_1'' &= x_1', \end{aligned} \tag{9.150}$$

and eventually

$$\begin{aligned} x_1''' &= x_1''\mathbf{E}_1'' \circ \mathbf{E}_1''' + x_3''\mathbf{E}_3'' \circ \mathbf{E}_1''', \\ x_3''' &= x_1''\mathbf{E}_1'' \circ \mathbf{E}_3''' + x_3''\mathbf{E}_3'' \circ \mathbf{E}_3''', \\ \mathbf{E}_2''' &= \mathbf{E}_2'', \end{aligned} \tag{9.151}$$

that is,

$$\begin{aligned} x_1''' &= x_1'' \cos\varphi_1 - x_3'' \sin\varphi_1, \\ x_3''' &= x_1'' \sin\varphi_1 + x_3'' \cos\varphi_1, \\ x_2''' &= x_2''. \end{aligned} \tag{9.152}$$

Note that

$$\boldsymbol{\omega} = \mathbf{E}_3\omega,$$
$$\dot{\boldsymbol{\varphi}}_1 = \mathbf{E}_2'''\dot{\varphi}_1,$$
$$\dot{\boldsymbol{\theta}}_1 = \mathbf{E}_1''\dot{\theta}_1. \tag{9.153}$$

Below we present the process of projecting vectors (9.153). We have successively

1. Vector $\boldsymbol{\omega}$:

$$\begin{aligned} &\omega_1' = 0, && \omega_2' = 0, && \omega_3' = \omega, \\ &\omega_2'' = \omega \sin\theta_1, && \omega_3'' = \omega\cos\theta_1, && \omega_1'' = 0, \\ &\omega_1''' = -\omega\cos\theta_1\sin\varphi_1, && \omega_3''' = \omega\cos\theta_1\cos\varphi_1, && \omega_2''' = \omega\sin\theta_1; \end{aligned} \tag{9.154}$$

2. Vector $\dot{\boldsymbol{\varphi}}_1$:

$$\dot{\varphi}_{13} = 0, \qquad \dot{\varphi}_{12}''' = \dot{\varphi}_1, \qquad \dot{\varphi}_{11}''' = 0; \tag{9.155}$$

3. Vector $\dot{\boldsymbol{\theta}}_1$:

$$\begin{aligned} &\dot{\theta}_{11}'' = \dot{\theta}_1, && \dot{\theta}_{12}'' = 0, && \dot{\theta}_{13}'' = 0, \\ &\dot{\theta}_{11}''' = \dot{\theta}_1\cos\varphi_1, && \dot{\theta}_{13}''' = \dot{\theta}_1\sin\varphi_1, && \dot{\theta}_{12}''' = 0. \end{aligned} \tag{9.156}$$

Taking into account the preceding calculations we obtain

$$\omega_{11} = -\omega\cos\theta_1\sin\varphi_1 + \dot{\theta}_1\cos\varphi_1,$$
$$\omega_{12} = \dot{\varphi}_1 + \omega\sin\theta_1,$$
$$\omega_{13} = \omega\cos\varphi_1\cos\theta_1 + \dot{\theta}_1\sin\varphi_1, \tag{9.157}$$

and eventually the vector of angular velocity of body 1 reads

$$\boldsymbol{\omega}_1 = \omega_{11}\mathbf{E}_1''' + \omega_{12}\mathbf{E}_2''' + \omega_{13}\mathbf{E}_3'''. \tag{9.158}$$

Let us now consider one of the connected bodies of number j. This body is connected to the body of number $j-1$ by means of a Cardan universal joint. This means that body j has two ways of moving relative to body $j-1$, which are described by angles θ_j and φ_j (Fig. 9.13) with respect to the system $O_jX_1'''X_2'''X_3'''$ of axes mutually parallel to the coordinate of body $j-1$ (Fig. 9.13a).

According to the calculations conducted earlier, by choosing an arbitrary point of body j of coordinates (x_1''', x_2''', x_3'''), we can describe its coordinates in the stationary system $O_1X_1X_2X_3$ using rotation matrix $\mathbf{A}_1$ introduced earlier and employing the previously described notion.

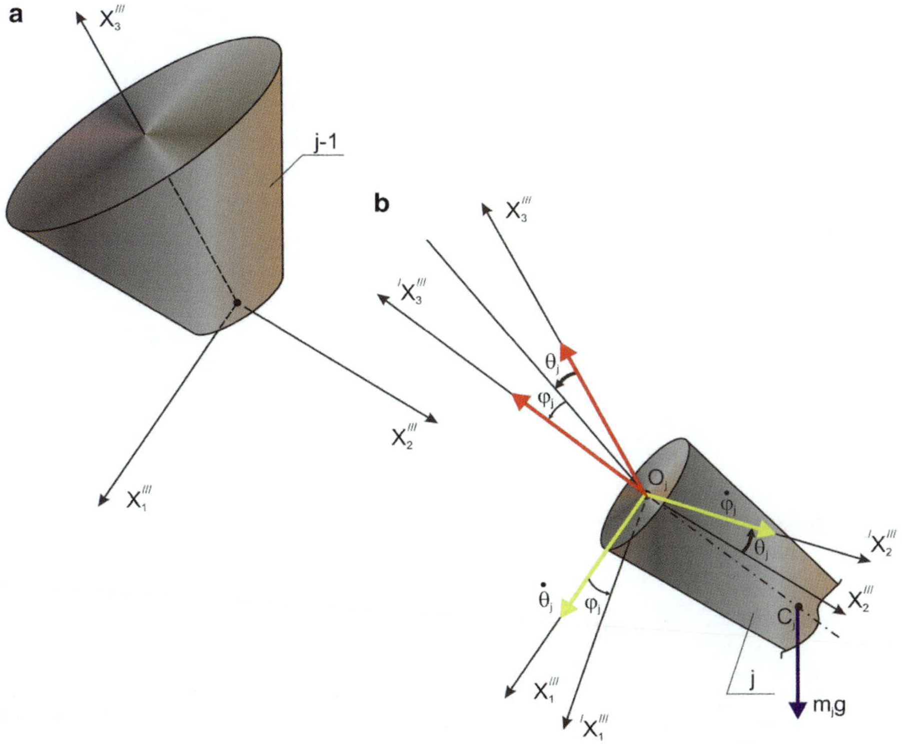

Fig. 9.13 The position of body j (**b**) with respect to body $j-1$ (**a**) is determined by two angles θ_j and φ_j

Our aim is to express the position of an arbitrary point belonging to an arbitrary body of number j and given in a coordinate system rigidly connected to body j in a stationary system, that is, $O_1X_1X_2X_3$. To this end it is necessary to apply the successive rotations of the bodies preceding body j, which is equivalent to the following notation:

$$\begin{aligned} &\mathbf{x}_{j-1} = \mathbf{A}_j\mathbf{x}_j, \qquad \mathbf{x}_{j-2} = \mathbf{A}_{j-1}\mathbf{x}_{j-1}, \\ &\mathbf{x}_{j-3} = \mathbf{A}_{j-2}\mathbf{x}_{j-2}, \quad \mathbf{x}_{j-k} = \mathbf{A}_{j-k}\mathbf{x}_{j-k}, \\ &\quad j \le N, j-k = 1, \end{aligned} \tag{9.159}$$

and hence we obtain

$$\mathbf{x}_{j-k} = \mathbf{A}_{j-k}\ldots\mathbf{A}_{j-2}\mathbf{A}_{j-1}\mathbf{A}_j\mathbf{x}_j. \tag{9.160}$$

For instance, in order to determine the coordinates of the point of the body N in a stationary system one should set $j = N$ and $k = N - 1$ in (9.160):

$$\mathbf{x}_j = \mathbf{A}_1\mathbf{A}_2\ldots\mathbf{A}_N\mathbf{x}_N, \tag{9.161}$$

where matrix $\mathbf{A}_1$ is the matrix that generates the motion described by (9.146), and therefore the angles describing the position of two bodies of numbers $j-1$ and j are valid for $2 \leq j \leq N$.

The matrix $\mathbf{A}_j$, $2 \leq j \leq N$, occurring in the preceding equations, can be easily obtained from matrix $\mathbf{A}_1$ after setting $\psi_1 = 0$:

$$\mathbf{A}_j = \begin{bmatrix} \cos\varphi_j & \sin\varphi_j \sin\theta_j & -\sin\varphi_j \cos\theta_j \\ 0 & \cos\theta_j & \sin\theta_j \\ \sin\varphi_j & -\cos\varphi_j \sin\theta_j & \cos\varphi_j \cos\theta_j \end{bmatrix}^{\mathrm{T}}. \tag{9.162}$$

Because the vector of angular velocity of body $j-1$ is known, we are able to determine the vector of angular velocity of body j in the system $O_j X_1''' X_2''' X_3'''$ rigidly connected to body j. According to Fig. 9.13b, where the known components of vector $\boldsymbol{\omega}_{j-1}$ are marked in red and those of vectors $\dot{\boldsymbol{\theta}}_j$ and $\dot{\boldsymbol{\varphi}}_j$ are marked in yellow, using matrix $\mathbf{A}_j$ we can express vector $\boldsymbol{\omega}_j$ in the coordinates of body j.

Vectors $\boldsymbol{\omega}_{j-1}$ and $\dot{\boldsymbol{\theta}}_j$ are projected onto the axes of the coordinate system of body j in the following way:

$$\begin{bmatrix} \cos\varphi_j & \sin\varphi_j \sin\theta_j & -\sin\varphi_j \cos\theta_j \\ 0 & \cos\theta_j & \sin\theta_j \\ \sin\theta_j & -\cos\varphi_j \sin\theta_j & \cos\varphi\_j \cos\theta_j \end{bmatrix}^{-1} \begin{bmatrix} \omega_{(j-1)_1} + \dot{\theta}_j \\ \omega_{(j-1)_2} \\ \omega_{(j-1)_3} \end{bmatrix}, \tag{9.163}$$

and additionally taking into account that $\dot{\boldsymbol{\varphi}}_j$ lies on the axis $O_j X_2'''$ we obtain

$$\begin{aligned} \omega_{j_1} &= \left(\omega_{(j-1)_1} + \dot{\theta}_j\right) \cos\varphi_j \\ &\quad + \omega_{(j-1)_2} \sin\varphi_j \sin\theta_j - \omega_{(j-1)_3} \sin\varphi_j \cos\theta_j, \\ \omega_{j2} &= \omega_{(j-1)_2} \cos\theta_j + \omega_{(j-1)_3} \sin\theta_j + \dot{\varphi}_j, \\ \omega_{j3} &= \left(\omega_{(j-1)_1} + \dot{\theta}_1\right) \sin\varphi_j \\ &\quad - \omega_{(j-1)_2} \cos\varphi_j \sin\theta_j + \omega_{(j-1)_3} \cos\varphi_j \cos\theta_j. \end{aligned} \tag{9.164}$$

In order to derive equations of motion we have to determine the positions of origins O_j of all previously used coordinate systems and the positions of mass centers C_j of all rigid bodies.

According to Fig. 9.11c the position of the center of joint O_2 can be determined from (9.145) because, substituting $\mathbf{x}''' = [0\ 0\ l_1]^{\mathrm{T}}$, we obtain

$$\begin{aligned} x_{1O_2} &= -l_1 \left(\sin\varphi_1 \cos\omega t + \cos\varphi_1 \sin\theta_1 \sin\omega t\right), \\ x_{2O_2} &= -l_1 \left(\sin\varphi_1 \sin\omega t - \cos\varphi_1 \sin\theta_1 \cos\omega t\right), \\ x_{3O_2} &= -l_1 \cos\varphi_1 \cos\theta_1, \end{aligned} \tag{9.165}$$

and the position of the mass center of body 1 is equal to

$$x_{1C_1} = -a_1 \left(\sin\varphi_1 \cos\omega t + \cos\varphi_1 \sin\theta_1 \sin\omega t\right),$$
$$x_{2C_1} = -a_1 \left(\sin\varphi_1 \sin\omega t - \cos\varphi_1 \sin\theta_1 \cos\omega t\right),$$
$$x_{3C_1} = -a_1 \cos\varphi_1 \cos\theta_1. \tag{9.166}$$

Recurrence formulas for the determination of positions of the remaining points O_j follow from the equation

$$\overrightarrow{O_{j-1}O_j} = -\mathbf{E}_3''' l_{j-1} = \left(x_{1Oj} - x_{1Oj-1}\right)\mathbf{E}_1 + \left(x_{2Oj} - x_{2Oj-1}\right)\mathbf{E}_2 + \left(x_{3Oj} - x_{3Oj-1}\right)\mathbf{E}_3. \tag{9.167}$$

Multiplying relation (9.159) in turn by $\mathbf{E}_i$, $i = 1, 2, 3$ we obtain

$$x_{1Oj} = x_{1Oj-1} - l_{j-1}\mathbf{E}_1 \circ \mathbf{E}_3''',$$
$$x_{2Oj} = x_{2Oj-1} - l_{j-1}\mathbf{E}_2 \circ \mathbf{E}_3''',$$
$$x_{3Oj} = x_{3Oj-1} - l_{j-1}\mathbf{E}_3 \circ \mathbf{E}_3''', \qquad j = 2, \ldots, N, \tag{9.168}$$

where vector $\mathbf{E}_3'''$ is associated with body $j - 1$.

In turn, the mass centers C_j of the rigid bodies are described by the equations

$$x_{1Cj} = x_{1Oj} - a_j\mathbf{E}_1 \circ \mathbf{E}_3''',$$
$$x_{2Cj} = x_{2Oj} - a_j\mathbf{E}_2 \circ \mathbf{E}_3''',$$
$$x_{3Cj} = x_{3Oj} - a_j\mathbf{E}_3 \circ \mathbf{E}_3''', \qquad j = 2, \ldots, N, \tag{9.169}$$

and vector $\mathbf{E}_3'''$ is associated with body j.

In order to exploit Lagrange's equations of the second kind, which serve to derive equations of motion of the connected rigid bodies, one should determine the kinetic energy T of the system of bodies, the potential energy V of the bodies in the gravitational force field, and the potential energy U accumulated in each of $j = 1, \ldots, N$ Cardan universal joints. The energies are given by the equations

$$T = \frac{1}{2}\sum_{j=1}^{N} m_j \left(\dot{x}_{1Cj}^2 + \dot{x}_{2Cj}^2 + \dot{x}_{3Cj}^2\right) + \frac{1}{2}\left(\sum_{j=1}^{N} I_{j_1}\omega_{j1}^2 + \sum_{j=1}^{N} I_{j_2}\omega_{j2}^2 + \sum_{j=1}^{N} I_{j_3}\omega_{j3}^2\right), \tag{9.170}$$

$$V = g \sum_{j=1}^{N} m_j x_{3Cj}, \tag{9.171}$$

$$U = \frac{1}{2} \sum_{j=1}^{N} k_j \left(\theta_j^2 + \varphi_j^2 \right), \tag{9.172}$$

where g is the acceleration of gravity and k_j denotes the stiffness coefficient of each of the Cardan universal joints j.

Lagrange's equations of the second kind in this case take the form

$$\frac{\mathrm{d}}{\mathrm{d}t} \frac{\partial T}{\partial \dot{q}_i} - \frac{\partial T}{\partial q_i} + \frac{\partial U}{\partial q_i} + \frac{\partial V}{\partial q_i} = 0, \qquad i = 1, \ldots, 6N, \tag{9.173}$$

where q_i are the generalized coordinates of the body.

Let us write the equations of motion that follow from relation (9.173) in the explicit form for the case of only one body $j = 1$.

From (9.166) we obtain

$$\begin{aligned}
\dot{x}_{1C_1} &= -a_1 \Big[\dot{\varphi}_1 \cos\varphi_1 \cos\omega t - \omega \sin\varphi_1 \sin\omega t - \dot{\varphi}_1 \sin\varphi_1 \sin\theta_1 \sin\omega t \\
&\quad + \dot{\theta}_1 \cos\varphi_1 \cos\theta_1 \sin\omega t + \omega \cos\varphi_1 \sin\theta_1 \cos\omega t \Big], \\
\dot{x}_{2C_1} &= -a_1 \Big[\dot{\varphi}_1 \cos\varphi_1 \sin\omega t + \omega \sin\varphi_1 \cos\omega t + \dot{\varphi}_1 \sin\varphi_1 \sin\theta_1 \cos\omega t \\
&\quad - \dot{\theta}_1 \cos\varphi_1 \cos\theta_1 \cos\omega t + \omega \cos\varphi_1 \sin\theta_1 \sin\omega t \Big], \\
\dot{x}_{3C_1} &= a_1 \Big[\dot{\varphi}_1 \sin\varphi_1 \cos\theta_1 + \dot{\theta}_1 \cos\varphi_1 \sin\theta_1 \Big].
\end{aligned} \tag{9.174}$$

Taking into account relations (9.174) and (9.149) in (9.170) and then formulas (9.166) in (9.171) from (9.173) we obtain the following equation of motion for the first body with two degrees of freedom with the adopted generalized coordinates θ_1 and φ_1:

$$\begin{aligned}
\ddot{\varphi}_1 &= [2(I_{12} + M_1)]^{-1} \{2 \left[\omega \dot{\Theta}_1 (C + I_{12}) - a_1 g m_1 \sin(\varphi_1) \right] \cos(\Theta_1) \\
&\quad - k_1 \varphi_1 + A \left(\omega^2 \cos^2(\Theta_1) - \dot{\Theta}_1^2 \right) \sin(\varphi_1) \}, \\
\ddot{\theta}_1 &= \left[4 (I_{11} + M_1) \cos^2(\varphi_1) + I_{13} \sin^2(\varphi_1) \right]^{-1} \cdot \omega^2 \{ (C - I_{11} \\
&\quad + 2I_{12} - I_{13}) \sin(2\Theta) - 4 \left[\left((C + I_{12}) \omega \cos(\Theta) + A \sin(2\varphi_1) \dot{\Theta}_1 \right) \dot{\varphi}_1 \right. \\
&\quad \left. - a_1 g m_1 \cos(\varphi_1) \sin(\Theta_1) - k_1 \Theta_1 \right] \},
\end{aligned} \tag{9.175}$$

where

$$M_1 = m_1 a_1^2, \qquad A = I_{11} - I_{13} + M_1,$$
$$B = A \cos(2\varphi_1), \qquad C = B + M_1.$$

9.8 Conservative Vibrations of a Rigid Body Supported Elastically in the Gravitational Field

Now we will make use of the previously introduced knowledge regarding the geometry of masses, the kinematics of a rigid body, and Lagrange's equations of the second kind to the derivation of equations of motion of a rigid body elastically supported at three points A_i, $i = 1, 2, 3$ (see [3]). The body has a mass M and moments of inertia with respect to principal centroidal axes of inertia $I_1 = I_{X_1'''}$, $I_2 = I_{X_2'''}$, and $I_3 = I_{X_3'''}$. Linear springs of stiffnesses k_i, $i = 1, 2, 3$ are connected with each of the points A_i. The springs carry the load only along their axes, that is, the spring of stiffness k_i carries the load along the axis OX_i, and their opposite ends B_i are subjected to a harmonic kinematic excitation ($b_i \cos \omega_i t$), which is illustrated in Fig. 9.14.

The considered body has six degrees of freedom, and the axes of the non-stationary system $CX_1'''X_2'''X_3'''$ situated at the mass center of the body C coincide with the axes of principal centroidal axes of inertia.

In the solution of this problem we will make certain assumptions. Introducing a stationary coordinate system at an arbitrary point O it is easy to notice that under the weight $M\mathbf{g}$ the rigid body goes down rotating simultaneously about a certain

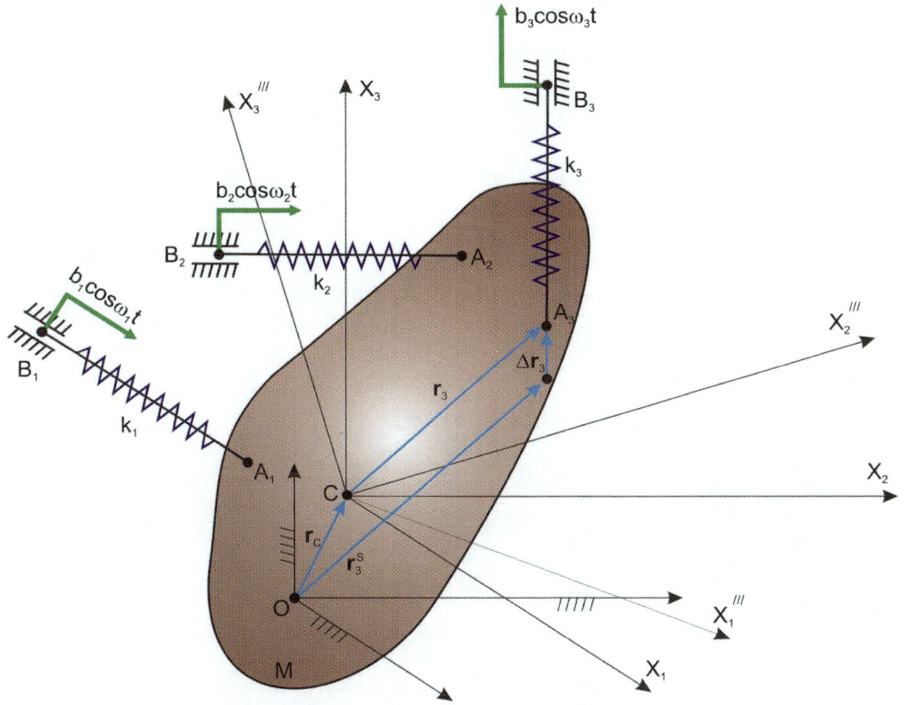

Fig. 9.14 A rigid body supported by springs and the introduced Cartesian coordinate systems (only vectors $\mathbf{r}_3$, $\Delta\mathbf{r}_3$, and $\mathbf{r}_3^S$ are shown)

unknown axis (it becomes skewed with respect to the Cartesian coordinate system of the origin O) assuming in this position a configuration of static equilibrium. Because later on we do not take into account the geometric non-linearities associated with the motion of springs, we can assume that in the static equilibrium position points O and C coincide. In a static equilibrium position, the gravity force $M\mathbf{g}$ is carried exclusively by a vertically situated spring, and a certain amount of potential energy will already be accumulated in it. Vibrations will be further measured from the static equilibrium position, and in this position all three Cartesian coordinate systems introduced in Fig. 9.14 are coincident.

From the calculations of Sect. 5.7 (see Chap. 5 of [1]) it follows that the general motion of a rigid body can be described by the translational motion of mass center C of this body and three rotations about the axes of the Cartesian coordinate system, where these axes coincide with the principal centroidal axes of inertia of the considered rigid body.

According to Fig. 9.14, this means that an arbitrary position of the body can be characterized by vector $\mathbf{r}_C$ (displacement of mass center C), and the system assumes a position described by the coordinate system $CX_1X_2X_3$. Next, it is possible to perform the rotation of the body about point C, measured, for example, by three of Euler's angles, in such a way that after three such rotations the axes of the system $CX_1X_2X_3$ are coincident with the axes $CX_1'''X_2'''X_3'''$.

Because the body has six degrees of freedom, the Cartesian coordinates of an arbitrary point that belongs to a rigid body can be described by the following six generalized coordinates.

1. The coordinates of the mass center $C(x_{1C}, x_{2C}, x_{3C})$, where $q_1 = x_{1C} \equiv x_1$, $q_2 = x_{2C} \equiv x_2$, $q_3 = x_{3C} \equiv x_3$, and the vector $\mathbf{r}_C = \mathbf{r}_C[x_{1C}, x_{2C}, x_{3C}]$ determines the translational motion of the body.
2. The Euler angles that represent the motion of a body about a point $q_4 = \theta_1$, $q_5 = \theta_2$, $q_6 = \theta_3$ (Fig. 9.15).

If we rotate the coordinate system $CX_1'''X_2'''X_3'''$ successively through angles θ_3, θ_2, and θ_1, then the radius vector of an arbitrary body point of coordinates (x_1''', x_2''', x_3'''), that is, $\mathbf{E}_1'''x_1''' + \mathbf{E}_2'''x_2''' + \mathbf{E}_3'''x_3'''$, is expressed in the coordinate system $CX_1X_2X_3$, which is illustrated in Fig. 9.16.

The rotation of the system $CX_1'''X_2'''X_3'''$ through the angle θ_3 (Fig. 9.16a) is equivalent to the following transformation:

$$\begin{aligned}\mathbf{E}_1''x_1'' + \mathbf{E}_2''x_2'' &= \mathbf{E}_1'''x_1''' + \mathbf{E}_2'''x_2''',\\ \mathbf{E}_3'' &= \mathbf{E}_3''',\end{aligned} \tag{9.176}$$

which after multiplying by $\mathbf{E}_1''$ and $\mathbf{E}_2''$ gives

$$\begin{aligned}x_1'' &= x_1'''\left(\mathbf{E}_1''' \circ \mathbf{E}_1''\right) + x_2'''\left(\mathbf{E}_2''' \circ \mathbf{E}_1''\right),\\ x_2'' &= x_1'''\left(\mathbf{E}_1''' \circ \mathbf{E}_2''\right) + x_2'''\left(\mathbf{E}_2''' \circ \mathbf{E}_2''\right),\\ x_3'' &= x_3''',\end{aligned} \tag{9.177}$$

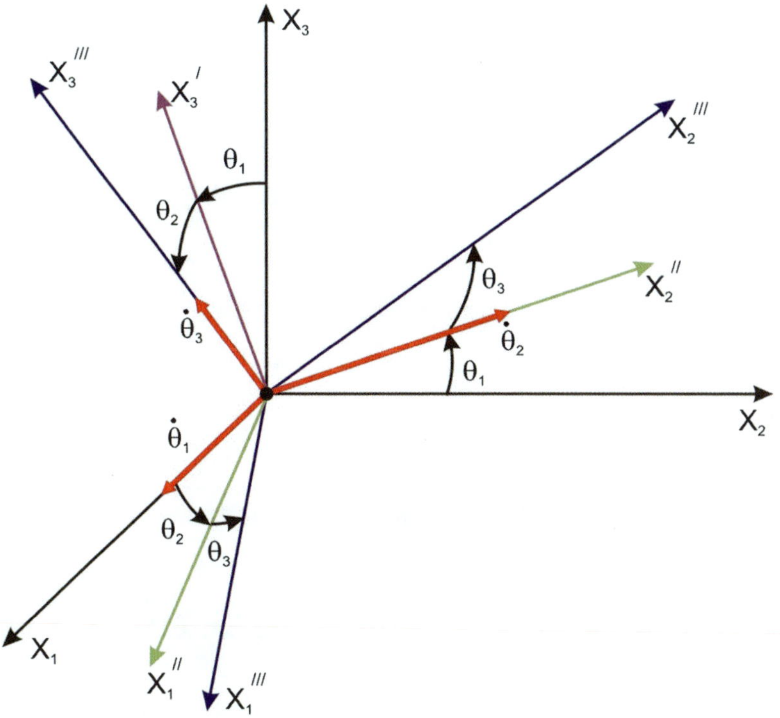

Fig. 9.15 Angles θ_i and angular velocities $\dot{\theta}_i$ introduced to analyze the part of rigid-body motion associated with motion about a point

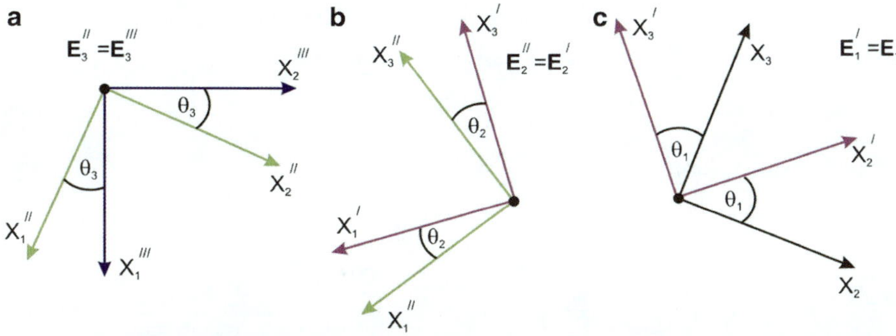

Fig. 9.16 Successive rotations of coordinate systems through angles θ_3 (a), θ_2 (b), and θ_1 (c)

and eventually we have

$$
\begin{aligned}
x_1'' &= x_1''' \cos\theta_3 - x_2''' \sin\theta_3, \\
x_2'' &= x_1''' \sin\theta_3 + x_2''' \cos\theta_3, \\
x_3'' &= x_3'''.
\end{aligned}
\tag{9.178}
$$

The next rotation through the angle θ_2 leads to the relationship (Fig. 9.16b)

$$\begin{aligned} \mathbf{E}'_1 x'_1 + \mathbf{E}'_3 x'_3 &= \mathbf{E}''_1 x''_1 + \mathbf{E}''_3 x''_3, \\ \mathbf{E}''_2 &= \mathbf{E}'_2, \end{aligned} \tag{9.179}$$

which after multiplying by $\mathbf{E}'_1$ and $\mathbf{E}'_3$ gives

$$\begin{aligned} x'_1 &= x''_1 \left(\mathbf{E}''_1 \circ \mathbf{E}'_1\right) + x''_3 \left(\mathbf{E}''_3 \circ \mathbf{E}'_1\right), \\ x'_3 &= x''_1 \left(\mathbf{E}''_1 \circ \mathbf{E}'_3\right) + x''_3 \left(\mathbf{E}''_3 \circ \mathbf{E}'_3\right), \\ x'_2 &= x''_2, \end{aligned} \tag{9.180}$$

and hence we have

$$\begin{aligned} x'_1 &= x''_1 \cos\theta_2 + x''_3 \sin\theta_2, \\ x'_3 &= -x''_1 \sin\theta_2 + x''_3 \cos\theta_2, \\ x'_2 &= x''_2. \end{aligned} \tag{9.181}$$

Then rotation through the angle θ_1 (Fig. 9.16c) is described by the relationships

$$\begin{aligned} \mathbf{E}_2 x_2 + \mathbf{E}_3 x_3 &= \mathbf{E}'_2 x'_2 + \mathbf{E}'_3 x'_3, \\ \mathbf{E}'_1 &= \mathbf{E}_1, \end{aligned} \tag{9.182}$$

from which we obtain

$$\begin{aligned} x_2 &= x'_2 \left(\mathbf{E}'_2 \circ \mathbf{E}_2\right) + x'_3 \left(\mathbf{E}'_3 \circ \mathbf{E}_2\right), \\ x_3 &= x'_2 \left(\mathbf{E}'_2 \circ \mathbf{E}_3\right) + x'_3 \left(\mathbf{E}'_3 \circ \mathbf{E}_3\right), \\ x_1 &= x'_1, \end{aligned} \tag{9.183}$$

or, in equivalent form,

$$\begin{aligned} x_2 &= x'_2 \cos\theta_1 - x'_3 \sin\theta_1, \\ x_3 &= x'_2 \sin\theta_1 + x'_3 \cos\theta_1, \\ x_1 &= x'_1. \end{aligned} \tag{9.184}$$

Taking into account the preceding relationships we obtain

$$\begin{aligned} x_1 &= x''_1 \cos\theta_2 + x''_3 \sin\theta_2 \\ &= \left(x'''_1 \cos\theta_3 - x'''_2 \sin\theta_3\right) \cos\theta_2 + x'''_3 \sin\theta_2 \\ &= x'''_1 \cos\theta_2 \cos\theta_3 - x'''_2 \sin\theta_3 \cos\theta_2 + x'''_3 \sin\theta_2, \end{aligned}$$

$$
\begin{aligned}
x_2 &= x_2' \cos\theta_1 - x_3' \sin\theta_1 \\
&= x_2'' \cos\theta_1 - \left(-x_1'' \sin\theta_2 + x_3'' \cos\theta_2\right) \sin\theta_1 \\
&= \left(x_1''' \sin\theta_3 + x_2''' \cos\theta_3\right) \cos\theta_1 \\
&\quad - \left[-\left(x_1''' \cos\theta_3 - x_2''' \sin\theta_3\right) \sin\theta_2 + x_3''' \cos\theta_2\right] \sin\theta_1 \\
&= x_1''' \left(\sin\theta_3 \cos\theta_1 + \cos\theta_3 \sin\theta_2 \sin\theta_1\right) \\
&\quad + x_2''' \left(\cos\theta_1 \cos\theta_3 - \sin\theta_3 \sin\theta_2 \sin\theta_1\right) - x_3''' \sin\theta_1 \cos\theta_2,
\end{aligned}
$$

$$
\begin{aligned}
x_3 &= x_2' \sin\theta_1 + x_3' \cos\theta_1 \\
&= x_2'' \sin\theta_1 + \left(-x_1'' \sin\theta_2 + x_3'' \cos\theta_2\right) \cos\theta_1 \\
&= \left(x_1''' \sin\theta_3 + x_2''' \cos\theta_3\right) \sin\theta_1 \\
&\quad + \left[-\left(x_1''' \cos\theta_3 - x_2''' \sin\theta_3\right) \sin\theta_2 + x_3''' \cos\theta_2\right] \cos\theta_1 \\
&= x_1''' \left(\sin\theta_1 \sin\theta_3 - \cos\theta_3 \sin\theta_2 \cos\theta_1\right) \\
&\quad + x_2''' \left(\cos\theta_3 \sin\theta_1 + \sin\theta_3 \sin\theta_2 \cos\theta_1\right) + x_3''' \cos\theta_1 \cos\theta_2,
\end{aligned} \tag{9.185}
$$

or, in the equivalent matrix form,

$$
\begin{bmatrix} x_1 \\ x_2 \\ x_3 \end{bmatrix} =
\begin{bmatrix}
\cos\theta_2 \cos\theta_3 & -\sin\theta_3 \cos\theta_2 & \sin\theta_2 \\
\begin{matrix} \sin\theta_3 \cos\theta_1 + \\ + \cos\theta_3 \sin\theta_2 \sin\theta_1 \end{matrix} & \begin{matrix} \cos\theta_1 \cos\theta_3 - \\ - \sin\theta_3 \sin\theta_2 \sin\theta_1 \end{matrix} & -\sin\theta_1 \cos\theta_2 \\
\begin{matrix} \sin\theta_1 \sin\theta_3 - \\ - \cos\theta_3 \sin\theta_2 \cos\theta_1 \end{matrix} & \begin{matrix} \cos\theta_3 \sin\theta_1 \\ + \sin\theta_3 \sin\theta_2 \cos\theta_1 \end{matrix} & \cos\theta_1 \cos\theta_2
\end{bmatrix}
\begin{bmatrix} x_1''' \\ x_2''' \\ x_3''' \end{bmatrix}. \tag{9.186}
$$

The change in time of each of the introduced angles $\theta_1 = \theta_1(t)$, $\theta_2 = \theta_2(t)$, and $\theta_3 = \theta_3(t)$ generates angular velocities $\dot{\theta}_1$, $\dot{\theta}_2$, and $\dot{\theta}_3$ of the vectors drawn in Fig. 9.15.

To determine the kinetic energy of a body it is necessary to know the components of the angular velocity vector

$$
\dot{\boldsymbol{\theta}} = \dot{\boldsymbol{\theta}}_1 + \dot{\boldsymbol{\theta}}_2 + \dot{\boldsymbol{\theta}}_3, \tag{9.187}
$$

expressed in the body coordinate system $CX_1'''X_2'''X_3'''$.

This time we will analyze successive rotations of the coordinate systems presented in Fig. 9.16 in the reverse order.

We have successively

$$x_1' = x_1,$$
$$x_2' = x_2 \cos\theta_1 + x_3 \sin\theta_1,$$
$$x_3' = -x_2 \sin\theta_1 + x_3 \cos\theta_1, \tag{9.188}$$

and then

$$x_1'' = x_1' \cos\theta_2 - x_3' \sin\theta_2,$$
$$x_2'' = x_2',$$
$$x_3'' = x_1' \sin\theta_2 + x_3' \cos\theta_2, \tag{9.189}$$

and eventually we obtain

$$x_1''' = x_1'' \cos\theta_3 + x_2'' \sin\theta_3,$$
$$x_2''' = -x_1'' \sin\theta_3 + x_2'' \cos\theta_3,$$
$$x_3''' = x_3''. \tag{9.190}$$

According to Fig. 9.15 we have

$$\dot{\boldsymbol{\theta}}_1 = \mathbf{E}_1'\dot{\theta}_1, \qquad \dot{\boldsymbol{\theta}}_2 = \mathbf{E}_2''\dot{\theta}_2, \qquad \dot{\boldsymbol{\theta}}_3 = \mathbf{E}_3'''\dot{\theta}_2. \tag{9.191}$$

We project each of the velocity vectors of (9.190) onto the axes of the system $CX_1'''X_2'''X_3'''$ according to (9.187)÷(9.190).

For vector $\dot{\boldsymbol{\theta}}_1$ we have

$$\dot{\theta}_{11}' = \dot{\theta}_1, \qquad \dot{\theta}_{12}' = 0, \qquad \dot{\theta}_{13}' = 0,$$
$$\dot{\theta}_{11}'' = \dot{\theta}_1 \cos\theta_2, \qquad \dot{\theta}_{12}'' = 0, \qquad \dot{\theta}_{13}'' = \dot{\theta}_1 \sin\theta_2,$$
$$\dot{\theta}_{11}''' = \dot{\theta}_1 \cos\theta_2 \cos\theta_3, \qquad \dot{\theta}_{12}''' = -\dot{\theta}_1 \cos\theta_2 \sin\theta_3, \qquad \dot{\theta}_{13}''' = \dot{\theta}_1 \sin\theta_2. \tag{9.192}$$

For vector $\dot{\boldsymbol{\theta}}_2$ we have

$$\dot{\theta}_{21}''' = \dot{\theta}_2 \sin\theta_3,$$
$$\dot{\theta}_{22}''' = \dot{\theta}_2 \cos\theta_3,$$
$$\dot{\theta}_{23}''' = 0. \tag{9.193}$$

Vector $\dot{\boldsymbol{\theta}}$ in the coordinate system has the form

$$\dot{\boldsymbol{\theta}} = \mathbf{E}_1''' \left(\dot{\theta}_1 \cos\theta_2 \cos\theta_3 + \dot{\theta}_2 \sin\theta_3\right) + \mathbf{E}_2''' \left(-\dot{\theta}_1 \cos\theta_2 \sin\theta_3 + \dot{\theta}_2 \cos\theta_3\right) + \mathbf{E}_3''' \left(\dot{\theta}_3 + \dot{\theta}_1 \sin\theta_2\right). \tag{9.194}$$

For the sake of simplification of calculation we will assume that the body is supported at three points A_1, A_2, and A_3, which are the ends of springs of linear stiffnesses that are in contact with the body (we neglect friction). As was already mentioned, the opposite ends B_1, B_2, B_3 of the springs are excited in a harmonic kinematic way, and the excitation has the form $x_{Bi} = b_i \cos \omega_i t$, $i = 1, 2, 3$. The springs are guided, and their axes are parallel to the corresponding axes of the coordinate system $OX_1X_2X_3$, and points A_i and B_i are allowed to move only along the axis OX_i. In this way, geometric non-linearities caused by the motion of the body and generated by the forces in springs do not appear, and the springs are assumed to be working in a linear range, which is expressed by their stiffness coefficients k_i.

In a static equilibrium position for $b_i = 0$ the coordinates of points B_n are described in the system $OX_1X_2X_3$, because points C and O are coincident ($C = O$), and the angles $\theta_n = 0$, $n = 1, 2, 3$. In turn, the coordinates of points A_i in the static equilibrium position are equal to

$$\begin{aligned} x^s_{A_1} &= r_{A_1x_1} = r'''_{A_1x_1}, \\ x^s_{A_2} &= r_{A_2x_2} = r'''_{A_2x_2}, \\ x^s_{A_3} &= r_{A_3x_3} = r'''_{A_3x_3}. \end{aligned} \tag{9.195}$$

Coordinates of points A_i associated with the direction of the corresponding spring and caused by the generalized displacements in the system $OX_1X_2X_3$ are given below:

$$\begin{aligned} x_{1A_1} &= x^s_{1A_1} + x_1 + \left[x'''_{1A_1} \cos\theta_2 \cos\theta_3 \right. \\ &\quad \left. - x'''_{2A_1} \sin\theta_3 \cos\theta_2 + x'''_{3A_1} \sin\theta_2\right], \\ x_{2A_2} &= x^s_{2A_2} + x_2 + \left[x'''_{1A_2} (\sin\theta_3 \cos\theta_1 + \cos\theta_3 \sin\theta_2 \sin\theta_1) \right. \\ &\quad \left. + x'''_{2A_2} (\cos\theta_1 \cos\theta_3 - \sin\theta_3 \sin\theta_2 \sin\theta_1) - x'''_{3A_2} \sin\theta_1 \cos\theta_2\right], \\ x_{3A_3} &= x^s_{3A_3} + x_3 + \left[x'''_{1A_3} (\sin\theta_1 \sin\theta_3 - \cos\theta_3 \sin\theta_2 \cos\theta_1) \right. \\ &\quad \left. + x'''_{2A_3} (\cos\theta_3 \sin\theta_1 + \sin\theta_3 \sin\theta_2 \cos\theta_1) + x'''_{3A_3} \cos\theta_1 \cos\theta_2\right], \end{aligned} \tag{9.196}$$

where now x_1, x_2, x_3 result from the translational displacement of the mass center of the rigid body.

The following forces are generated in the springs k_i:

$$\begin{aligned} P_1 &= -k_1\lambda_1 = -k_1 \left[x_1 + x'''_{1A_1} \cos\theta_2 \cos\theta_3 - x'''_{2A_1} \sin\theta_3 \cos\theta_2 \right. \\ &\quad \left. + x'''_{3A_1} \sin\theta_2 - x_{1B_1}(t)\right], \\ P_2 &= -k_2\lambda_2 = -k_2 \left[x_2 + x'''_{1A_2} (\sin\theta_3 \cos\theta_1 + \cos\theta_3 \sin\theta_2 \sin\theta_1) \right. \\ &\quad + x'''_{2A_2} (\cos\theta_1 \cos\theta_3 - \sin\theta_3 \sin\theta_2 \sin\theta_1) \\ &\quad \left. - x'''_{3A_2} \sin\theta_1 \cos\theta_2 - x_{2B_2}(t)\right], \end{aligned}$$

$$P_3 = -k_3\lambda_3 = -k_3\left[x_3 + x'''_{1A_3}\left(\sin\theta_1\sin\theta_3 - \cos\theta_3\sin\theta_2\cos\theta_1\right)\right.$$
$$+ x'''_{2A_3}\left(\cos\theta_3\sin\theta_1 - \sin\theta_3\sin\theta_2\cos\theta_1\right)$$
$$\left.+ x'''_{3A_3}\cos\theta_1\cos\theta_2 - x_{3B_3}(t)\right], \qquad (9.197)$$

where λ_1, λ_2, λ_3 denote deflections of the springs situated in parallel to the axes OX_1, OX_2, and OX_3 measured from a static equilibrium position.

Because the vibrations are observed from a static equilibrium position, potential energy is accumulated only in the springs and is equal to

$$V = \frac{1}{2}\left(k_1\lambda_1^2 + k_2\lambda_2^2 + k_3\lambda_3^2\right). \qquad (9.198)$$

Using (9.198), the kinetic energy of the considered rigid body is equal to

$$T = \frac{1}{2}M\left(\dot{x}_1^2 + \dot{x}_2^2 + \dot{x}_3^2\right) + \frac{I_1}{2}\left(\dot{\theta}_1\cos\theta_2\cos\theta_3 + \dot{\theta}_2\sin\theta_3\right)^2$$
$$+ \frac{I_2}{2}\left(-\dot{\theta}_1\cos\theta_2\sin\theta_3 + \dot{\theta}_2\cos\theta_3\right)^2 + \frac{I_3}{2}\left(\dot{\theta}_3 + \dot{\theta}_1\sin\theta_2\right)^2. \qquad (9.199)$$

Lagrange's equations of the second kind for the considered case have the following form:

$$\frac{\mathrm{d}}{\mathrm{d}t}\frac{\partial T}{\partial \dot{q}_j} - \frac{\partial T}{\partial q_j} + \frac{\partial V}{\partial q_j} = 0, \qquad j = 1,\ldots,6, \qquad (9.200)$$

where $q_1 = x_1, q_2 = x_2, q_3 = x_3, q_4 = \theta_1, q_5 = \theta_2, q_6 = \theta_3$.

We calculate successively

$$\frac{\partial T}{\partial \dot{x}_1} = M\dot{x}_1, \qquad \frac{\partial T}{\partial \dot{x}_2} = M\dot{x}_2, \qquad \frac{\partial T}{\partial \dot{x}_3} = M\dot{x}_3,$$

$$\frac{\partial T}{\partial \dot{\theta}_1} = I_1\left(\dot{\theta}_1\cos^2\theta_2\cos^2\theta_3 + \dot{\theta}_2\cos\theta_2\cos\theta_3\sin\theta_3\right)$$
$$+ I_2\left(\dot{\theta}_1\cos^2\theta_2\sin^2\theta_3 - \dot{\theta}_2\cos\theta_2\sin\theta_3\cos\theta_3\right)$$
$$+ I_3\left(\dot{\theta}_1\sin^2\theta_2 + \dot{\theta}_3\sin\theta_2\right),$$

$$\frac{\partial T}{\partial \dot{\theta}_2} = I_1\left(\dot{\theta}_2\sin^2\theta_3 + \frac{1}{2}\dot{\theta}_1\cos\theta_2\sin 2\theta_3\right)$$
$$+ I_2\left(\dot{\theta}_1\cos^2\theta_3 - \frac{1}{2}\dot{\theta}_1\cos\theta_2\sin 2\theta_3\right),$$

$$\frac{\partial T}{\partial \dot{\theta}_3} = I_3\left(\dot{\theta}_3 + \dot{\theta}_1 \sin\theta_2\right),$$

$$\frac{\partial T}{\partial \theta_1} = 0,$$

$$\begin{aligned}\frac{\partial T}{\partial \theta_2} &= -\frac{1}{2}I_1\left(\dot{\theta}_1^2\cos^2\theta_3\sin 2\theta_2 + \dot{\theta}_1\dot{\theta}_2\sin\theta_2\sin 2\theta_3\right)\\ &\quad + \frac{1}{2}I_2\left(-\dot{\theta}_1^2\sin^2\theta_3\sin 2\theta_2 + \dot{\theta}_1\dot{\theta}_2\sin\theta_2\sin 2\theta_3\right)\\ &\quad + I_3\left(\frac{1}{2}\dot{\theta}_1^2\sin 2\theta_2 + \dot{\theta}_1\dot{\theta}_3\cos\theta_2\right),\end{aligned}$$

$$\begin{aligned}\frac{\partial T}{\partial \theta_3} &= I_1\left(-\frac{1}{2}\dot{\theta}_1^2\cos^2\theta_2\sin 2\theta_3 + \dot{\theta}_1\dot{\theta}_2\cos\theta_2\cos 2\theta_3 + \frac{1}{2}\dot{\theta}_2^2\sin 2\theta_3\right)\\ &\quad + I_2\left(\frac{1}{2}\dot{\theta}_1^2\cos^2\theta_2\sin 2\theta_3 - \dot{\theta}_1\dot{\theta}_2\cos\theta_2\cos 2\theta_3 - \frac{1}{2}\dot{\theta}_2^2\sin 2\theta_3\right),\end{aligned}$$

$$\frac{\partial V}{\partial x_1} = k_1\left[x_1 + x'''_{1A_1}\cos\theta_2\cos\theta_3 - x'''_{2A_1}\sin\theta_3\cos\theta_2 + x'''_{3A_1}\sin\theta_2 - x'''_{1B_1}(t)\right],$$

$$\begin{aligned}\frac{\partial V}{\partial x_2} &= k_2\left[x_2 + x'''_{1A_2}(\sin\theta_3\cos\theta_1 + \cos\theta_3\sin\theta_2\sin\theta_1)\right.\\ &\quad \left.+ x'''_{2A_2}(\cos\theta_1\cos\theta_3 - \sin\theta_3\sin\theta_2\sin\theta_1) - x'''_{3A_2}\sin\theta_1\cos\theta_2 - x_{2B_2}(t)\right],\end{aligned}$$

$$\begin{aligned}\frac{\partial V}{\partial x_3} &= k_3\left[x_3 + x'''_{1A_3}(\sin\theta_1\sin\theta_3 - \cos\theta_3\sin\theta_2\cos\theta_1)\right.\\ &\quad \left.+ x'''_{2A_3}(\cos\theta_3\sin\theta_1 + \sin\theta_3\sin\theta_2\cos\theta_1) + x'''_{3A_3}\cos\theta_1\cos\theta_2 - x_{3B_3}(t)\right],\end{aligned}$$

$$\begin{aligned}\frac{\partial V}{\partial \theta_1} &= k_2\left\{x_2\left[x'''_{1A_2}(\cos\theta_3\sin\theta_2\cos\theta_1 - \sin\theta_3\sin\theta_1)\right.\right.\\ &\quad - x'''_{2A_2}(\sin\theta_1\cos\theta_3 + \sin\theta_3\sin\theta_2\cos\theta_1)\\ &\quad - x'''_{3A_2}\cos\theta_1\cos\theta_3] + [x'''_{1A_2}(\sin\theta_3\cos\theta_1 + \cos\theta_3\sin\theta_2\sin\theta_1)\\ &\quad + x'''_{2A_2}(\cos\theta_1\cos\theta_3 - \sin\theta_3\sin\theta_2\sin\theta_1)\\ &\quad - x'''_{3A_2}\sin\theta_1\cos\theta_2 - x_{2B_2}(t)\big]\cdot\big[x'''_{1A_2}(-\sin\theta_3\sin\theta_1 + \cos\theta_3\sin\theta_2\cos\theta_1)\\ &\quad \left.\left.+ x'''_{2A_2}(-\sin\theta_1\cos\theta_3 - \sin\theta_3\sin\theta_2\cos\theta_1) - x'''_{3A_2}\cos\theta_1\cos\theta_2\right]\right\}\\ &\quad + k_3\big\{x_3\big[x'''_{1A_3}(\cos\theta_1\sin\theta_3 + \cos\theta_3\sin\theta_2\sin\theta_1)\\ &\quad + x'''_{2A_3}(\cos\theta_3\cos\theta_1 - \sin\theta_3\sin\theta_2\sin\theta_1) - x'''_{3A_3}\sin\theta_1\cos\theta_2\big]\\ &\quad + \big[x'''_{1A_3}(\sin\theta_1\sin\theta_3 - \cos\theta_3\sin\theta_2\cos\theta_1)\\ &\quad + x'''_{2A_3}(\cos\theta_3\sin\theta_1 + \sin\theta_3\sin\theta_2\cos\theta_1)\end{aligned}$$

$$
\begin{aligned}
&+ x'''_{3A_3} \cos\theta_1 \cos\theta_2 - x_{3B_3}(t)\big] \cdot \big[x'''_{1A_3} (\cos\theta_1 \sin\theta_3 + \cos\theta_3 \sin\theta_2 \sin\theta_1) \\
&+ x'''_{2A_3} (\cos\theta_3 \cos\theta_1 - \sin\theta_3 \sin\theta_2 \sin\theta_1) - x'''_{3A_3} \sin\theta_1 \cos\theta_2\big]\big\},
\end{aligned}
$$

$$
\begin{aligned}
\frac{\partial V}{\partial \theta_2} &= k_1\big\{x_1 \left(-x'''_{1A_1} \sin\theta_2 \cos\theta_3 + x'''_{2A_1} \sin\theta_3 \sin\theta_2 + x'''_{3A_1} \cos\theta_2\right) \\
&+ \big[x'''_{1A_1} \cos\theta_2 \cos\theta_3 - x'''_{2A_1} \sin\theta_3 \cos\theta_2 \\
&+ x'''_{3A_1} \sin\theta_2 - x_{1B_1}(t)\big] \cdot \big[-x'''_{1A_1} \sin\theta_2 \cos\theta_3 \\
&+ x'''_{2A_1} \sin\theta_3 \sin\theta_2 + x'''_{3A_1} \cos\theta_2\big]\big\} + k_2\big\{x_2\big[x'''_{1A_2} \cos\theta_3 \cos\theta_2 \sin\theta_1 \\
&- x'''_{2A_2} \sin\theta_3 \cos\theta_2 \sin\theta_1 + x'''_{3A_2} \sin\theta_1 \sin\theta_2\big] \\
&+ \big[x'''_{1A_2} (\sin\theta_3 \cos\theta_1 + \cos\theta_3 \sin\theta_2 \sin\theta_1) \\
&+ x'''_{2A_2} (\cos\theta_1 \cos\theta_3 - \sin\theta_3 \sin\theta_2 \sin\theta_1) \\
&- x'''_{3A_2} \sin\theta_1 \cos\theta_2 - x_{2B_2}(t)\big] \cdot \big[x'''_{1A_2} \cos\theta_3 \cos\theta_2 \sin\theta_1 \\
&- x'''_{2A_2} \sin\theta_3 \cos\theta_2 \sin\theta_1 + x'''_{3A_2} \sin\theta_1 \sin\theta_2\big]\big\} \\
&+ k_3\big\{x_3\big[-x'''_{1A_3} \cos\theta_3 \cos\theta_2 \cos\theta_1 + x'''_{2A_3} \sin\theta_3 \cos\theta_2 \cos\theta_1 \\
&- x'''_{3A_3} \cos\theta_1 \sin\theta_3\big] + \big[x'''_{1A_3} (\sin\theta_1 \sin\theta_3 - \cos\theta_3 \sin\theta_2 \cos\theta_1) \\
&+ x'''_{2A_3} (\cos\theta_3 \sin\theta_1 + \sin\theta_3 \sin\theta_2 \cos\theta_1) + x'''_{3A_3} \cos\theta_1 \cos\theta_2 \\
&- x_{3B_3}(t)\big] \cdot \big[-x'''_{1A_3} \cos\theta_3 \cos\theta_2 \cos\theta_1 \\
&+ x'''_{2A_3} \sin\theta_3 \cos\theta_2 \cos\theta_1 - x'''_{3A_3} \cos\theta_1 \sin\theta_2\big]\big\},
\end{aligned}
$$

$$
\begin{aligned}
\frac{\partial V}{\partial \theta_3} &= k_1\big\{x_1 \left(-x'''_{1A_1} \cos\theta_2 \sin\theta_3 - x'''_{2A_1} \cos\theta_3 \cos\theta_2 + x'''_{3A_1} \cos\theta_2\right) \\
&+ \big[x'''_{1A_1} \cos\theta_2 \cos\theta_3 - x'''_{2A_1} \sin\theta_3 \cos\theta_2 \\
&+ x'''_{3A_1} \sin\theta_2 - x_{1B_1}(t)\big] \cdot \big[-x'''_{1A_1} \cos\theta_2 \sin\theta_3 - x'''_{2A_1} \cos\theta_3 \cos\theta_2\big]\big\} \\
&+ k_2\big\{x_2\big[x'''_{1A_2} (\cos\theta_3 \cos\theta_1 - \sin\theta_3 \sin\theta_2 \sin\theta_1) \\
&- x'''_{2A_2} (\cos\theta_1 \sin\theta_3 + \cos\theta_3 \sin\theta_2 \sin\theta_1)\big] \\
&+ \big[x'''_{1A_2} (\sin\theta_3 \cos\theta_1 + \cos\theta_3 \sin\theta_2 \sin\theta_1) \\
&+ x'''_{2A_2} (\cos\theta_1 \cos\theta_3 - \sin\theta_3 \sin\theta_2 \sin\theta_1) \\
&- x'''_{3A_2} \sin\theta_1 \cos\theta_2 - x_{2B_2}(t)\big] \cdot \big[x'''_{1A_2} (\cos\theta_3 \cos\theta_1 - \sin\theta_3 \sin\theta_2 \sin\theta_1) \\
&- x'''_{2A_2} (\cos\theta_1 \sin\theta_3 + \cos\theta_3 \sin\theta_2 \sin\theta_1)\big]\big\} \\
&+ k_3\big\{2x_3\big[x'''_{1A_3} (\sin\theta_1 \cos\theta_3 + \sin\theta_3 \sin\theta_2 \sin\theta_1) \\
&+ x'''_{2A_3} (-\sin\theta_3 \sin\theta_1 + \cos\theta_3 \sin\theta_2 \cos\theta_1)\big]
\end{aligned}
$$

$$+ 2\big[x'''_{1A_3}(\sin\theta_1\sin\theta_3 - \cos\theta_3\sin\theta_2\cos\theta_1)$$
$$+ x'''_{2A_3}(\cos\theta_3\sin\theta_1 + \sin\theta_3\sin\theta_2\cos\theta_1)$$
$$+ x'''_{3A_3}\cos\theta_1\cos\theta_2 - x_{3B_3}(t)\big]\cdot\big[x'''_{1A_3}(\sin\theta_1\cos\theta_3 + \sin\theta_3\sin\theta_2\sin\theta_1)$$
$$+ x'''_{2A_3}(-\sin\theta_3\sin\theta_1 + \cos\theta_3\sin\theta_2\cos\theta_1)\big]\big\}. \tag{9.201}$$

Substituting the results of calculations (9.201) into Lagrange's equations (9.200) we obtain six non-linear ordinary differential equations that describe the conservative dynamics of the considered rigid body subjected to kinematic excitation at three points and supported with linear springs. For small vibrations the problem can be simplified by the introduction of approximations $\cos\theta_i \cong 1$ and $\sin\theta_i \cong \theta_i$.

In the general case in various places of the body we can apply external forces $\mathbf{F}_S = \mathbf{F}_S(t)$ whose points of application are determined by vectors $\mathbf{r}_S$, $s = 1, 2, \ldots, S$, and whose amount can be smaller or greater than the number of the introduced generalized coordinates. In this case equations of motion of the system can be derived using Lagrange's equations of the second kind of the form

$$\frac{\mathrm{d}}{\mathrm{d}t}\frac{\partial T}{\partial \dot{q}_j} - \frac{\partial T}{\partial q_j} + \frac{\partial V}{\partial q_j} = Q_j, \qquad j = 1, \ldots, 6, \tag{9.202}$$

where Q_j is the generalized force corresponding to the generalized coordinate q_j and should be expressed in terms of forces $\mathbf{F}_S$.

We determine the generalized forces based on the principle that the sum of works δW of forces $\mathbf{F}_S$ $(s = 1, 2, \ldots, S)$ during virtual displacements $\delta\mathbf{r}_S$ is equal to the sum of works of generalized forces Q_j during virtual displacements δq_j, that is,

$$\delta W = \mathbf{F}_s \circ \delta\mathbf{r}_s = Q_j\delta q_j, \tag{9.203}$$

where the Einstein summation applies. Coordinates of the vectors that determine the positions of force vectors must be expressed in terms of the generalized coordinates, that is, we have

$$\mathbf{r}_s = \mathbf{r}_s(q_1, \ldots, q_6, t), \tag{9.204}$$

hence we obtain

$$\delta\mathbf{r}_s = \frac{\partial\mathbf{r}_s}{\partial q_j}\delta q_j, \quad j = 1, \ldots, 6, \quad s = 1, \ldots, S. \tag{9.205}$$

Substituting relations (9.205) into (9.203) we obtain

$$\delta W = \sum_{s=1}^{S}\mathbf{F}_s \circ \left(\sum_{j=1}^{6}\frac{\partial\mathbf{r}_s}{\partial q_j}\delta q_j\right)$$
$$= \sum_{j=1}^{6}\left(\sum_{s=1}^{S}\mathbf{F}_s \circ \frac{\partial\mathbf{r}_s}{\partial q_j}\right)\delta q_j = \sum_{j=1}^{6}Q_j\delta q_j, \tag{9.206}$$

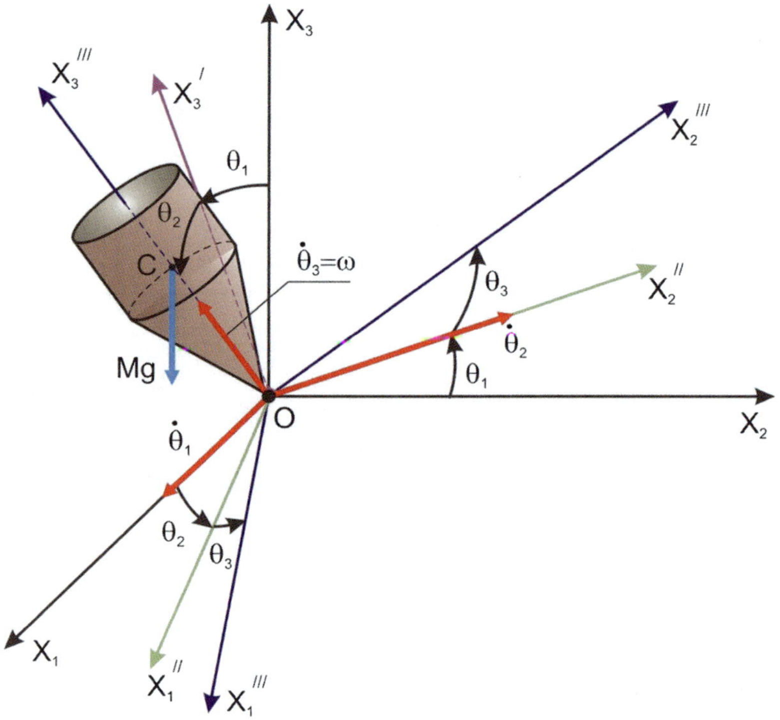

Fig. 9.17 Axisymmetrical body in motion about a point and the introduced systems of coordinates

from which follows

$$Q_j(t) = \sum_{s=1}^{S} \mathbf{F}(t) \circ \frac{\partial \mathbf{r}_s}{\partial q_j}, \tag{9.207}$$

which was the aim of our calculations.

Below we will consider small vibrations of a free axisymmetrical body in motion about a point, which was the subject of analysis in [4]. As distinct from the previous general calculations, here we introduce several simplifications leading to a system of linear differential equations.

Example 9.1. Determine the critical angular velocity of the top shown in Fig. 9.17, on the assumption that the angular velocity with respect to its symmetry axis OX_3''' is equal to $\dot{\theta}_3 = \omega = \text{const}$, and the mass center M of the body is situated at distance a from point O.

From (9.199), and setting $I_1 = I_2 = I$ and $I_3 = I_O$, we obtain

$$
\begin{aligned}
T &= \frac{I}{2}\left(\dot{\theta}_1 \cos\theta_2 \cos\theta_3 + \dot{\theta}_2 \sin\theta_3\right)^2 \\
&\quad + \frac{I}{2}\left(-\dot{\theta}_1 \cos\theta_2 \sin\theta_3 + \dot{\theta}_2 \cos\theta_3\right)^2 + \frac{I_O}{2}\left(\dot{\theta}_3 + \dot{\theta}_1 \sin\theta_2\right)^2 \\
&= \frac{I}{2}\left(\dot{\theta}_1^2 \cos^2\theta_2 + \dot{\theta}_2^2\right) + \frac{I_O}{2}\left(\omega + \dot{\theta}_1 \sin\theta_2\right)^2. \qquad (*)
\end{aligned}
$$

We assume that the generalized coordinates θ_1 and θ_2 are small and that $\omega \gg \dot{\theta}_1$ and $\omega \gg \dot{\theta}_2$. On these assumptions $(*)$ takes the form

$$
T = \frac{I}{2}\left(\dot{\theta}_1^2 \cos^2\theta_2 + \dot{\theta}_2^2\right) + \frac{I_O\omega^2}{2} + I_O\omega\dot{\theta}_1\theta_2.
$$

We calculate successively

$$
\begin{aligned}
\frac{\mathrm{d}}{\mathrm{d}t}\frac{\partial T}{\partial \dot{\theta}_2} &= I\ddot{\theta}_2, \\
\frac{\partial T}{\partial \theta_2} &= I_O\omega\dot{\theta}_1 + 2\frac{I}{2}\dot{\theta}_1^2 \sin\theta_2\cos\theta_2 \cong I_O\omega\dot{\theta}_1, \\
\frac{\mathrm{d}}{\mathrm{d}t}\frac{\partial T}{\partial \dot{\theta}_1} &= I_O\omega\dot{\theta}_2 + \frac{\mathrm{d}}{\mathrm{d}t}I\dot{\theta}_1\cos^2\theta_2 \\
&= I_O\omega\dot{\theta}_2 + I\ddot{\theta}_1\cos^2\theta_2 - 2I\dot{\theta}_1\dot{\theta}_2\sin\theta_2\cos\theta_2 \\
&\cong I_O\omega\dot{\theta}_2 + I\ddot{\theta}_1, \\
\frac{\partial T}{\partial \theta_1} &= 0.
\end{aligned}
$$

Because of the small values of the angles θ_1 and θ_2 we neglect the potential energy of a body in the gravitational field, and the generalized forces are equal to $Q_1 = Mga\sin\theta_1 \cong Mga\theta_1$ and $Q_2 = Mga\sin\theta_2 \cong Mga\theta_2$.

From Lagrange's equations we obtain

$$
\begin{aligned}
I\ddot{\theta}_1 + I_O\omega\dot{\theta}_2 &= Mga\theta_1, \\
I\ddot{\theta}_2 - I_O\omega\dot{\theta}_1 &= Mga\theta_2,
\end{aligned}
$$

and dividing them by I we have

$$
\begin{aligned}
\ddot{\theta}_1 - \alpha^2\theta_1 + \beta\dot{\theta}_2 &= 0, \\
\ddot{\theta}_2 - \alpha^2\theta_2 + \beta\dot{\theta}_1 &= 0, \qquad (**)
\end{aligned}
$$

where $\alpha^2 = \frac{Mga}{I}$, $\beta = \frac{I_O\omega}{I}$.

Finally, the system of two ordinary differential equations of the form $(**)$ is left for the analysis. We seek their solutions in the form of the functions

$$\theta_1 = \theta_{1O} e^{i\omega_O t}, \qquad \theta_2 = \theta_{2O} e^{i\omega_O t}, \qquad i^2 = -1.$$

Inserting the preceding solutions into $(**)$ we obtain

$$\begin{bmatrix} -\left(\omega_O^2 + \alpha^2\right) & i\omega_O\beta \\ -i\omega_O\beta & -\left(\omega_O^2 + \alpha^2\right) \end{bmatrix} \begin{bmatrix} \theta_{1O} \\ \theta_{2O} \end{bmatrix} = \begin{bmatrix} 0 \\ 0 \end{bmatrix},$$

and the desired frequencies of free vibrations are determined from the characteristic equation

$$\begin{vmatrix} -\left(\omega_O^2 + \alpha^2\right) & i\omega_O\beta \\ -i\omega_O\beta & -\left(\omega_O^2 + \alpha^2\right) \end{vmatrix} = 0.$$

From the preceding equation we obtain

$$\left(\omega_O^2 + \alpha^2\right)^2 - \omega_O^2\beta^2 = \omega_O^4 - \omega_O^2\left(\beta^2 - 2\alpha^2\right) + \alpha^4 = 0.$$

The roots of the preceding equation determine the frequencies of free vibrations, which are equal to

$$\omega_O^2 = \frac{1}{2}\left(\beta^2 - 2\alpha^2\right) \pm \sqrt{\beta^2 - 4\alpha^2}.$$

We determine the critical value of velocity for which the assumed small vibrations about point O occur from the condition $\beta = 2\alpha$, and it is equal to $\omega = \frac{2}{I_O}\sqrt{IMga}$. □

9.9 A Wobblestone Dynamics

9.9.1 Coulomb–Contensou Friction Model

Since the time of the ancient Celts it has been known that certain bodies having a center of mass not coincident with their centroid and having principal centroidal axes of inertia not coincident with their geometric axes exhibit certain interesting dynamical behaviors. An example of such a rigid body is the so-called *wobblestone* (a half-ellipsoid solid having many other names, e.g., *rattleback* or *celt*), which lies on a flat horizontal surface, sets in rotational motion about the vertical axis, and rotates in only one direction. The imposition of an initial velocity in the opposite direction leads to a quick cessation of the rotation in this direction, and subsequently the stone starts its transverse vibrations and rotation in the opposite direction.

The Celts believed that all bodies (objects) possessed consciousness and spirit. Wobblestones came from meteorites, which were readily found in Ireland and, as they provided a *yes* or *no* answer, were used by Celts for fortune telling.

The first scientific work concerning the dynamics of a wobblestone was published by Walker [5] in 1896. Walker observed that the amazing dynamics of the wobblestone was the result of the asymmetry of the principal centroidal axes of inertia with respect to the symmetrical axes of the stone. Analyzing this problem was not easy because of the absence of linear terms in equations of dynamics and because of the need for an adequate friction model. In 1986 Hermann returned to this topic [6]. He pointed to the possibility of the return motion of the stone in either one or two directions depending on the geometric and inertial parameters of a given rigid body. The lack of inclusion of slip by Hermann led to a contradiction with the earlier assumptions of Magnus [7], where a linear dependence was assumed between the friction force and the velocity of the contact point of the stone and horizontal surface on which the stone moved. The problem was taken up by Caughey [8], but his model deviated considerably from reality. In 1982 Kane and Levinson [9] introduced a model that was closer to reality and took into account rolling but in the absence of slip. The model allowed for a demonstration of several changes in the direction of rotation, and the introduction of viscous damping into the system allowed for their reduction to one or two changes.

One year later Lindberg and Longman [10] observed that it was impossible to decouple the linearized equations of motion of a wobblestone due to the presence of gyroscopic terms.

A more complex model of stone dynamics was proposed by Garcia and Hubbard [11] that included aerodynamic dissipation, dry friction, and slip; in addition, numerical analysis was enhanced by experimental investigations.

The lack of a complete explanation, including the lack of an adequate mathematical model of the dynamics of a wobblestone, also represented a challenge for leading Russian mechanicians. In 2002 Markeev [12] conducted an analysis of the dynamics of a wobblestone in the neighborhood of static and dynamic equilibrium positions on the assumption of an absence of friction. However, the derived perturbation equations included only the local dynamics of the stone. On the other hand, they were verified in an experimental way as well.

In 2006 Borisov et al. [13] undertook a mathematical modeling of a heavy unbalanced ellipsoid rolling without slip on a horizontal surface pointing to phenomena similar to those observed during the motion of a wobblestone. In 2008, Zhuravlev and Klimov [14] presented a model of a wobblestone that was the closest to reality and that introduced the possibility of slip of the contact point between the stone and a horizontal surface and, additionally, included the CCZ friction model, discussed earlier in Sect. 2.11.2 of [1]. The aforementioned friction leads to mutual coupling of rotations and slips of the body, and the derived equations of motion allow one to carry out a global analysis of the dynamics of a wobblestone with the aid of numerical methods.

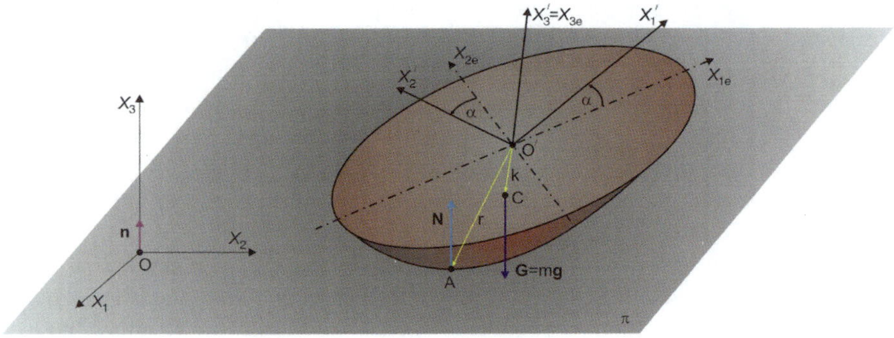

Fig. 9.18 Wobblestone on a horizontal plane π and axes of a body coordinate system and absolute coordinate system

Figure 9.18 shows a view of a wobblestone that includes the axes of the absolute system $OX_1X_2X_3$ and systems of principal centroidal axes of inertia rotated through angle α with respect to the principal axes of the ellipsoid.

The position of point of contact A between the stone and plane π in the body system located at the mass center of the stone is given by the vector $\mathbf{k}+\mathbf{r}$, where $\mathbf{k}=\overrightarrow{CO}=(0,0,-k)^{\mathrm{T}}$ and $\mathbf{r}=\overrightarrow{O'A}=(x_1', x_2', x_3')^{\mathrm{T}}$.

The principal geometric axes of an ellipsoid are denoted by a, b, c, where $a > b > c$. Moreover, in an absolute system the unit vector was introduced such that $\mathbf{n}\circ\mathbf{g}<0$. Equations of motion of the wobblestone in a body system have the form

$$
\begin{aligned}
m\frac{\mathrm{d}\mathbf{v}}{\mathrm{d}t}+\boldsymbol{\omega}\times(m\mathbf{v}) &= -m\mathbf{g}\circ\mathbf{n}+N\mathbf{n}-\mathbf{T},\\
I_C\frac{\mathrm{d}\boldsymbol{\omega}}{\mathrm{d}t}+\boldsymbol{\omega}\times(I_C\boldsymbol{\omega}) &= N(\mathbf{k}+\mathbf{r})\times\mathbf{n}-(\mathbf{k}+\mathbf{r})\times\mathbf{T},\\
\frac{\mathrm{d}\mathbf{n}}{\mathrm{d}t}+\boldsymbol{\omega}\times\mathbf{n} &= \mathbf{0},
\end{aligned}
\tag{9.208}
$$

where the third equation of system (9.208) is the Poisson equation.

Above, m is the mass of the wobblestone; I_c is the diagonal inertia matrix with respect to the mass center C of non-zero elements B_1, B_2, and B_3; $\mathbf{T}$ is the friction force directed against the velocity of point A, where $\mathbf{v}_A=\mathbf{v}+\omega\times(\mathbf{k}+\mathbf{r})$; and $\mathbf{N}=N\mathbf{n}$ is a normal force at the point of contact (reaction at the contact point is equal to $\mathbf{R}=\mathbf{N}+\mathbf{T}$).

Three vector equations (9.208) have ten unknowns: $\mathbf{n}=(n_1,n_2,n_3)^{\mathrm{T}}$, $\mathbf{v}=(v_1,v_2,v_3)^{\mathrm{T}}$ and $\omega=(\omega_1,\omega_2,\omega_3)^{\mathrm{T}}$ and the scalar N. An additional scalar equation necessary to solve the problem results from the observation that during the motion of a wobblestone, it remains in contact the entire time with the horizontal plane OX_1X_2 whose normal vector is $\mathbf{n}$ (only this case will be considered). The perpendicularity condition generates the tenth scalar equation of the form

$$
(\mathbf{v}+\boldsymbol{\omega}\times(\mathbf{k}+\mathbf{r}))\circ\mathbf{n}=0. \tag{9.209}
$$

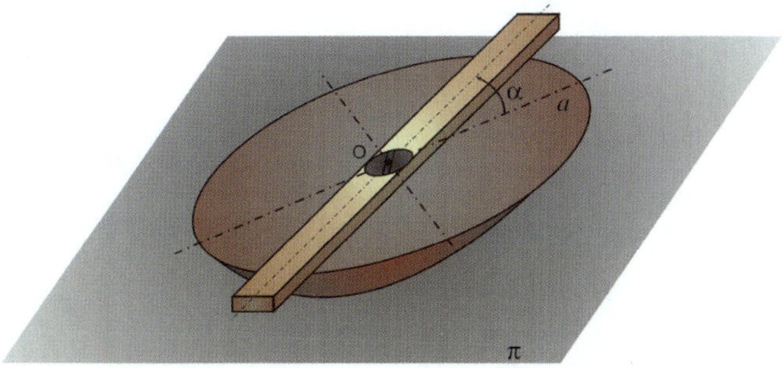

Fig. 9.19 Construction of a wobblestone

The friction force occurring in the first two equations of (9.208) has the form

$$\mathbf{T} = \frac{\mu N(\mathbf{v} + \boldsymbol{\omega} \times (\mathbf{k} + \mathbf{r}))}{|\mathbf{v} + \boldsymbol{\omega} \times (\mathbf{k} + \mathbf{r})| + \frac{8\rho}{3\pi}|\boldsymbol{\omega} \circ \mathbf{n}|}, \tag{9.210}$$

and, as can be seen, it depends on $\mathbf{v}$, $\boldsymbol{\omega}$, $\mathbf{n}$, and $\mathbf{r} + \mathbf{k}$. Above, μ is the coefficient of kinetic friction, ρ is the radius of the circular contact path between two bodies. On the assumption that ρ is small, the friction moment M_T can be neglected.

In Fig. 9.19 it is shown how one may easily construct a wobblestone.

On one half of a homogeneous ellipsoid along its axis a we place a homogeneous rod fixed by a screw to the ellipsoid at point O. Such an arrangement results in a change to the position of the mass center. Next, we rotate the rod through angle α, changing in this way both the moments of inertia and the positions of the principal centroidal axes of inertia.

To simplify the numerical simulation of systems (9.208) and (9.209), let us choose a coordinate system so as to simplify the second of vector equations (9.208) by diagonalization of matrix I_C. The Cartesian coordinate system $O'X_{1e}X_{2e}X_{3e}$ will be called an ellipsoid system. The second coordinate system $O'X_1'X_2'X_3'$ is rotated with respect to the previous one through angle α about the axis $O'X_3' = O'X_{3e}$. The aforementioned coordinate systems have their corresponding axes parallel to those of systems $CX_{1e}X_{2e}X_{3e}$ and $CX_1'X_2'X_3'$. The system $CX_1'X_2'X_3'$ is a system of principal centroidal axes of inertia (since axis $CX_3' = CX_{3e}$ is the principal centroidal axis, the products of inertia $I_{CX_{1e}X_{3e}} = I_{CX_{2e}X_{3e}} = I_{CX_1'X_3'} = I_{CX_2'X_3'} = 0$).

If we know the moments of inertia in the system $CX_{1e}X_{2e}X_{3e}$, then after rotation through appropriate angle α the inertia matrix assumes a diagonal form $I_C = \mathrm{diag}[I_1, I_2, I_3]$.

In order to determine the moments of inertia $I_1 \equiv I_{1'1'}$, $I_2 \equiv I_{2'2'}$, we will make use of the results of calculations from Example 3.9 in [1].

We have accordingly

$$I_{CX_1'} = I_{1'1'} \equiv I_1 = I_{CX_{1e}}\cos^2\alpha + I_{CX_{2e}}\sin^2\alpha - I_{CX_{1e}X_{2e}}\sin 2\alpha,$$
$$I_{CX_2'} = I_{2'2'} \equiv I_2 = I_{CX_{1e}}\sin^2\alpha + I_{CX_{2e}}\cos^2\alpha + I_{CX_{1e}X_{2e}}\sin 2\alpha,$$
$$I_{CX_3'} = I_{OX_3'} = I_3. \tag{9.211}$$

The product of inertia in the coordinate system $CX_1'X_2'X_3'$ is equal to

$$I_{CX_1'X_2'} = \frac{I_{CX_{1e}} - I_{CX_{2e}}}{2}\sin 2\alpha + I_{CX_{1e}X_{2e}}\cos 2\alpha = 0, \tag{9.212}$$

because we choose the angle α, so that $I_{CX_1'X_2'} = 0$. On the basis of (9.212) we determine the value of the unknown angle

$$\alpha = \frac{1}{2}\arctan\frac{2I_{CX_{1e}X_{2e}}}{I_{CX_{2e}} - I_{CX_{1e}}}. \tag{9.213}$$

Having determined the angle α, we can now express an arbitrary vector $\mathbf{r}$ described by the coordinates $O'X_{1e}X_{2e}X_{3e}$ in terms of the coordinates obtained after rotation of this system through the angle α, that is, coordinates of the system $O'X_1'X_2'X_3'$, and vice versa. According to (5.176) we obtain

$$\begin{bmatrix} x_{1e} \\ x_{2e} \\ x_{3e} \end{bmatrix} = \begin{bmatrix} \cos\alpha & -\sin\alpha & 0 \\ \sin\alpha & \cos\alpha & 0 \\ 0 & 0 & 1 \end{bmatrix} \begin{bmatrix} x_1' \\ x_2' \\ x_3' \end{bmatrix}. \tag{9.214}$$

An equation of an ellipsoid in an ellipsoid system has the canonical form

$$\phi(\mathbf{r}) = \frac{x_{1e}^2}{a^2} + \frac{x_{2e}^2}{b^2} + \frac{x_{3e}^2}{c^2} - 1 = 0, \tag{9.215}$$

and according to (9.215) in the system $O'X_1'X_2'X_3'$ this equation takes the form

$$\phi(\mathbf{r}) = \frac{(-x_1'\cos\alpha + x_2'\sin\alpha)^2}{a^2} + \frac{(x_1'\sin\alpha + x_2'\cos\alpha)^2}{b^2} + \frac{x_3'}{c^2} - 1 = 0. \tag{9.216}$$

Next, we express the function $\phi(\mathbf{r})$ with the aid of matrix $\mathbf{R}$ in the following way:

$$\phi(\mathbf{r}) + 1 = \mathbf{r} \circ \mathbf{R}\mathbf{r} \tag{9.217}$$

or, in expanded form,

$$
\begin{aligned}
&x_1'^2(b^2\cos^2\alpha + a^2\sin^2\alpha) + x_2'^2(b^2\sin^2\alpha + a^2\cos^2\alpha)\\
&+ \frac{2x_1'x_2'(a^2-b^2)\sin\alpha\cos\alpha}{a^2b^2} + \frac{x_3'^2}{c^2} = R_{11}x_1'^2 + R_{22}x_2'^2\\
&+ R_{33}x_3'^2 - (R_{12}+R_{21})x_1'x_2' - (R_{13}+R_{31})x_1'x_3'\\
&- (R_{23}+R_{32})x_2'x_3'. \qquad (9.218)
\end{aligned}
$$

Equating terms at the same powers of variables x_1' and x_2' and their combinations, we obtain the desired elements of matrix $\mathbf{R}$:

$$
\begin{aligned}
R_{11} &= \frac{(b^2\cos^2\alpha + a^2\sin^2\alpha)}{a^2b^2},\\
R_{22} &= \frac{(b^2\sin^2\alpha + a^2\cos^2\alpha)}{a^2b^2},\\
R_{33} &= \frac{1}{c^2},\\
R_{12} &= R_{21} = \frac{b^2-a^2}{a^2b^2}\sin\alpha\cos\alpha,\\
R_{13} &= R_{31} = R_{23} = R_{32} = 0. \qquad (9.219)
\end{aligned}
$$

Note that

$$\frac{1}{2}\frac{\mathrm{d}\phi}{\mathrm{d}\mathbf{r}} = \mathbf{Rr}. \qquad (9.220)$$

A normal unit vector at the point of contact $\mathbf{n}$ is defined as

$$\mathbf{n} = -\frac{\frac{\mathrm{d}\phi}{\mathrm{d}\mathbf{r}}}{\left|\frac{\mathrm{d}\phi}{\mathrm{d}\mathbf{r}}\right|}, \qquad (9.221)$$

where

$$n_1^2(t) + n_2^2(t) + n_3^2(t) = 1. \qquad (9.222)$$

The normal unit vector $\mathbf{n}$ and vector $\frac{\mathrm{d}\phi}{\mathrm{d}\mathbf{r}}$ are proportional to one another, that is,

$$\mathbf{n} = f\frac{\mathrm{d}\phi}{\mathrm{d}\mathbf{r}}, \qquad f < 0. \qquad (9.223)$$

The coefficient f and the coordinates of the point of contact in the ellipsoid system are determined from (9.215) and (9.223), that is, one has to solve the following system of equations:

$$\frac{x_{1e}^2}{a^2} + \frac{x_{2e}^2}{b^2} + \frac{x_{3e}^2}{c^2} - 1 = 0,$$

$$n_{x_{1e}} = \frac{2f x_{1e}}{a^2}, \quad n_{x_{2e}} = \frac{2f x_{2e}}{b^2}, \quad n_{x_{3e}} = \frac{2f x_{3e}}{c^2}. \tag{9.224}$$

Solutions of system (9.224) have the form

$$x_{1e} = -\frac{a^2 n_{x_{1e}}(t)}{\sqrt{a^2 n_{x_{1e}}^2(t) + b^2 n_{x_{2e}}^2(t) + c^2 n_{x_{3e}}^2(t)}},$$

$$x_{2e} = -\frac{b^2 n_{x_{2e}}(t)}{\sqrt{a^2 n_{x_{1e}}^2(t) + b^2 n_{x_{2e}}^2(t) + c^2 n_{x_{3e}}^2(t)}},$$

$$x_{3e} = -\frac{c^2 n_{x_{3e}}(t)}{\sqrt{a^2 n_{x_{1e}}^2(t) + b^2 n_{x_{2e}}^2(t) + c^2 n_{x_{3e}}^2(t)}},$$

$$f = -\frac{1}{2}\sqrt{a^2 n_{x_{1e}}^2(t) + b^2 n_{x_{2e}}^2(t) + c^2 n_{x_{3e}}^2(t)}. \tag{9.225}$$

Because, according to (9.215), $\phi(\mathbf{r}) = 0$, from (9.217) we obtain

$$\mathbf{r} \circ \mathbf{R}\mathbf{r} = 1. \tag{9.226}$$

Vector $\mathbf{r}$ in the preceding equation will be expressed using vector $\mathbf{n}$. According to (9.220) and (9.223) we have

$$\mathbf{n} = 2f\,\mathbf{R} \circ \mathbf{r}. \tag{9.227}$$

Taking into account (9.227) in (9.226) we obtain

$$\left(\frac{1}{2f}\right)^2 \mathbf{R}^{-1}\mathbf{n} \circ \mathbf{R}\mathbf{R}^{-1}\mathbf{n} = 1, \tag{9.228}$$

which makes it possible to determine one of the desired unknowns

$$f = -\frac{1}{2}\sqrt{R^{-1}\mathbf{n} \circ \mathbf{n}}. \tag{9.229}$$

The desired vector $\mathbf{r}$, according to (9.227), is expressed in terms of unit vector $\mathbf{n}$ in the following way:

$$\mathbf{r} = -\frac{R^{-1}\mathbf{n}}{\sqrt{R^{-1}\mathbf{n} \circ \mathbf{n}}}. \tag{9.230}$$

The preceding calculations generalize the results obtained earlier [see (9.225)].

In order to numerically solve the problem, that is, the system of algebraic differential equations (9.208) and (9.209), let us differentiate algebraic equation (9.209) of the form

$$\mathbf{v}_A \circ \mathbf{n} = \mathbf{0}, \tag{9.231}$$

which ensures the perpendicularity of vectors $\mathbf{n}$ and $\mathbf{v}_A$, which means that the velocity of the point of contact $\mathbf{v}_A$ lies on the plane π.

Differentiating (9.231) in the absolute system $OX_1X_2X_3$ we obtain

$$\frac{\mathrm{d}\mathbf{v}_A}{\mathrm{d}t} \circ \mathbf{n} + \mathbf{v}_A \circ \frac{\mathrm{d}\mathbf{n}}{\mathrm{d}t} = \frac{\mathrm{d}\mathbf{v}_A}{\mathrm{d}t} \circ \mathbf{n} = 0, \tag{9.232}$$

because $\frac{\mathrm{d}\mathbf{n}}{\mathrm{d}t} = 0$, where $\frac{\mathrm{d}}{\mathrm{d}t}$ is the derivative in the absolute coordinate system.

Since

$$\begin{aligned}
\frac{\mathrm{d}\mathbf{v}_A}{\mathrm{d}t} &= \frac{\tilde{\mathrm{d}}\mathbf{v}_A}{\mathrm{d}t} + \boldsymbol{\omega} \times \mathbf{v}_A \\
&= \frac{\tilde{\mathrm{d}}}{\mathrm{d}t}(\mathbf{v} + \boldsymbol{\omega} \times (\mathbf{r} + \mathbf{k})) + \boldsymbol{\omega} \times (\mathbf{v} + (\boldsymbol{\omega} \times (\mathbf{r} + \mathbf{k}))) \\
&= \dot{\tilde{\mathbf{v}}} + \dot{\tilde{\boldsymbol{\omega}}} \times (\mathbf{r} + \mathbf{k}) + \boldsymbol{\omega} \times (\dot{\tilde{\mathbf{r}}} + \dot{\tilde{\mathbf{k}}}) + \boldsymbol{\omega} \times (\mathbf{v} + (\boldsymbol{\omega} \times (\mathbf{r} + \mathbf{k}))),
\end{aligned} \tag{9.233}$$

condition (9.232) takes the form

$$\left[\dot{\tilde{\mathbf{v}}} + \dot{\tilde{\boldsymbol{\omega}}} \times (\mathbf{r} + \mathbf{k}) + \boldsymbol{\omega} \times \dot{\tilde{\mathbf{r}}} + \boldsymbol{\omega} \times (\mathbf{v} + \boldsymbol{\omega} \times (\mathbf{r} + \mathbf{k}))\right] \circ \mathbf{n} = 0. \tag{9.234}$$

Differentiating condition (9.231) in the local system $CX_1'X_2'X_3'$ we have

$$\begin{aligned}
&\frac{\tilde{\mathrm{d}}\mathbf{v}_A}{\mathrm{d}t} \circ \mathbf{n} + \mathbf{v}_A \circ \frac{\tilde{\mathrm{d}}\mathbf{n}}{\mathrm{d}t} \\
&= (\dot{\tilde{\mathbf{v}}} + \dot{\tilde{\boldsymbol{\omega}}} \times (\mathbf{r} + \mathbf{k}) + \boldsymbol{\omega} \times \dot{\tilde{\mathbf{r}}}) \circ \mathbf{n} + (\mathbf{v} + (\boldsymbol{\omega} \times (\mathbf{r} + \mathbf{k})) \circ \dot{\tilde{\mathbf{n}}} = \mathbf{0}.
\end{aligned} \tag{9.235}$$

In this way the problem eventually is reduced to a solution of differential equations (9.208) and (9.234) or (9.235).

The problem is solved in the following way. The system of differential equations (9.208) and (9.235) was solved numerically. Algebraic calculations were conducted based on equations (9.214)–(9.230). An illustrative result of calculations is presented in Fig. 9.20.

In Fig. 9.20b the change in direction of rotation $\omega_3(t)$ is clearly seen, in Fig. 9.20c it is seen that at the beginning and end of motion the normal force is equal to the weight of the stone, whereas from Fig. 9.20a it follows that all components of the vector of velocity of the center of mass of the stone are characterized by oscillatory changes.

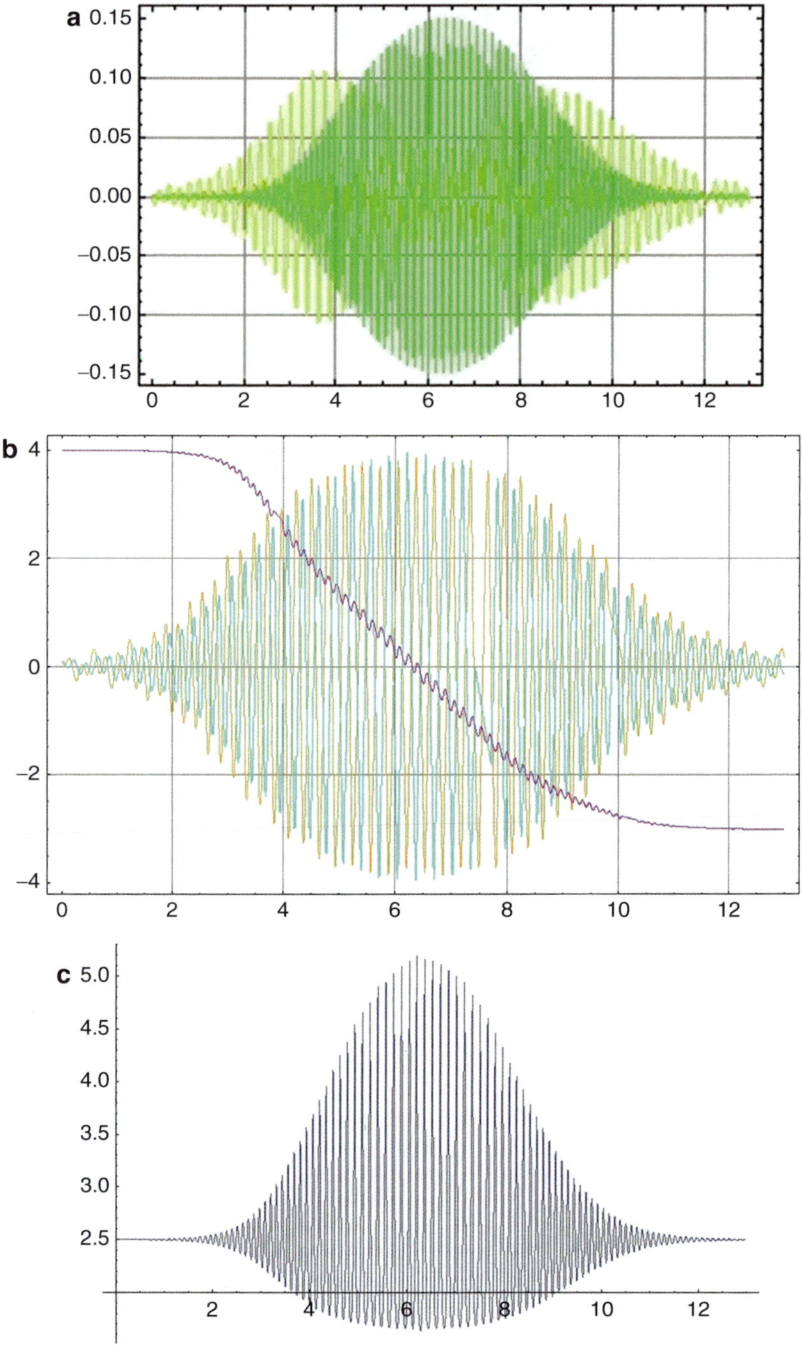

Fig. 9.20 Dynamics of the wobblestone: (**a**) plots of $v_1(t)$, $v_2(t)$ and $v_3(t)$; (**b**) plots of $\omega_1(t)$, $\omega_2(t)$, and $\omega_3(t)$; (**c**) plot of $N(t)$

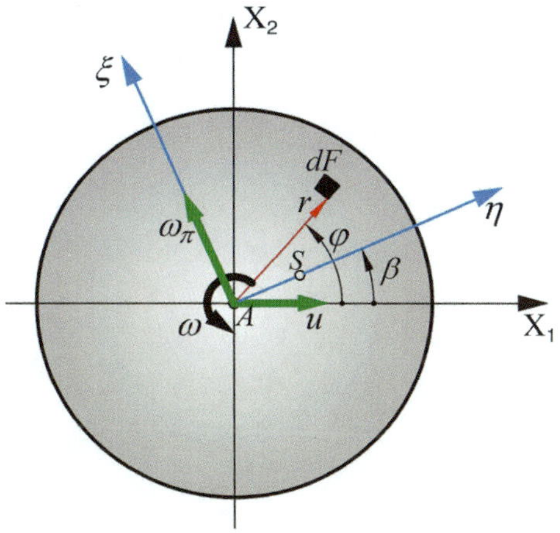

Fig. 9.21 Contact patch and coordinate systems

9.9.2 *Tangens Hyperbolicus Approximations of the Spatial Model of Friction*

In Fig. 9.21 is presented a non-dimensional circular contact area (of radius equal to one) with the center at point A, with a relative translational non-dimensional velocity of magnitude $u = v_A/\rho$ (where v_A is the length of the real sliding velocity of point A and ρ is the real radius of the contact surface) and relative angular velocity ω. Without loss of generality we assume that the velocity u is directed along the X_1 axis of the introduced coordinate system AX_1X_2. We assume that in the contact pressure distribution, initially possessing a central symmetry, distortion appears due to rolling resistance, and the final stress distribution is symmetric with respect to the η axis of the $A\eta\xi$ coordinate system. The resultant normal force is applied at point S, and the rolling resistance vector is opposite to the ξ axis (see [14–16] for more details).

The resultant non-dimensional friction force components and friction torque can be expressed as follows:

$$\begin{aligned} T_{X_1}(u,\omega,\beta) &= T_{OX_1} + T_{rX_1}, \\ T_{X_2}(u,\omega,\beta) &= T_{OX_2} + T_{rX_2}, \\ M(u,\omega,\beta) &= M_O + M_r, \end{aligned} \tag{9.236}$$

where T_{OX_1}, T_{OX_2}, and M_O are the corresponding friction force components along the X_1 and X_2 axes and friction torque for case where there is no rolling resistance, while T_{rX_1}, T_{rX_2}, and M_r are the corresponding components of friction force and torque related to rolling resistance. Assuming that Coulomb's law holds true on an arbitrary surface element dF, the corresponding non-dimensional elements of the friction model (with the non-dimensional friction coefficient equal to one) have the following integral form in a polar coordinate system:

$$T_{OX_1}(u,\omega) = \int_0^{2\pi}\int_0^1 \sigma_O\, t_{X_1} \mathrm{d}r\mathrm{d}\varphi, \quad T_{rX_1}(u,\omega,\beta) = \int_0^{2\pi}\int_0^1 \sigma_r\, t_{X_1} \mathrm{d}r\mathrm{d}\varphi,$$

$$T_{OX_2}(u,\omega) = \int_0^{2\pi}\int_0^1 \sigma_O\, t_{X_2} \mathrm{d}r\mathrm{d}\varphi, \quad T_{rX_2}(u,\omega,\beta) = \int_0^{2\pi}\int_0^1 \sigma_r\, t_{X_2} \mathrm{d}r\mathrm{d}\varphi,$$

$$M_O(u,\omega) = r\int_0^{2\pi}\int_0^1 \sigma_O (t_{X_2}\cos\varphi - t_{X_1}\sin\varphi)\mathrm{d}r\mathrm{d}\varphi,$$

$$M_r(u,\omega,\beta) = r\int_0^{2\pi}\int_0^1 \sigma_r (t_{X_2}\cos\varphi - t_{X_1}\sin\varphi)\mathrm{d}r\mathrm{d}\varphi, \tag{9.237}$$

where

$$t_{X_1}(r,\varphi) = \frac{r\,(u-\omega\, r\sin\varphi)}{\sqrt{u^2 - 2\,\omega\, u\, r\sin\varphi + \omega^2 r^2}},$$

$$t_{X_2}(r,\varphi) = \frac{\omega\, r^2\cos\varphi}{\sqrt{u^2 - 2\,\omega\, u\, r\sin\varphi + \omega^2 r^2}},$$

while σ_O and σ_r are components of the non-dimensional contact stress distribution $\sigma(r,\varphi,\beta) = \sigma_O(r) + \sigma_r(r,\varphi,\beta)$. For σ_O having central symmetry, $T_{OX_2} = 0$.

The distortion in the stress distribution related to the rolling resistance is assumed to be a linear function with one parameter $0 \le k_r \le 1$ [15]:

$$\sigma_r(r,\varphi,\beta) = \sigma_O(r)\, k_r r\cos(\varphi-\beta), \tag{9.238}$$

where σ_0 is the non-dimensional (for the non-dimensional surface element dF and additionally related to the real resultant normal reaction) contact stress distribution for cases without rolling resistance, which for the Hertz law takes the form

$$\sigma_O(r) = \frac{3}{2\pi}\sqrt{1-r^2}. \tag{9.239}$$

The integral model (9.236)÷(9.237) is not convenient in direct application to real problems of modeling and simulation. Moving the origin of the polar coordinate system to the instantaneous center of velocities one can obtain exact analytical expressions of the components (9.237) in terms of elementary functions [15]. But they are still inconvenient to use because of their complexity. One way to avoid this problem is to construct suitable approximations. One of the simplest approximations is the first-order Padé one proposed by Kireenkov [15] for the complete combined model of sliding and rolling resistance in the following form:

$$\begin{aligned}
T_{OX_1(\mathrm{P1})} &= \frac{u}{u + a_{11}\,|\omega|},\\
T_{rX_1(\mathrm{P1})} &= a_{21}\frac{\omega\,k_r\,\sin\beta}{u + a_{11}\,|\omega|},\\
T_{rX_2(\mathrm{P1})} &= b_{O1}\frac{\omega\,k_r\,\cos\beta}{|\omega| + b_{11}u},\\
M_{O(\mathrm{P1})} &= c_{O1}\frac{\omega}{|\omega| + c_{11}u},\\
M_{r(\mathrm{P1})} &= c_{21}\frac{u\,k_r\,\sin\beta}{|\omega| + c_{11}u}.
\end{aligned} \tag{9.240}$$

The coefficients of model (9.240) are determined by the following conditions:

$$\begin{aligned}
\left.\frac{\partial T_{OX_1(\mathrm{P1})}}{\partial u}\right|_{u=0} &= \left.\frac{\partial T_{OX_1}}{\partial u}\right|_{u=0},\\
T_{rX_1(\mathrm{P1})}\big|_{u=0} &= T_{rX_1}\big|_{u=0},\\
T_{rX_2(\mathrm{P1})}\big|_{u=0} &= T_{rX_2}\big|_{u=0},\\
\left.\frac{\partial T_{rX_2(\mathrm{P1})}}{\partial \omega}\right|_{\omega=0} &= \left.\frac{\partial T_{rX_2}}{\partial \omega}\right|_{\omega=0},\\
M_{O(\mathrm{P1})}\big|_{u=0} &= M_O\big|_{u=0},\\
\left.\frac{\partial M_{O(\mathrm{P1})}}{\partial \omega}\right|_{\omega=0} &= \left.\frac{\partial M_O}{\partial \omega}\right|_{\omega=0},\\
M_{r(\mathrm{P1})}\big|_{\omega=0} &= M_r\big|_{\omega=0},
\end{aligned}$$

and for the Hertz case (9.239) we have $a_{11} = 8/(3\pi)$, $a_{21} = -1/4$, $b_{O1} = 3\pi/32$, $b_{11} = 15\pi/32$, $c_{O1} = 3\pi/16$, $c_{11} = 15\pi/16$, $c_{21} = -3\pi/16$. The approximations (9.240) preserve the values but do not completely satisfy all first partial derivatives of the functions (9.237) at $u = 0$ or $\omega = 0$. To satisfy all first partial derivatives it is necessary to use a second-order Padé approximation [15]:

$$\begin{aligned}
T_{OX_1(\mathrm{P2})} &= \frac{u^2 + a_{12}u\,|\omega|}{u^2 + a_{12}u\,|\omega| + \omega^2},\\
T_{rX_1(\mathrm{P2})} &= a_{22}\frac{|\omega|}{u^2 + \omega^2}\omega\,k_r\,\sin\beta,\\
T_{rX_2(\mathrm{P2})} &= b_{O2}\frac{|\omega| + b_{12}u}{u^2 + b_{12}u\,|\omega| + \omega^2}\omega\,k_r\,\cos\beta,
\end{aligned}$$

$$M_{O(\mathrm{P2})} = c_{O2}\frac{|\omega| + c_{12}u}{u^2 + c_{12}u\,|\omega| + \omega^2}\omega,$$

$$M_{r(\mathrm{P2})} = c_{22}\frac{u^2}{u^2 + \omega^2}\,k_r\sin\beta. \tag{9.241}$$

Coefficients of the model (9.241) are determined in an analogous way to the model (9.240) constants. For the Hertz case (9.239) we have $a_{12} = 8\pi/8$, $a_{22} = -3\pi/32$, $b_{O2} = 3\pi/32$, $b_{12} = 32/(15\pi)$, $c_{O2} = 3\pi/16$, $c_{12} = 16/(15\pi)$, $c_{22} = -1/5$. Approximations (9.241) completely satisfy the values and all first partial derivatives of the functions (9.237) at $u = 0$ or $\omega = 0$.

Here the complete set of tangens hyperbolicus approximations of a coupled model of dry friction and rolling resistance for a circular contact area between interacting bodies is used in the following form:

$$T_{OX_1(\mathrm{th})} = \tanh\left(h_1\,u\,|\omega|^{-1}\right),$$

$$T_{rX_1(\mathrm{th})} = f_1 k_r\sin\beta\,,$$

$$T_{rX_2(\mathrm{th})} = f_2 k_r\cos\beta,$$

$$M_{O(\mathrm{th})} = f_2 - f_1,$$

$$M_{r(\mathrm{th})} = h_5\left(1 - \tanh\left(s_2 u^{-q_2}|\omega|^{q_2}\right)\right)\,k_r\sin\beta\,, \tag{9.242}$$

where $\quad f_1 = h_2\left(1 - \tanh\left(s_1 u^{q_1}|\omega|^{-q_1}\right)\right)\,\mathrm{sign}\,(\omega)\,,\quad f_2 = h_3\tanh\left(h_4 u^{-1}\omega\right).$

The coefficients of model (9.242) are determined analogously to the case of models (9.240) and (9.241). The number of constants is smaller because of the use of certain relation between functions T_{rX_1}, T_{rX_2}, and M_O [the same relation can also be used for models (9.240)–(9.241), but we present them in the form proposed by Kireenkov [15]]. For the Hertz case (9.239) we have $h_1 = 3\pi/8$, $h_2 = -3\pi/32$, $h_3 = 3\pi/32$, $h_4 = 32/(15\pi)$, $h_5 = -1/5$. The approximations (9.242) completely satisfy the values and all first partial derivatives of functions (9.237) at $u = 0$ or $\omega = 0$. Coefficients q_1, q_2, s_1, s_2 do not affect values and first partial derivatives at $u = 0$ or $\omega = 0$, but they are chosen to be $q_1 = 1.75$, $q_2 = 2.5$, $s_1 = 1.25$, $s_2 = 0.275$ for the best fitting of the integral model (9.237) for different values of u and ω.

Figure 9.22 presents a comparison of three approximate models (9.240)–(9.242) and an exact integral model (9.237). The tangens hyperbolicus approximation is closest to the exact integral model. It is significantly more accurate than the second-order Padé approximation.

The wobblestone as a semi-ellipsoid rigid body with its mass center at point C, touching a rigid, flat, and fixed horizontal surface π (parallel to the X_1X_2 plane of the global stationary coordinate system $X_1X_2X_3$) at point A is presented in Fig. 9.23.

The equations of motion in the non-stationary coordinate system $OX_1X_2X_3$ (with axes parallel to the principal centroidal axes of inertia; we assume that the geometrical axis X_{3e} of the ellipsoid is parallel to one of them) are as follows:

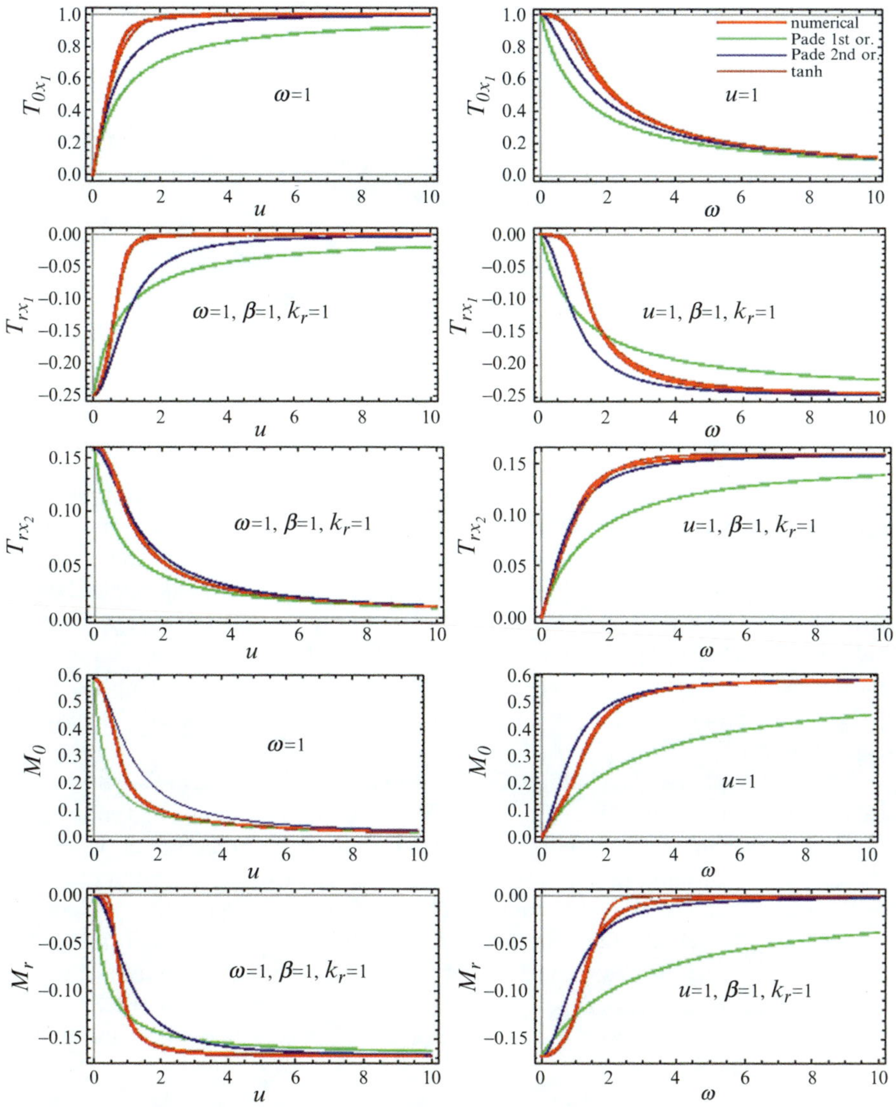

Fig. 9.22 Exact and approximated components of friction models

$$
\begin{aligned}
m\frac{\mathrm{d}\mathbf{v}}{\mathrm{d}t} + \boldsymbol{\omega} \times (m\mathbf{v}) &= -mg\mathbf{n} + N\mathbf{n} + \mathbf{T}, \\
\mathbf{B}\frac{\mathrm{d}\boldsymbol{\omega}}{\mathrm{d}t} + \boldsymbol{\omega} \times (\mathbf{B}\mathbf{v}) &= (\mathbf{r} - \mathbf{k}) \times (N\mathbf{n} + \mathbf{T}) + \mathbf{M}_t + \mathbf{M}_r, \\
\frac{\mathrm{d}\mathbf{n}}{\mathrm{d}t} + \boldsymbol{\omega} \times \mathbf{n} &= 0,
\end{aligned}
\tag{9.243}
$$

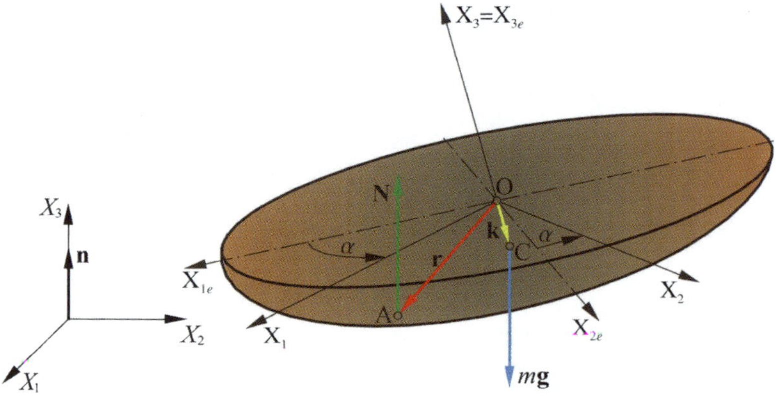

Fig. 9.23 The wobblestone on a horizontal plane π

where m is the mass of the wobblestone, $\mathbf{B} = \mathrm{diag}(B_1, B_2, B_3)$ is the tensor of inertia of the solid, $\mathbf{v}$ is the absolute velocity of the mass center C, $\boldsymbol{\omega}$ is the absolute angular velocity of the body, N is the value of the normal reaction of the horizontal plane, $\mathbf{n}$ is the unit vector normal to the plane X_1X_2, $\mathbf{T}$ (not included in Fig. 9.23) is the sliding friction force at point of contact A, and $\mathbf{M}_t$ and $\mathbf{M}_r$ (not included in Fig. 9.23) are respectively the dry friction and the rolling resistance torques applied to the body. Vector $\mathbf{r}$ indicates the actual contact point position and vector $\mathbf{k}$ determines the mass center position [16].

The combined models of sliding friction and rolling resistance cannot be directly used in the form presented previously for the wobblestone modeling and simulations with the use of standard numerical methods of integration. One reason is that the expressions for friction forces and torques have singularity for $u = 0$ and $\omega = 0$. Another problem arises from the fact that for $u = 0$ the directions of the components T_{rX_1} and T_{rX_2} are indefinite. Due to that reason we will express them in the $A\eta\xi$ coordinate system. But the angle β will still be indefinite due to the lack of sliding velocity. Similar problems appear due to the absence of rolling. In order to avoid these difficulties we propose the following specific approximations of the friction and rolling resistance models:

$$\mathbf{T} = -\mu N \, \mathbf{T}_{O(a)} - \mu N \left(T_{rX_1(a)} c_\beta + T_{rX_2(a)} s_\beta \right) \frac{\boldsymbol{\omega}_\beta}{\|\boldsymbol{\omega}_\beta\| + \varepsilon}$$

$$- \mu N \left(T_{rX_1(a)} s_\beta - T_{rX_2(a)} c_\beta \right) \frac{\boldsymbol{\omega}_\pi}{\|\boldsymbol{\omega}_\pi\| + \varepsilon},$$

$$\mathbf{M}_t = -\mu\rho N \, M_{t(a)} \mathbf{n}, \quad \mathbf{M}_r = -\frac{f_r N \boldsymbol{\omega}_\pi}{\|\boldsymbol{\omega}_\pi\| + \varepsilon}, \tag{9.244}$$

where (a) at the end of an index stands for some kind of approximation.

For a linear Padé approximation we have

$$\mathbf{T}_{O(\mathrm{P1}\varepsilon)} = \frac{\mathbf{u}}{\|\mathbf{u}\| + a_{11}\,|\omega_n| + \varepsilon}, \quad T_{rX_1(\mathrm{P1}\varepsilon)} = a_{21}\frac{\omega_n\, k_r s_\beta}{\|\mathbf{u}\| + a_{11}\,|\omega_n| + \varepsilon},$$
$$T_{rX_2(\mathrm{P1}\varepsilon)} = b_{O1}\frac{\omega_n\, k_r c_\beta}{|\omega_n| + b_{11}\,\|\mathbf{u}\| + \varepsilon}, \quad M_{t(\mathrm{P1}\varepsilon)} = \frac{c_{O1}\omega_n + c_{21}\,\|\mathbf{u}\|\, k_r s_\beta}{|\omega_n| + c_{11}\,\|\mathbf{u}\| + \varepsilon}, \tag{9.245}$$

and the second-order Padé approximation model takes the form

$$\mathbf{T}_{O(\mathrm{P2}\varepsilon)} = \frac{\|\mathbf{u}\| + a_{12}\,|\omega_n|}{\mathbf{u}^2 + a_{12}\,\|\mathbf{u}\|\,|\omega_n| + \omega_n^2 + \varepsilon}\mathbf{u},$$
$$T_{rX_1(\mathrm{P2}\varepsilon)} = a_{22}\frac{|\omega_n|\;\omega_n k_r s_\beta}{\mathbf{u}^2 + \omega_n^2 + \varepsilon},$$
$$T_{rX_2(\mathrm{P2}\varepsilon)} = b_{O2}\frac{|\omega_n| + b_{12}\,\|\mathbf{u}\|}{\mathbf{u}^2 + b_{12}\,\|\mathbf{u}\|\,|\omega_n| + \omega_n^2 + \varepsilon}\omega_n\, k_r c_\beta,$$
$$M_{t(\mathrm{P2}\varepsilon)} = \frac{|\omega_n| + c_{12}\,\|\mathbf{u}\|}{\mathbf{u}^2 + c_{12}\,\|\mathbf{u}\|\,|\omega_n| + \omega_n^2 + \varepsilon}\omega_n + c_{22}\frac{\mathbf{u}^2 k_r s_\beta}{\mathbf{u}^2 + \omega_n^2 + \varepsilon}, \tag{9.246}$$

while the tangens hyperbolicus model will be as follows:

$$\mathbf{T}_{O(\mathrm{th}\varepsilon)} = \tanh\left(h_1\frac{\|\mathbf{u}\|}{|\omega_n| + \varepsilon}\right)\frac{\mathbf{u}}{\|\mathbf{u}\| + \varepsilon},$$
$$T_{rX_1(\mathrm{th}\varepsilon)} = f_{1\varepsilon} k_r s_\beta, \qquad T_{rX_2(\mathrm{th}\varepsilon)} = f_{2\varepsilon} k_r c_\beta,$$
$$M_{t(\mathrm{th}\varepsilon)} = f_{2\varepsilon} - f_{1\varepsilon} + h_5\left(1 - \tanh\left(s_2\left(\frac{|\omega_n| + \varepsilon}{\|\mathbf{u}\|}\right)^{q_2}\right)\right) k_r s_\beta, \tag{9.247}$$

where $f_{1\varepsilon} = h_2\left(1 - \tanh\left(s_1\left(\frac{\|\mathbf{u}\|}{|\omega_n|+\varepsilon}\right)^{q_1}\right)\right)\operatorname{sign}(\omega_n)$, $f_{2\varepsilon} = h_3\tanh\left(h_4\frac{\omega_n}{\|\mathbf{u}\|+\varepsilon}\right)$, where μ is the dry friction coefficient, ρ is the radius of the contact path (we assume that the circular contact path between bodies with constant radius is independent of the normal force), $f_r = \rho\int_0^{2\pi}\int_0^1 \sigma_r r^2\cos(\varphi - \sigma)\,\mathrm{d}r\mathrm{d}\varphi$ is the rolling resistance coefficient and for the Hertz case $f_r = \rho k_r/5$, $\mathbf{u}$ is the normalized velocity $\mathbf{v}_A$ of the body point in contact with the horizontal surface, ω_n is the projection of the angular velocity onto the X_3 axis, $\boldsymbol{\omega}_\pi$ is the component of the angular velocity lying on the π plane, ω_β is the vector lying on the π plane of the same length as ω_π but perpendicular to it, and c_β and s_β are approximated sine and cosine functions of the angle β (angle between the sliding and rolling directions):

$$\mathbf{v}_A = \mathbf{v} + \boldsymbol{\omega} \times (\mathbf{r} - \mathbf{k}), \qquad \mathbf{u} = \frac{\mathbf{v}_A}{\rho},$$
$$\omega_n = \boldsymbol{\omega} \cdot \mathbf{n}, \qquad \boldsymbol{\omega}_\pi = \boldsymbol{\omega} - \omega_n \mathbf{n}, \qquad \boldsymbol{\omega}_\beta = \boldsymbol{\omega}_\pi \times \mathbf{n},$$
$$c_\beta = \frac{u_{\eta\varepsilon}}{\sqrt{u_{\eta\varepsilon}^2 + u_\xi^2}}, \qquad s_\beta = \frac{u_\xi}{\sqrt{u_{\eta\varepsilon}^2 + u_\xi^2}}, \tag{9.248}$$
$$u_{\eta\varepsilon} = \mathbf{u} \cdot \boldsymbol{\omega}_\beta + \varepsilon, \qquad u_\xi = -\mathbf{u} \cdot \boldsymbol{\omega}_\pi.$$

The $\mathbf{M}_r$ vector is constructed on the assumption that the rolling resistance torque opposes the angular velocity component lying on the π plane (it is equivalent to assuming a rigid π plane and deformable wobblestone). The parameter ε is introduced in order to smooth the equations and avoid numerical problems around some singularities. The differential equations of motion (9.243) are supplemented with the following algebraic equation:

$$(\mathbf{v} + \boldsymbol{\omega} \times (\mathbf{r} - \mathbf{k})) \circ \mathbf{n} = 0, \tag{9.249}$$

which follows from the fact that the velocity $\mathbf{v}_A$ lies on the plane π. Equations (9.243) and (9.249) now form the differential-algebraic equation set. One way to solve them is to differentiate condition (9.249) with respect to time and then treat it as an additional equation while solving the governing equations algebraically with respect to the corresponding derivatives and the normal reaction N.

To complete the model, the relation between vectors $\mathbf{r}$ and $\mathbf{n}$ should be given. Taking the ellipsoid equation

$$\frac{r_{1e}^2}{a^2} + \frac{r_{2e}^2}{b^2} + \frac{r_{3e}^2}{c^2} = 1, \tag{9.250}$$

(where a, b, and c are the semiaxes of the ellipsoid) and the condition of tangent contact between the ellipsoid and the horizontal plane

$$n_{1e} = \frac{\lambda r_{1e}}{a^2}, \qquad n_{2e} = \frac{\lambda r_{2e}}{b^2}, \qquad n_{3e} = \frac{\lambda r_{3e}}{c^2}, \tag{9.251}$$

we can find the relation between the components of vectors $\mathbf{r}$ and $\mathbf{n}$ in the $OX_{1e}X_{2e}X_{3e}$ coordinate system. Since the $OX_1X_2X_3$ coordinate system is obtained by rotation of the $OX_{1e}X_{2e}X_{3e}$ system around the X_{3e} axis through the angle α, the corresponding relation in the $OX_1X_2X_3$ coordinate system can be found easily.

All the presented results were obtained for the following parameters and initial conditions: $m = 0.25\,\mathrm{kg}$, $g = 10\,\mathrm{m/s^2}$, $\alpha = 0.3\,\mathrm{rad}$, $B_1 = 10^{-4}\,\mathrm{kg \cdot m^2}$, $B_2 = 8 \cdot 10^{-4}\,\mathrm{kg \cdot m^2}$, $B_3 = 10^{-3}\,\mathrm{kg \cdot m^2}$, $a = 0.08\,\mathrm{m}$, $b = 0.016\,\mathrm{m}$, $c = 0.012\,\mathrm{m}$, $k_1 = k_2 = 0$, $k_3 = 0.002\,\mathrm{m}$, $\mu = 0.5$, $\rho = 6 \cdot 10^{-4}\,\mathrm{m}$, $k_r = 1$, $\epsilon = 10^{-4}\,\mathrm{rad/s}$, $v_{1O} = v_{2O} = v_{3O} = 0\,\mathrm{m/s}$, $n_{1O} = n_{2O} = 0$, $n_{3O} = 1$.

Figure 9.24 shows the results of simulation of the wobblestone initially spinning at $\omega_{3O} = 20\,\mathrm{rad/s}$ but also wobbling at $\omega_{2O} = 1\,\mathrm{rad/s}$ ($\omega_{1O} = 0$) for three

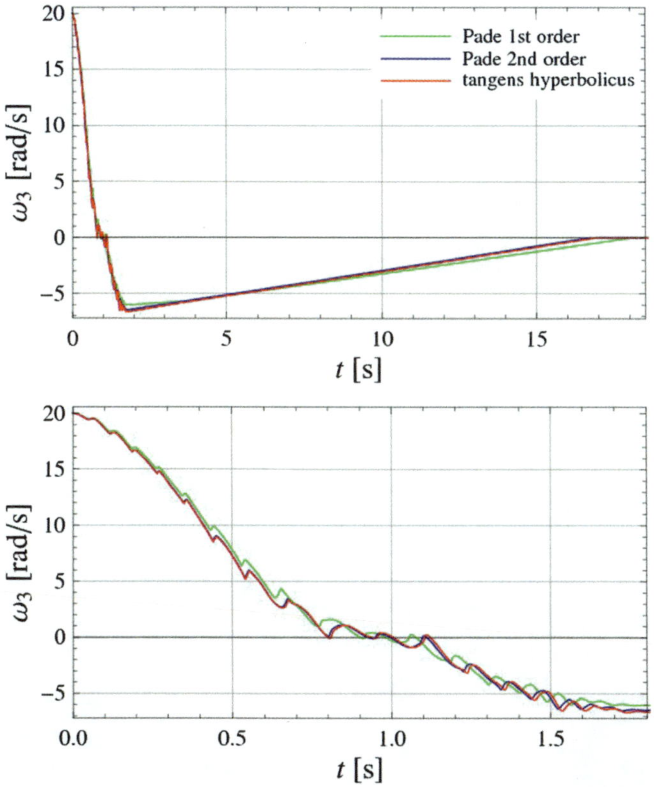

Fig. 9.24 Wobblestone response with initial conditions $\omega_{1O} = 0$, $\omega_{2O} = 1$, $\omega_{3O} = 20$ (rad/s) for different approximations of the friction model

different approximations: the first-order Padé, the second-order Padé, and tangens hyperbolicus approximation. The wobblestone exhibits typical behavior for that kind of solid, that is, we can observe that after some time the spin changes sign and the motion vanishes. The differences between the three solutions are seen, especially between the solution using the first-order Padé approximation and the others. The significance of that difference depends on the kind of application of the developed model and simulation. The rest of the presented results were obtained using of a tangens hyperbolicus approximation.

Figure 9.25 presents similar results for initial spin $\omega_{30} = 4$ rad/s and wobbling at $\omega_{2O} = 1$ rad/s ($\omega_{1O} = 0$). We can also see the corresponding behavior of a system with friction torque M_t switched off, where the motion ends with the wobblestone spinning with constant velocity without wobbling, but the initial portion of motion does not differ significantly from the motion of the wobblestone with the friction torque. The corresponding normal force history is also presented in Fig. 9.25.

In Fig. 9.26 the final ω_3 angular velocity for a wobblestone without dry friction torque M_t for different initial conditions is presented in the form of contour plots. One can observe that for one direction of the spin ($\omega_{3O} = 2$ rad/s in Fig. 9.26a)

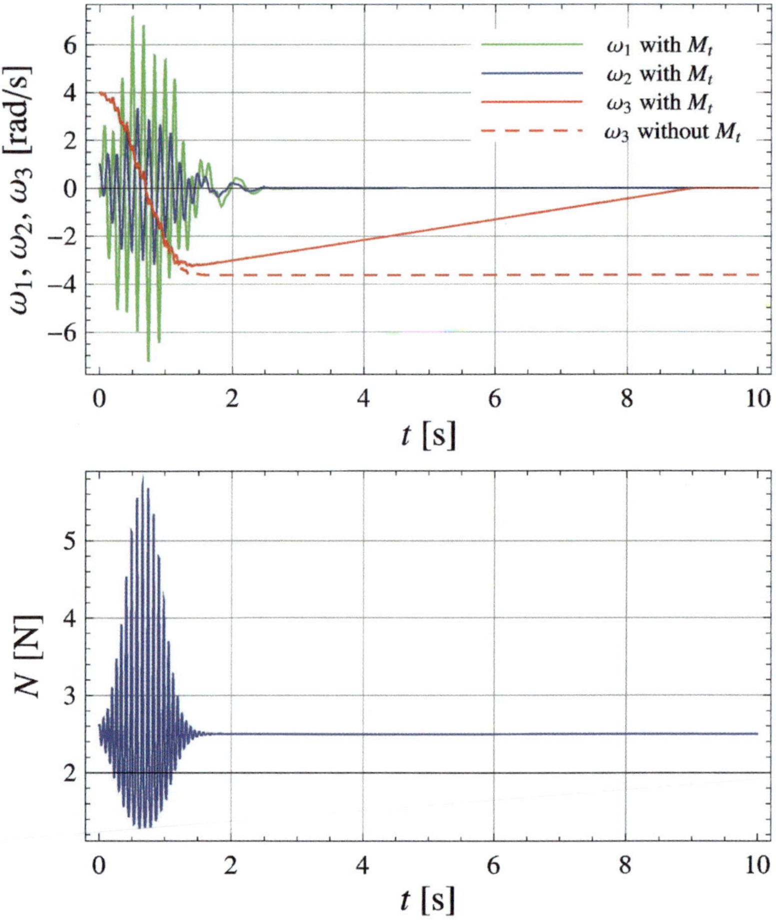

Fig. 9.25 Wobblestone response and normal force history with initial conditions $\omega_{1O} = 0, \omega_{2O} = 1, \omega_{30} = 4$ (rad/s) for tangens hyperobolicus approximation

there is some area around the point (0,0) on the plane of the initial conditions ω_{1O}–ω_{2O} for which there are no reversals. The changes in the sign of spin take place for initial conditions outside that area, that is, for strong enough initial wobbling. For the opposite sign of the initial spin ($\omega_{3O} = -3$ rad/s in Fig. 9.26b) it is difficult to observe the spin reversal. The most interesting plot, however, is the that in Fig. 9.26c, where the section along the ω_3 axis in the initial-condition space is shown. The ω_3 axis is stable in the sense that a small enough perturbation of the wobblestone spinning with any value and any sign will decay after some time, and the stone will continue spinning with (almost) the same velocity. To observe the reversals, the initial wobbling must be large enough and only for the proper sign of the spin.

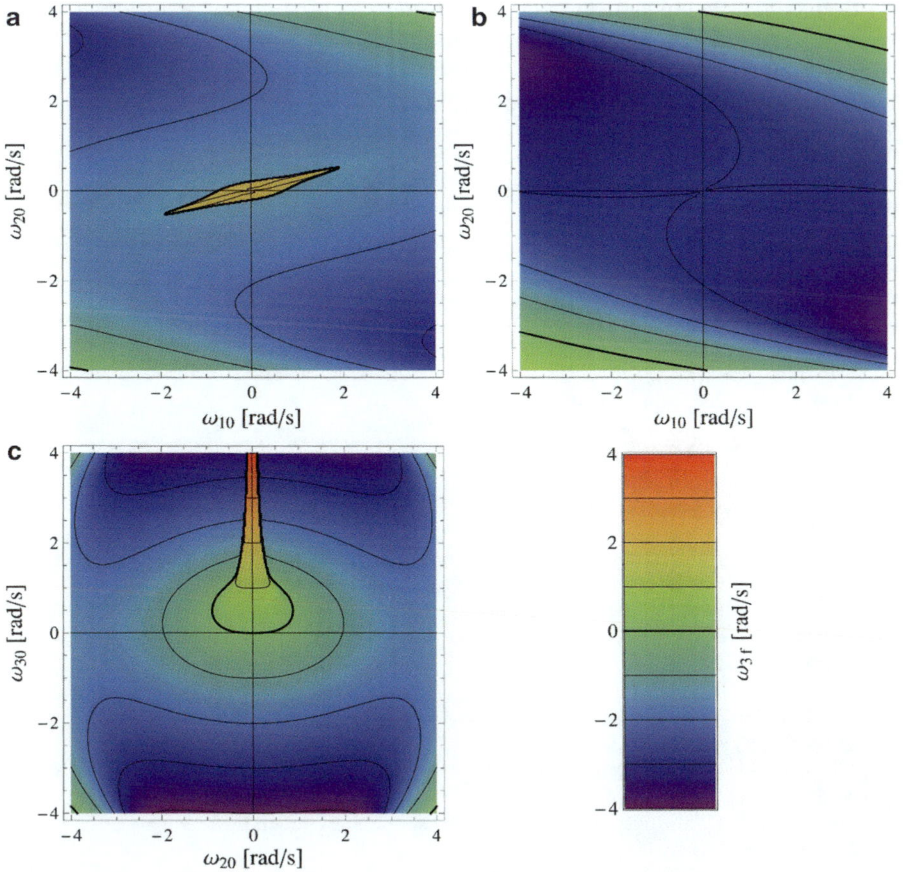

Fig. 9.26 Wobblestone final ω_3 angular velocity (rad/s) (without dry friction torque M_t) for different initial conditions. (**a**) $\omega_{3O} = 2$ rad/s. (**b**) $\omega_{3O} = -3$ rad/s. (**c**) $\omega_{1O} = 0$ rad/s

To conclude this discussion of our results, also presented in [16], a complete set of tangens hyperbolicus approximations of the spatial friction model coupled with the rolling resistance for the circular contact area between interacting bodies was developed and then compared with corresponding Padé approximations of the first and second order, well known from the literature, as well as with the numerical solution of the exact integral model. It was shown that tangens hyperbolicus approximations are the closest to the exact solution. Applying three different approximations to the wobblestone model the differences in simulation results were shown. In applications where high accuracy of simulation is required, a model with a second-order Padé or tangens hyperbolicus approximation should be used. Taking into account that the complexity of both approximations is comparable and that the tangens hyperbolicus approximation is actually closer to the exact integral model of friction, it seems reasonable to use the second one.

Both the presented model of the wobblestone and its simulations are very realistic as compared with most earlier works on the celt since the correct spatial friction model, coupled with rolling resistance torque, has been applied, however with the significant simplifying assumption of a circular contact area between stone and table with a constant radius independent of the normal force.

References

1. J. Awrejcewicz, *Classical Mechanics: Statics and Kinematics* (Springer, New York, 2012)
2. A.D. Markeev, *Theoretical Mechanics* (Nauka, Moscow, 1990) (in Russian)
3. Z. Koruba, J.W. Osiecki, *Elements of Advanced Mechanics* (Kielce University of Technology Press, Kielce, 2007) (in Polish)
4. V.A. Svietlitskiy, I.V. Stasenko, *Collection of Problems on Vibrations Theory* (Vyshaia Shkola, Moscow, 1979) (in Russian)
5. G.T. Walker, On a dynamical top. Q. J. Pure Appl. Math. **28**, 175–184 (1896)
6. B. Sir Hermann, The rigid body dynamics of unidirectional spin. Proc. R. Soc. A **405**, 265–279 (1986)
7. K. Magnus, On the theory of the Celtic Stone. Zeitschrift für angewandte Mathematik und Mechanik **54**, 54–55 (1974) (in German)
8. T.K. Caughey, A mathematical model of the rattleback. Int. J. Non-Linear Mech. **15**, 293–302 (1980)
9. T.R. Kane, D.A. Levinson, Realistic mathematical modelling of a rattleback. Int. J. Non-Linear Mech. **17**, 175–186 (1982)
10. R.E. Lindberg, R.W. Longman, On the dynamic behavior of the wobblestone. Acta Mech. **49**, 81–94 (1983)
11. A. Garcia, M. Hubbard, Spin reversal of the rattleback: theory and experiment. Proc. R. Soc. A **418**, 165–197 (1988)
12. A.P. Markeev, On the dynamics of a solid on an absolutely rough plane. Regular Chaot. Dyn. **7**(2), 153–160 (2002)
13. A.V. Borisov, A.A. Kilin, I.S. Mamaev, New effects of rattlebacks. Dokl. Phys. **51**(5), 272–275 (2006)
14. V.Ph. Zhuravlev, D.M. Klimov, Global motion of the celt. Mech. Solids **43**(3), 320–327 (2008)
15. A.A. Kireenkov, Combined model of sliding and rolling friction in dynamics of bodies on a rough surface. Mech. Solids **43**(3), 116–131 (2008)
16. G. Kudra, J. Awrejcewicz, Tangens hyperbolicus approximations of the spatial model of friction coupled with rolling resistance. Int. J. Bifurcat. Chaos **21**(10), 2905–2917 (2011)

Chapter 10
Stationary Motions of a Rigid Body and Their Stability

10.1 Stationary Conservative Dynamics

Let us consider a scleronomic mechanical system described by the system of differential equations

$$\dot{\mathbf{x}} = \mathbf{f}(\mathbf{x}), \tag{10.1}$$

where $\mathbf{x} \in \mathbf{R}^N$ and for an arbitrary choice of initial conditions $\mathbf{x}(t_0) = \mathbf{x}^0$ there exists a unique solution $\mathbf{x} = \mathbf{x}(\mathbf{x}^0, t)$, where $\cdot = \frac{\mathrm{d}}{\mathrm{d}t}$, and time $t \in [0, +\infty)$. Owing to the autonomous nature of (10.1) we will further assume $t_0 = 0$. From the considerations of Chap. 9 it follows that in the case of conservative dynamics of a rigid body first integrals exist that lead to simplification of such systems.

Thus, let us assume that system (10.1) has M first integrals of the forms

$$U_0(\mathbf{x}(t)) = C_0, U_1(\mathbf{x}(t)) = C_1, \ldots, U_M = (\mathbf{x}(t)) = C_M, \quad M < N - 1, \tag{10.2}$$

where C_i, $i = 0, 1, \ldots, M$, are constants.

Theorem 10.1. *If one of the first integrals of system* (10.1) *has a non-singular stationary (constant) value for certain fixed (constant) values of the remaining first integrals of this system at a certain point* $\mathbf{x}^0$, *then the solution* $\mathbf{x} \equiv \mathbf{x}^0$ *describes the actual motion of the dynamical system governed by* (10.1).

Proof. see [1] Let $\mathbf{C} = [C_1, \ldots, C_M]^{\mathrm{T}}$, and the first integral $U_0(\mathbf{x}(t)) = C_0$ has a constant (stationary) value for certain fixed values of the remaining first integrals $\mathbf{U}(\mathbf{x}^0) = \mathbf{C}^0$, $\mathbf{U}(\mathbf{x}) = [U_1(\mathbf{x}), \ldots, U_M(\mathbf{x})]^{\mathrm{T}}$. □

Let us introduce M values of undetermined Lagrange multipliers $\boldsymbol{\lambda} = [\lambda_1, \ldots, \lambda_M]^{\mathrm{T}}$ and the function

$$W(\mathbf{x}, \boldsymbol{\lambda}) = U_0(\mathbf{x}) + \boldsymbol{\lambda}^{\mathrm{T}}(\mathbf{U}(\mathbf{x}) - \mathbf{C}^0). \tag{10.3}$$

J. Awrejcewicz, *Classical Mechanics: Dynamics*, Advances in Mechanics and Mathematics 29, DOI 10.1007/978-1-4614-3740-6_10,

Because by assumption $\mathbf{x}^0$ is non-singular, we have

$$\det\left(\frac{\partial^2 W(\mathbf{x},\boldsymbol{\lambda})}{\partial \mathbf{x}^2}\right)_{(\mathbf{x}^0,\boldsymbol{\lambda}^0)} \neq 0. \tag{10.4}$$

Function (10.3) is a linear combination of the first integrals of system (10.1), and after its substitution into (10.1) we obtain

$$\frac{\mathrm{d}W}{\mathrm{d}t} = \left(\frac{\partial W}{\partial \mathbf{x}}\right)^{\mathrm{T}} \frac{\mathrm{d}\mathbf{x}}{\mathrm{d}t} = \left(\frac{\partial W}{\partial \mathbf{x}}\right)^{\mathrm{T}} \mathbf{f}(\mathbf{x}) \equiv 0, \tag{10.5}$$

which means that W is also a first integral of (10.1), where $1, \lambda_1, \lambda_2, \ldots, \lambda_M$ are constant coefficients. The second derivative of (10.5) with respect to $\mathbf{x}$ has the form

$$\left(\frac{\partial^2 W}{\partial \mathbf{x}^2}\right)^{\mathrm{T}} \mathbf{f}(\mathbf{x}) + \frac{\partial \mathbf{f}}{\partial \mathbf{x}}\left(\frac{\partial W}{\partial \mathbf{x}}\right) \equiv 0. \tag{10.6}$$

Stationary values $\mathbf{x} = \mathbf{x}^0$ and $\boldsymbol{\lambda} = \boldsymbol{\lambda}^0$ are determined from algebraic equations of the form

$$\frac{\partial W}{\partial \mathbf{x}} = \mathbf{0}, \qquad \frac{\partial W}{\partial \boldsymbol{\lambda}} = \mathbf{0}. \tag{10.7}$$

Taking (10.7) into account in (10.6) we obtain

$$\left(\frac{\partial^2 W}{\partial \mathbf{x}^2}\right)_{\mathbf{x}^0} \mathbf{f}(\mathbf{x})^0 = \mathbf{0}. \tag{10.8}$$

Taking into account condition (10.4) we obtain

$$\mathbf{f}(\mathbf{x})^0 = \mathbf{0}, \tag{10.9}$$

which we set out to demonstrated.

The classic approach [2] consists in the determination of solutions of non-linear algebraic equations (10.9) in order to determine $\mathbf{x}^0$. Provided that we know the first integrals of the analyzed system, the problem becomes simpler because we are dealing with quadratic forms (10.7), which often may be just linear algebraic equations.

Stationary solutions $U_0(\mathbf{x}) = C_0$ depend on the values of the remaining fixed constants $\mathbf{C}$, constituting in this way a family of solutions $\mathbf{x} = \mathbf{x}^0(\mathbf{C})$ in the space $R^N \times R^M$ $(\mathbf{x} \in R^N, \mathbf{C} \in R^M)$.

Theorem 10.2. *If one of the first integrals of system* (10.1) *has its local extremum (minimum or maximum) for the fixed values of the remaining first integrals at a certain point* $\mathbf{x}^0$, *then the solution* $\mathbf{x} \equiv \mathbf{x}^0$ *is a stable stationary motion of dynamical system* (10.1).

Proof. Let the integral $U_0(\mathbf{x}) = C_0$ attain the extremum C_0^0 for the fixed values $\mathbf{C}^0$ of the remaining first integrals at point $\mathbf{x}_0$.

Let us consider the solution $\mathbf{x}^0(t) = \mathbf{x}(\mathbf{x}^0, t)$ of system (10.1), for $U_0(\mathbf{x}^0(t)) \equiv U_0(\mathbf{x}^0) = C_0^0$, $U(\mathbf{x}^0(t)) \equiv U(\mathbf{x}^0) = \mathbf{C}^0$. The quantity C_0^0 is the extremum value of the function $U_0(\mathbf{x})$ for the fixed values $\mathbf{C}^0$ of the first integrals $U(\mathbf{x}) = \mathbf{C}$ attained at point $\mathbf{x}^0$. This means that $\mathbf{x}^0(t) \equiv \mathbf{x}^0$ is the stationary motion of the analyzed dynamical system.

Let us now introduce a quadratic form as follows:

$$V(\mathbf{x}) = (U_0(\mathbf{x})) - C_0^0)^2 + (\mathbf{U}(\mathbf{x}) - \mathbf{C}^0)^T(\mathbf{U}(\mathbf{x}) - \mathbf{C}^0). \tag{10.10}$$

The function $\frac{\mathrm{d}V}{\mathrm{d}t} \equiv 0$; moreover, $V(\mathbf{x}_0) = 0$ and $V(\mathbf{x}) > 0$ for $\mathbf{x} - \mathbf{x}^0 \neq 0$. Since if in the neighborhood of point $\mathbf{x}^0$ variables $\mathbf{x}(t)$ satisfy the condition $U(\mathbf{x}(t)) = \mathbf{C}^0$ for $\mathbf{x} \neq \mathbf{x}_0$, then $V(\mathbf{x}) = (U_0(\mathbf{x}) - C_0^0) > 0$. In turn, for $\mathbf{x} = \mathbf{x}_0$ we have $V(\mathbf{x}) = (\mathbf{U}(\mathbf{x}) - \mathbf{C}^0)^{\mathrm{T}}(\mathbf{U}(\mathbf{x}) - \mathbf{C}^0) > 0$. This means that $V(\mathbf{x})$ is a positive-definite function ($V(\mathbf{x}) > 0$) and $\frac{\mathrm{d}V}{\mathrm{d}t} \equiv 0$ in the neighborhood of $\mathbf{x}^0$. According to the Lyapunov[1] stability theorem [2], the stationary motion $\mathbf{x}(t) \equiv \mathbf{x}^0$ is stable. □

The conditions for the function $U_0(\mathbf{x})$ to attain its extremum at point $\mathbf{x}^0$ for additional conditions $U(\mathbf{x}) = \mathbf{C}$ are equivalent to the determinacy with regard to the sign (positive- or negative-definite) of a quadratic form of the form

$$\delta^2 W = \frac{1}{2}(\mathbf{x} - \mathbf{x}^0)^T \left(\frac{\partial^2 W}{\partial \mathbf{x}^2}\right)_{\mathbf{x}^0} (\mathbf{x} - \mathbf{x}^0) \tag{10.11}$$

on the linear manifold

$$\delta \mathbf{U} = \left(\frac{\partial \mathbf{U}}{\partial \mathbf{x}}\right)_{\mathbf{x}^0} (\mathbf{x} - \mathbf{x}^0) = \mathbf{0}. \tag{10.12}$$

Let us consider the equation of disturbances in the neighborhood $\mathbf{x} \equiv \mathbf{x}^0$ of the form

$$\delta \dot{\mathbf{x}} = \left[\frac{\partial \mathbf{f}}{\partial \mathbf{x}}\right]_{\mathbf{x}^0} \delta \mathbf{x}, \tag{10.13}$$

that is, in the Taylor series expansion about $\mathbf{x}^0$ we retain only linear terms. Equation of disturbances (10.13) possesses an integral of the form

$$\delta^2 W = \frac{1}{2} \delta \mathbf{x}^{\mathrm{T}} \left[\frac{\partial^2 W}{\partial \mathbf{x}^2}\right]_{\mathbf{x}^0} \delta \mathbf{x} = \gamma_0 \tag{10.14}$$

---

[1] Alexandr Lyapunov (1857–1918), Russian mathematician who made a great contribution to the theory of stability, working mainly in Kharkiv and Saint Petersburg.

and additionally M linear integrals of the form

$$\delta \mathbf{U} = \left[\frac{\partial U}{\partial \mathbf{x}}\right]_{\mathbf{x}^0} \delta \mathbf{x} = \boldsymbol{\gamma}. \tag{10.15}$$

A characteristic equation corresponding to the matrix in (10.13) of the form

$$\det\left[\sigma \mathbf{E}_n - \left(\frac{\partial \mathbf{f}}{\partial \mathbf{x}}\right)_{\mathbf{x}^0}\right] = 0, \tag{10.16}$$

which is a polynomial of the Nth order, has exactly M zero roots (eigenvalues) $\sigma_0 = 0, \sigma_1 = 0, \ldots, \sigma_M = 0$ if $\operatorname{rank}(\frac{\partial \mathbf{f}}{\partial \mathbf{x}})_{x^0} = N - M$.

Theorem 10.3. *The stationary motion* $\mathbf{x} \equiv \mathbf{x}^0$ *of dynamical system* (10.1) *is unstable if the determinant*

$$(-1)^M \det\begin{bmatrix} [\mathbf{0}_{M\times N}], & \left[\frac{\partial \mathbf{U}}{\partial \mathbf{x}}\right] \\ \left[\frac{\partial \mathbf{U}}{\partial \mathbf{x}}\right]^T, & \left[\frac{\partial^2 W}{\partial \mathbf{x}^2}\right] \end{bmatrix}_{\mathbf{0}^0} < 0 \tag{10.17}$$

and the rank of the matrix $[\frac{\partial \mathbf{f}}{\partial \mathbf{x}}]_{\mathbf{x}^0}$ *is equal to* $N - M$, *that is,*

$$\operatorname{rank}\left[\frac{\partial \mathbf{f}}{\partial \mathbf{x}}\right]_{\mathbf{x}^0} = N - M.$$

The proof of the preceding theorem is given in [1].

Example 10.1. Investigate the stability of the stationary motions of a rigid body about a fixed point (Sect. 9.2) in the Euler case. Equations of motion (9.29) have two first integrals, (9.32) and (9.34), of the form

$$\begin{aligned} T &= \frac{1}{2}\left(I_1\omega_1'^2 + I_2\omega_2'^2 + I_3\omega_3'^2\right) = T^* \equiv \text{const}, \\ K_0^2 &= I_1\omega_1'^2 + I_2\omega_2'^2 + I_3\omega_3'^2 = K_0^{*2} \equiv \text{const}, \end{aligned} \tag{10.1.1}$$

and we assume $I_3 > I_2 > I_1$.

In this case function (10.3) takes the form

$$W(\boldsymbol{\omega}', \lambda) = T - \frac{1}{2}\lambda(K_0^2 - K_0^{*2}). \tag{10.1.2}$$

The preceding function is stationary if the following conditions are satisfied:

$$\frac{\partial W}{\partial \omega_1'} = I_1\omega_1'(1-\lambda I_1) = 0, \qquad \frac{\partial W}{\partial \omega_2'} = I_2\omega_2'(1-\lambda I_2) = 0,$$
$$\frac{\partial W}{\partial \omega_3'} = I_3\omega_3'(1-\lambda I_3) = 0, \qquad K_0^2 = K_0^{*2} \equiv \text{const.} \tag{10.1.3}$$

System (10.1.3) has the following solutions:

$$\omega_1' = \omega_2' = \omega_3' = 0, \qquad \lambda \in R \quad \text{and} \quad K_0^* = 0, \tag{10.1.4}$$
$$\omega_1' = \omega, \quad \omega_2' = \omega_3' = 0, \qquad \lambda = I_1^{-1} \quad \text{and} \quad K_0^{*2} = \omega^2 I_1^2, \tag{10.1.5}$$
$$\omega_1' = \omega_3' = 0, \quad \omega_2' = \omega, \qquad \lambda = I_2^{-1} \quad \text{and} \quad K_0^{*2} = \omega^2 I_2^2, \tag{10.1.6}$$
$$\omega_1' = \omega_2' = 0, \quad \omega_3' = \omega, \qquad \lambda = I_3^{-1} \quad \text{and} \quad K_0^{*2} = \omega^2 I_3^2. \tag{10.1.7}$$

Solution (10.1.4) corresponds to a state of equilibrium, and it is stable with respect to ω_1', ω_2', and ω_3', because in this case the energy integral T [see (10.1.1)] attains its global minimum.

The three remaining stationary solutions correspond to body rotations about principal centroidal axes of inertia. Let us conduct an analysis of solution (10.1.5). According to (10.11), and taking into account (10.1.3), we have

$$\delta^2 W = \frac{1}{2}[\delta\omega_1', \delta\omega_2', \delta\omega_3']\left[\frac{\partial^2 W}{\partial \boldsymbol{\omega}^2}\right]_{\boldsymbol{\omega}^0}\begin{bmatrix}\delta\omega_1'\\ \delta\omega_2'\\ \delta\omega_3'\end{bmatrix},$$
$$\left[\frac{\partial^2 W}{\partial \boldsymbol{\omega}^2}\right] = \begin{bmatrix} I_1(1-\lambda I_1) & 0 & 0\\ 0 & I_2(1-\lambda I_2) & 0\\ 0 & 0 & I_3(1-\lambda I_3)\end{bmatrix}, \tag{10.1.8}$$

that is,

$$\delta^2 W = \frac{1}{2}\left[I_2\left(1-\frac{I_2}{I_1}\right)(\delta\omega_2')^2 + I_3\left(1-\frac{I_3}{I_1}\right)(\delta\omega_3')^2\right]. \tag{10.1.9}$$

Linear manifold (10.12) in this case takes the form

$$\delta(K_0^2) = \left[\frac{\partial K_0^2}{\partial \boldsymbol{\omega}}\right]_{\boldsymbol{\omega}^0}\delta\boldsymbol{\omega}$$
$$= \begin{bmatrix} 2I_1^2\omega_1' & 0 & 0\\ 0 & 2I_2^2\omega_2' & 0\\ 0 & 0 & 2I_3^2\omega_3'\end{bmatrix}_{\boldsymbol{\omega}^0}\begin{bmatrix}\delta\omega_1'\\ \delta\omega_2'\\ \delta\omega_3'\end{bmatrix} = 2I_1^2\omega\delta\omega_1 = 0, \tag{10.1.10}$$

hence for $\omega \neq 0$ we obtain $\delta\omega_1 = 0$.

Because by assumption $I_3 > I_2 > I_1$, we have $\delta^2 W < 0$. It follows that for $K_0^2 = \text{const}$ the function T on solution (10.1.5) attains its maximum.

In the case of solution (10.1.7) we have

$$\delta^2 W = \frac{1}{2}\left[I_1\left(1 - \frac{I_1}{I_3}\right)(\delta\omega_1)^2 + I_2\left(1 - \frac{I_2}{I_3}\right)(\delta\omega_2)^2\right],$$

$$\delta(K_0^2) = 2I_3^2\omega\delta\omega_3 = 0. \tag{10.1.11}$$

In this case $\delta^2 W > 0$ and for $K_0^2 = \text{const}$ the function T on solution (10.1.7) attains its minimum.

From the preceding analysis it follows that, by Theorem 10.2, a rigid body with fixed mass center, after the introduction of the initial rotations about the smallest or largest axis of inertia, rotates infinitely long, even after small disturbances of motion, because its stationary rotations about these axes are stable.

In the end, let us consider stationary rotational motion about the average axis of an ellipsoid, that is, solution (10.1.6).

In this case we have

$$\delta^2 W = \frac{1}{2}\left[I_1\left(1 - \frac{I_1}{I_2}\right)(\delta\omega_1)^2 + I_3\left(1 - \frac{I_3}{I_2}\right)(\delta\omega_3)^2\right],$$

$$\delta(K_0^2) = 2I_2^2\omega\delta\omega_2 = 0. \tag{10.1.12}$$

The first term of the first equation is positive, whereas the second is negative, and in order to estimate the stability we will make use of Theorem 10.3. The matrix corresponding to the quadratic form $\delta^2 W$ in this case takes the form

$$\left[\frac{\partial^2 W}{\partial \boldsymbol{\omega}^2}\right] = \begin{bmatrix} I_1\left(1 - \frac{I_1}{I_2}\right) & 0 \\ 0 & I_3\left(1 - \frac{I_3}{I_2}\right) \end{bmatrix}. \tag{10.1.13}$$

The determinant of the preceding matrix is equal to

$$\begin{vmatrix} I_1\left(1 - \frac{I_1}{I_2}\right) & 0 \\ 0 & I_3\left(1 - \frac{I_3}{I_2}\right) \end{vmatrix} = I_1 I_3\left(1 - \frac{I_1}{I_2}\right)\left(1 - \frac{I_3}{I_2}\right) < 0. \tag{10.1.14}$$

According to Theorem 10.3 one should now determine the rank of the matrix $[\frac{\partial \mathbf{f}}{\partial \boldsymbol{\omega}}]_{\omega^0}$.

After the linearization of equations of motion (9.29) we obtain

$$I_1(\delta\dot{\omega}_1') + (I_3 - I_2)\omega_3'\delta\omega_2' + (I_3 - I_2)\omega_2'\delta\omega_3' = 0,$$
$$I_2(\delta\dot{\omega}_2') + (I_1 - I_3)\omega_1'\delta\omega_3' + (I_1 - I_3)\omega_3'\delta\omega_1' = 0,$$
$$I_3(\delta\dot{\omega}_3') + (I_2 - I_1)\omega_1'\delta\omega_2' + (I_2 - I_1)\omega_2'\delta\omega_1' = 0, \tag{10.1.15}$$

hence after taking into account (10.1.6) we have

$$\begin{aligned}
&I_1(\delta\dot{\omega}_1') + (I_3 - I_2)\omega\delta\omega_3' = 0,\\
&I_2(\delta\dot{\omega}_2') = 0,\\
&I_3(\delta\dot{\omega}_3') + (I_2 - I_1)\omega\delta\omega_1' = 0,
\end{aligned} \tag{10.1.16}$$

that is, the matrix

$$\left[\frac{\partial \mathbf{f}}{\partial \boldsymbol{\omega}}\right]_{\omega^0} = \begin{bmatrix} 0 & 0 & \frac{I_3-I_2}{I_1}\omega \\ 0 & 0 & 0 \\ \frac{I_2-I_1}{I_3}\omega & 0 & 0 \end{bmatrix}. \tag{10.1.17}$$

The rank of this matrix is equal to 2, and because $M = 1$ and $N = 3$, the second condition of Theorem 10.3 is also satisfied. Eventually, we come to the conclusion that rotation about the average axis of an inertia ellipsoid is unstable. □

10.2 Invariant Sets of Conservative Systems and Their Stability

We will start our calculations with the definition of an invariant set.

Definition 10.1. The set **M** is called an *invariant set (positive invariant)* of dynamical system (10.1) if for $\mathbf{x}^0 \in M$ (start of motion) we have $\mathbf{x}(\mathbf{x}^0, t) \in \mathbf{M}$ (motion of the system) for all $t > 0$ and all $\mathbf{x}^0 \in M$.

Because in our case we consider systems that satisfy Theorem 10.1, we say that the manifold **M** generates the functions $U_0(\mathbf{x})$ for the previously mentioned additional conditions for the remaining first integrals to be constant, that is, $\mathbf{U}(\mathbf{x}) = \mathbf{C} \equiv const$:

1. A non-singular stationary value on the set **M**, where

$$\delta U_0|_{\delta\mathbf{U}=\mathbf{0}} = 0, \qquad \delta^2 U_0|_{\delta\mathbf{U}=\mathbf{0}} \neq 0. \tag{10.18}$$

2. A local minimum (maximum) value if $U_0(\mathbf{M}) = const$ and there exists $\delta > 0$ such that for any **x** satisfying the condition

$$0 < \mathrm{dist}(\mathbf{x}, \mathbf{M}) < \delta, \qquad \mathbf{U}(\mathbf{x}) = \mathbf{C}, \tag{10.19}$$

also the following condition is satisfied:

$$\mathbf{U}_0(\mathbf{x}) > (< 0)\mathbf{U}_0(\mathbf{M}),$$

where

$$\mathrm{dist}(\mathbf{x}, \mathbf{M}) = \min_{\mathbf{x}^* \in \mathbf{M}} ||\mathbf{x} - \mathbf{x}^*||. \tag{10.20}$$

Theorem 10.4. *If a certain set* $\mathbf{M}_0$ *generates a non-singular stationary value of one of the first integrals of system* (10.1) *for fixed values of the remaining first integrals, then* $\mathbf{M}_0$ *is an invariant set of dynamical system* (10.1).

The proof of the preceding theorem is given in [1], and it is carried out in an analogous way to the proof of Theorem 10.1 since if $\dim \mathbf{M}_0 = 0$, then the invariant manifold $\mathbf{M}_0$ reduces to the points (positions) of equilibrium of the considered dynamical system (10.1). If $\dim \mathbf{M}_0 > 0$, then the stationary value generated by this set corresponds to stationary motions of system (10.1).

In the general case the manifold $\mathbf{M}_0$ is filled up with stationary motions of system (10.1), which can be periodic, quasiperiodic, or chaotic.

Theorem 10.5. *If a compact set* $\mathbf{M}_0$ *generates a local non-singular stationary extremum (minimum or maximum) value for fixed values of the remaining first integrals of system* (10.1), *then* $\mathbf{M}_0$ *is a stable invariant set of dynamical system* (10.1).

The proof of this theorem is provided in [1]. From the theorem it follows that the open set $\mathbf{M}_0$ is invariant (one should prove this), and subsequently it should be proved that this invariant set is stable. Let us emphasize that the proof of the theorem does not require the assumption of differentiability of first integrals. Let us note that stationary motions $\mathbf{x}^0(\mathbf{x}^0, t)$ lying on the invariant set $\mathbf{M}_0$ are stable in reference to some of the variables that characterize the distance of the disturbed motion from the set $\mathbf{M}_0$, but in general these motions are unstable with respect to disturbances $\mathbf{x} - \mathbf{x}^0(\mathbf{x}^0, t)$.

Example 10.2. Determine the invariant set corresponding to a stationary motion of a rigid body about a fixed point in the Euler case about the average axis of an inertia ellipsoid.

As was already shown in Example 10.1, the rotation of a body about the average axis of an ellipsoid of inertia is unstable, and according to (10.1.1) in this case we have $T = \frac{1}{2} I_2 \omega^2$ and $K_0^2 = I_2^2 \omega^2$.

Equations of motion (9.29) are satisfied for the invariant set determined by the equations

$$\begin{aligned} I_1 \omega_1'^2 + I_2 \omega_2'^2 + I_3 \omega_3'^2 &= I_2 \omega^2, \\ I_1^2 \omega_1'^2 + I_2^2 \omega_2'^2 + I_3^2 \omega_3'^2 &= I_2^2 \omega^2, \end{aligned} \tag{$*$}$$

where $\omega \in R$. From the preceding equations we determine

$$\begin{aligned} \omega_1'^2 &= \frac{I_2(I_2 - I_3)}{I_1(I_1 - I_3)}(\omega_2^2 - \omega_2'^2), \\ \omega_3'^2 &= \frac{I_2(I_1 - I_2)}{I_3(I_1 - I_3)}(\omega_2^2 - \omega_2'^2), \end{aligned} \tag{$**$}$$

and after substitution into the second of the equations of motion (9.29) we obtain

$$\dot{\omega}_2' = \pm\sqrt{\frac{(I_2 - I_3)(I_1 - I_2)}{I_3 I_3}}(\omega^2 - \omega_2'^2). \tag{10.21}$$

Equation (∗∗) describes the stationary motion of a rigid body about a fixed point in the Euler case about the average axis of an inertia ellipsoid on the manifold given by (∗). □

References

1. A.V. Karapetian, *Stationary Motion Stability* (Editorial URSS, Moscow, 1998) (in Russian)
2. J. Awrejcewicz, *Oscillations of Lumped Deterministic Systems* (WNT, Warsaw, 1996) (in Polish)

Chapter 11
Geometric Dynamics

11.1 Introduction

A classical approach to the dynamics of Hamiltonian systems (or dynamical systems in general) is based on the notion of a *phase space* (Chaps. 2 and 3). It turns out that the phase space of a Hamiltonian system possesses certain geometric properties [1]. One of the first scientists to notice that was H. Poincaré[1] [2]. By exploiting the properties of the phase space of Hamiltonian systems and Hamilton's equations themselves, it is possible to formulate mechanics in the language of symplectic geometry [1, 3]. The phase space plays here the role of a manifold on which is defined a certain quantity called a *symplectic form*, which relates mechanics to geometry. In turn, the subject of analysis of this chapter is the geometric approach to the dynamics of mechanical systems, henceforth called *geometric dynamics*. As the name implies, it deals with the geometric formulation of dynamics with aid of the formalism of differential geometry. It turns out that the dynamics of Hamiltonian systems can be formulated in the language of geometry of a Riemann[2]–Finsler[3] space [4, 5]. The object of investigation of the present chapter is the geometric description of dynamics in a Riemannian space. In order to use the aforementioned formalism of a Riemannian space we have to be able to describe a differentiable manifold and define on it a metric tensor [6, 7].

In other words, we need to have at our disposal the Riemannian space "obtained from the investigated dynamical system." As far as the potential manifolds are concerned we can choose from several spaces naturally occurring in dynamics such as, for example, a configuration space, an extended configuration space, a

[1]Henri Poincaré (1854–1912), a great French physicist and mathematician being i.a. a pioneer of chaos theory.

[2]Bernhard Riemann (1826–1866), influential German mathematician who made essential contributions to analysis and differential geometry.

[3]Paul Finsler (1894–1970), German and Swiss mathematician.

J. Awrejcewicz, *Classical Mechanics: Dynamics*, Advances in Mechanics and Mathematics 29, DOI 10.1007/978-1-4614-3740-6_11,

configuration time-space, an extended configuration time-space, a tangent bundle of the configuration space. All those manifolds can be equipped with a *metric tensor*, that is, a *metric*.[4] The following ways of choosing a metric exist:

1. A configuration space Q + the Jacobi metric.
2. An extended configuration space $Q \times R$ + the Eisenhart metric.
3. A configuration time-space $Q \times R$ + the Finsler metric.
4. An extended configuration time-space $Q \times R^2$ + the Eisenhart metric.
5. A tangent bundle of a configuration space TQ + the Sasaki metric.

The choice of one of the preceding alternative descriptions is a matter of convenience and depends on the class of the dynamical system under investigation. In other words, not every dynamical system can be described (geometrized) in each of the previously mentioned cases. In the case of the configuration manifold with the Jacobi metric we can describe only Hamiltonian systems with an energy integral. In the case of the extended configuration time-space with the Eisenhart metric [8] we can describe non-autonomous systems. On the other hand, the Finsler geometry allows for the description of systems with a velocity-dependent potential, which, in turn, is not possible within Riemannian geometry. In the present chapter we will consider one of the aforementioned descriptions, namely, the configuration space with the Jacobi metric. The class of systems that can be described within the framework of this formalism are systems with Lagrangians of the form

$$L = \frac{1}{2} a_{ij}(\mathbf{q}) \dot{q}^i \dot{q}^j - V(\mathbf{q}), \tag{11.1}$$

where $a_{ij}(\mathbf{q})$ are the components of a kinetic energy matrix. The essence of geometric dynamics is the fact that the motion of the whole system can be identified with the motion of a certain virtual point along a geodesic in a Riemannian space. Therefore, we have to somehow associate a metric tensor with the dynamics of the investigated dynamical system, which is one of the fundamental problems of the geometric approach. A solution of this problem[5] is derived from the possibility of a variational formulation of many mechanical problems. This is possible because geodesic equations can be obtained also as a result of the variation of a certain expression A that is the length of the geodesic between two points. In other words, the length of arc of the geodesic has to assume a stationary value, which is expressed by the equation

$$\delta A = \int_{P_1}^{P_2} \mathrm{d}s = 0. \tag{11.2}$$

This means that from all paths connecting points P_1 and P_2 we choose that one for which (11.2) holds true. An elementary square whose length is equal to the

---

[4]We will often use both terms in an interchangeable way.

[5]We emphasize that it is not the only way to obtain the metric tensor.

length between two points lying infinitesimally close to each other is defined in a Riemannian space as

$$ds^2 = g_{ij}(\mathbf{q})dq^i dq^j, \tag{11.3}$$

where $g_{ij}(\mathbf{q})$ are the components of a metric tensor that, in general, can depend on the coordinates $\mathbf{q}$ [9]. Quantity (11.3) can be written in a different way using the matrix form of a metric tensor $\mathbf{g}$, namely,

$$ds^2 = \begin{bmatrix} dq^1 \\ dq^2 \\ \vdots \\ dq^N \end{bmatrix}^T \begin{bmatrix} g_{11} & g_{12} & \cdots & g_{1N} \\ g_{21} & g_{22} & \cdots & g_{2N} \\ \vdots & \vdots & \ddots & \vdots \\ g_{N1} & g_{N2} & \cdots & g_{NN} \end{bmatrix} \begin{bmatrix} dq^1 \\ dq^2 \\ \vdots \\ dq^N \end{bmatrix}, \tag{11.4}$$

where the symmetry of a metric tensor was used ($g_{ij}(\mathbf{q}) = g_{ji}(\mathbf{q})$). Now, substituting the previously described length between the aforementioned two points, that is, (11.3), into (11.2), we obtain

$$\delta \int_{P_1}^{P_2} ds = \delta \int_{P_1}^{P_2} \sqrt{g_{ij} dq^i dq^j} = \delta \int_{u_1}^{u_2} \sqrt{w} du, \tag{11.5}$$

where

$$w = g_{ij} p^i p^j, \quad p^i = \frac{dq^i}{ds}.$$

Moving the variation inside the integral and then carrying out the variation of the integrand we obtain

$$\int_{u_1}^{u_2} \left(\frac{\partial \sqrt{w}}{\partial q^r} \delta q^r + \frac{\partial \sqrt{w}}{\partial p^r} \delta p^r \right) du = 0. \tag{11.6}$$

Because

$$\delta p^r = \delta \left(\frac{dq^r}{du} \right) = \frac{d}{du} (\delta q^r),$$

integrating by parts the second term in (11.6) we obtain

$$\int_{u_1}^{u_2} \frac{\partial \sqrt{w}}{\partial p^r} \delta p^r du = \frac{\partial \sqrt{w}}{\partial p^r} \delta q^r \Big|_{u_1}^{u_2} - \int_{u_1}^{u_2} \frac{d}{du} \left(\frac{\partial \sqrt{w}}{\partial p^r} \right) \delta q^r du. \tag{11.7}$$

Because the variations δq^r at the ends of the integration interval vanish, we have

$$\int_{u_1}^{u_2} \frac{\partial \sqrt{w}}{\partial p^r} \delta p^r du = - \int_{u_1}^{u_2} \frac{d}{du} \left(\frac{\partial \sqrt{w}}{\partial p^r} \right) \delta q^r du. \tag{11.8}$$

Substituting the obtained result into (11.6) we find

$$\int_{u_1}^{u_2} \left(\frac{\partial \sqrt{w}}{\partial q^r} - \frac{\mathrm{d}}{\mathrm{d}u} \left(\frac{\partial \sqrt{w}}{\partial p^r} \right) \right) \delta q^r \, \mathrm{d}u = 0. \tag{11.9}$$

Because in the integration interval variations δq^r are arbitrary, the preceding integral vanishes when the integrand becomes zero [4]. Hence we obtain Euler–Lagrange equations

$$\frac{\mathrm{d}}{\mathrm{d}u} \left(\frac{\partial \sqrt{w}}{\partial p^r} \right) - \frac{\partial \sqrt{w}}{\partial q^r} = 0. \tag{11.10}$$

Carrying out the differentiation in the preceding expression we obtain

$$\begin{aligned} \frac{\mathrm{d}}{\mathrm{d}u} \left(\frac{\partial \sqrt{w}}{\partial p^r} \right) - \frac{\partial \sqrt{w}}{\partial q^r} &= \frac{\mathrm{d}}{\mathrm{d}u} \left(\frac{1}{2\sqrt{w}} \frac{\partial w}{\partial p^r} \right) - \frac{1}{2\sqrt{w}} \frac{\partial w}{\partial q^r} \\ &= \frac{\mathrm{d}}{\mathrm{d}u} \left(\frac{1}{2\sqrt{w}} \right) \frac{\partial w}{\partial p^r} + \frac{1}{2\sqrt{w}} \frac{\mathrm{d}}{\mathrm{d}u} \left(\frac{\partial w}{\partial p^r} \right) - \frac{1}{2\sqrt{w}} \frac{\partial w}{\partial q^r} \\ &= -\frac{1}{4w^{3/2}} \frac{dw}{du} \frac{\partial w}{\partial p^r} + \frac{1}{2\sqrt{w}} \frac{\mathrm{d}}{\mathrm{d}u} \left(\frac{\partial w}{\partial p^r} \right) - \frac{1}{2\sqrt{w}} \frac{\partial w}{\partial q^r} = 0, \end{aligned} \tag{11.11}$$

and then, multiplying through by $2\sqrt{w}$,

$$\frac{\mathrm{d}}{\mathrm{d}u} \left(\frac{\partial \sqrt{w}}{\partial p^r} \right) - \frac{\partial \sqrt{w}}{\partial q^r} = \frac{1}{2w} \frac{\mathrm{d}w}{\mathrm{d}u} \frac{\partial w}{\partial p^r}. \tag{11.12}$$

Because the parameter u was arbitrary, we can set $u = s$, that is, now we take as the parameter the length of arc of a geodesic. Then, using relation (11.5) we find

$$w = g_{ij} p^i p^j = g_{ij} \frac{\mathrm{d}q^i}{\mathrm{d}s} \frac{\mathrm{d}q^j}{\mathrm{d}s} = 1, \tag{11.13}$$

hence

$$\frac{\mathrm{d}w}{\mathrm{d}u} = 0.$$

Equation (11.12) then takes the form

$$\frac{\mathrm{d}}{\mathrm{d}u} \left(\frac{\partial \sqrt{w}}{\partial p^r} \right) - \frac{\partial \sqrt{w}}{\partial q^r} = 0. \tag{11.14}$$

Because

$$\frac{\partial w}{\partial p^r} = g_{ij} \delta^i_r p^j + g_{ij} \delta^j_r p^i = 2 g_{ir} p^i, \tag{11.15}$$

(11.14) takes the form

$$\frac{\mathrm{d}}{\mathrm{d}s}\left(2g_{ir}p^i\right) - \frac{\partial g_{ij}}{\partial q^r}p^i p^j = 0. \tag{11.16}$$

Differentiating with respect to s we have

$$2\frac{\partial g_{ir}}{\partial q^j}\frac{\mathrm{d}q^j}{\mathrm{d}s}p^i + 2g_{ir}\frac{\mathrm{d}p^i}{\mathrm{d}s} - \frac{\partial g_{ij}}{\partial q^r}p^i p^j = 0. \tag{11.17}$$

Now, using the definition $p^i = \frac{\mathrm{d}q^i}{\mathrm{d}s}$, we obtain

$$2g_{ir}\frac{\mathrm{d}^2 q^i}{\mathrm{d}s^2} + 2\frac{\partial g_{ir}}{\partial q^j}\frac{\mathrm{d}q^i}{\mathrm{d}s}\frac{\mathrm{d}q^j}{\mathrm{d}s} - \frac{\partial g_{ij}}{\partial q^r}\frac{\mathrm{d}q^i}{\mathrm{d}s}\frac{\mathrm{d}q^j}{\mathrm{d}s} = 0. \tag{11.18}$$

The second term in the preceding equation can be written in the form

$$2\frac{\partial g_{ir}}{\partial q^j}\frac{\mathrm{d}q^i}{\mathrm{d}s}\frac{\mathrm{d}q^j}{\mathrm{d}s} = \frac{\partial g_{ir}}{\partial q^j}\frac{\mathrm{d}q^i}{\mathrm{d}s}\frac{\mathrm{d}q^j}{\mathrm{d}s} + \frac{\partial g_{jr}}{\partial q^i}\frac{\mathrm{d}q^i}{\mathrm{d}s}\frac{\mathrm{d}q^j}{\mathrm{d}s}. \tag{11.19}$$

This is possible because the term $\frac{\partial g_{ir}}{\partial q^j}$ is summed with the term $\frac{\mathrm{d}q^i}{\mathrm{d}s}\frac{\mathrm{d}q^j}{\mathrm{d}s}$, which is symmetrical with respect to indices i and j, and thus (11.19) represents a symmetrization of the expression $\frac{\partial g_{ir}}{\partial q^j}$. Using this fact in relation (11.18) we obtain

$$2g_{ir}\frac{\mathrm{d}^2 q^i}{\mathrm{d}s^2} + \left(\frac{\partial g_{ir}}{\partial q^j} + \frac{\partial g_{jr}}{\partial q^i} - \frac{\partial g_{ij}}{\partial q^r}\right)\frac{\mathrm{d}q^i}{\mathrm{d}s}\frac{\mathrm{d}q^j}{\mathrm{d}s} = 0. \tag{11.20}$$

Multiplying the preceding equation by g^{nr} and summing with respect to the index r we obtain

$$g^{nr}g_{ir}\frac{\mathrm{d}^2 q^i}{\mathrm{d}s^2} + \frac{1}{2}g^{nr}\left(\frac{\partial g_{ir}}{\partial q^j} + \frac{\partial g_{jr}}{\partial q^i} - \frac{\partial g_{ij}}{\partial q^r}\right)\frac{\mathrm{d}q^i}{\mathrm{d}s}\frac{\mathrm{d}q^j}{\mathrm{d}s} = 0, \tag{11.21}$$

which defines *equations of a geodesic* [10]

$$\frac{\mathrm{d}^2 q^n}{\mathrm{d}s^2} + \Gamma^n_{ij}\frac{\mathrm{d}q^i}{\mathrm{d}s}\frac{\mathrm{d}q^j}{\mathrm{d}s} = 0, \quad n = 1, 2, \ldots, N, \tag{11.22}$$

where

$$\Gamma^n_{ij} := \frac{1}{2}g^{nr}\left(\frac{\partial g_{ir}}{\partial q^j} + \frac{\partial g_{jr}}{\partial q^i} - \frac{\partial g_{ij}}{\partial q^r}\right), \tag{11.23}$$

and N is a dimension of a Riemannian space. Geodesic equations (11.22) are determined completely by a metric tensor $\mathbf{g}$. Thus, having found the relation between the dynamics of the examined system and the metric tensor, we obtain geodesics corresponding to the dynamics of the investigated system.

Equations of motion of a dynamical system can also be obtained using the variational calculus. This is possible by virtue of the variational formulation of Lagrangian mechanics exploiting the principle of least action (see [1, 3, 11])

$$\delta S = \delta \int_{t_1}^{t_2} L(\mathbf{q}, \dot{\mathbf{q}})\,\mathrm{d}t = 0, \tag{11.24}$$

where S is the action functional. Here we are dealing with a situation analogous to the one in the case of a Riemannian space where geodesic equations were obtained by demanding that the variation of a functional of the action A vanishe [see (11.2)]. The difference consists in the fact that now the integration variable is time t. Thus, as before, from the principle of least action and substituting into (11.10) the Lagrangian L instead of $\sqrt{w}$ we obtain Euler–Lagrange equations:

$$\frac{\mathrm{d}}{\mathrm{d}t}\left(\frac{\partial L}{\partial \dot{q}^n}\right) - \frac{\partial L}{\partial q^n} = 0, \quad n = 1, 2, \ldots, N. \tag{11.25}$$

Substituting Lagrangian (11.1) into (11.24) we find

$$\frac{\mathrm{d}}{\mathrm{d}t}\left(a_{in}\dot{q}^i\right) - \frac{1}{2}\frac{\partial a_{ij}}{\partial q^n}\dot{q}^i\dot{q}^j + \frac{\partial V}{\partial q^n} = 0. \tag{11.26}$$

Differentiating, we keep in mind that quantities a^{ij} are dependent on q^i and obtain

$$\frac{\partial a_{in}}{\partial q^j}\dot{q}^i\dot{q}^j + a_{in}\ddot{q}^i - \frac{1}{2}\frac{\partial a_{ij}}{\partial q^n}\dot{q}^i\dot{q}^j + \frac{\partial V}{\partial q^n} = 0. \tag{11.27}$$

Next, we multiply relation (11.27) through by g^{kr}, and because $g^{kn}g_{in} = \delta^k_i$, we obtain the equations of motion

$$\ddot{q}^k + a^{kn}\left(\frac{\partial a_{in}}{\partial q^j} - \frac{1}{2}\frac{\partial a_{ij}}{\partial q^n}\right)\dot{q}^i\dot{q}^j + a^{kn}\frac{\partial V}{\partial q^n} = 0. \tag{11.28}$$

The fundamental demand during the geometrization of dynamical systems is that geodesic equations in the given Riemannian space projected onto the configuration space Q give equations of motion of the considered dynamical system. Whether the geometrization is possible or not is determined by the following theorem, stated in a way appropriate for the case of the Jacobi metric [4].

Theorem 11.1. *For a given conservative dynamical system of total energy E whose Lagrangian has the form*

$$L = \frac{1}{2}a_{ij}(\mathbf{q})\dot{q}^i\dot{q}^j - V(\mathbf{q})$$

it is always possible to find the transformation of the metric

$$g_{ij} = e^{\varphi(\mathbf{q})}a_{ij}, \qquad \text{where} \qquad \varphi(\mathbf{q}) = \ln 2(E - V)$$

such that geodesics in a Riemannian space are the trajectories of the dynamical system.

Thus determining the correspondence between dynamics and a Riemannian space in this way, we can shift the investigation and analysis of dynamical systems to the investigation of the behavior of geodesics in the given Riemannian space, which is precisely the essence of geometric dynamics. The basic tool for such analysis is the Jacobi–Levi-Civita[6] equation (JLC), also known as the *equation of geodesic deviation*. Subsequently, we will derive this equation and analyze it for the presented model of a Riemannian space (a configuration space and the Jacobi metric) and specifically for mechanical systems with two degrees of freedom.

11.2 The Jacobi Metric on Q

In the case of conservative systems in a quite simple way it is possible to obtain a metric that is provided by the kinetic energy itself. Let us consider then a conservative dynamical system with N degrees of freedom described by the following Lagrangian:

$$L = \frac{1}{2}a_{ij}(\mathbf{q})\dot{q}^i\dot{q}^j - V(\mathbf{q}). \tag{11.29}$$

Because our considered mechanical system is conservative, the total energy E is an integral of motion. In that case Hamilton's variational principle [1]

$$\delta\int_{t_1}^{t_2} L(\mathbf{q}, \dot{\mathbf{q}})\mathrm{d}t = 0 \tag{11.30}$$

is reduced to the Maupertuis principle of least action

$$\delta\int_{t_1}^{t_2} 2T\mathrm{d}t = 0, \tag{11.31}$$

[6]Tullio Levi-Civita (1873–1941), Italian mathematician of Jewish origin who investigated celestial mechanics, the three-body problem, and hydrodynamics.

where

$$T = \frac{1}{2} a_{ij}(\mathbf{q}) \dot{q}^i \dot{q}^j$$

is the kinetic energy of the system. If the total energy is the constant of motion, we have $E = T + V$, and substituting $V = E - T$ instead of V into the expression for Lagrangian (11.29) we find $L = 2T - E$. Substituting the quantity obtained in this way into (11.30) we obtain

$$\delta \int_{t_1}^{t_2} (2T - E)\,\mathrm{d}t = \delta \int_{t_1}^{t_2} 2T\mathrm{d}t - \delta \int_{t_1}^{t_2} E\mathrm{d}t = \delta \int_{t_1}^{t_2} 2T\mathrm{d}t = 0. \tag{11.32}$$

Exploiting the analogy between formulas (11.2) and (11.31) we obtain

$$\mathrm{d}s = 2T\mathrm{d}t = 2\,(E - V)\,\mathrm{d}t, \tag{11.33}$$

where we make use of the fact that $E = T + V$. Next, squaring equation (11.33) by sides we obtain

$$\mathrm{d}s^2 = 4\,(E - V)\,T\mathrm{d}t^2. \tag{11.34}$$

Then substituting

$$T = \frac{1}{2} a_{ij}(\mathbf{q}) \dot{q}^i \dot{q}^j$$

into (11.34) we find

$$\begin{aligned} \mathrm{d}s^2 &= 2\,(E - V)\,a_{ij}(\mathbf{q}) \dot{q}^i \dot{q}^j \mathrm{d}t^2 \\ &= 2\,(E - V)\,a_{ij}(\mathbf{q}) \mathrm{d}q^i \mathrm{d}q^j. \end{aligned} \tag{11.35}$$

On the other hand,

$$\mathrm{d}s^2 = g_{ij} \mathrm{d}q^i \mathrm{d}q^j, \tag{11.36}$$

and taking the preceding relation into account we obtain a metric of the form

$$g_{ij} = 2W a_{ij}, \qquad \text{where} \qquad W = E - V. \tag{11.37}$$

The preceding metric is called the *Jacobi metric* and will be denoted by g. The quantity W introduced in formula (11.37) assumes the same values as the kinetic energy T of the investigated system because, by definition, it is the difference of the total and potential energy. However, the adopted formalism does not allow for the explicit introduction of the kinetic energy into the metric because of the occurrence of velocities in T. Therefore, the introduction of T instead of W into (11.37) would imply that the metric is dependent on velocity, and such a case is described not by the geometry of a Riemannian space but by the geometry of a Finsler space, which is not a subject of interest of this chapter.

The considered Riemannian space in this case is a configuration space Q defined as

$$Q = \{q \in R^N : E - V \geq 0\}. \tag{11.38}$$

It is the manifold with a boundary

$$\partial Q = \{q \in R^N : E - V = 0\}. \tag{11.39}$$

Let us note that a metric of form (11.37) is degenerate when $E = V$ (in other words, when the kinetic energy of the system becomes zero).

A configuration space Q of a dynamical system with N degrees of fredom has the structure of a differentiable manifold in which the role of local coordinates is played by generalized coordinates. This manifold, together with a metric $\mathbf{g}$ defined on it, forms the desired Riemannian space $M = (Q, \mathbf{g})$. According to the assumptions of the theory, geodesic equations projected onto a configuration space should lead to the determination of equations of motion. Because in our case a Riemannian manifold is the configuration manifold, geodesic equations should be identical with the equations of motion. In order to obtain these equations we make use of Euler–Lagrange equations for a geodesic Lagrangian,[7] but then *Christoffel*[8] *symbols* have to be found, which is more laborious than the application of Euler–Lagrange equations to a geodesic Lagrangian. The geodesic Lagrangian is obtained from relation (11.36) after "dividing" a linear element by sides by ds. Because we obtained equations of geodesics from (11.14), substituting

$$g_{ij} p^i p^j = g_{ij} \frac{\mathrm{d}q^i}{\mathrm{d}s} \frac{\mathrm{d}q^j}{\mathrm{d}s} = L \tag{11.40}$$

into (11.14), we obtain Euler–Lagrange equations for the geodesic Lagrangian

$$\frac{\mathrm{d}}{\mathrm{d}s}\left(\frac{\partial L}{\partial\left(\frac{dq^k}{\mathrm{d}s}\right)}\right) - \frac{\partial L}{\partial q^k} = 0. \tag{11.41}$$

The quantities occurring in the preceding equations are equal to

$$\frac{\partial L}{\partial\left(\frac{dq^k}{\mathrm{d}s}\right)} = 2g_{ik}\frac{\mathrm{d}q^i}{\mathrm{d}s}, \qquad \frac{\partial L}{\partial q^k} = \frac{\partial g_{ij}}{\partial q^k}\frac{\mathrm{d}q^i}{\mathrm{d}s}\frac{\mathrm{d}q^j}{\mathrm{d}s}. \tag{11.42}$$

---

[7]Obviously, we can obtain geodesic equations from formula (10.1.5).

[8]Elwin Bruno Christoffel (1829–1900), German mathematician and physicist working mainly at the University of Strasbourg.

Differentiating the first of the expressions we obtain

$$\frac{\mathrm{d}}{\mathrm{d}s}\left(\frac{\partial L}{\partial\left(\frac{\mathrm{d}q^k}{\mathrm{d}s}\right)}\right) = 2\frac{\partial g_{ik}}{\partial q^j}\frac{\mathrm{d}q^i}{\mathrm{d}s}\frac{\mathrm{d}q^j}{\mathrm{d}s} + 2g_{ik}\frac{\mathrm{d}^2q^i}{\mathrm{d}s^2}. \tag{11.43}$$

Then, substituting relations (11.42) and (11.43) into (11.41) we find

$$g_{ik}\frac{\mathrm{d}^2q^i}{\mathrm{d}s^2} + \left(\frac{\partial g_{ik}}{\partial q^j} - \frac{1}{2}\frac{\partial g_{ij}}{\partial q^n}\right)\frac{\mathrm{d}q^i}{\mathrm{d}s}\frac{\mathrm{d}q^j}{\mathrm{d}s} = 0. \tag{11.44}$$

Using formula (11.37) we obtain

$$\frac{\partial g_{ik}}{\partial q^j} = 2W\frac{\partial a_{ik}}{\partial q^j} - 2a_{ik}\frac{\partial V}{\partial q^j}. \tag{11.45}$$

Substituting the preceding result into (11.44) and applying formula (11.33) we find

$$\begin{aligned} &a_{ik}\frac{\mathrm{d}^2q^i}{\mathrm{d}s^2} + \frac{a_{ik}}{W}\frac{\mathrm{d}V}{\mathrm{d}t}\frac{dq^i}{\mathrm{d}t} + \left(\frac{\partial a_{ik}}{\partial q^j} - \frac{1}{2}\frac{\partial a_{ij}}{\partial q^n}\right)\frac{\mathrm{d}q^i}{\mathrm{d}t}\frac{\mathrm{d}q^j}{\mathrm{d}t} \\ &- \frac{a_{ik}}{W}\frac{\partial V}{\partial q^j}\frac{\mathrm{d}q^i}{\mathrm{d}t}\frac{\mathrm{d}q^j}{\mathrm{d}t} + \frac{\partial V}{\partial q^k} = 0. \end{aligned}$$

Because

$$\frac{\mathrm{d}V}{\mathrm{d}t} = \frac{\partial V}{\partial q^j}\frac{\mathrm{d}q^j}{\mathrm{d}t},$$

we obtain

$$\ddot{q}^n + a^{kn}\left(\frac{\partial a_{ik}}{\partial q^j} - \frac{1}{2}\frac{\partial a_{ij}}{\partial q^k}\right)\dot{q}^i\dot{q}^j + a^{kn}\frac{\partial V}{\partial q^n} = 0, \tag{11.46}$$

where

$$\dot{q}^j = \frac{\mathrm{d}q^j}{\mathrm{d}t}.$$

These are equations of motion (11.28). Thus the metric g is properly defined. Here the Riemannian space has the same dimension as the configuration space of the system. This dimension is equal to the number of degrees of freedom of the investigated mechanical system.

11.3 The Jacobi–Levi-Civita Equation

As mentioned previously, the fundamental tool of geometric dynamics is the equation of geodesic deviation called the *Jacobi–Levi-Civita equation* or the *JLC equation* for short. In order to derive this equation let us consider a one-parameter family of geodesics V_2 in N-dimensional space V_N. Let s (the length of a geodesic) be the parameter varying along each of the geodesic lines of this family. By u let us denote a parameter that is constant for each of the geodesic lines and changes while switching between geodesics. In other words, the parameter u numbers geodesics in the family

$$V_2 = \{q \in V_N : \gamma_u(s) = q\}, \tag{11.47}$$

where γ is a geodesic. The family V_2 forms a two-dimensional space. Let us denote by $T_p V_2$ a space tangent to V_2 at point p. This space is spanned by the vectors

$$\left\{ \frac{\partial}{\partial u}, \frac{\partial}{\partial s} \right\}. \tag{11.48}$$

Let us introduce the following designations:

$$\boldsymbol{\partial}_s = \frac{\partial}{\partial s}, \quad \boldsymbol{\partial}_u = \frac{\partial}{\partial u}, \tag{11.49}$$

where vector $\boldsymbol{\partial}_u$ is called a *Jacobi vector*. Because vector $\boldsymbol{\partial}_s$ is tangent to a geodesic, from the definition of the geodesic [7] we obtain

$$\nabla_s \boldsymbol{\partial}_s = 0. \tag{11.50}$$

Moreover, the following equality holds true:

$$\nabla_s \boldsymbol{\partial}_u = \nabla_u \boldsymbol{\partial}_s. \tag{11.51}$$

Because the preceding vectors belong to the space V_N, they can be expanded in terms of the basis of this space:

$$\boldsymbol{\partial}_s = \frac{\partial q^i}{\partial s}\mathbf{e}_i, \qquad \boldsymbol{\partial}_u = \frac{\partial q^j}{\partial u}\mathbf{e}_j. \tag{11.52}$$

On the left-hand side of formula (11.51) we have

$$\begin{aligned}
\nabla_s \boldsymbol{\partial}_u &= \nabla_{\frac{\partial q^i}{\partial s}\mathbf{e}_i}\left(\frac{\partial q^j}{\partial u}\mathbf{e}_j\right) = \frac{\partial q^j}{\partial s}\nabla_i\left(\frac{\partial q^j}{\partial u}\mathbf{e}_j\right) \\
&= \frac{\partial q^j}{\partial s}\frac{\partial q^i}{\partial u}\nabla_i(\mathbf{e}_j) + \frac{\partial q^j}{\partial s}\mathbf{e}_i\left(\frac{\partial q^j}{\partial u}\right)\mathbf{e}_j.
\end{aligned}$$

Now, using relation (11.52) we obtain

$$\nabla_s \boldsymbol{\partial}_u = \frac{\partial q^j}{\partial s}\frac{\partial q^i}{\partial u}\Gamma_{ij}^k \mathbf{e}_k + \frac{\partial^2 q^j}{\partial s \partial u}\mathbf{e}_j = \left(\frac{\partial^2 q^k}{\partial s \partial u} + \frac{\partial q^i}{\partial s}\frac{\partial q^j}{\partial u}\Gamma_{ij}^k\right)\mathbf{e}_k .$$

In an analogous way we obtain the right-hand side of formula (11.51) in the form

$$\nabla_u \boldsymbol{\partial}_s = \left(\frac{\partial^2 q^k}{\partial u \partial s} + \frac{\partial q^i}{\partial s}\frac{\partial q^j}{\partial u}\Gamma_{ji}^k\right)\mathbf{e}_k . \tag{11.53}$$

Because Christoffel symbols are symmetrical and

$$\frac{\partial^2 q^k}{\partial u \partial s} = \frac{\partial^2 q^k}{\partial s \partial u}, \tag{11.54}$$

relation (11.51) is satisfied. Let us act on the proven relationship (11.51) from the left with an affine connection. We obtain

$$\nabla_s^2 \boldsymbol{\partial}_u = \nabla_s \nabla_u \boldsymbol{\partial}_s . \tag{11.55}$$

Let us introduce a Riemann tensor [7]

$$R\left(\boldsymbol{\partial}_s, \boldsymbol{\partial}_u\right)\boldsymbol{\partial}_s = \nabla_s \nabla_u \boldsymbol{\partial}_s - \nabla_u \nabla_s \boldsymbol{\partial}_s - \nabla_{[\boldsymbol{\partial}_s, \boldsymbol{\partial}_u]}\boldsymbol{\partial}_s , \tag{11.56}$$

where in our case $[\boldsymbol{\partial}_s, \boldsymbol{\partial}_u] = 0$. Indeed, using relation (11.52) we find

$$\begin{aligned} [\boldsymbol{\partial}_s, \boldsymbol{\partial}_u] &= \left[\frac{\partial q^i}{\partial s}\mathbf{e}_i , \frac{\partial q^j}{\partial u}\mathbf{e}_j\right] = \frac{\partial q^i}{\partial s}\mathbf{e}_i \left(\frac{\partial q^j}{\partial u}\mathbf{e}_j\right) \\ &- \frac{\partial q^j}{\partial u}\mathbf{e}_j \left(\frac{\partial q^i}{\partial s}\mathbf{e}_i\right) = \frac{\partial q^i}{\partial s}\frac{\partial q^j}{\partial u}\left(\mathbf{e}_i \mathbf{e}_j - \mathbf{e}_j \mathbf{e}_i\right) = 0 \end{aligned} \tag{11.57}$$

because mixed partial derivatives are commutative. Hence the Riemann tensor (11.56) is reduced to the form

$$R\left(\boldsymbol{\partial}_s, \boldsymbol{\partial}_u\right)\boldsymbol{\partial}_s = \nabla_s \nabla_u \boldsymbol{\partial}_s - \nabla_u \nabla_s \boldsymbol{\partial}_s . \tag{11.58}$$

Taking into account formula (11.50) we find

$$R\left(\boldsymbol{\partial}_s, \boldsymbol{\partial}_u\right)\boldsymbol{\partial}_s = \nabla_s \nabla_u \boldsymbol{\partial}_s . \tag{11.59}$$

Substituting the obtained result into (11.55) and exploiting the antisymmetry of the Riemann tensor we obtain an equation of the form

$$\nabla_s^2 \boldsymbol{\partial}_u + R\left(\boldsymbol{\partial}_s, \boldsymbol{\partial}_u\right)\boldsymbol{\partial}_s = 0. \tag{11.60}$$

This equation is called the *Jacobi–Levi-Civita equation* (JLC) or the *equation of geodesic deviation*. As can be seen, it describes the evolution of vector $\boldsymbol{\partial}_u$ along a geodesic. Another derivation of the equation of geodesic deviation can be found in [10].

In order to carry out any calculations at all, we have to pass from the tensor form (i.e., independent of the specific coordinate system) of the JLC equation (11.60) to an equation in local coordinates.[9] Because in a Riemannian space the basis consists of vectors $\{\mathbf{e}_1, \mathbf{e}_2, \ldots, \mathbf{e}_N\}$, let us resolve vectors $\boldsymbol{\partial}_u$ and $\boldsymbol{\partial}_s$ into this basis

$$\boldsymbol{\partial}_u = \mathbf{J} = J^i \mathbf{e}_i, \qquad \boldsymbol{\partial}_s \to \frac{\mathrm{d}}{\mathrm{d}s} = \frac{\mathrm{d}q^j}{\mathrm{d}s}\mathbf{e}_j = Y^j \mathbf{e}_j. \tag{11.61}$$

Let us note that the partial derivative ∂_s is replaced by an ordinary derivative because we are interested in the evolution of the Jacobi vector $\mathbf{J}$ along a geodesic that is parameterized by s. At first, let us calculate the quantity

$$\begin{aligned}
\nabla_s^2 \mathbf{J} &= \nabla_s \left(\nabla_s \left(J^i \mathbf{e}_i\right)\right) = \nabla_s \left(J^i \nabla_s \mathbf{e}_i + \frac{\mathrm{d}J^i}{\mathrm{d}s}\mathbf{e}_i\right) \\
&= \nabla_s \left(J^i Y^j \nabla_j \mathbf{e}_i + \frac{\mathrm{d}J^i}{\mathrm{d}s}\mathbf{e}_i\right) = \nabla_s \left(J^i Y^j \Gamma^k_{ji} \mathbf{e}_k + \frac{\mathrm{d}J^i}{\mathrm{d}s}\mathbf{e}_i\right) \\
&= \nabla_s \left(\left(\frac{\mathrm{d}J^k}{\mathrm{d}s} + J^i Y^j \Gamma^k_{ji}\right)\mathbf{e}_k\right).
\end{aligned} \tag{11.62}$$

Repeating analogous steps we further obtain

$$\begin{aligned}
\nabla_s^2 \mathbf{J} &= Y^n \nabla_n \left(\left(\frac{\mathrm{d}J^k}{\mathrm{d}s} + J^i Y^j \Gamma^k_{ji}\right)\mathbf{e}_k\right) \\
&= Y^n \left(\frac{\mathrm{d}J^k}{\mathrm{d}s} + J^i Y^j \Gamma^k_{ji}\right) \nabla_n \mathbf{e}_k + Y^n \mathbf{e}_n \left(\frac{\mathrm{d}J^k}{\mathrm{d}s} + J^i Y^j \Gamma^k_{ji}\right)\mathbf{e}_k \\
&= Y^n \left(\frac{\mathrm{d}J^k}{\mathrm{d}s} + J^i Y^j \Gamma^k_{ji}\right) \Gamma^l_{nk} \mathbf{e}_l + \frac{\mathrm{d}}{\mathrm{d}s}\left(\frac{\mathrm{d}J^k}{\mathrm{d}s} + J^i Y^j \Gamma^k_{ji}\right)\mathbf{e}_k \\
&= \left(Y^n \left(\frac{\mathrm{d}J^k}{\mathrm{d}s} + J^i Y^j \Gamma^k_{ji}\right) \Gamma^l_{nk} + \frac{\mathrm{d}}{\mathrm{d}s}\left(\frac{\mathrm{d}J^l}{\mathrm{d}s} + J^i Y^j \Gamma^l_{ji}\right)\right)\mathbf{e}_l.
\end{aligned}$$

When we introduce the following notation[10]

$$\frac{\delta J^k}{\delta s} = \frac{\mathrm{d}J^k}{\mathrm{d}s} + J^i Y^j \Gamma^k_{ji}, \tag{11.63}$$

[9] Since, by definition, a Riemannian space has the structure of a differentiable manifold, in general there is no single global coordinate system but many so-called local coordinate systems.

[10] Often this quantity is called the absolute derivative of a tensor [8].

equation (11.62) can be written in the form

$$\nabla_s^2 \mathbf{J} = \left(\frac{\mathrm{d}}{\mathrm{d}s} \left(\frac{\delta J^l}{\delta s} \right) + \Gamma^l_{nk} Y^n \frac{\delta J^k}{\delta s} \right) \mathbf{e}_l . \tag{11.64}$$

Using relationship (11.63) again, we obtain

$$\nabla_s^2 \mathbf{J} = \frac{\delta}{\delta s} \left(\frac{\delta J^k}{\delta s} \right) \mathbf{e}_k = \frac{\delta^2 J^k}{\delta s^2} \mathbf{e}_k . \tag{11.65}$$

Let us calculate the second term in equation (11.60)

$$\begin{aligned} R(\mathbf{J}, \boldsymbol{\partial}_s)\boldsymbol{\partial}_s &= R(J^i \mathbf{e}_i, Y^j \mathbf{e}_j) Y^k \mathbf{e}_k \\ &= J^i Y^j Y^k R(\mathbf{e}_i, \mathbf{e}_j)\mathbf{e}_k = J^i Y^j Y^k R^n_{kij} \mathbf{e}_n , \end{aligned}$$

where we used the linearity of a Riemann tensor with respect to each of its arguments. Substituting the obtained result and relation (11.65) into (11.60) we obtain

$$\left(\frac{\delta^2 J^n}{\delta s^2} + J^i Y^j Y^k R^n_{kij} \right) \mathbf{e}_n = 0. \tag{11.66}$$

Hence the JLC equation in the local coordinates takes the form

$$\frac{\delta^2 J^n}{\delta s^2} + J^i Y^j Y^k R^n_{kij} = 0. \tag{11.67}$$

As can be seen, these are N second-order differential equations with respect to the parameter s. They are a starting point for any further calculations. Let us note that in these equations there occur quantities Y^k, which are the components of vector $\boldsymbol{\partial}_s$ (11.61), or, to put it differently, they are the components of a velocity vector in Riemannian space. They satisfy the equations

$$\begin{aligned} \nabla_s \boldsymbol{\partial}_s &= Y^i \nabla_i (Y^j \mathbf{e}_j) = Y^i Y^j \Gamma^k_{ij} \mathbf{e}_k + \frac{\mathrm{d}Y^k}{\mathrm{d}s} \mathbf{e}_k \\ &= \left(Y^i Y^j \Gamma^k_{ij} + \frac{\mathrm{d}Y^k}{\mathrm{d}s} \right) \mathbf{e}_k = 0. \end{aligned} \tag{11.68}$$

Taking into account relation (11.61) we obtain the geodesic equations

$$\frac{\mathrm{d}^2 q^k}{\mathrm{d}s^2} + \Gamma^k_{ij} \frac{\mathrm{d}q^i}{\mathrm{d}s} \frac{\mathrm{d}q^j}{\mathrm{d}s} = \frac{\delta Y^k}{\delta s} = 0, \tag{11.69}$$

where $k = 1, 2, \ldots, N$.

11.4 The JLC Equation in Geodesic Coordinates

The obtained equations (11.67) can be already directly applied to the investigated system. However, their number can be diminished by choosing a certain basis in the considered Riemannian space. That is, let us consider the system of vectors E_i that satisfy the following conditions:

$$\mathbf{E}_1 = \boldsymbol{\partial}_s, \qquad g(E_i, E_j) = \delta_{ij}, \qquad \nabla_s \mathbf{E}_i = 0. \tag{11.70}$$

Having defined the basis already in the Riemannian space, let us resolve the vectors occurring in the JLC equation (11.60) on this basis

$$\boldsymbol{\partial}_s = \mathbf{E}_1, \quad \boldsymbol{\partial}_u = I^n \mathbf{E}_n. \tag{11.71}$$

Let us now substitute expansions (11.70) into the JLC equation (11.60). As a result we obtain

$$\nabla_s^2 (I^n \mathbf{E}_n) + R(I^n \mathbf{E}_n, \mathbf{E}_1)\mathbf{E}_1 = 0. \tag{11.72}$$

The first of the terms in the preceding expression takes the form

$$\nabla_s^2 (I^n \mathbf{E}_n) = \nabla_s (\nabla_s (I^n \mathbf{E}_n)) = \frac{\mathrm{d}^2 I^n}{\mathrm{d}s^2} \mathbf{E}_n, \tag{11.73}$$

where we twice used the fact that vectors $\mathbf{E}_n$ undergo parallel translation along the geodesic. Substituting the obtained result into (11.72) we obtain

$$\frac{\mathrm{d}^2 I^n}{\mathrm{d}s^2} \mathbf{E}_n + R(I^n \mathbf{E}_n, \mathbf{E}_1)\mathbf{E}_1 = 0. \tag{11.74}$$

Now, let us calculate the second term in the preceding equation. Let us resolve vector $\mathbf{E}_n$ on the basis $\{\mathbf{e}_1, \mathbf{e}_2, \ldots, \mathbf{e}_N\}$

$$\begin{aligned} R(I^n \mathbf{E}_n, \mathbf{E}_1)\mathbf{E}_1 &= R(I^n \Psi_n^j \mathbf{e}_j, \Psi_1^k \mathbf{e}_k)\Psi_1^l \mathbf{e}_l \\ &= I^n \Psi_n^j \Psi_1^k \Psi_1^l R(\mathbf{e}_j, \mathbf{e}_k)\mathbf{e}_l = I^n \Psi_n^j \Psi_1^k \Psi_1^l R^r_{ljk} \mathbf{e}_r. \end{aligned} \tag{11.75}$$

Because the preceding expression is a vector, it can be represented as a linear combination of the vectors of the basis E_i in the form

$$\begin{aligned} R(I^n \mathbf{E}_n, \mathbf{E}_1)\mathbf{E}_1 &= \sum_{q=1}^{N} g(I^n \Psi_n^j \Psi_1^k \Psi_1^l R^r_{ljk} \mathbf{e}_r, \Psi_q^i \mathbf{e}_i)\mathbf{E}_q \\ &= \sum_{q=1}^{N} I^n \Psi_n^j \Psi_1^k \Psi_1^l \Psi_q^i R^r_{ljk} g(\mathbf{e}_r, \mathbf{e}_i)\mathbf{E}_q \\ &= \sum_{q=1}^{N} I^n \Psi_n^j \Psi_1^k \Psi_1^l \Psi_q^i R_{iljk} \mathbf{E}_q. \end{aligned} \tag{11.76}$$

We insert here a sigma sign because the index of summation q occurs at the same level, and consequently the Einstein summation convention does not apply. Substituting relation (11.76) into (11.74) we obtain

$$\sum_{q=1}^{N}\left(\frac{\mathrm{d}^2 I^q}{\mathrm{d}s^2}+I^n\Psi_n^j\Psi_1^k\Psi_1^l\Psi_q^i R_{iljk}\right)\mathbf{E}_q=0, \tag{11.77}$$

which leads to an equation of the form

$$\frac{\mathrm{d}^2 I^q}{\mathrm{d}s^2}+I^n\Psi_n^j\Psi_1^k\Psi_1^l\Psi_q^i R_{iljk}=0. \tag{11.78}$$

The quantities $\Psi_1^k\Psi_1^l$ occurring in the preceding expression are components of the vector $\mathbf{E}_1=\boldsymbol{\partial}_s$. On the other hand, from relation (11.61) we have $\boldsymbol{\partial}_s=Y^j\mathbf{e}_j$, so we obtain $Y^k=\Psi_1^k$ and $Y^l=\Psi_1^l$. Substituting this into (11.78) we find

$$\frac{\mathrm{d}^2 I^q}{\mathrm{d}s^2}+I^n\Psi_n^j Y^k Y^l\Psi_q^i R_{iljk}=0. \tag{11.79}$$

For $q=1$ (i.e., the component along the geodesic) the preceding equation assumes a simple form:

$$\frac{\mathrm{d}^2 I^1}{\mathrm{d}s^2}=0, \tag{11.80}$$

because

$$I^n\Psi_n^j Y^k Y^l\Psi_1^i R_{iljk}=I^n\Psi_n^j Y^k Y^l Y^i R_{iljk}=0. \tag{11.81}$$

The preceding quantity vanishes because it is the sum of products of quantities R_{iljk} and Y^iY^l, which are respectively antisymmetric and symmetric with respect to the exchange of the indices il. Because of relationship (11.80), it suffices to consider only the equations

$$\frac{\mathrm{d}^2 I^q}{\mathrm{d}s^2}+I^n\Psi_n^j Y^k Y^l\Psi_q^i R_{iljk}=0, \tag{11.82}$$

where $q=2,3,\ldots,N$. In the preceding equations quantities occur that are dependent on the solutions of geodesic equations (or, which is equivalent, of equations of motion). This means that in order to solve the JLC equation one should simultaneously solve equations of motion.[11] However, for systems with many degrees of freedom certain procedures enable the solution of the JLC equation without having to know the solutions of the equations of motion of the investigated system [4]. This approach facilitates, on certain assumptions, the calculation of

[11] A similar situation occurs in the case of calculation of Lyapunov exponents.

Lyapunov exponents by means of an analytical method. However, it requires the application of a mathematical apparatus that is beyond the scope of this book.

11.5 The JLC Equation for the Jacobi Metric

The JLC equation (11.67) for the Jacobi metric has the form

$$\frac{\delta^2 J^i}{\delta s^2} + J^k Y^j Y^l R^i_{jkl} = 0. \tag{11.83}$$

Using formula (11.63) we find

$$\frac{\delta^2 J^i}{\delta s^2} = \frac{\mathrm{d}}{\mathrm{d}s}\left(\frac{\mathrm{d}J^i}{\mathrm{d}s} + \Gamma^i_{jk} J^j Y^k\right) + \Gamma^i_{nl}\left(\frac{\mathrm{d}J^n}{\mathrm{d}s} + \Gamma^n_{jk} J^j Y^k\right) Y^l. \tag{11.84}$$

Next, we differentiate and group the terms, obtaining

$$\frac{\delta^2 J^i}{\delta s^2} = \frac{\mathrm{d}^2 J^i}{\mathrm{d}s^2} + 2\frac{\mathrm{d}J^n}{\mathrm{d}s}\Gamma^i_{nl} Y^l + J^n \left(\frac{\partial \Gamma^i_{nk}}{\partial q^j} Y^j Y^k + \Gamma^i_{nk}\frac{\mathrm{d}Y^k}{\mathrm{d}s} + \Gamma^i_{jl}\Gamma^j_{nk} Y^l Y^k\right). \tag{11.85}$$

Taking into account formula (11.68) we have

$$\frac{\mathrm{d}Y^k}{\mathrm{d}s} = -\Gamma^k_{rl} Y^r Y^k.$$

Substituting the obtained relationship into (11.85) we obtain

$$\frac{\delta^2 J^i}{\delta s^2} = \frac{\mathrm{d}^2 J^i}{\mathrm{d}s^2} + 2\frac{\mathrm{d}J^n}{\mathrm{d}s}\Gamma^i_{nl} Y^l + J^n \left(\frac{\partial \Gamma^i_{nr}}{\partial q^l} - \Gamma^i_{nk}\Gamma^k_{rl} + \Gamma^i_{kl}\Gamma^k_{nr}\right) Y^l Y^r.$$

Substituting the preceding expression into the JLC equation (11.83) we obtain

$$\frac{\mathrm{d}^2 J^i}{\mathrm{d}s^2} + 2\frac{\mathrm{d}J^n}{\mathrm{d}s}\Gamma^i_{nl} Y^l + J^n \left(\frac{\partial \Gamma^i_{nr}}{\partial q^l} - \Gamma^i_{nk}\Gamma^k_{rl} + \Gamma^i_{kl}\Gamma^k_{nr} + R^i_{rnl}\right) Y^l Y^r = 0.$$

Next, we make use of the form of the Riemann tensor, and we have

$$\frac{\mathrm{d}^2 J^i}{\mathrm{d}s^2} + 2\frac{\mathrm{d}J^n}{\mathrm{d}s}\Gamma^i_{nl} Y^l + J^n \frac{\partial \Gamma^i_{rl}}{\partial q^l} Y^l Y^r = 0. \tag{11.86}$$

In order to use the preceding equation efficiently it has to be reduced to the differential equation with respect to time t. Using relation (11.33) we obtain

$$\frac{\mathrm{d}}{\mathrm{d}s} = \frac{1}{2W}\frac{\mathrm{d}}{\mathrm{d}t},$$

$$\frac{\mathrm{d}^2}{\mathrm{d}s^2} = \frac{1}{4W^2}\frac{\mathrm{d}^2}{\mathrm{d}t^2} - \frac{1}{4W^3}\frac{\mathrm{d}W}{\mathrm{d}t}\frac{\mathrm{d}}{\mathrm{d}t}.$$

Substitutitng the preceding relations into (11.86) we obtain the JLC equation for the Jacobi metric in the form

$$\ddot{J}^i - \frac{\dot{W}}{W}\dot{J}^i + 2\dot{J}^n \Gamma^i_{nl} X^l + J^n \frac{\partial \Gamma^i_{rl}}{\partial q^l} X^l X^r = 0, \tag{11.87}$$

where

$$X^r = \frac{\mathrm{d}q^r}{\mathrm{d}t}.$$

If the dimension of a Riemannian space (which in the present case is determined by the configuration space) is equal to N, then we obtain a system of N ordinary differential equations with respect to time.

11.6 Mechanical Systems with Two Degrees of Freedom

Already in systems with two degrees of freedom it is possible to observe chaotic behavior because then the phase space is four-dimensional [12, 13]. Since in the case of conservative systems descriptions by means of the Jacobi metric and the Eisenhart metric are equivalent, we can choose either of them in an arbitrary way. In the case of the Jacobi metric a Riemannian space is two-dimensional because it is a configuration space whose dimension is equal to the number of degrees of freedom of the investigated system. On the other hand, in the case of the Eisenhart metric the space is four-dimensional. Let us consider the Jacobi metric (11.37) in the matrix form

$$\mathbf{G} = 2W\begin{bmatrix} a_{11} & a_{12} \\ a_{12} & a_{22} \end{bmatrix}, \qquad W = E - V(\mathbf{q}), \tag{11.88}$$

which corresponds to a Lagrangian of the form

$$L = \frac{1}{2}a_{11}(\dot{q}^1)^2 + \frac{1}{2}a_{22}(\dot{q}^2)^2 + a_{12}\dot{q}^1\dot{q}^2 - V(\mathbf{q}). \tag{11.89}$$

Our aim is to obtain the JLC equation for the conservative mechanical system with two degrees of freedom whose dynamics is described by Lagrangian (11.89). To this end we make use of previously obtained results (11.82) and write the JLC equation for a system with two degrees of freedom in the form

$$\frac{\mathrm{d}^2 I^q}{\mathrm{d}s^2} + I^n \Psi_n^j Y^k Y^l \Psi_q^i R_{iljk} = 0.$$

The preceding equations are equations in a geodesic coordinate system. The first equation ($q = 1$), according to (11.80), reduces to the form

$$\frac{\mathrm{d}^2 I^1}{\mathrm{d}s^2} = 0,$$

whereas the second one takes the form

$$\frac{\mathrm{d}^2 I^2}{\mathrm{d}s^2} + I^2 \Psi_2^j Y^k Y^l \Psi_2^i R_{iljk} = 0. \tag{11.90}$$

The unknown quantities in the preceding equation are the quantities Ψ_2^i, which, in turn, are the components of basis vector $\mathbf{E}_2$. Those components are determined from condition (11.70). Let us write the orthonormality condition in the matrix form

$$\boldsymbol{\Psi}^{\mathrm{T}} \mathbf{G} \boldsymbol{\Psi} = \mathbf{I}, \tag{11.91}$$

where $\mathbf{I}$ is an identity matrix and $\boldsymbol{\Psi}$ is a matrix of the form

$$\boldsymbol{\Psi} = \begin{bmatrix} \Psi_1^1 & \Psi_2^1 \\ \Psi_1^2 & \Psi_2^2 \end{bmatrix} = \begin{bmatrix} Y^1 & \Psi_2^1 \\ Y^2 & \Psi_2^2 \end{bmatrix}.$$

Following transformation of condition (11.91) we obtain the equation

$$\boldsymbol{\Psi}^{\mathrm{T}} \boldsymbol{\Psi} = \mathbf{G}^{-1},$$

which, in turn, leads to equations of the form

$$\left(Y^1\right)^2 + \left(\Psi_2^1\right)^2 = \frac{G_{22}}{\det \mathbf{G}},$$
$$\left(Y^2\right)^2 + \left(\Psi_2^2\right)^2 = \frac{G_{11}}{\det \mathbf{G}}.$$

As solutions we take

$$\Psi_2^1 = \frac{1}{\sqrt{\det \mathbf{G}}}\left(G_{12} Y^1 + G_{22} Y^2\right),$$
$$\Psi_2^2 = -\frac{1}{\sqrt{\det \mathbf{G}}}\left(G_{11} Y^1 + G_{12} Y^2\right). \tag{11.92}$$

Taking into account the fact that in a two-dimensional space there exists only one non-vanishing component of the Riemann tensor and that there are symmetries of components, (11.90) takes the form

$$\frac{\mathrm{d}^2 I^2}{\mathrm{d}s^2} + I^2 R_{2121}\left(\Psi_2^2 Y^1 - \Psi_2^1 Y^2\right) = 0.$$

Next, using (11.92), we obtain

$$\frac{\mathrm{d}^2 I^2}{\mathrm{d}s^2} + I^2 \frac{R_{2121}}{\det \mathbf{G}} = 0$$

or, in equivalent form,

$$\frac{\mathrm{d}^2 I^2}{\mathrm{d}s^2} + \frac{1}{2} R I^2 = 0,$$

where R denotes a scalar of curvature. The preceding equation in time domain takes the form

$$\ddot{I} - \frac{\dot{W}}{W}\dot{I} + 2RW^2 I = 0, \qquad I = I^2.$$

Now, applying the substitution

$$I = \Lambda e^{\frac{1}{2}\int \frac{\dot{W}}{W}\mathrm{d}t} = \Lambda\sqrt{W},$$

we obtain

$$\ddot{\Lambda} + \Omega(t)\,\Lambda = 0, \tag{11.93}$$

where

$$\Omega(t) = \frac{1}{2}\left(\frac{\ddot{W}}{W} - \frac{1}{2}\left(\frac{\dot{W}}{W}\right)^2 + 4RW^2\right). \tag{11.94}$$

The obtained equation serves as a basis for the analysis of dynamical systems with two degrees of freedom in a Riemannian space. It should be emphasized that the function $\Omega(t)$ does not depend on time explicitly but only implicitly through solutions to equations of motion. This means that in order to carry out its analysis we have to solve the JLC equation simultaneously with equations of motion.

Example 11.1. As an example of a mechanical system with two degrees of freedom we will consider a double physical pendulum. The dynamics of such a pendulum is described by a Lagrangian of the form

$$L = \frac{1}{2}\left(m_1 c_1^2 + J_1 + m_2 l_1^2\right)\dot{\varphi}_1^2 + \frac{1}{2}\left(m_2 c_2^2 + J_2\right)\dot{\varphi}_2^2$$
$$+ m_2 c_2 l_1 \dot{\varphi}_1 \dot{\varphi}_2 \cos\phi + g\left(m_1 c_1 + m_2 l_1\right)\left(\cos\varphi_1 - 1\right)$$
$$+ m_2 g c_2 \left(\cos\varphi_2 - 1\right), \qquad (*)$$

where $\phi = \varphi_1 - \varphi_2$; $m_1,\ m_2$ denote the masses of particular links of the pendulum; $J_1,\ J_2$ are the moments of inertia of the links about their mass centers; $c_1,\ c_2$ denote the positions of the mass centers of the links; and $l_1,\ l_2$ denote the lengths of the links.

In order to write the preceding Lagrangian in non-dimensional form we introduce the following scaling:

$$\tau = t\sqrt{\frac{m_1 g c_1 + m_2 g l_1}{J_1 + m_1 c_1^2 + m_2 l_1^2}},$$
$$\beta = \frac{J_2 + m_2 g l_1}{J_1 + m_1 c_1^2 + m_2 l_1^2},$$
$$\kappa = \frac{m_2 c_2 l_1}{J_1 + m_1 c_1^2 + m_2 l_1^2},$$
$$\mu = \frac{m_2 c_2}{m_1 c_1 + m_2 l_1}.$$

Lagrangian $(*)$ takes the following non-dimensional form:

$$L = \frac{1}{2}\dot{\varphi}_1^2 + \frac{\beta}{2}\dot{\varphi}_2^2 + \kappa\dot{\varphi}_1\dot{\varphi}_2\cos\phi + \left(\cos\varphi_1 - 1\right) + \left(\cos\varphi_2 - 1\right).$$

Equations of dynamics obtained from Euler–Lagrange equations for Lagrangian $(*)$ take the form

$$\ddot{\boldsymbol{\varphi}} + \boldsymbol{a}^{-1}\boldsymbol{b}\dot{\boldsymbol{\varphi}}^2 + \boldsymbol{a}^{-1}\boldsymbol{\eta} = \mathbf{0}, \qquad (**)$$

where

$$\boldsymbol{\varphi} = \begin{bmatrix}\varphi_1\\ \varphi_2\end{bmatrix}, \qquad \dot{\boldsymbol{\varphi}}^2 = \begin{bmatrix}\dot{\varphi}_1^2\\ \dot{\varphi}_2^2\end{bmatrix}, \qquad \boldsymbol{a} = \begin{bmatrix}1 & \kappa\cos\phi\\ \kappa\cos\phi & \beta\end{bmatrix},$$
$$\boldsymbol{b} = \begin{bmatrix}0 & -\kappa\sin\phi\\ -\kappa\sin\phi & 0\end{bmatrix}, \qquad \boldsymbol{\eta} = \begin{bmatrix}\sin\varphi_1\\ \mu\sin\varphi_2\end{bmatrix}.$$

Geometrization will be carried out in a configuration space with the Jacobi metric, which in this case has the form

$$\mathbf{g} = 2W \begin{bmatrix} 1 & \kappa \cos\phi \\ \kappa \cos\phi & \beta \end{bmatrix},$$

$$W = E - 1 - \mu + \cos\varphi_1 + \mu \cos\varphi_2.$$

At first, let us calculate the Christoffel symbols after taking (11.23) into account:

$$\begin{aligned}
\Gamma^1_{11} &= \frac{1}{2W \det a} \left(2\kappa^2 \sin\varphi_1 \cos^2\phi \right.\\
&\qquad \left. + W\kappa^2 \sin(2\phi) - \mu\kappa \sin\varphi_2 \cos\phi - \beta \sin\varphi_1\right),\\
\Gamma^2_{22} &= \frac{1}{2W \det a} \left(2\mu\kappa^2 \sin\varphi_2 \cos^2\phi \right.\\
&\qquad \left. - W\kappa^2 \sin(2\phi) - \beta\kappa \sin\varphi_1 \cos\phi - \beta\mu \sin\varphi_2\right),\\
\Gamma^2_{11} &= \frac{1}{2W \det a} \left(\mu \sin\varphi_2 - 2W\kappa \sin\phi - \kappa \sin\varphi_1 \cos\phi\right),\\
\Gamma^1_{22} &= \frac{1}{2W \det a} \left(\beta \sin\varphi_1 + 2W\kappa \sin\phi - \mu\kappa \sin\varphi_2 \cos\phi\right),\\
\Gamma^2_{12} &= \frac{1}{2W \det a} \left(\kappa\mu \sin\varphi_2 \cos\phi - \beta \sin\varphi_1\right),\\
\Gamma^1_{12} &= \frac{\beta}{2W \det a} \left(\kappa \sin\varphi_1 \cos\phi - \mu \sin\varphi_2\right).
\end{aligned}$$

Because the system has two degrees of freedom, the configuration space, and consequently the Riemannian space, is a two-dimensional space. In two-dimensional Riemannian space we have only one non-zero component of the Riemann tensor, which in our case takes the form

$$\begin{aligned}
R_{2121} &= \mu \cos\varphi_2 + 2W\kappa \cos\phi + \beta \cos\varphi_1\\
&\quad + \frac{1}{W}(\kappa \sin\varphi_1 \cos\phi - \mu \sin\varphi_2)^2 + \frac{\sin^2\varphi_1 \det a}{W}\\
&\quad - \frac{\kappa \sin\phi}{\det a}(\kappa\beta \sin\varphi_1 \cos\phi - \mu\kappa \sin\varphi_2 \cos\phi\\
&\quad - \beta\mu \sin\varphi_2 + \beta \sin\varphi_1) - \frac{2W\kappa^3 \sin^2\phi \cos\phi}{\det a},
\end{aligned}$$

where the components of the Riemann tensor are found from the formula[12]

[12]Obviously we apply here the Einstein summation convention, that is, we carry out the summation with respect to repeating indices.

$$R^l_{ijk} = \frac{\partial \Gamma^l_{ik}}{\partial q^j} - \frac{\partial \Gamma^l_{ij}}{\partial q^k} + \Gamma^n_{ik}\Gamma^l_{nj} - \Gamma^n_{ij}\Gamma^l_{nk}.$$

Next, we obtain a curvature scalar R of the form

$$\begin{aligned} R = {} & \frac{\kappa \cos\phi}{W \det a} - \frac{\kappa^3 \sin^2\phi \cos\phi}{W \det a^2} \\ & + \frac{1}{2W^3 \det a}(\kappa \sin\varphi_1 \cos\phi - \mu \sin\varphi_2)^2 + \frac{\sin^2\varphi_1}{2W^3} \\ & - \frac{\kappa \sin\phi}{2W^2 \det a^2}(\kappa\beta \sin\varphi_1 \cos\phi - \mu\kappa \sin\varphi_2 \cos\phi \\ & - \beta\mu \sin\varphi_2 + \beta \sin\varphi_1) + \frac{\mu \cos\varphi_2 + \beta \cos\varphi_1}{2W^2 \det a}. \end{aligned}$$

We substitute the last obtained form of the curvature scalar into (11.94) and solve the resulting JLC equation (11.93) simultaneously with equations of motion (∗∗). Further details and numerical examples can be found in [14, 15]. □

References

1. V.I. Arnold, *Mathematical Methods of Classical Mechanics* (Springer, Berlin, 1978)
2. H. Poincaré, *Les Méthodes Nouvelles de la Méchanique Céleste* (Gauthier-Villars, Paris, 1892) (in French)
3. J.V. Jose, E.J. Saletan, *Classical Dynamics: A Contemporary Approach* (Cambridge University Press, Cambridge, 1998)
4. L. Casetti, M. Pettini, E.D.G. Cohen, *Geometric approach to hamiltonian dynamics and statistical mechanics*. Phys. Rep. **337**(3), 237–342 (2000)
5. P. Finsler, *Über Kurven und Flachen in allegemeinen Raumen*. Dissertation, Göttingen, 1918 (Birkhauser, Basel, 1951) (in German)
6. M.P. do Carmo, *Riemannian Geometry* (Birkhauser, Boston, 1993)
7. M. Nakahara, *Geometry, Topology and Physics* (Adam Hilger, Bristol, 1990)
8. L.U. Eisenhart, *Riemannian Geometry* (Princeton University Press, Princeton, 1925)
9. I.M. Gelfand, S.W. Fomin, *Calculus of Variations* (Prentice Hall, Englewood Cliffs, 1963)
10. J.L. Synge, A. Schild, *Tensor Calculus* (Dover, New York, 1978)
11. G. Białkowski, *Classical Mechanics* (PWN, Warsaw, 1975) (in Polish)
12. J. Awrejcewicz, *Classical Mechanics: Statics and Kinematics* (Springer, New York, 2012)
13. J. Awrejcewicz, *Bifurcation and Chaos in Simple Dynamical Systems* (World Scientific, Singapore, 1989)
14. J. Awrejcewicz, D. Sendkowski, M. Kazmierczak, Geometrical approach to the swinging pendulum dynamics. Comput. Struct. **84**(24–25), 1577–1583 (2006)
15. J. Awrejcewicz, D. Sendkowski, Geometric analysis of the dynamics of a double pendulum. J. Mech. Mater. Struct. **2**(8), 1421–1430 (2007)

Printed by Printforce, the Netherlands